Chaos Theory Tamed

Chaos Theory Tamed

Garnett P. Williams

US Geological Survey (Ret.)

Taylor & Francis Limited,1 Gunpowder Square, London EC4A 3DE

British Library Cataloguing-in-Publication Data
A CIP catalogue record for this book is available from the British Library.
ISBN: 0-7484-0749-9

Library of Congress Cataloging-in-Publication Data are available

Cover design by Youngs Design in Production.
Typeset in Times New Roman
Printed in Great Britain by T. J. International, Padstow, Cornwall.

Contents

Preface

Virtually every branch of the sciences, engineering, economics, and related fields now discusses or refers to chaos. James Gleick's 1987 book, *Chaos: making a new science* and a 1988 one-hour television program on chaos aroused many people's curiosity and interest. There are now quite a few books on the subject. Anyone writing yet another book, on any topic, inevitably goes through the routine of justifying it. My justification consists of two reasons:

- Most books on chaos, while praiseworthy in many respects, use a high level of math. Those books have been written by specialists for other specialists, even though the authors often label them "introductory." Amato (1992) refers to a "cultural chasm" between "the small group of mathematically inclined initiates who have been touting" chaos theory, on the one hand, and most scientists (and, I might add, "everybody else"), on the other. There are relatively few books for those who lack a strong mathematics and physics background and who might wish to explore chaos in a particular field. (More about this later in the Preface.)
- Most books, in my opinion, don't provide understandable derivations or explanations of many key concepts, such as Kolmogorov–Sinai entropy, dimensions, Fourier analysis, Lyapunov exponents, and others. At present, the best way to get such explanations is either to find a personal guru or to put in gobs of frustrating work studying the brief, condensed, advanced treatments given in technical articles.

Chaos is a mathematical subject and therefore isn't for everybody. However, to understand the fundamental concepts, you don't need a background of anything more than introductory courses in algebra, trigonometry, geometry, and statistics. That's as much as you'll need for this book. (More advanced work, on the other hand, does require integral calculus, partial differential equations, computer programming, and similar topics.)

In this book, I assume no prior knowledge of chaos, on your part. Although chaos covers a broad range of topics, I try to discuss only the most important ones. I present them hierarchically. Introductory background perspective takes up the first two chapters. Then come seven chapters consisting of selected important material (an auxiliary toolkit) from various fields. Those chapters provide what I think is a good and necessary foundation—one that can be arduous and time-consuming to get from other sources. Basic and simple chaos-related concepts follow. They, in turn, are prerequisites for the slightly more advanced concepts that make up the later chapters. (That progression means, in turn, that some chapters are on a very simple level, others on a more advanced level.) In general, I try to present a plain-vanilla treatment, with emphasis on the idealized case of low-dimensional, noise-free chaos. That case is indispensable for an introduction. Some real-world data, in contrast, often require sophisticated and as-yet-developing methods of analysis. I don't discuss those techniques.

The absence of high-level math of course doesn't mean that the reading is light entertainment. Although there's no way to avoid some specialized terminology, I define such terms in the text as well as in a Glossary. Besides, learning and using a new vocabulary (a new language) is fun and exciting. It opens up a new world.

My goal, then, is to present a basic, semitechnical introduction to chaos. The intended audience consists of chaos nonspecialists who want a foothold on the fundamentals of chaos theory, regardless of their academic level. Such nonspecialists may not be comfortable with the more formal mathematical approaches that some books follow. Moreover, many readers (myself included) often find a formal writing style more difficult to understand. With this wider and less mathematically inclined readership in mind, I have deliberately kept the writing informal—"we'll" instead of "we will," "I'd" instead of "I would," etc. Traditionalists who are used to a formal style may be uneasy with this. Nonetheless, I hope it will help reduce the perceived distance between subject and reader.

I'm a geologist/hydrologist by training. I believe that coming from a peripheral field helps me to see the subject differently. It also helps me to understand—and I hope answer—the types of questions a nonspecialist has. Finally, I hope it will help me to avoid using excessive amounts of specialized jargon.

In a nutshell, this is an elementary approach designed to save you time and trouble in acquiring many of the fundamentals of chaos theory. It's the book that I wish had been available when I started looking into chaos. I hope it'll be of help to you.

In regard to units of measurement, I have tried to compromise between what I'm used to and what I suspect most readers are used to. I've used metric units (kilometers, centimeters, etc.) for length because that's what I have always used in the scientific field. I've used Imperial units (pounds, Fahrenheit, etc.) in most other cases.

I sincerely appreciate the benefit of useful conversations with and/or help from A. V. Vecchia, Brent Troutman, Andrew Fraser, Ben Mesander, Michael Mundt, Jon Nese, William Schaffer, Randy Parker, Michael Karlinger, Leonard Smith, Kaj Williams, Surja Sharma, Robert Devaney, Franklin Horowitz, and John Moody. For critically reading parts of the manuscript, I thank James Doerner, Jon Nese, Ron Charpentier, Brent Troutman, A. V. Vecchia, Michael Karlinger, Chris Barton, Andrew Fraser, Troy Shinbrot, Daniel Kaplan, Steve Pruess, David Furbish, Liz Bradley, Bill Briggs, Dean Prichard, Neil Gershenfeld, Bob Devaney, Anastasios Tsonis, and Mitchell Feigenbaum. Their constructive comments helped reduce errors and bring about a much more readable and understandable product. I also thank Anthony Sanchez, Kerstin Williams, and Sebastian Kuzminsky for their invaluable help on the figures.

I especially want to thank Roger Jones for his steadfast support, perseverance, hard work, friendly advice and invaluable expertise as editor. He has made the whole process a rewarding experience for me. Other authors should be so fortunate.

Symbols

a A constant

b A constant or scalar

c (a) A constant; (b) a component dimension in deriving the Hausdorff–Besicovich dimension

c_a Intercept of line $\Delta R = c_a - H'_{KS}m$, and taken as a rough indicator of the accuracy of the measurements

d Component dimension in deriving Hausdorff–Besicovich dimension; elsewhere, a derivative

e Base of natural logarithms, with a value equal to 2.718 . . .

f A function

h Harmonic number

i A global counter, often representing the *i*th bin of the group of bins into which we divide values of x

j (a) A counter, often representing the *j*th bin of the group of bins into which we divide values of y; (b) the imaginary number $(-1)^{0.5}$

k Control parameter

k_n *k* value at time *t* or observation *n*

k_∞ *k* value (logistic equation) at which chaos begins

m A chosen lag, offset, displacement, or number of intervals between points or observations

n Number (position) of an iteration, observation or period within a sequence (e.g. the *n*th observation)

r Scaling ratio

s Standard deviation

s^2 Variance (same as power)

t Time, sometimes measured in actual units and sometimes just in numbers of events (no units)

u An error term

v_i Special variable

w A variable

x Indicator variable (e.g. time, population, distance)

x_i (a) The *i*th value of *x*, *i*th point, or *i*th bin; (b) all values of *x* as a group

x_j A trajectory point to which distances from point x_i are measured in calculating the correlation dimension

x_n Value of the variable *x* at the *n*th observation.

x_t Value of the variable x at time or observation t (hence x_0, x_1, x_2, etc.)
x^* Attractor (a value of x)
y Dependent variable or its associated value
y_h Height of the wave having harmonic number h
y_0 Value of dependent variable y at the origin
z A variable
A Wave amplitude
C_ε Correlation sum
D A multipurpose or general symbol for dimension, including embedding dimension
D_c Capacity (a type of dimension)
D_H Hausdorff (or Hausdorff–Besicovich) dimension
D_I Information dimension, numerically equal to the slope of a straight line on a plot of I_ε (arithmetic scale) versus $1/\varepsilon$ (log scale)
$\boldsymbol{E}$ An observed vector, usually not perpendicular to any other observed vectors
F Wave frequency
G Labeling symbol in definition of correlation dimension
H Entropy (sometimes called information entropy)
H_t Entropy at time t
H_w Entropy computed as a weighted sum of the entropies of individual phase space compartments
H_{KS} Kolmogorov–Sinai (K–S) entropy
H'_{KS} Kolmogorov–Sinai (K–S) entropy as estimated from incremental redundancies
H_X Self-entropy of system X
$H_{X,Y}$ Joint entropy of systems X and Y
$H_{X|Y}$ Conditional entropy for system X
H_Y Self-entropy of system Y
$H_{Y|X}$ Conditional entropy for system Y
$H_{\Delta t}$ Entropy computed over a particular duration of time t
I Information
I_i Information contributed by compartment i
I_w Total information contributed by all compartments
I_X Information for dynamical system X
$I_{X;Y}$ Mutual information of coupled systems X and Y
I_Y Information of dynamical system Y
$I_{Y;X}$ Mutual information of coupled systems X and Y
I_ε Information needed to describe an attractor or trajectory to within an accuracy ε
K Boltzmann's constant
L Length or distance
L_ε Estimated length, usually by approximations with small, straight increments of length ε
L_w Wavelength
M_ε An estimate of a measure (a determination of length, area, volume, etc.)
M_{tr} A true value of a measure
N Total number of data points or observations
N_d Total number of dimensions or variables
N_r Total number of possible bin-routes a dynamical system can take during its evolution from an arbitrary starting time to some later time
N_s Total number of possible or represented states of a system

N_ε Number of points contained within a circle, sphere, or hypersphere of a given radius
P Probability
P_i (a) Probability associated with the *i*th box, sphere, value, etc.; (b) all probabilities of a distribution, as a group
P_s Sequence probability
$P(x_i)$ (a) Probability of class x_i from system X; (b) all probabilities of the various classes of x, as a group
$P(x_i,y_j)$ Joint probability that system X is in class x_i when system Y is in class y_j
$P(y_j)$ (a) Probability of class y_j from system Y; (b) all probabilities of the various classes of y, as a group
$P(y_j|x_i)$ Conditional probability that system Y will be in class y_j, given that system X is in class x_i
R Redundancy
R_m Autocorrelation at lag m
T Wave period
$\boldsymbol{U}$ A unit vector representing any of a set of mutually orthogonal vectors
$\boldsymbol{V}$ A vector constructed from an observed vector so as to be orthogonal to similarly constructed vectors of the same set
X A system or ensemble of values of random variable x and its probability distribution
Y A system or ensemble of values of random variable y and its probability distribution
α Fourier cosine coefficient
β Fourier sine coefficient
δ Difference between two computed values of a trajectory, for a given iteration number
δ_a Orbit difference obtained by extrapolating a straight line back to $n = 0$ on a plot of orbit difference versus iteration n
δ_0 Difference between starting values of two trajectories
ε Characteristic length of scaling device (ruler, box, sphere, etc.)
ε_0 Largest length of scaling device for which a particular relation holds
θ (a) Central or inclusive angle, such as the angle subtended during a rotating-disk experiment (excluding phase angle) or the angle between two vectors; (b) an angular variable or parameter
λ Lyapunov exponent (global, not local)
ν Correlation dimension (correlation exponent)
$\hat{\nu}$ Estimated correlation dimension
π 3.1416 . . .
ϕ phase angle
Σ summation symbol
Δ an interval, range, or difference
ΔR incremental redundancy (redundancy at a given lag minus redundancy at the previous lag)
| "given," or "given a value of"

PART I
BACKGROUND

What is this business called "chaos"? What does it deal with, and why do people think it's important? Let's begin with those and similar questions.

Chapter 1
Introduction

The concept of **chaos** is one of the most exciting and rapidly expanding research topics of recent decades. Ordinarily, chaos is disorder or confusion. In the scientific sense, chaos does involve some disarray, but there's much more to it than that. We'll arrive at a more complete definition in the next chapter.

The chaos that we'll study is a particular class of *how something changes over time*. In fact, *change* and *time* are the two fundamental subjects that together make up the foundation of chaos. The weather, Dow–Jones industrial average, food prices, and the size of insect populations, for example, all change with time. (In chaos jargon, these are called **systems**. A "system" is an assemblage of interacting parts, such as a weather system. Alternatively, it is a group or sequence of elements, especially in the form of a chronologically ordered **set** of data. We'll have to start speaking in terms of systems from now on.) Basic questions that led to the discovery of chaos are based on change and time. For instance, what's the qualitative long-term behavior of a changing system? Or, given nothing more than a record of how something has changed over time, how much can we learn about the underlying system? Thus, "behavior over time" will be our theme.

The next chapter goes over some reasons why chaos can be important to you. Briefly, if you work with numerical measurements (data), chaos can be important because its presence means that long-term predictions are worthless and futile. Chaos also helps explain irregular behavior of something over time. Finally, whatever your field, it pays to be familiar with new directions and new interdisciplinary topics (such as chaos) that play a prominent role in many subject areas. (And, by the way, the only kind of data we can analyze for chaos are rankable numbers, with clear intervals and a zero point as a standard. Thus, data such as "low, medium, or high" or "male/female" don't qualify.)

The easiest way to see how something changes with time (a **time series**) is to make a graph. A baby's weight, for example, might change as shown in Figure 1.1a; Figure 1.1b is a hypothetical graph showing how the price of wheat might change over time.

Even when people don't have any numerical measurements, they can simulate

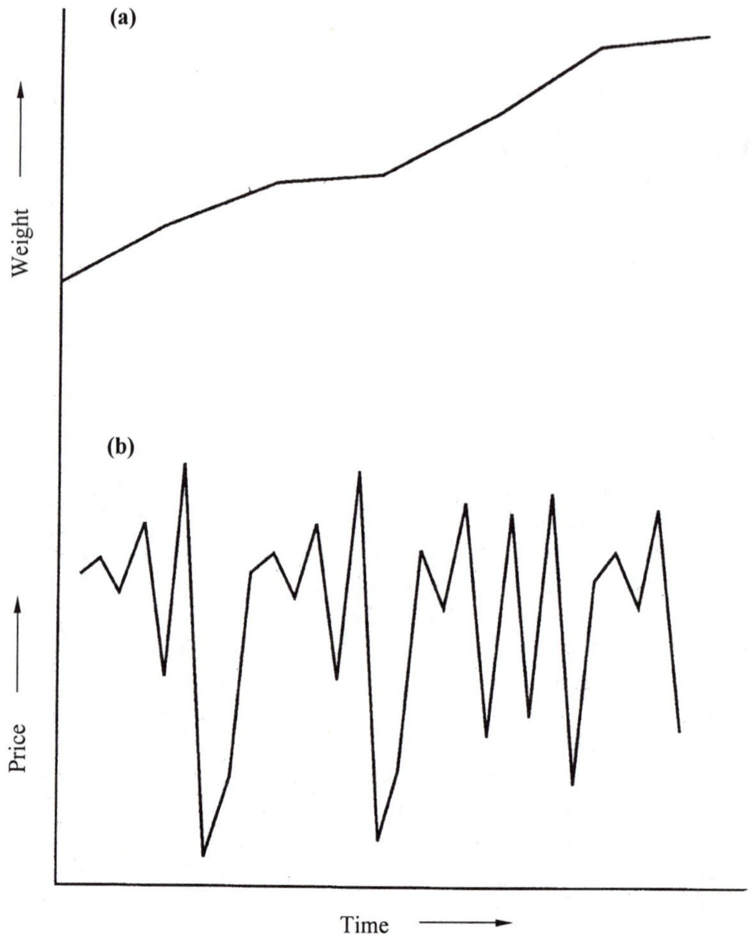

Figure 1.1 Hypothetical time series: (a) change of a baby's weight with time; (b) change in price of wheat over time.

a time series using some specified rule, usually a mathematical equation. The equation describes how a quantity changes from some known beginning state. Figure 1.1b—a pattern that happens to be chaotic—is an example. I generated the pattern with the following special but simple equation (from Grebogi et al. 1983):

$$x_{t+1} = 1.9 - x_t^2 . \quad (1.1)$$

Here x_t (spoken as "x of t") is the value of x at a time t, and x_{t+1} ("x of t plus one") is the value of x at some time interval (day, year, century, etc.) later. That shows one of the requirements for chaos: the value at any time depends in part on the previous value. (The price of a loaf of bread today isn't just a number pulled out of a hat; instead, it depends largely on yesterday's price.) To generate a chaotic time series with Equation 1.1, I first assigned (arbitrarily) the value 1.0 for x_t and used the equation to compute x_{t+1}. That gave $x_{t+1} = 1.9-1^2 = 0.9$. To simulate the idea that the next value depends on the previous one, I then fed back into the equation the x_{t+1} just computed (0.9), but put it in the position of the given x_t. Solving for the new x_{t+1} gave $x_{t+1} = 1.9-0.9^2 = 1.09$. (And so time here is represented by repeated calculations of the equation.) For the next time increment, the computed x_{t+1} (1.09) became the new x_t, and so on, as indicated in the following table:

Input value (x_t)	New value (x_{t+1})
1.0	0.9
0.9	1.09
1.09	0.712
0.712	1.393
etc.	

Repeating this process about 30 times produced a record of widely fluctuating values of x_{t+1} (the time series of Figure 1.1b).

Just looking at the time series of Figure 1.1b, nobody can tell whether it is chaotic. In other words, erratic-looking temporal behavior is just a superficial indicator of possible chaos. Only a detailed analysis of the data, as explained in later chapters, can reveal whether the time series is chaotic.

The simulated time series of Figure 1.1b has several key traits:

- It shows complex, unsystematic motion (including large, sudden qualitative changes), rather than some simple curve, trend, cycle, or equilibrium. (A possible analogy is that many evolving systems in our world show instability, upheaval, surprise, perpetual novelty, and radical events.)
- The indiscriminate-looking pattern didn't come from a haphazard process, such as plucking numbered balls out of a bowl. Quite the contrary: it came from a specific equation. Thus, a chaotic sequence looks haphazard but really is **deterministic**, meaning that it follows a rule. That is, some law, equation, or fixed procedure determines or specifies the results. Furthermore, for given values of the **constants** and input, future results are predictable. For instance, given the constant 1.9 and a value for x_t in Equation 1.1, we can compute x_{t+1} exactly. (Because of that deterministic origin, some people refer to chaos as "deterministic chaos.")
- The equation that generated the chaotic behavior (Eq. 1.1) is simple. Therefore, complex behavior doesn't necessarily have a complex origin.
- The chaotic behavior came about with just one variable (x). (A **variable** is a quantity that can have different numerical values.) That is, chaos doesn't have to come from the interaction of many variables. Instead, just one variable can do it.
- The pattern is entirely self-generated. In other words, aside from any influence of the constant (explored in later chapters), the chaos develops without any external influences whatsoever.

- The irregular evolution came about without the direct influence of sampling or measurement **error** in the calculations. (There aren't any error terms in the equation.)

The revelation that disorganized and complex-looking behavior can come from an elementary, deterministic equation or simple underlying cause was a real surprise to many scientists. Curiously, various fields of study many years earlier accepted a related idea: collections of small entities (particles or whatever) behave haphazardly, even though physical laws govern the particles individually.

Equation 1.1 shows why many scientists are attracted to chaos: behavior that looks complex and even impossible to decipher and understand can be relatively easy and comprehensible. Another attraction for many of us is that many basic concepts of chaos don't require advanced mathematics, such as calculus, differential equations, complex variables, and so on. Instead, you can grasp much of the subject with nothing more than basic algebra, plane geometry, and maybe some rudimentary statistics. Finally, an unexpected and welcome blessing is that, to analyze for chaos, we don't have to know the underlying equation or equations that govern the system.

Chaos is a young and rapidly developing field. Indeed, much of the information in this book was only discovered since the early 1970s. As a result, many aspects of chaos are far from understood or resolved. The most important unresolved matter is probably this: *at present, chaos is extremely difficult to identify in real-world data*. It certainly appears in mathematical (computer) exercises and in some laboratory experiments. (In fact, as we'll see later, once we introduce the idea of nonlinearity into theoretical models, chaos is unavoidable.) However, there's presently a big debate as to whether anyone has clearly identified chaos in field data. (The same difficulty, of course, accompanies the search for any kind of complex structure in field data. Simple patterns we can find and approximate; complex patterns are another matter.) In any event, we can't just grab a nice little set of data, apply a simple test or two, and declare "chaos" or "no chaos."

The reason why recognizing chaos in real-world data is such a monumental challenge is that the analysis methods aren't yet perfected. The tools we have right now look attractive and enticing. However, they were developed for highly idealized conditions, namely:

- systems of no more than two or three variables
- very big datasets (*typically many thousands of observations*, and in some cases millions)
- unrealistically high **accuracy** in the data measurements
- data having negligible amounts of **noise** (unwanted disturbance superimposed on, or unexplainable variability in, useful data).

Problems arise when data don't fulfil those four criteria. Ordinary data (mine and very possibly yours) rarely fulfil them. For instance, datasets of 50–100 values (the kind I'm used to) are way too small. One of the biggest problems is that, when applied to ordinary data, the present methods often give plausible but misleading results, suggesting chaos when in fact there isn't any.

Having said all that, here's something equally important on the positive side: applying chaos analysis to a set of data (even if those data aren't ideal) can reveal many important features that other, more traditional tools might not disclose.

Description and theory of chaos are far ahead of identifying chaos in real-world data. However, with the present popularity of chaos as a research topic, new and improved methods are emerging regularly.

Summary

Chaos (deterministic chaos) deals with long-term evolution—how something changes over a long time. A chaotic time series looks irregular. Two of chaos's important practical implications are that long-term predictions under chaotic conditions are worthless and complex behavior can have simple causes. Chaos is difficult to identify in real-world data because the available tools generally were developed for idealistic conditions that are difficult to fulfil in practice.

Chapter 2
Chaos in perspective

Where chaos occurs

Chaos, as mentioned, deals mostly with how something evolves over time. Space or distance can take the place of time in many instances. For that reason, some people distinguish between "temporal chaos" and "spatial chaos."

What kinds of processes in the world are susceptible to chaos? Briefly, chaos happens only in **deterministic**, **nonlinear**, **dynamical** systems. (I'll define "non-linear" and "dynamical" next. At the end of the book there is a glossary of these and other important terms.) Based on those qualifications, here's an admittedly imperfect but nonetheless reasonable definition of chaos:

> Chaos is sustained and disorderly-looking long-term evolution that satisfies certain special mathematical criteria and that occurs in a deterministic non-linear system.

Chaos theory is the principles and mathematical operations underlying chaos.

Nonlinearity

Nonlinear means that output isn't directly proportional to input, or that a change in one variable doesn't produce a proportional change or reaction in the related variable(s). In other words, a system's values at one time aren't proportional to the values at an earlier time. An alternate and sort of "cop-out" definition is that nonlinear refers to anything that isn't linear, as defined below. There are more formal, rigid, and complex mathematical definitions, but we won't need such detail. (In fact, although the meaning of "nonlinear" is clear intuitively, the experts haven't yet come up with an all-inclusive definition acceptable to everyone. Interestingly, the same is true of other common mathematical terms, such as number, system, set, point, infinity, random, and certainly chaos.) A **nonlinear equation** is an equation involving two variables, say x and y, and two **coefficients**, say b and c, in some form that doesn't plot as a straight line on ordinary graph paper.

The simplest nonlinear response is an all-or-nothing response, such as the freezing of water. At temperatures higher than 0°C, nothing happens. At or below that threshold temperature, water freezes. With many other examples, a nonlinear relation describes a curve, as opposed to a threshold or straight-line relation, on arithmetic (uniformly scaled) graph paper.

Most of us feel a bit uncomfortable with nonlinear equations. We prefer a **linear** relation any time. A **linear equation** has the form $y = c + bx$. It plots as a straight line on ordinary graph paper. Alternatively, it is an equation in which the variables are directly proportional, meaning that no variable is raised to a power other than one. Some of the reasons why a linear relation is attractive are:

- the equation is easy
- we're much more familiar and comfortable with the equation
- extrapolation of the line is simple
- comparison with other linear relations is easy and understandable
- many software packages are commercially available to provide extensive statistical analyses.

Classical mathematics wasn't able to analyze nonlinearity effectively, so linear approximations to curves became standard. Most people have such a strong preference for straight lines that they usually try to make nonlinear relations into straight lines by transforming the data (e.g. by taking **logarithms** of the data). (To "**transform**" data means to change their numerical description or scale of measurement.) However, that's mostly a graphical or analytical convenience or gimmick. It doesn't alter the basic nonlinearity of the physical process. Even when people don't transform the data, they often fit a straight line to points that really plot as a curve. They do so because either they don't realize it is a curve or they are willing to accept a linear approximation.

Campbell (1989) mentions three ways in which linear and nonlinear phenomena differ from one another:

- *Behavior over time* Linear processes are smooth and regular, whereas nonlinear ones may be regular at first but often change to erratic-looking.
- *Response to small changes in the environment or to stimuli* A linear process changes smoothly and in proportion to the stimulus; in contrast, the response of a **nonlinear system** is often much greater than the stimulus.
- *Persistence of local pulses* Pulses in **linear systems** decay and may even die out over time. In nonlinear systems, on the other hand, they can be highly coherent and can persist for long times, perhaps forever.

A quick look around, indoors and out, is enough to show that Nature doesn't produce straight lines. In the same way, processes don't seem to be linear. These days, voices are firmly declaring that many—possibly most—actions that last over time are nonlinear. Murray (1991), for example, states that "If a mathematical model for any biological phenomenon is linear, it is almost certainly irrelevant from a biological viewpoint." Fokas (1991) says: "The laws that govern most of the phenomena that can be studied by the physical sciences, engineering, and social sciences are, of course, nonlinear." Fisher (1985) comments that nonlinear motions make up by far the most common class of things in the universe. Briggs & Peat (1989) say that linear systems seem almost the exception rather than the rule and refer to an "ever-sharpening picture of universal nonlinearity." Even more radically, Morrison (1988) says simply that "linear systems don't exist in nature." Campbell et al. (1985) even object to the term nonlinear, since it implies that linear relations are the norm or standard, to which we're supposed to compare other types of relations. They attribute to Stanislaw Ulam the comment that using the term "nonlinear science" is like referring to the bulk of zoology as the study of non-elephant animals.

Dynamics

The word **dynamics** implies force, energy, motion, or change. A **dynamical system** is anything that moves, changes, or evolves in time. Hence, chaos deals with what the experts like to refer to as **dynamical-systems theory** (the study of phenomena that vary with time) or **nonlinear dynamics** (the study of nonlinear movement or evolution).

Motion and change go on all around us, every day. Regardless of our particular specialty, we're often interested in understanding that motion. We'd also like to forecast how something will behave over the long run and its eventual outcome.

Dynamical systems fall into one of two categories, depending on whether the system loses energy. A **conservative** dynamical system has no friction; it doesn't lose energy over time. In contrast, a **dissipative** dynamical system has friction; it loses energy over time and therefore always approaches some asymptotic or limiting condition. That asymptotic or limiting state, under certain conditions, is where chaos occurs. Hence, dissipative systems are the only kind we'll deal with in this book.

Phenomena happen over time in either of two ways. One way is at **discrete** (separate or distinct) intervals. Examples are the occurrence of earthquakes, rainstorms, and volcanic eruptions. The other way is **continuously** (air temperature and humidity, the flow of water in perennial rivers, etc.). Discrete intervals can be spaced evenly in time, as implied for the calculations done for Figure 1.1b, or irregularly in time. Continuous phenomena might be measured continuously, for instance by the trace of a pen on a slowly moving strip of paper. Alternatively, we might measure them at discrete intervals. For example, we might measure air temperature only once per hour, over many days or years.

Special types of equations apply to each of those two ways in which phenomena happen over time. Equations for discrete time changes are **difference equations** and are solved by **iteration**, explained below. In contrast, equations based on a continuous change (continuous measurements) are **differential equations**.

You'll often see the term "**flow**" with differential equations. To some authors (e.g. Bergé et al. 1984: 63), a flow is a system of differential equations. To others (e.g. Rasband 1990: 86), a flow is the solution of differential equations.

Differential equations are often the most accurate mathematical way to describe a smooth continuous evolution. However, some of those equations are difficult or impossible to solve. In contrast, difference equations usually can be solved right away. Furthermore, they are often acceptable approximations of differential equations. For example, a baby's growth is continuous, but measurements taken at intervals can approximate it quite well. That is, it is a continuous development that we can adequately represent and conveniently analyze on a discrete-time basis. In fact, Olsen & Degn (1985) say that difference equations are the most powerful vehicle to the understanding of chaos. We're going to confine ourselves just to discrete observations in this book. The physical process underlying those discrete observations might be discrete or continuous.

Iteration is a mathematical way of simulating discrete-time evolution. To iterate means to repeat an operation over and over. In chaos, it usually means to solve or apply the same equation repeatedly, often with the outcome of one solution fed back in as input for the next, as we did in Chapter 1 with Equation 1.1. It is a standard method for analyzing activities that take place in equal, discrete time steps or continuously, but whose equations can't be solved exactly, so that we have to settle for successive discrete approximations (e.g. the time-behavior of materials and fluids). I'll use "number of iterations" and "time" synonymously and interchangeably from now onward.

Iteration is the mathematical counterpart of **feedback**. Feedback in general is any response to something sent out. In mathematics, that translates as "what goes out comes back in again." It is output that returns to serve as input. In temporal processes, feedback is that part of the past that influences the present, or that part of the present that influences the future. Positive feedback amplifies or accelerates the output. It causes an event to become magnified over time. Negative feedback dampens or inhibits output, or causes an event to die away over time. Feedback shows up in climate, biology, electrical engineering, and probably in most other fields in which processes continue over time.

The time frame over which chaos might occur can be as short as a fraction of a second. At the other extreme, the time series can last over hundreds of thousands of years, such as from the Pleistocene Epoch (say about 600000 years ago) to the present.

The multidisciplinary nature of chaos

In theory, virtually anything that happens over time could be chaotic. Examples are epidemics, pollen production, populations, incidence of forest fires or droughts, economic changes, world ice volume, rainfall rates or amounts, and so on. People have looked for (or studied) chaos in physics, mathematics, communications, chemistry, biology, physiology, medicine, ecology, hydraulics, geology, engineering, atmospheric sciences, oceanography, astronomy, the solar system, sociology, literature, economics, history, international relations, and in other fields. That makes it truly interdisciplinary. It forms a common ground or builds bridges between different fields of study.

Uncertain prominence

Opinions vary widely as to the importance of chaos in the real world. At one extreme are those scientists who dismiss chaos as nothing more than a mathematical curiosity. A middle group is at least receptive. Conservative members of the middle group feel that chaos may well be real but that the evidence so far is more illusory than scientific (e.g. Berryman & Millstein 1989). They might say, on the one hand, that chaos has a rightful place as a study topic and perhaps ought to be included in college courses (as indeed it is). On the other hand, they feel that it doesn't necessarily pervade everything in the world and that its relevance is probably being oversold. This sizeable group (including many chaos specialists) presently holds that, although chaos appears in mathematical experiments (iterations of nonlinear equations) and in tightly controlled laboratory studies, nobody has yet conclusively found it in the physical world.

Moving the rest of the way across the spectrum of attitudes brings us to the exuberant, ecstatic fringe element. That euphoric group holds that chaos is the third scientific revolution of the twentieth century, ranking right up there with relativity and quantum mechanics. Such opinions probably stem from reports that researchers find or claim to find chaos in chemical reactions, the weather, asteroid movement, the motion of atoms held in an electromagnetic field, lasers, the electrical activity of the heart and brain, population fluctuations of plants and animals, the stock market, and even in the cries of newborn babies (Pool 1989a, Mende et al. 1990). One enthusiastic proponent believes there's now substantial though nonrigorous evidence that chaos is the rule rather than the exception in Newtonian dynamics. He also says that "few observers doubt that chaos is ubiquitous throughout nature" (Ford 1989). Verification or disproval of such a claim awaits further research.

Causes of chaos

Chaos, as Chapter 1 showed, can arise simply by iterating mathematical equations. Chapter 13 discusses some factors that might cause chaos in such iterations.

The conditions required for chaos in our physical world (Bergé et al. 1984: 265), on the other hand, aren't yet fully known. In other words, if chaos does develop in Nature, the reason generally isn't clear. Three possible causes have been proposed:

- An increase in a control factor to a value high enough that chaotic, disorderly behavior sets in. Purely mathematical calculations and controlled laboratory experiments show this method of initiating chaos. Examples are iterations of equations of various sorts, the selective heating of fluids in small containers (known as Rayleigh–Benard experiments), and oscillating chemical reactions (Belousov–Zhabotinsky experiments). Maybe that same cause operates with natural physical systems out in the real world (or maybe it doesn't). Berryman & Millstein (1989), for example, believe that even though natural ecosystems normally don't become chaotic, they could if humans interfered. Examples of such interference are the extermination of predators or an increase in an organism's growth rate through biotechnology.

- The nonlinear interaction of two or more separate physical operations. A popular classroom example is the double pendulum—one pendulum dangling from the lower end of another pendulum, constrained to move in a plane. By regulating the upper pendulum, the teacher makes the lower pendulum flip about in strikingly chaotic fashion.
- The effect of ever-present environmental noise on otherwise regular motion (Wolf 1983). At the least, such noise definitely hampers our ability to analyze a time series for chaos.

Benefits of analyzing for chaos

Important reasons for analyzing a set of data for chaos are:

- Analyzing data for chaos can help indicate whether haphazard-looking fluctuations actually represent an orderly system in disguise. If the sequence is chaotic, there's a discoverable law involved, and a promise of greater understanding.
- Identifying chaos can lead to greater accuracy in short-term predictions. Farmer & Sidorowich (1988a) say that "most forecasting is currently done with linear methods. Linear dynamics cannot produce chaos, and linear methods cannot produce good forecasts for chaotic time series." Combining the principles of chaos, including nonlinear methods, Farmer & Sidorowich found that their forecasts for short time periods were roughly 50 times more accurate than those obtained using standard linear methods.
- Chaos analysis can reveal the time-limits of reliable predictions and can identify conditions where long-term forecasting is largely meaningless. If something is chaotic, knowing when reliable predictability dies out is useful, because predictions for all later times are useless. As James Yorke said, "it's worthwhile knowing ahead of time when you can't predict something" (Peterson 1988).
- Recognizing chaos makes modeling easier. (A **model** is a simplified representation of some process or phenomenon. *Physical* or *scale* models are miniature replicas [e.g. model airplanes] of something in real life. *Mathematical* or *statistical* models explain a process in terms of equations or statistics. *Analog* models simulate a process or system by using one-to-one "analogous"

physical quantities [e.g. lengths, areas, ohms, or volts] of another system. Finally, *conceptual* models are qualitative sketches or mental images of how a process works.) People often attribute irregular evolution to the effects of many external factors or variables. They try to model that evolution statistically (mathematically). On the other hand, just a few variables or deterministic equations can describe chaos.

Although chaos theory does provide the advantages just mentioned, it doesn't do everything. One area where it is weak is in revealing details of a particular underlying physical law or governing equation. There are instances where it might do so (see, for example, Farmer & Sidorowich 1987, 1988a,b, Casdagli 1989, and Rowlands & Sprott 1992). At present, however, we usually don't look for (or expect to discover) the rules by which a system evolves when analyzing for chaos.

Contributions of chaos theory

Some important general benefits or contributions from the discovery and development of chaos theory are the following.

Randomness

Realizing that a simple, deterministic equation can create a highly irregular or unsystematic time series has forced us to reconsider and revise the long-held idea of a clear separation between determinism and randomness. To explain that, we've got to look at what "**random**" means. Briefly, there isn't any general agreement. Most people think of "random" as disorganized, haphazard, or lacking any apparent order or pattern. However, there's a little problem with that outlook: a long list of random events can show streaks, clumps, or patterns (Paulos 1990: 59–65). Although the individual events are not predictable, certain aspects of the clumps are.

Other common definitions of "random" are:

- every possible value has an equal chance of selection
- a given observation isn't likely to recur
- any subsequent observation is unpredictable
- any or all of the observations are hard to compute (Wegman 1988).

My definition, modified from that of Tashman & Lamborn (1979: 216) is: random means based strictly on a chance mechanism (the luck of the draw), with negligible deterministic effects. That definition implies that, in spite of what most people probably would assume, anything that is random really has some inherent determinism, however small. Even a flip of a coin varies according to some influence or associated forces. A strong case can be made that there isn't any such thing as true randomness, in the sense of no underlying determinism or outside influence

(Ford 1983, Kac 1983, Wegman 1988). In other words, the two terms "random" and "deterministic" aren't mutually exclusive; anything random is also deterministic, and both terms can characterize the same sequence of data. That notion also means that there are degrees of randomness (Wegman 1988). In my usage, "random" implies negligible determinism.

People used to attribute apparent randomness to the interaction of complex processes or to the effects of external unmeasured forces. They routinely analyzed the data statistically. Chaos theory shows that such behavior can be attributable to the nonlinear nature of the system rather than to other causes.

Dynamical systems technology

Recognition and study of chaos has fostered a whole new technology of dynamical systems. The technology collectively includes many new and better techniques and tools in nonlinear dynamics, time-series analysis, short- and long-range prediction, quantifying complex behavior, and numerically characterizing non-Euclidean objects. In other words, studying chaos has developed procedures that apply to many kinds of complex systems, not just chaotic ones. As a result, chaos theory lets us describe, analyze, and interpret temporal data (whether chaotic or not) in new, different, and often better ways.

Nonlinear dynamical systems

Chaos has brought about a dramatic resurgence of interest in nonlinear dynamical systems. It has thereby helped accelerate a new approach to science and to numerical analysis in general. (In fact, the International Federation of Nonlinear Analysts was established in 1991.) In so doing, it has diminished the apparent role of linear processes. For instance, scientists have tended to think of Earth processes in terms of Newton's laws, that is, as reasonably predictable if we know the appropriate laws and present condition. However, the nonlinearity of many processes, along with the associated sensitive dependence on initial conditions (discussed in a later chapter), makes reliable predictability very difficult or impossible. Chaos emphasizes that basic impossibility of making accurate long-term predictions. In some cases, it also shows how such a situation comes about. In so doing, chaos brings us a clearer perspective and understanding of the world as it really is.

Controlling chaos

Studying chaos has revealed circumstances under which we might want to avoid chaos, guide a system out of it, design a product or system to lead into or against it, stabilize or control it, encourage or enhance it, or even exploit it. Researchers are actively pursuing those goals. For example, there's already a vast literature on controlling chaos (see, for instance, Abarbanel et al. 1993, Shinbrot 1993, Shinbrot et al. 1993, anonymous 1994). Berryman (1991) lists ideas for avoiding chaos in ecology. A field where we might want to encourage chaos is physiology; studies of diseases, nervous disorders, mental depression, the brain, and the heart suggest that many physiological features behave chaotically in healthy individuals and more regularly in unhealthy ones (McAuliffe 1990). Chaos reportedly brings about greater efficiency in mixing processes (Pool 1990b, Ottino et al. 1992). Finally, it might help encode electronic messages (Ditto & Pecora 1993).

Historical development

The word "chaos" goes back to Greek mythology, where it had two meanings:

- the primeval emptiness of the universe before things came into being
- the abyss of the underworld.

Later it referred to the original state of things. In religion it has had many different and ambiguous meanings over many centuries. Today in everyday English it usually means a condition of utter confusion, totally lacking in order or organization. Robert May (1974) seems to have provided the first written use of the word in regard to deterministic nonlinear behavior, but he credits James Yorke with having coined the term.

Foundations

Various bits and snatches of what constitutes deterministic chaos appeared in scientific and mathematical literature at least as far back as the nineteenth century. Take, for example, the chaos themes of **sensitivity to initial conditions** and long-term unpredictability. The famous British physicist James Clerk Maxwell reportedly said in an 1873 address that "...When an infinitely small variation in the present state may bring about a finite difference in the state of the system in a finite time, the condition of the system is said to be unstable ... [and] renders impossible the prediction of future events, if our knowledge of the present state is only approximate, and not accurate" (Hunt & Yorke 1993). French mathematician Jacques Hadamard remarked in an 1898 paper that an error or discrepancy in initial conditions can render a system's long-term behavior unpredictable (Ruelle 1991: 47–9). Ruelle says further that Hadamard's point was discussed in 1906 by French physicist Pierre Duhem, who called such long-term predictions "forever unusable." In 1908, French mathematician, physicist, and philosopher Henri Poincaré contributed a discussion along similar lines. He emphasized that slight differences in initial conditions eventually can lead to large differences, making prediction for all practical purposes "impossible." After the early 1900s the general theme of sensitivity to initial conditions receded from attention until Edward Lorenz's work of the 1960s, reviewed below.

Another key aspect of chaos that was first developed in the nineteenth century is entropy. Eminent players during that period were the Frenchman Sadi Carnot, the German Rudolph Clausius, the Austrian Ludwig Boltzmann, and others. A third important concept of that era was the Lyapunov exponent. Major contributors were the Russian–Swedish mathematician Sofya Kovalevskaya and the Russian mathematician Aleksandr Lyapunov.

During the twentieth century many mathematicians and scientists contributed parts of today's chaos theory. Jackson (1991) and Tufillaro et al. (1992: 323–6) give good skeletal reviews of key historical advances. May (1987) and Stewart (1989a: 269) state, without giving details, that some biologists came upon chaos in the 1950s. However, today's keen interest in chaos stems largely from a 1963 paper by meteorologist Edward Lorenz[1] of the Massachusetts Institute of Technology. Modeling weather on his computer, Lorenz one day set out to duplicate a pattern he had derived previously. He started the program on the computer, then left the room for a cup of coffee and other business. Returning a couple of hours later to inspect the "two months" of weather forecasts, he was astonished to see that the predictions for the later stages differed radically from those of the original run. The reason turned out to be that, on his "duplicate" run, he hadn't specified his input data to the usual number of decimal places. These were just the sort of potential discrepancies first mentioned around the turn of the century. Now recognized as one of the chief features of "sensitive dependence on initial conditions," they have emerged as a main characteristic of chaos.

Chaos is truly the product of a team effort, and a large team at that. Many contributors have worked in different disciplines, especially various branches of mathematics and physics.

The coming of age

The scattered bits and pieces of chaos began to congeal into a recognizable whole in the early 1970s. It was about then that fast computers started becoming more available and affordable. It was also about then that the fundamental and crucial importance of *nonlinearity* began to be appreciated. The improvement in and accessibility to fast and powerful computers was a key development in studying

1. **Lorenz, Edward N. (1917–)** Originally a mathematician, Ed Lorenz worked as a weather forecaster for the US Army Air Corps during the Second World War. That experience apparently hooked him on meteorology. His scientific interests since that time have centered primarily on weather prediction, atmospheric circulation, and related topics. He received his Master's degree (1943) and Doctor's degree (1948) from Massachusetts Institute of Technology (MIT) and has stayed at MIT for virtually his entire professional career (now being Professor Emeritus of Meteorology). His major (but not only) contribution to chaos came in his 1963 paper "Deterministic nonperiodic flow." In that paper he included one of the first diagrams of what later came to be known as a strange attractor.

nonlinearity and chaos. Computers "love to" iterate and are good at it—much better and faster than anything that was around earlier. Chaos today is intricately, permanently and indispensably welded to computer science and to the many other disciplines mentioned above, in what Percival (1989) refers to as a "thoroughly modern marriage."

From the 1970s onward, chaos's momentum increased like the proverbial locomotive. Many technical articles have appeared since that time, especially in such journals as *Science*, *Nature*, *Physica D*, *Physical Review Letters*, and *Physics Today*. As of 1991, the number of papers on chaos was doubling about every two years. Some of the new journals or magazines devoted largely or entirely to chaos and nonlinearity that have been launched are *Chaos*, *Chaos, Solitons, and Fractals*, *International Journal of Bifurcation and Chaos*, *Journal of Nonlinear Science*, *Nonlinear Science Today*, and *Nonlinearity*. In addition, entire courses in chaos now are taught in colleges and universities. In short, chaos is now a major industry.

Along with the articles and journals, many books have emerged. Gleick's (1987) general nonmathematical introduction has been widely acclaimed and has sold millions of copies. Examples of other nonmathematical introductions are Briggs & Peat (1989), Stewart (1989a), Peters (1991), and Çambel (1993). Peitgen et al. (1992) give very clear and not overly mathematical explanations of many key aspects of chaos. A separate reference list at the end of this book includes a few additional books. Interesting general articles include those by Pippard (1982), Crutchfield et al. (1986), Jensen (1987), Ford (1989), and Pool (1989a–f). For good (but in some cases a bit technical) review articles, try Olsen & Degn (1985), Grebogi et al. (1987), Gershenfeld (1988), Campbell (1989), Eubank & Farmer (1990), and Zeng et al. (1993). Finally, the 1992 edition of the *McGraw-Hill encyclopedia of science and technology* includes articles on chaos and related aspects of chaos.

Summary

Chaos occurs only in deterministic, nonlinear, dynamical systems. The prominence of nonlinear (as compared to linear) systems in the world has received more and more comment in recent years. A dynamical system can evolve in either of two ways. One is on separate, distinct occasions (discrete intervals). The other is continuously. If it changes continuously, we can measure it discretely or continuously. Equations based on discrete observations are called difference equations, whereas those based on continuous observations are called differential equations. Iteration is a mathematical way of simulating regular, discrete intervals. Iteration also incorporates the idea that any object's value at a given time depends on its value at the previous time. Chaos is a truly interdisciplinary topic in that specialists in many subjects study it. Just how prominent it is remains to be determined.

Proposed causes of chaos in the natural world are a critical increase in a control factor, the nonlinear interaction of two or more processes, and environmental noise. Analyzing data for chaos can reveal whether an erratic-looking time series is actually deterministic, lead to more accurate short-term predictions, tell us whether long-term predicting for the system is at all feasible, and simplify modeling. On the other hand, chaos analysis doesn't reveal underlying physical laws. Important general benefits that chaos theory has brought include reconsidered or revised concepts about determinism versus randomness, many new and improved techniques in nonlinear analysis, the realization that nonlinearity is much more common and important than heretofore assumed, and identifying conditions under which chaos might be detrimental or desirable.

Chaos theory (including sensitivity to initial conditions, long-term unpredictability, entropy, Lyapunov exponents, and probably other aspects) goes back at least as far as the late nineteenth century. However, most of present-day chaos theory as a unified discipline has developed since the late 1960s or early 1970s, largely because computers became more available around that time. The number of publications on chaos began increasing sharply in the early 1990s.

PART II
THE AUXILIARY TOOLKIT

Chaos theory uses many tools, mostly from mathematics, physics, and statistics. Much of the secret to understanding chaos theory is being familiar with those tools right from the start. They provide an essential foundation. Chapters 3–9 cover the more important ones.

Chapter 3

Phase space—the playing field

We'll begin by setting up the arena or playing field. One of the best ways to understand a dynamical system is to make those dynamics visual. A good way to do that is to draw a graph. Two popular kinds of graph show a system's dynamics. One is the ordinary time-series graph that we've already discussed (Fig. 1.1). Usually, that's just a two-dimensional plot of some variable (on the vertical **axis**, or ordinate) versus time (on the horizontal axis, or abscissa).

Right now we're going to look at the other type of graph. It doesn't plot time directly. The axis that normally represents time therefore can be used for some other variable. In other words, the new graph involves more than one variable (besides time). A point plotted on this alternate graph reflects the **state** or **phase** of the system at a particular time (such as the phase of the Moon). Time shows up but in a relative sense, by the sequence of plotted points (explained below).

The space on the new graph has a special name: **phase space** or **state space** (Fig. 3.1). In more formal terms, phase space or state space is an abstract mathematical space in which **coordinates** represent the variables needed to specify the phase (or state) of a dynamical system. The phase space includes all the instantaneous states the system can have. (Some specialists draw various minor technical distinctions between "phase space" and "state space," but I'll go along with the majority and treat the two synonymously.)

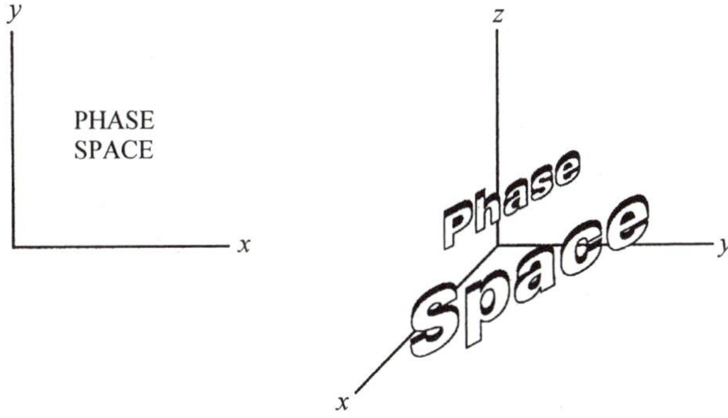

Figure 3.1 Two-dimensional (left diagram) and three-dimensional phase space.

As a complement to the common time-series plot, a phase space plot provides a different view of the evolution. Also, whereas some time series can be very long and therefore difficult to show on a single graph, a phase space plot condenses all the data into a manageable space on a graph. Thirdly, as we'll see later, structure we might not see on a time-series plot often comes out in striking fashion on the phase space plot. For those reasons, most of chaos theory deals with phase space.

A graph ordinarily can accommodate only three variables or fewer. However, chaos in real-world situations often involves many variables. (In some cases, there are ways to discount the effects of most of those variables and to simplify the analysis.) Although no one can draw a graph for more than three variables while still keeping the axes at right angles to one another, the idea of phase space holds for any number of variables. How do you visualize a phase space with more than three variables or dimensions? Perhaps the easiest way is to stop thinking in terms of a physical or graphical space (such as three mutually perpendicular axes) and just think in terms of number of variables.

Systems having more than three variables often can be analyzed only mathematically. (However, some tools enable us to simplify or condense the information contained in many variables and thus still use graphs advantageously.) Mathematical analysis of systems with more than three variables requires us to buy the idea that it is legitimate to extend certain mathematical definitions and relations valid for three or fewer dimensions to four or more dimensions. That assumption is largely intuitive and sometimes wrong. Not much is known about the mathematics of higher-dimensional space (Cipra 1993).

In chaos jargon, phase space having two axes is called "two-space." For three variables, it is three-space, and so on.

Chaos theory deals with two types of phase space: **standard phase space** (my term) and **pseudo phase space**. The two types differ in the number of independent physical features they portray (e.g. temperature, wind velocity, humidity, etc.) and in whether a plotted point represents values measured at the same time or at successive times.

Standard phase space

Standard phase space (hereafter just called phase space) is the phase space defined above: an abstract space in which coordinates represent the variables needed to specify the state of a dynamical system at a particular time. On a graph, a plotted point neatly and compactly defines the system's condition for some measuring occasion, as indicated by the point's coordinates (values of the variables). For example, we might plot a baby's height against its weight. Any plotted point represents the state of the baby (a dynamical system!) at a particular time, in terms of height and weight (Fig. 3.2). The next plotted point is the same baby's height and weight at one time interval later, and so on. Thus, the succession of plotted points

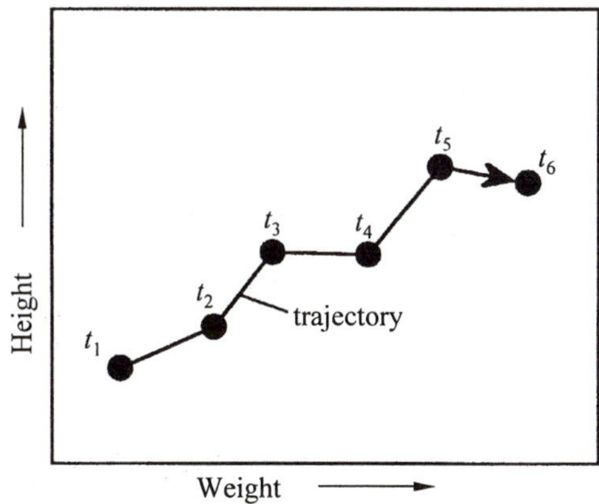

Figure 3.2 Standard phase space graph of baby's height and weight progression. The first measuring occasion ("time one") is labeled t_1; the second is t_2, etc.

shows how the baby grew over time. That is, comparing successive points shows how height has changed relative to weight, over time, *t*.

A line connecting the plotted points in their chronological order shows temporal evolution more clearly on the graph. The complete line on the graph (i.e. the sequence of measured values or list of successive **iterates** plotted on a phase space graph) describes a time path or **trajectory**. A trajectory that comes back upon itself to form a closed loop in phase space is called an **orbit**. (The two terms are often used synonymously.) Each plotted point along any trajectory has evolved directly from (or is partly a result of) the preceding point. As we plot each successive point in phase space, the plotted points migrate around. Orbits and trajectories therefore reflect the *movement* or *evolution* of the dynamical system, as with our baby's weights and heights. That is, an orbit or trajectory moves around in the phase space with time. The trajectory is a neat, concise geometric picture that describes part of the system's history. When drawn on a graph, a trajectory isn't necessarily smooth, like the arcs of comets or cannonballs; instead, it can zigzag all over the phase space, at least for discrete data.

The phase space plot is a world that shows the trajectory and its development. Depending on various factors, different trajectories can evolve for the same system. The phase space plot and such a family of trajectories together are a **phase space portrait**, **phase portrait**, or **phase diagram**.

The phase space for any given system isn't limitless. On the contrary, it has rigid boundaries. The minimum and maximum possible values of each variable define the boundaries. We might not know what those values are.

A phase space with plotted trajectories ideally shows the complete set of all possible states that a dynamical system can ever be in. Usually, such a full portrayal is possible only with hypothetical data; a phase space graph of real-world data usually doesn't cover all of the system's possible states. For one thing, our measuring device might not have been able to measure the entire range of possible values, for each variable. For another, some combinations of values might not have occurred during the study. In addition, some relevant variables might not be on the graph,

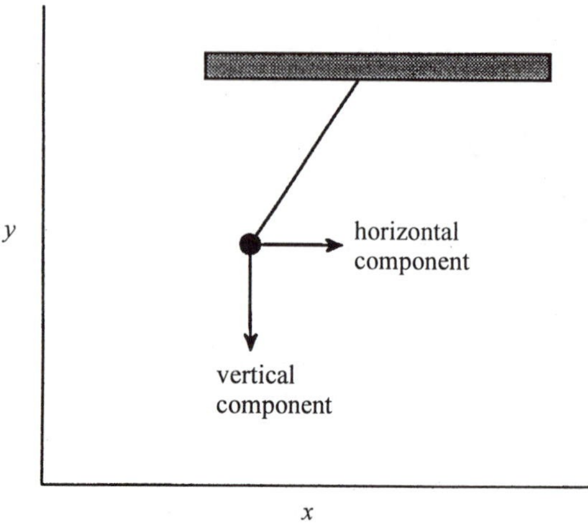

Figure 3.3 Pendulum and its complete description at any time by means of two position coordinates (*x*,*y*) and two velocity components.

either because we don't know they are relevant or we've already committed the three axes.

For any system, there are often several ways to define a standard phase space and its variables. In setting up a phase space, one major goal is to describe the system by using the fewest possible number of variables. That simplifies the analysis and makes the system easier to understand. (And besides, if we want to draw a graph we're limited to three variables.) A pendulum (Fig. 3.3) is a good example. Its velocity changes with its position or displacement along its arc. We could describe any such position in terms of two variables—the *x*,*y* coordinates. Similarly, we could describe the velocity at any time in terms of a horizontal component and a vertical component. That's four variables needed to describe the state of the system at any time. However, the two velocity components could be combined into one variable, the angular velocity. That simplifies our analysis by reducing the four variables to three. Furthermore, the two position variables *x* and *y* could be combined into one variable, the angular position. That reduces the number of variables to two—a situation much more preferable than four.

Phase space graphs might contain hundreds or even thousands of chronologically plotted points. Such a graph is a global picture of many or all of the system's possible states. It can also reveal whether the system likes to spend more time in certain areas (certain combinations of values of the variables).

The trajectory in Figure 3.2 (the baby's growth) is a simple one. Phase space plots in nonlinear dynamics and chaos produce a wide variety of patterns—some simple and regular (e.g. arcs, loops and doughnuts), others with an unbelievable mixture of complexity and beauty.

Pseudo (lagged) phase space

Maps

One of the most common terms in the chaos literature is **map**. Mathematically, a map (defined more completely below) is a **function**. And what's a function? As applied to a variable, a function is a "dependent" or output variable whose value is uniquely determined by one or more input ("independent") variables. Examples are the variable or function y in the relations $y = 3x$, $y = 5x^3-2x$, and $y = 4x^2+3.7z^{-2.66}$. In a more general sense, a function is an equation or relation between two groups, A and B, such that at least one member of group A is matched with one member of group B (a "single-valued" function) or is matched with two or more members of group B (a "multi-valued" function). An example of a single-valued function is $y = 5x$; for any value of x, there is only one value of y, and vice versa. An example of a multi-valued function is $y = x^2$; for $y = 4$, x can be +2 or −2, that is, one value of y is matched with two values of x. "Function" often means "single-valued function."

It's customary to indicate a function with parentheses. For instance, to say that "x is a function f of time t," people write $x = f(t)$. From the context, it is up to us to see that they aren't using parentheses here to mean multiplication.

In chaos, a map or function is an equation or rule that specifies how a dynamical system evolves forward in time. It turns one number into another by specifying how x, usually (but not always) via a discrete step, goes to a new x. An example is $x_{t+1} = 1.9-x_t^2$ (Eq. 1.1). Equation 1.1 says that x_t evolves one step forward in time (acquires a new value, x_{t+1}) by an amount equal to $1.9-x_t^2$. In general, given an initial or beginning value x_0 (pronounced "x naught") and the value of any constants, the map gives x_1; given x_1, the map gives x_2; and so on. Like a geographic map, it clearly shows a route and accurately tells us how to get to our next point. Figuratively, it is an input–output machine. The machine knows the equation, rule, or correspondence, so we put in x_0 and get out x_1; input x_1, output x_2; and so on. Sometimes you'll also see "map" used as a verb to indicate the mathematical process of assigning one value to another, as in Equation 1.1.

As you've probably noticed, computing a map follows an orderly, systematic series of steps. Any such orderly series of steps goes by the general name of an **algorithm**. An algorithm is any recipe for solving a problem, usually with the aid of a computer. The "recipe" might be a general plan, step-by-step procedure, list of rules, set of instructions, systematic program, or set of mathematical equations. An algorithm can have many forms, such as a loosely phrased group of text statements, a picture called a flowchart, or a rigorously specified computer program.

Because a map in chaos theory is often an equation meant to be iterated, many authors use "map" and "equation" interchangeably. Further, they like to show the iterations on a graph. In such cases, they might call the graph a "map." "Map" (or its equivalent, "**mapping**") can pertain to either form—equation or graph.

A **one-dimensional map** is a map that deals with just one physical feature, such as temperature. It's a rule (such as an iterated equation) relating that feature's value at one time to its value at another time. Graphs of such data are common. By convention, the input or older value goes on the horizontal axis, and the output value or function goes on the vertical axis. That general plotting format leaves room to label the axes in various ways. Four common labeling schemes for the ordinate and abscissa, respectively, are:

- "output *x*" and "input *x*"
- "x_{t+1}" and "x_t"
- "next *x*" and "*x*"
- "*x*" and "previous *x*."

(Calling the graph "one-dimensional" is perhaps unfortunate and confusing, because it has two axes or coordinates. A few authors, for that reason, call it "two dimensional." It's hard to reverse the tide now, though.)

Plotting a one-dimensional map is easy (Fig. 3.4). The first point has an abscissa value of the input *x* ("x_t") and an ordinate value of the output *x* ("x_{t+1}"). We therefore go along the horizontal axis to the first value of *x*, then vertically to the second value of *x*, and plot a point. For the second point, everything moves up by one. That means the abscissa's new value becomes observation number two (which is the old x_{t+1}), and the ordinate's new value is observation number three, and so on. (Hence, each measurement participates twice—first as the ordinate value for one plotted point, then as the abscissa value for the next point.)

Each axis on a standard phase space graph represents a different variable (e.g. Fig. 3.2). In contrast, our graph of the one-dimensional map plots two successive

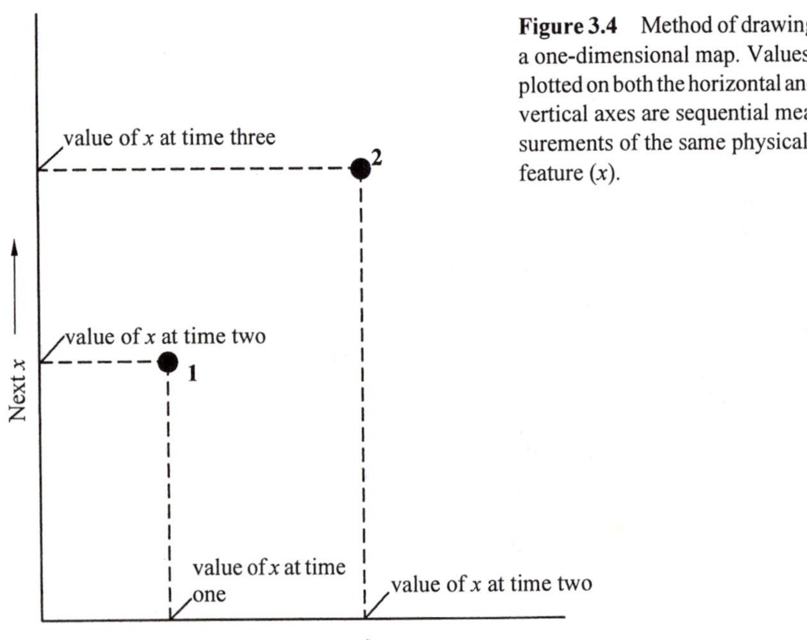

Figure 3.4 Method of drawing a one-dimensional map. Values plotted on both the horizontal and vertical axes are sequential measurements of the same physical feature (*x*).

measurements (x_{t+1} versus x_t) of one measured feature, x. Because x_t and x_{t+1} each have a separate axis on the graph, **chaologists** (those who study chaos) think of x_t and x_{t+1} as separate variables ("time-shifted variables") and their associated plot as a type of phase space. However, it's not a real phase space because the axes all represent the same feature (e.g. stock price) rather than different features. Also, each plotted point represents sequential measurements rather than a concurrent measurement. Hence, the graphical space for a one-dimensional map is really a **pseudo phase space**. Pseudo phase space is an imaginary graphical space in which the axes represent values of just one physical feature, taken at different times.

In the most common type of pseudo phase space, the different temporal measurements of the variable are taken at a constant time interval. In other cases, the time interval isn't necessarily constant. Examples of pseudo phase space plots representing such varying time intervals are most of the so-called **return maps** and **next-amplitude maps**, discussed in later chapters. The rest of this chapter deals only with a constant time interval.

Our discussion of sequential values so far has involved only two axes on the graph (two sequential values). In the same way, we can draw a graph of three successive values (x_t, x_{t+1}, x_{t+2}). Analytically, in fact, we can consider any number of sequential values. (Instead of starting with one value and going forward in time for the next values, some chaologists prefer to start with the latest measurement and work back in time. Labels for the successive measurements then are x_t, x_{t-1}, x_{t-2}, and so on.)

Lag

Pseudo phase space is a graphical arena or setting for comparing a time series to later measurements within the same data (a subseries). For instance, a plot of x_{t+1} versus x_t shows how each observation (x_t) compares to the next one (x_{t+1}). In that comparison, we call the group of x_t values the basic series and the group of x_{t+1} values the subseries. By extending that idea, we can also compare each observation to the one made two measurements later (x_{t+2} versus x_t), three measurements later (x_{t+3} versus x_t), and so on. The displacement or amount of offset, in units of number of events, is called the **lag**. Lag is a selected, constant interval in time (or in number of iterations) between the basic time series and any subseries we're comparing to it. It specifies the rule or basis for defining the subseries. For instance, the subseries x_{t+1} is based on a lag of one, x_{t+2} is based on a lag of two, and so on.

A table clarifies lagged data. Let's take a simplified example involving x_t, x_{t+1}, and x_{t+2}. Say the air temperature outside your window at noon on six straight days is 5, 17, 23, 13, 7, and 10 degrees Centigrade, respectively. The first column of Table 3.1 lists the day or time, t. The second column is the basic time series, x_t (the value of x, here temperature, at the associated t). The third column lists, for each x_t, the next x (that is, x_{t+1}). Column four gives, for each x_t, the corresponding value of x_{t+2}.

The columns for x_{t+1} and x_{t+2} each define a different subset of the original time series of column two. Specifically, the column for x_{t+1} is one observation removed from the basic data, x_t. That is, the series x_{t+1} represents a lag of one (a "lag-one series"). The column for x_{t+2} is offset from the basic data by two observations (a lag of two, or a "lag-two series"). The same idea lets us make up subsets based on a lag of three, four, and so on.

Table 3.1 Hypothetical time series of six successive noon temperatures and lagged values.

(1)		(2)	(3)	(4)
Time		Variable one = noon temp. = basic series	Variable two = lag-one series	Variable three = lag-two series
t		x_t	x_{t+1}	x_{t+2}
1		5	17	23
2		17	23	13
3		23	13	7
4		13	7	10
5		7	10	
6		10		
	Sum	75	70	53
	Average	12.5	14.0	13.3

One way to compare the various series is statistically. That is, we might compute certain statistics for each series. Table 3.1, for example, shows the average value of each series. Or, we might compute the autocorrelation coefficient (Ch. 7) for each series. The statistical approach compares the two series on a group basis. Another way to compare the two series is graphically (e.g. Fig. 3.4). In the following paragraphs, we'll deal with the graphical comparison. Both approaches play a key role in *many* chaos analyses.

Two simple principles in defining and plotting lagged data are:

- The first axis or coordinate usually (but not always) is the original time series, x_t. Unless we decide otherwise, that original or basic series advances by one increment or measurement, irrespective of the lag we choose. That is, each successive value in the x_t series usually is the next actual observation in the time series (e.g. Table 3.1, col. 2).
- The only role of lag is to define the values that make up the subseries. Thus, for any value of x_t the associated values plotted on the second and third axes are offset from the abscissa value by the selected lag. Lag doesn't apply to values in the x_t series.

The number of observations in a lagged series is $N-m$, where N is the total number of observations in the original time series and m is the lag. For instance, in Table 3.1, the lag-one series (col. 3) has $N-m=6-1=5$ observations. The lag-two series (col. 4) has 6–2 or 4 observations.

To analyze the data of Table 3.1 using a lag of two, we consider the first event with the third, the second with the fourth, and so on. ("Consider" here implies using the two events together in some mathematical or graphical way.) On a pseudo phase space plot, the abscissa represents x_t and the ordinate x_{t+2}. The value of the lag (here two) tells us how to get the y coordinate of each plotted point. Thus, for the noon temperatures, the first plotted point has an abscissa value of 5 (the first value in our measured series) and an ordinate value of 23 (the third value in the basic data, i.e. the second value after 5, since our specified lag is two). For the second plotted point, the abscissa value is 17 (the second value in the time series); the ordinate value is two observations later, or 13 (the fourth value of the original series). The third point consists of values three and five. Thus, x_t is 23 (the third measurement), and x_{t+2} is 7 (the fifth value). And so on. The overall plot shows how values two observations apart relate to each other.

The graphical space on the pseudo phase space plot of the data of Table 3.1 is a **lagged phase space** or, more simply, a **lag space. Lagged phase space** is a special type of pseudo phase space in which the coordinates represent lagged values of one physical feature. Such a graph is long-established and common in time-series analysis. In chaos theory as well, it's a basic, important and standard tool.

Some authors like to label the first observation as corresponding to "time zero" rather than time one. They do so because that first value is the starting or original value in the list of measurements. The next value from that origin (our observation number two in Table 3.1) then becomes the value at time one, and so on. Both numbering schemes are common.

Embedding dimension

A pseudo (lagged) phase space graph such as Figure 3.4 can have two or three axes or dimensions. Furthermore, as mentioned earlier in the chapter, chaologists often extend the idea of phase space to more than three dimensions. In fact, we can analyze mathematically—and compare—any number of subseries of the basic data. Each such subseries is a dimension, just as if we restricted ourselves to three of them and plotted them on a graph. Plotting lagged values of one feature in that manner is a way of "putting the basic data to bed" (or **embedding** them) in the designated number of phase space axes or dimensions. Even when we're analyzing them by crunching numbers rather than by plotting, the total number of such dimensions (subseries) that we analyze or plot is called the **embedding dimension**, an important term. The embedding dimension is the total number of separate time series (including the original series, plus the shorter series obtained by lagging that series) included in any one analysis.

The "analysis" needn't involve every possible subseries. For example, we might set the lag at one and compare x_t with x_{t+1} (an embedding dimension of two). The next step might be to add another subgroup (x_{t+2}) and consider groups x_t, x_{t+1}, and x_{t+2} together. That's an embedding dimension of three. Comparing groups x_t, x_{t+1}, x_{t+2}, and x_{t+3} means an embedding dimension of four, and so on.

From a graphical point of view, the embedding dimension is the number of axes on a pseudo phase space graph. Analytically, it's the number of variables (x_t, x_{t+1}, etc.) in an analysis. In prediction, we might use a succession of lagged values to predict the next value in the lagged series. In other words, from a prediction point of view, the embedding dimension is the number of successive points (possibly lagged points) that we use to predict each next value in the series.

The preceding comments point out something worth remembering about the embedding dimension: it isn't a fixed characteristic of the original dataset. Rather, it's a number that you control. You'll often systematically increase it in a typical analysis, as explained in later chapters. It represents the number of selected parts you choose to analyze, within a measured time series.

The embedding dimension has nothing to do with lag. Suppose, for example, that you decide to use an embedding dimension of two. That means only that you are going to compare two series from the basic measurements. One series usually is the basic data themselves, namely x_t. The other series (subseries) can be any part of that original series. For instance, designating the lag as one means comparing x_t with x_{t+1}, a lag of five involves comparing x_t with x_{t+5}, a lag of 12 means comparing x_t with x_{t+12}, and so on, and they'd all be for an embedding dimension of *two*.

Embedding a time series is one of the most important tools in chaos theory. I'll mention here a few applications, even though their meaning won't be clear until we've defined the terms in future chapters. The major application of embedding is in reconstructing an attractor. It's also useful for taking Poincaré sections, identifying determinism in disorderly data, estimating dimensions, measuring Lyapunov exponents, and making short-term predictions. More recent applications include reducing noise and approximating nonmeasured vectors in lag space (Kostelich & Yorke 1990).

Summary

Phase space or state space is an abstract mathematical space in which coordinates represent the variables needed to specify the phase (or state) of a dynamical system at a particular time. If the system has three or fewer variables, measured values of the variables can be plotted on a graph; in that case, the phase space is the imaginary space between the graph's axes. Two types of phase space are standard phase space and pseudo phase space. On a graph (three coordinates or fewer) of standard phase space, each axis stands for a key variable (e.g. temperature, price, weight, etc.). A plotted point represents the state of the system at one time. A sequence of plotted points shows how that system varies over time. On a pseudo phase space plot, in contrast, the axes or coordinates represent successive values of the same physical feature. The most common type of pseudo phase space plot uses a constant time interval (a "lag") between successive measurements. It's therefore called a lag-space plot. A good example of a lag-space plot is a graphed version of a one-dimensional map. (A map is a rule that specifies how a dynamical system evolves in time.) A one-dimensional map is a function that deals only with one measured feature, say x; it specifies how an input value x_t (plotted on the horizontal axis) goes in discrete fashion to an output value x_{t+1} (plotted on the vertical axis). The sequence of points on either type of phase space graph is a trajectory in the phase space. The concepts of phase space and pseudo phase space apply to any number of coordinates or dimensions, not just to three or fewer. The number of dimensions in any one pseudo phase space analysis is called the embedding dimension. Embedding a time series is a basic step in most types of chaos analysis.

Chapter 4

Distances and lines in space

Straight-line distance between two points

Several aspects of chaos theory require computing the distance between two points in phase space. Here's a brief review of that easy computation.

Two-dimensional case

In a standard two-dimensional phase space, all distances lie in a plane (Fig. 4.1a). Ordinarily, you'll have the coordinates (x,y) of any two points for which you need the spanning distance. The distance from one point to the other is the length of the hypotenuse of a right triangle that we draw between the two points and lines parallel to the two axes. We can compute that hypotenuse length (L in Fig. 4.1a) from the standard **Pythagorean theorem**. The theorem says that the square of the hypotenuse of a right triangle is the sum of (length of side one)^2+(length of side two)^2. Taking the square root of both sides of the equation gives the distance between the two points (the length of the hypotenuse L) as ([length of side one]^2 + [length of side two]^2)$^{0.5}$. If the x and y coordinates for point A are x_1, y_1 and those for point B are x_2, y_2 (Fig. 4.1a), then:

length of side one = x coordinate for point B minus x coordinate for point A
= $x_2 - x_1$, and
length of side two = y value for point B minus y value for point A
= $y_2 - y_1$.

The desired distance (length L of the hypotenuse of the right triangle in Fig. 4.1a) then is:

$$L = ([x_2 - x_1]^2 + [y_2 - y_1]^2)^{0.5}. \qquad (4.1)$$

As an example, say point A is at $x_1 = 1$ and $y_1 = 2$ (written 1,2) and point B is at 8,6. The distance between the two points then is:

$$
\begin{aligned}
L &= ([8-1]^2+[6-2]^2)^{0.5}\\
&= (49+16)^{0.5}\\
&= 8.1.
\end{aligned}
$$

Three-dimensional case

The Pythagorean theorem also is valid in three dimensions (Fig. 4.1b). It's written just as it is in the two-dimensional case but with the third variable (z) also included:

$$L = ([x_2-x_1]^2+[y_2-y_1]^2+[z_2-z_1]^2)^{0.5}. \tag{4.2}$$

As in two dimensions, the components (here three) are distances measured along the three axes.

(a) Two-dimensional space

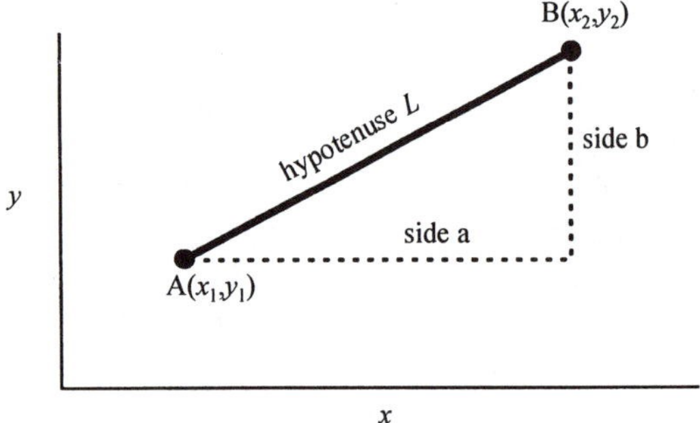

(b) Three-dimensional space

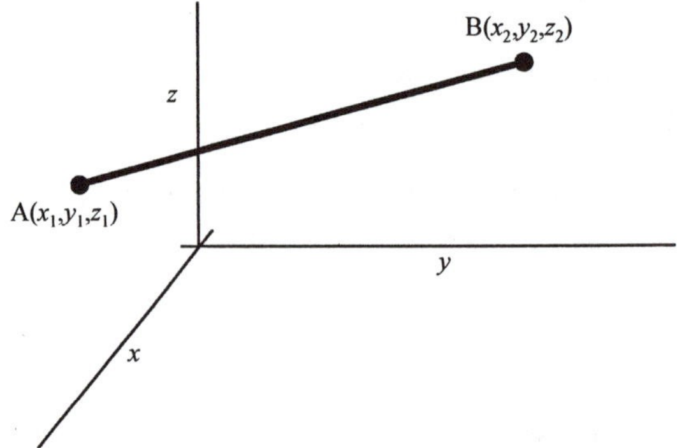

Figure 4.1 Distance between two points in phase space.

If the coordinates of the two points in Figure 4.1b are (–2,2,3) for A and (4,3,2) for B, then the distance between the two points is

$$\begin{aligned} L &= ([4-(-2)]^2+[3-2]^2+[2-3]^2)^{0.5} \\ &= (6^2+1^2+[-1]^2)^{0.5} \\ &= 6.2. \end{aligned}$$

Multidimensional case

"Multi" in this book means "more than two." Hence, the three-dimensional case covers part of the multidimensional case. Let's extend that to even more dimensions. Visualizing such a case isn't easy. It requires a leap of imagination. Also, we have to assume that the same principles that apply to the two- and three-dimensional cases also apply to higher dimensions. (Actually, we just define the higher-dimensional case that way.) Therefore, the distance L between any two points in multidimensional space is

$$L = ([x_2-x_1]^2+[y_2-y_1]^2+[z_2-z_1]^2+ \ldots +[w_2-w_1]^2)^{0.5} \tag{4.3}$$

where w is the last variable of the group.

Equation 4.3 applies to any number of dimensions. It's quite common in chaos analysis, including vector analysis. Some authors call it the "**distance formula**."

An important case in chaos analysis is that for which the first point is at the **origin** of a set of graphical axes. In that case, x_1, y_1, and so on, are all zero. The distance formula (Eq. 4.3) then reduces to:

$$L = (x_2{}^2+y_2{}^2+z_2{}^2+ \ldots +w_2{}^2)^{0.5}. \tag{4.4}$$

Getting the parameters for a straight-line equation

Another common task is to find the general equation for a straight line between any two phase space points, usually in two dimensions. Any straight line has the standard equation $y = \mathrm{c}+\mathrm{b}x$, in which x and y are variables and b (slope of line) and c (intercept, i.e. value of y at $x = 0$) are *constants*. Constants b and c are **parameters**. The task is to find their values. As a prelude, we need to talk about parameters.

Parameters are important in chaos and in research in all fields. However, people often use the term improperly, namely as a synonym for "variable." "Variable" is a general term denoting a characteristic or property that can take on any of a range of values. In ordinary usage, a variable (sometimes called a state variable) can, and often does, change at any time. A "parameter" is a special kind of variable. It is a variable that either changes extremely slowly or that remains *constant*. "Parameter" has several meanings:

- In physics, a parameter is a controllable quantity that physicists keep constant for one or more experiments but which they then change as desired. It usually reflects the intensity of the force that drives the system. A physical system can have more than one such parameter at a time.
- In mathematics, a parameter is an arbitrary constant in an equation, placed in such a position that changing it gives various cases of whatever the equation represents. Example: the equation of a straight line, $y = c+bx$, as just defined. In that equation, x and y are variables; the coefficients (constants) b and c are parameters. Changing the parameters b and c gives different versions of the straight line (i.e. straight lines having different slopes and intercepts).
- Also in mathematics, a parameter is a special variable in terms of which two or more other variables can be written. For instance, say you throw a tennis ball horizontally from a rooftop. The ball's horizontal progress (x) varies with time t (say, according to $x = 50t$). Its vertical drop (y) also varies with t (say, $y = -t^2$). Here t is a parameter because we can express both x and y in terms of t, in separate equations. The two equations ($x = 50t$ and $y = -t^2$) are **parametric equations**.
- In statistics, a parameter is a fixed numerical characteristic of a population. For example, the average diameter of all the apples in the world. Usually we have only a *sample* of a population, so in the statistics sense we have only a parameter *estimate* rather than a parameter's true value.

There are also some less common mathematical definitions of "parameter." *None of the definitions justify using it as a synonym for "variable."* Chaos theory uses "parameter" in either of the first two senses listed above. Where an author uses the term correctly, its meaning is clear from the context.

So much for the meaning of "parameter." Now let's look at three easy ways to get the parameters (b and c) of a straight-line equation.

- Take a ruler and measure the line's slope and intercept directly, on the two-dimensional graph, as follows:
  1. The slope of any line is the tangent of the angle that it makes with the horizontal axis. That slope or tangent is the opposite side over the adjacent side of any right triangle that has the straight line as hypotenuse. In the example of Figure 4.2, the ratio of those two distances (i.e. the slope of the line) is 0.67.
  2. Find the y intercept (parameter c) by extrapolating the line to where it intersects the y axis and reading the value of y. For the line in Figure 4.2, that intersection is at $y = 1.0$.

  With b = 0.67 and c = 1.0, the equation of the straight line in Figure 4.2 is $y = 1.0+0.67x$. This method is relatively crude.
- Given the x,y coordinates of any two points on the line, compute slope as a y distance divided by an x distance, then substitute that slope value, along with the coordinates of any point, into the model equation to compute the intercept, c. For instance, say point A is at $x = 0.6$ and $y = 1.4$ (i.e. point A is at 0.6,1.4) and point B is at 3,3. Then the y distance associated with those points is y_2-y_1

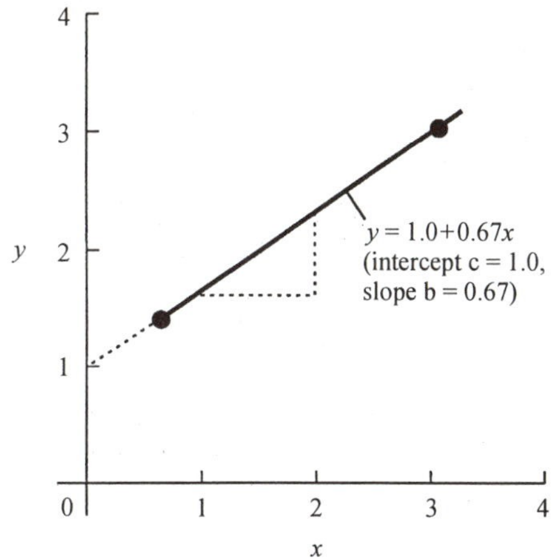

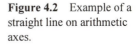

Figure 4.2 Example of a straight line on arithmetic axes.

= 3−1.4 = 1.6. The x distance is $x_2-x_1 = 3-0.6 = 2.4$. The slope b therefore is y distance/x distance = $(y_2-y_1)/(x_2-x_1) = 1.6/2.4 = 0.67$.

The second and final step is to choose either point and insert its coordinates along with the value for b into the model equation to solve for c. Let's take point A, for which $x = 0.6$ and $y = 1.4$. From $y = c+bx$, we have $1.4 = c+0.67(0.6)$. Rearranging to solve for c gives c = 1.0.

- Insert the coordinates of each point into the model equation, in turn, to get two equations with two unknowns. Next, rearrange each equation to define parameter b or c. Then equate one equation to the other to solve for the other unknown. Finally, substitute the value of that parameter along with the coordinates of one point into the model equation to get the remaining parameter. For instance, the model equation $y = c+bx$ using the coordinates of point A is $1.4 = c+0.6b$. Rearrangement gives $c = 1.4-0.6b$. Similarly, the model equation with the coordinates of point B (for which $x = 3$ and $y = 3$) is $3 = c+3b$, or $c = 3-3b$. Equating the two different expressions for c gives $c = 1.4-0.6b = 3-3b$. Rearranging and solving for b gives b = 0.67. Finally, substituting that value of b and the coordinates of either point (say, point A) into the model equation: $1.4 = c+0.67\ (0.6)$, c = 1.0.

Interpolation

Interpolation is the estimation of one or more values between two known values. In chaos theory, someone might interpolate to process irregularly measured basic data, to calculate Poincaré sections (Ch. 18), and for other purposes. We'll discuss four common situations that involve interpolation in dynamical systems: linear interpolation, cubic splines, the intersection of two straight lines, and the intersection of a line with a plane (see also Wilkes 1966, Davis 1986). In general, those methods apply only to three or fewer dimensions.

Linear interpolation

Linear interpolation assumes a straight-line relation between two or more adjacent points. It usually involves just one variable, as in a time series. The easiest case uses just two points. For example, suppose you measure variable y at two successive times. At time t_1 you get value y_1 and at t_2 you get y_2. To estimate that variable (say y') at an intermediate time (t'), just set up simple proportions (Fig. 4.3):

$$\frac{(y_2 - y_1)}{(t_2 - t_1)} = \frac{(y' - y_1)}{(t' - t_1)}.$$

Cross-multiplying and rearranging to solve for y' gives

$$y' = \frac{(y_2 - y_1)\,(t' - t_1)}{(t_2 - t_1)} + y_1. \qquad (4.5)$$

Linear interpolation between two points is relatively safe when four conditions are satisfied:

- the measured points are close together
- the number of points being estimated isn't too different from the number of original points

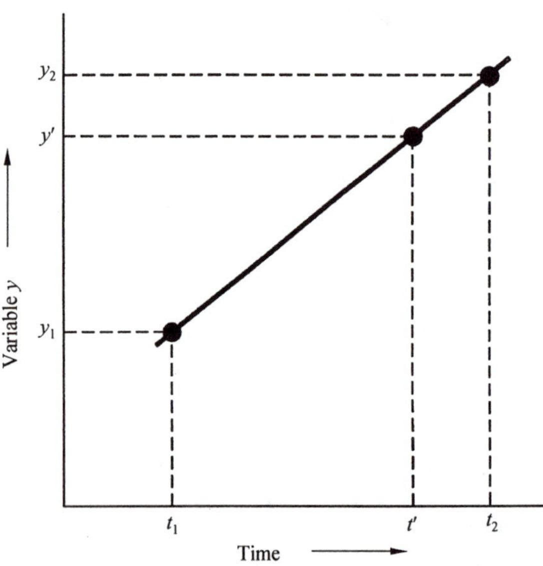

Figure 4.3 Simple linear interpolation between two points (y_1,y_2) along a time series.

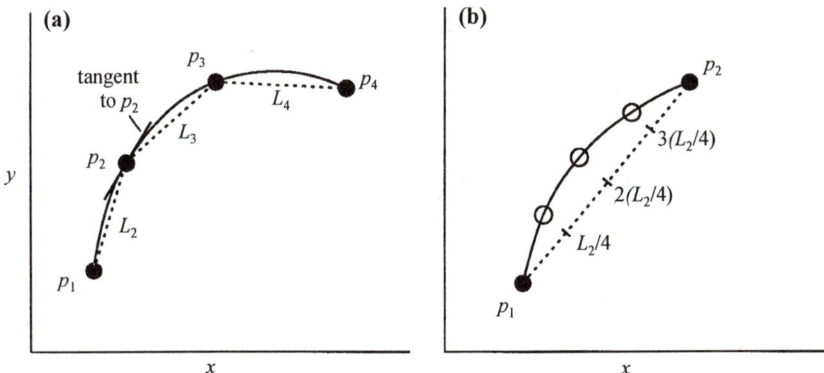

Figure 4.4 Interpolation by a cubic spline: (a) four hypothetical data points (p_1, p_2, p_3, p_4) connected by three chords (L_2, L_3, L_4) and a spline (the smooth curve); (b) subdivision of a chord with associated interpolated points (open circles).

- the original data are accurate
- the local relation really is linear.

Conversely, it can be risky if:

- the measured points are far apart
- you try to generate many more artificial points than original points
- the original data have much noise
- the local relation isn't approximately linear.

One way of reducing the error associated with such problems is to fit a straight line to several (e.g. three or four) adjacent points and to use that line for interpolation.

Cubic splines

The method of **cubic splines**, a second common interpolation technique, allows for a curved (nonlinear) relation between two points. It also has certain mathematical advantages when several successive segments are considered as a unit. A **spline** (rhymes with wine) is a drafting tool—a flexible strip, anchored at various points but taking on the shape of a smooth curve connecting all those points. It's used as a guide in drawing curved lines. A mathematical spline follows the same principles—anchored at data points and having the shape of a smooth curve connecting them.

Figure 4.4a (based on Davis 1986: 206) shows four hypothetical data points. The first point, p_1, is located at x_1,y_1, the second (p_2) is at x_2,y_2, and so on. A smooth curve (spline) connects the four points. In addition, a chord or straight line connects each successive pair of points; chord L_2 connects points p_1 and p_2, chord L_3 connects points p_2 and p_3, and so on.

To construct a spline segment between two known data points, the chord serves as a sort of base line from which we estimate the nearby spline. We divide each chord into arbitrary equal subintervals. Then we interpolate spline points, using the end of a nearby chord subinterval as a basis. To see the basic idea, let's divide the straight-line distance L_2 between known points p_1 and p_2 into four equal parts (Fig. 4.4b). The length of each part is $L_2/4$. Cumulative distances from p_1 along the chord then are $L_2/4$ for the first point to be interpolated, $2(L_2/4)$ for the second point, and $3(L_2/4)$ for the third point. That scheme lets us interpolate three spline points—one corresponding to the end of chordal distance $L_2/4$, one for $2(L_2/4)$, and one for $3(L_2/4)$, as explained below. Each spline point is near but not directly on the associated chord (Fig. 4.4b). The final spline results from connecting the interpolated points. Ordinarily, we'd use a straight line to connect each successive pair of points, so the crude example just described doesn't give a smooth curve (and hence I've cheated in drawing a smooth curve in Fig. 4.4b). In practice, we'd use many more subintervals to get acceptably close to a smooth curve.

With two variables or dimensions (x and y), an x coordinate and a y coordinate need to be computed to define each interpolated point. The same general equation (spline equation) works for both coordinates. That general equation is:

$$\hat{p}_c = a_1 + a_2 L_c + a_3 L_c^{\,2} + a_4 L_c^{\,3} \tag{4.6}$$

in which $\hat{p}_c$ is the x or y coordinate of an interpolated point, L_c is a cumulative distance (expressed as a proportion, such as $L_2/4$) along the chord between the two known points, and the a_i's are constants or coefficients. One set of coefficients a_i is required to compute the x values of interpolated points within each spline segment, and a different set is needed for the y values. Equation 4.6 is a **cubic polynomial**—cubic because the highest power to which the **independent variable** is raised is 3, and **polynomial** because the entire expression has more than one term. (The Glossary has a more complete definition of "polynomial.")

Values needed for the computations are the coordinates of known points and slopes of tangent lines at the interior points (p_2 and p_3 of Fig. 4.4a). Simultaneously solving a set of matrix algebra relations gives the slopes of the tangent lines. Also, each variable requires a separate solution. That is, two-dimensional (x,y) data require first going through the procedure using x coordinates, then repeating it using y coordinates. That, in turn, allows a determination of the four coefficients a_i to use in Equation 4.6, for a particular dimension (x or y), within a particular spline segment. Davis (1986: 204–211) gives further details.

Intersection of two straight lines

Another interpolation situation involves two known points in a two-dimensional phase space, an assumed straight line between them, and another straight line that intersects the first line (Fig. 4.5). The job is to estimate the coordinates of the point

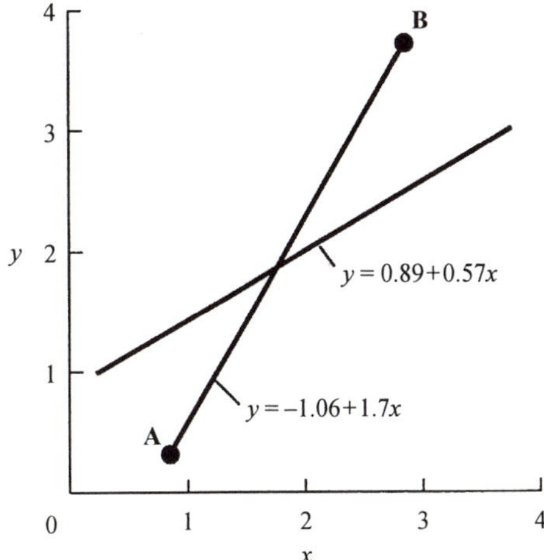

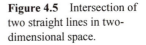

Figure 4.5 Intersection of two straight lines in two-dimensional space.

where the two lines intersect. The two points (e.g. A and B in Fig. 4.5) were measured at successive times, so the problem again boils down to estimating coordinate values at some intermediate time.

The solution is straightforward, although it requires knowing the parameter values in the equations of both lines. Any of the three methods discussed above can give those parameters. Obtaining them leaves two equations (one for each line), with unknown x and y values in each equation. Exactly at (and only at) the point of intersection, the two equations give the same x and y values. Hence, for that particular point the x and y values (or equivalent expressions of each) are interchangeable between the two equations. We can then get the desired coordinates by equating one equation to the other, that is, by solving them simultaneously.

To illustrate with Figure 4.5, say the equations of the two intersecting straight lines (with parameters determined by any of the methods explained earlier) are: $y = -1.06+1.7x$ and $y = 0.89+0.57x$. To find the unique pair of x and y values that are common to both lines, we solve the equations simultaneously. We can do that either by equating the two or by subtracting the lower equation from the upper. Either of those approaches eliminates y. For our example, that leaves $0 = -1.95+1.13x$, or $x = 1.73$. We then use the x value and the equation of either line to get y. For instance, with the second equation we have $y = 0.89+0.57\ (1.73)$ or $y = 1.88$. Hence, the two lines cross at $x = 1.73$, $y = 1.88$.

Intersection of a line with a plane

The preceding example was two-dimensional. Another common situation is that of a trajectory piercing a plane in three-dimensional space (Fig. 4.6). For example, chaologists place an imaginary reference plane between measurements made at two successive times. They then estimate (interpolate) the coordinates at the point where the trajectory pierces that plane. Basic data are the values of the three variables—x_1, y_1, and z_1 at time 1 and x_2, y_2, and z_2 at time 2. Between those two phase space points, we can put the plane wherever we want. That is, we dictate the equation for the plane. One way of writing a plane's equation is $(1/a_x)x+(1/a_y)y+(1/a_z)z=1$, where a_x, a_y, and a_z are the plane's intercepts with the x, y, and z axes, respectively. Let's take the easiest case, namely that for which we assume that the trajectory between the two measured points is a straight line. The x, y, and z coordinates of any point along that trajectory are **linear functions** of two features, namely time (t) and the coordinates of the measured variables:

$$x = x_1+t(x_2-x_1) \tag{4.7}$$
$$y = y_1+t(y_2-y_1) \tag{4.8}$$
$$z = z_1+t(z_2-z_1). \tag{4.9}$$

Using the measured values of the variables for each of the two times, the general procedure consists of three simple steps:

1. Plug the measured values of the coordinates into Equations 4.7–9 to get x, y, and z as functions of time. (Those equations then describe the straight-line trajectory in terms of time, between our two measured points.)
2. Substitute those definitions of the variables into the equation for the plane and solve for t, the time at which the trajectory intersects the plane.
3. Plug that intersection time into Equations 4.7–9 to get the values of x, y, and z for that time.

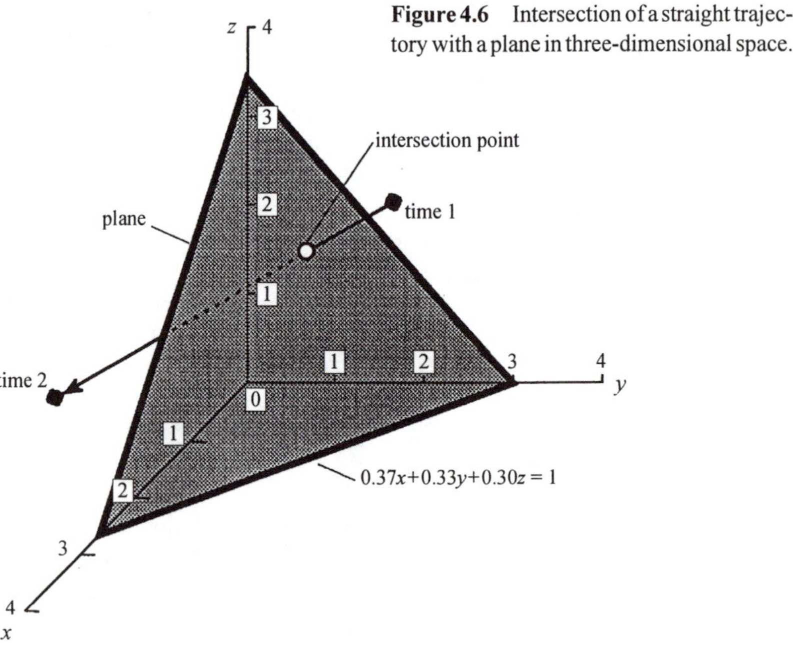

Figure 4.6 Intersection of a straight trajectory with a plane in three-dimensional space.

Let's run through a quick, hypothetical example. Suppose we're studying a continuous system that has three variables. We measure (sample) that system at two discrete times. At time 1 we get 0.5,2,2 for x, y, and z, respectively, and at time 2 we get 0,–2,0.2. Now we insert a plane between them, defined by $0.37x+0.33y+0.30z = 1$ (Fig. 4.6). The job is to interpolate values for x, y, and z at the point where the trajectory intersects the plane. Armed with the measured data, we follow the three steps just mentioned:

1. Inserting the observed values of x, y, and z into Equations 4.7–9:

$$\begin{aligned} x &= 0.5+t\,(0-0.5) \\ &= 0.5+t\,(-0.5) \\ &= 0.5-0.5t \end{aligned} \tag{4.7a}$$

$$\begin{aligned} y &= 2+t\,(-2-2) \\ &= 2+t\,(-4) \\ &= 2-4t \end{aligned} \tag{4.8a}$$

and

$$\begin{aligned} z &= 2+t\,(0.2-2) \\ &= 2+t\,(-1.8) \\ &= 2-1.8t. \end{aligned} \tag{4.9a}$$

Those three equations define the straight line between our two points. They do so in terms of time, t.

2. Substitute those three definitions of the variables into the equation for the plane to get the time at which the trajectory intersects the plane:

$$\begin{aligned} &0.37x+0.33y+0.30z = 1 \text{ (our plane)} \\ &0.37\,(0.5-0.5t)+0.33\,(2-4t)+0.30\,(2-1.8t) = 1 \\ &0.19-0.19t+0.66-1.32t+0.60-0.54t = 1 \\ &1.45-2.05t = 1 \\ &t = 0.22. \end{aligned}$$

3. Knowing t for the time of intersection, plug t into Equations 4.7a–9a to get the values of x, y, and z for that particular t:

$$\begin{aligned} x &= 0.5-0.5t (4.7a) \\ &= 0.5-0.5\,(0.22) \\ &= 0.39 \end{aligned}$$

$$\begin{aligned} y &= 2-4t (4.8a) \\ &= 2-4\,(0.22) \\ &= 1.12 \end{aligned}$$

and

$$\begin{aligned} z &= 2-1.8t(4.9\text{a}) \\ &= 2-1.8\,(0.22) \\ &= 1.60. \end{aligned}$$

All the explanations discussed in this chapter have been couched in terms of standard phase space, with variables x, y, and z. However, the same concepts also apply to pseudo (lagged) phase space. For instance, chaos analyses might require the straight-line distance between two points in lagged phase space, interpolation between two or more points in lagged phase space, and so on. All we do in those cases is substitute x_t, x_{t+1}, and x_{t+2} (if the lag is one), or more generally the appropriately lagged values of the time series, in place of x, y, and z.

Summary

Chaos theory often requires the straight-line distance between two points in phase space, parameters for straight-line equations, and several types of interpolation. Regardless of the number of dimensions (variables), a straight-line distance (L) between two points can be computed with the generalized "distance formula" (a rearrangement of the Pythagorean theorem). For any number of variables (x, y, z, . . . w), that formula is

$$L = ([x_2-x_1]^2+[y_2-y_1]^2+[z_2-z_1]^2+\ldots+[w_2-w_1]^2)^{0.5}.$$

There are several easy ways to find the parameters of a straight-line equation, given the coordinates for any two points on the line. The methods all involve short, simple algebraic steps using the coordinates and the model equation, $y = c+bx$. Between two known points, for any selected intermediate time, linear interpolation is easy: just set up a proportion involving the values of the variable at the two times. A more sophisticated method of interpolating between two known points—cubic splines—assumes a curved (cubic polynomial) relation between the two points. A third common type of interpolation is to get the point of intersection of two straight lines. The way to do that interpolation, once you've found the parameters of those straight-line equations, is to solve the two equations simultaneously. The fourth popular type of required interpolation is to estimate the three-dimensional point of intersection of a straight line with a plane, given the plane's equation and the coordinates of two discrete measurements on each side of the plane. That involves three steps:

1. Use the coordinate values to get the equations that define the line in terms of time.
2. Substitute those definitions into the plane's equation to get the time of intersection.

3. Insert that intersection time into the equations that define the line, to get x, y, and z at the time of intersection.

All the above concepts apply not only to standard phase space but also to lagged phase space.

Chapter 5
Vectors

Chaologists speak of phase space points as "vectors." Also, many aspects of vectors are used in connection with Lyapunov exponents (Ch. 25) and with other mathematical features of chaos.

Scalars and vectors

Magnitudes alone are enough to describe quantities such as length, area, volume, time, temperature, and speed. Such quantities are called **scalars**, because a simple scale can indicate their values. Other quantities, in contrast, are slightly more complex in that they involve not only magnitude but also direction. Examples are force (including weight), velocity, and acceleration. (The force of gravity acts *downward*; the automobile's *speed* is 50 kilometers per hour but its *velocity* is 50 kilometers per hour *in a specified direction*; and so on.) These more comprehensive quantities are called **vectors**.

Graphically, a vector has a beginning point (A or C in Fig. 5.1a) and a terminal point (B or D in Fig. 5.1a). The length of the straight line drawn from the starting point to the terminal point represents the vector's *magnitude* (also called its *length* or **norm**). An arrowhead at the terminal point indicates the vector's *direction.* The letters identifying the vector itself (e.g. AB, CD) are written in bold italic type (***AB***, ***CD***).

Two vectors, such as ***AB*** and ***CD*** in Figure 5.1a, are equal if (and only if) they have the same magnitude and direction. That lets us plot many vectors, all of them equal, at different locations on a graph. Their different starting and finishing points make no difference, as long as the vectors' magnitude and direction are equal.

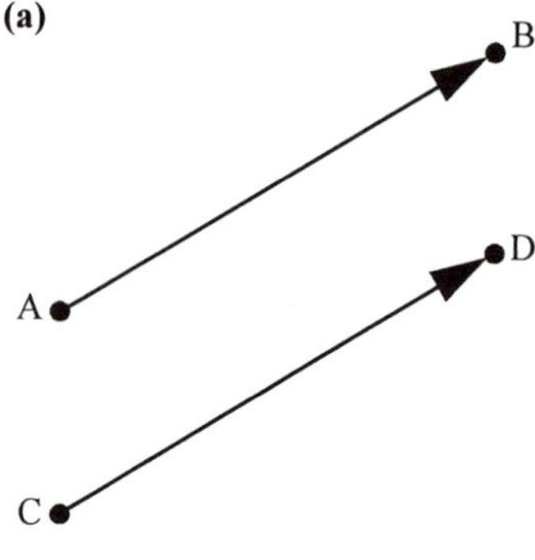

Figure 5.1 Equality of vectors. The two vectors in (a) are equal. The two vectors in (b) are equal, and the lower one is a coordinate vector because it starts at the origin of the graph.

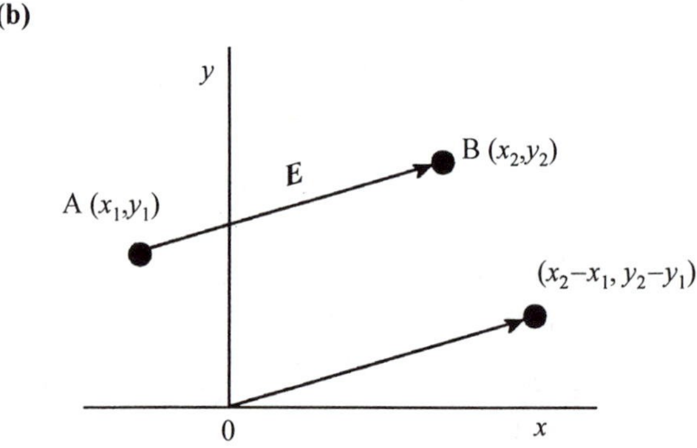

Coordinate vectors

In Figure 5.1b, the vector ***E*** is defined by its two endpoints A (at x_1,y_1) and B (at x_2,y_2). Say we reposition that vector, keeping the same orientation and magnitude, to where its starting point is at the origin of the graph (0,0). The terminal point then is at x_2-x_1, y_2-y_1. (The easiest way to verify those coordinates is to draw an example yourself, on a piece of graph paper.) A vector whose starting point is at the origin of the graph is a **coordinate vector**. The coordinates of its terminal point define such a vector. For example, say a vector has one endpoint (A) at –3,4 and the other (B) at 3,5. The associated coordinate vector (which, by definition, starts at the origin of the graph) then has an endpoint located at $x = x_2-x_1 = 3-(-3) = 6$ and $y = y_2-y_1 = 5-4 = 1$.

Chaos theory extends those ideas to data plotted in phase space. Any measurement of x and its associated y plots as a point in phase space. Chaologists call that point a vector, in the sense that they imagine a straight line from the origin to the point. (Technically, it's a *coordinate* vector, but for convenience they drop "coordinate.") In other words, chaologists define a vector just in terms of the coordinates of its terminal point, the implication being that it starts at the origin of the graph.

The idea of vectors in phase space also applies to pseudo (lagged) phase space (a plot of sequential values of the same variable). For instance, a simple listing of measurements of x (with values x_1, x_2, etc. . . .) is a **scalar time series**. The phase space has only one axis. However, a chaos analysis often involves a lag-space graph of x_t versus x_{t+m}, where m is the lag. That graph has two axes, as in Figure 5.2a. Each pair of values (i.e. each plotted point) on a lag-space graph is a **lag vector, delay vector**, or **state-space vector**. It defines a point's location (reflecting the direction and magnitude from the origin) on the graph. In other words, we've taken our scalar time series and from it created a **vector time series**. The entire list of paired values used to plot the points is a **vector array**.

The same terms also apply to lag-space graphs of three axes (Fig. 5.2b) or indeed to any number of coordinates. For instance, with three axes and a lag of two, the axes stand for x_t, x_{t+2}, and x_{t+4}. The first plotted point—the first vector or lag vector—is the point for x_1, x_3, and x_5; the second vector consists of x_2, x_4, and x_6; the third vector is made up of x_3, x_5, and x_7; and so on. In general, we can construct a vector time series with any number of coordinates.

Every vector described in terms of its coordinates has a **dimension**. The

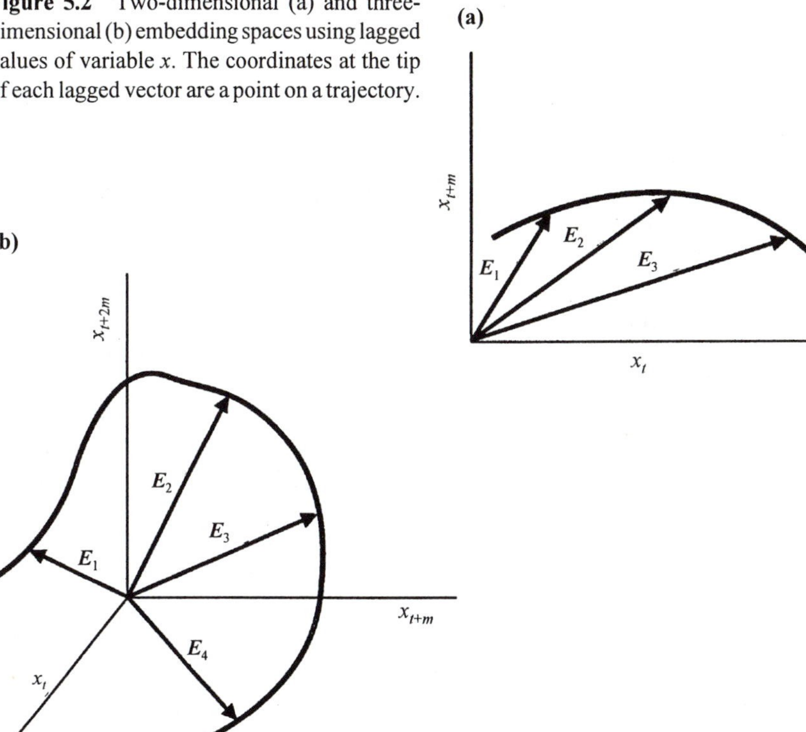

Figure 5.2 Two-dimensional (a) and three-dimensional (b) embedding spaces using lagged values of variable x. The coordinates at the tip of each lagged vector are a point on a trajectory.

vector's dimension is the number of components (axes or coordinates) it has. If it has only two or three coordinates (and hence can be drawn on a graph), the dimension equals the number of axes. For instance, in lag space that has two axes, each plotted point has two components (x_t and x_{t+m}), so each point is a *two-dimensional* vector. (Since we're in lag space, the two dimensions are embedding dimensions, and the vector is a lag vector.) Similarly, a *three-dimensional vector* is any point on a graph of x_t and two lagged values, such as x_{t+1} and x_{t+2}. Chaos analyses deal with vectors of any number of dimensions.

The technical literature often defines a lag vector in symbol form using embedding dimension D and possibly lag. Let's look first at the case for a lag of one. A typical expression for a lag of one is: lag vector = $x_t, x_{t+1}, x_{t+2}, \ldots, x_{t+(D-1)}$. That's "mathspeak" for saying that the final embedding dimension corresponds to x at $t+(D-1)$. Here's why that's true mathematically. For two axes or dimensions, $D = 2$. If the lag is one, the plotted data are x_t and x_{t+1}. To see that the last axis, x_{t+1}, equals $t+(D-1)$, we just plug in 2 for D in the expression $t+(D-1)$. That gives $t+(2-1) = t+1$. Similarly, with three axes ($D = 3$) and a lag of one, we plot x_t, x_{t+1}, and x_{t+2}. Here the last axis represents x at $t+2$. That value ($t+2$) again equals $t+(D-1)$, as we find by inserting 3 for D.

Now let's generalize that symbolism for a lag m that is greater than one. A typical expression for that case is: lag vector = $x_{t+(D-1)m}$. Again, let's look at the reason for the notation $t+(D-1)\,m$. A lag m of zero means we're analyzing only the basic time series, x_t. Setting $m = 0$ in the expression $t+(D-1)m$, we're left with just t. Therefore, $x_{t+(D-1)m}$ is just x_t, so the expression works for a lag of zero. When m is 1, $t+(D-1)m$ reduces to $t+(D-1)$, so the final axis again is $x_{t+(D-1)}$, as above. Let's also check our new expression for a lag of two. Using a lag of two, we'd plot x_t and x_{t+2} for two dimensions; we'd plot x_t, x_{t+2}, and x_{t+4} for three dimensions; we'd have x_t, x_{t+2}, x_{t+4}, and x_{t+6} for four dimensions (manageable only mathematically, not graphically); and so on. In each case, the final axis corresponds to $t+(D-1)m$. For instance, at $m = 2$ and $D = 4$, the last dimension is for x_{t+6}, and x_{t+6} also equals $x_{t+(D-1)m}$.

To reduce possible confusion, some authors write out the first one or two components of a generalized vector (and then tack on the generalized expression), rather than using only the generalized expression. That is, they will define a generalized vector as $x_t, x_{t+m}, \ldots, x_{t+(D-1)m}$. Also, those who prefer to start their analysis with the last measurement and proceed backward in time have a minus sign in place of the plus. Thus, their notation is $x_{t-m}, \ldots, x_{t-(D-1)m}$. There are also other notation schemes for lag-space vectors.

Vector addition and subtraction

Let's review how to add two vectors, such as $\boldsymbol{E}_1$ and $\boldsymbol{E}_2$ in Figure 5.3a. We first place the beginning point of one vector ($\boldsymbol{E}_2$ in Fig. 5.3a) at the terminal point of the other, maintaining of course the same orientation and magnitude for the one that we shifted (Fig. 5.3b). The straight line from the starting point of the stationary vector $\boldsymbol{E}_1$ to the terminal point of the shifted vector $\boldsymbol{E}_2$ then is the **resultant** or sum of the two vectors. Thus, if $\boldsymbol{E}_1$ and $\boldsymbol{E}_2$ are the two original vectors and $\boldsymbol{E}_3$ is the resultant, we write $\boldsymbol{E}_1+\boldsymbol{E}_2=\boldsymbol{E}_3$. That method of graphical addition is known as the **triangle law** (Fig. 5.3b).

The same resultant emerges via an alternative method called the **parallelogram law** (Fig. 5.3c). With that procedure, we shift one vector (here $\boldsymbol{E}_2$), such that its starting point coincides with that of the stationary vector ($\boldsymbol{E}_1$) to form two sides of a parallelogram. Next, we complete the parallelogram by drawing in its two other sides. The diagonal drawn from the common (starting) point of the two vectors then is the resultant, $\boldsymbol{E}_1+\boldsymbol{E}_2$.

The resultant $\boldsymbol{E}_3$ ($=\boldsymbol{E}_1+\boldsymbol{E}_2$) has a couple of noteworthy features. First, it's a vector itself, with both magnitude and direction. Secondly, both the equals sign and the idea of addition are used in a special sense here, because placing a ruler on Figure 5.3 and measuring the lengths of $\boldsymbol{E}_1$, $\boldsymbol{E}_2$ and $\boldsymbol{E}_3$ shows that the length of $\boldsymbol{E}_1$ added to the length of $\boldsymbol{E}_2$ doesn't equal the length of the resultant, $\boldsymbol{E}_3$. Hence, adding absolute magnitudes (lengths) isn't the same as adding vectors. In vector addition, the equals sign means "results in the new vector . . .".

In practice, we add vectors by adding their respective coordinates. For example, let's add vector $\boldsymbol{E}_1$ (defined, say, as 2,1) and vector $\boldsymbol{E}_2$ (3,4) (Fig. 5.4, upper right quadrant). The x coordinate of the resultant is the x value of vector $\boldsymbol{E}_1$ plus the x value of vector $\boldsymbol{E}_2$. Here that's 2+3 or 5. Similarly, the y value of the resultant is at the y value of $\boldsymbol{E}_1$ plus the y value of $\boldsymbol{E}_2$. Here that's 1+4 = 5. In general, if $\boldsymbol{E}_1$ is (x_1,y_1) and $\boldsymbol{E}_2$ is (x_2,y_2), then

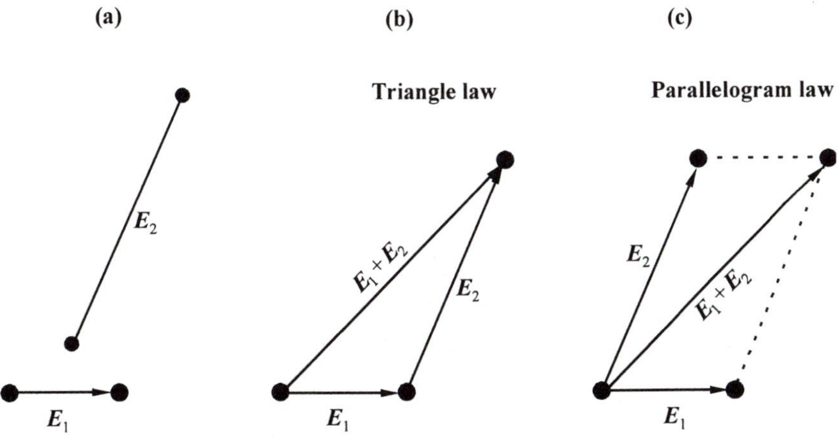

Figure 5.3 Triangle and parallelogram laws of vector addition.

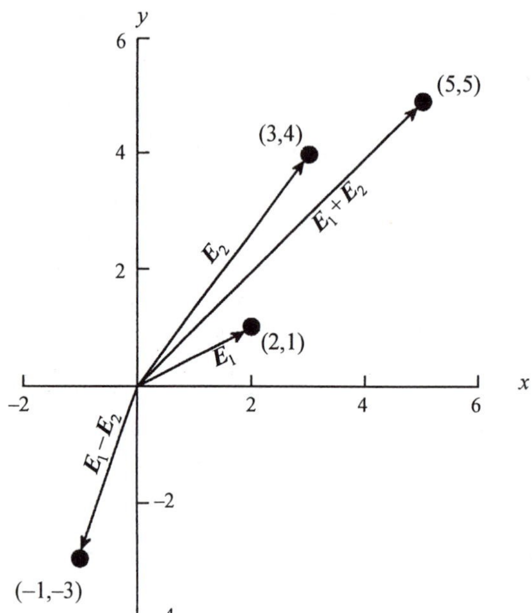

Figure 5.4 Addition (upper right quadrant) and subtraction (lower left quadrant) of two vectors, $\boldsymbol{E}_1$ and $\boldsymbol{E}_2$.

$$\boldsymbol{E}_1+\boldsymbol{E}_2=(x_1+x_2, y_1+y_2). \tag{5.1}$$

In the same way, we subtract one vector from another by subtracting the respective coordinates. For the two-dimensional case,

$$\boldsymbol{E}_1-\boldsymbol{E}_2=(x_1-x_2, y_1-y_2). \tag{5.2}$$

For example, subtracting $\boldsymbol{E}_2$ (3,4) from $\boldsymbol{E}_1$ (2,1) gives a new vector defined as –1,–3 (Fig. 5.4, lower left quadrant).

Scalar multiplication

If you have \$100 and by some miracle triple your money, you end up with \$300. That process in equation form is \$100×3 = \$300. The point here is that, mathematically, you got the \$300 by multiplying the \$100 by 3, not by \$3. The dollar values (\$100 and \$300) in that example (from Hoffmann 1975) are counterparts to vectors; the factor 3 is a counterpart to a scalar. In other words, multiplying a vector by a scalar (a constant or pure number) yields another vector.

Figures 5.5a and 5.5b show one- and two-dimensional examples, respectively. In general, if a vector $\boldsymbol{E}$ has coordinates x,y,z, and so on, then the relation for multiplying that vector by a scalar b is

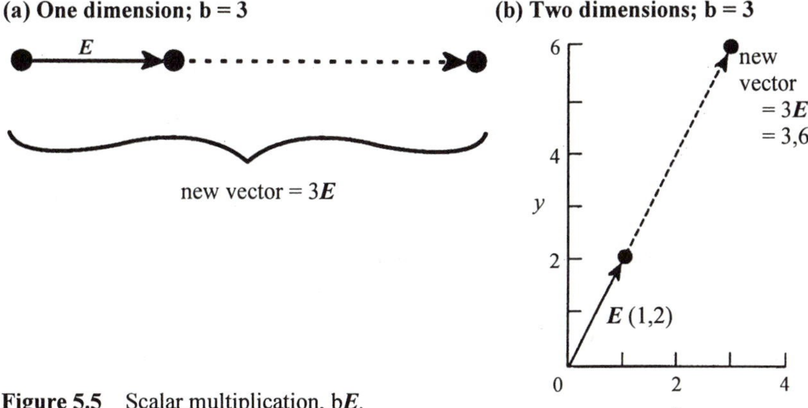

Figure 5.5 Scalar multiplication, b$\boldsymbol{E}$.

$$b\boldsymbol{E} = (bx, by, bz, \text{etc.}). \tag{5.3}$$

For example, suppose two-dimensional vector $\boldsymbol{E}$ is the point located at (1,2). Multiplying that vector by the scalar 3 results in a new vector three times as long, located at $x = 3\cdot1 = 3$ and $y = 3\cdot2 = 6$ (Fig. 5.5b).

The dot product

A certain multiplication process recurs frequently in dealing with two vectors. That process takes the product of the two vectors' x coordinates (in two dimensions, $x_1 \cdot x_2$), the product of their y coordinates ($y_1 \cdot y_2$), and then sums those individual products ($x_1 \cdot x_2 + y_1 \cdot y_2$). In three or more dimensions, we also add on $z_1 \cdot z_2$, and so on. Thus, there are as many items to add as there are dimensions or axes. In mathematical shorthand, say with vectors $\boldsymbol{E}_1$ and $\boldsymbol{E}_2$ you'll see the process written as $\boldsymbol{E}_1 \cdot \boldsymbol{E}_2$ (pronounced "E one dot E two"). Capsulizing the general idea in symbols for two dimensions:

$$\boldsymbol{E}_1 \cdot \boldsymbol{E}_2 = x_1 \cdot x_2 + y_1 \cdot y_2. \tag{5.4}$$

The resulting number isn't a vector. Instead, it's simply a scalar and goes by any of three names—the **scalar product**, **dot product**, or **inner product**. The dot, in other words, symbolizes multiplication whether we're multiplying scalars (pure numbers) or vectors.

As an example, suppose we've got two vectors in three-dimensional space. The first vector is 4,1,3 (i.e. the vector going from the graph's origin to the point at $x = 4$, $y = 1$, and $z = 3$). The second vector is 2,–3,2. The dot product then is $(4,1,3)\cdot(2,-3,2)$. It's computed (Eq. 5.4 for three dimensions) as $x_1 \cdot x_2 + y_1 \cdot y_2 + z_1 \cdot z_2 = (4\cdot2)+(1\cdot[-3])+(3\cdot2) = 8-3+6 = 11$.

The angle between two vectors

Computing one of the key indicators of chaos, **Lyapunov exponents**, can require the angle between two vectors. An example might be to determine the angle θ between vectors ***OA*** and ***OB***, as sketched in Figure 5.6. You'll feel more comfortable using the required formula if we take a few seconds to derive it here. The derivation follows a general pattern or outline that occurs very often in science and mathematics, as well as in this book. It's a three-step procedure for getting a symbol definition of a new quantity. Those three steps are:

1. Define the new quantity conceptually (in words), using distinct components. In many cases (but not with the angle between two vectors) this first step (conceptualization) includes dividing a raw measure by some standard or reference measure.
2. Substitute symbol definitions for the individual ingredients.
3. Simplify that symbol definition algebraically, if possible. Some ways to do that are to cancel out common terms, rearrange terms from one side of the equation to the other, combine terms, and to use global symbols to replace individual ones. Where simplification is possible, the original components are often unrecognizable in the resulting streamlined or economized definition.

In short: conceptualize, symbolize, and simplify.

As a preliminary move toward applying those three steps to the angle between two vectors, we draw a line connecting the ends of the two vectors. For instance, having vectors ***OA*** and ***OB*** in Figure 5.6, we draw line AB to complete a triangle. We'll call vector ***OA*** side one of the triangle, ***OB*** side two, and AB side three. Now let's go through the three steps.

1. The first step of the procedure—conceptualizing—is easy in this case because we don't have to come up with an original idea. The idea is already established in the form of the well known law of cosines. That law relates the three sides of a triangle and the inclusive angle between sides one and two as follows:

Length of side 3 squared
= length of side 1 squared

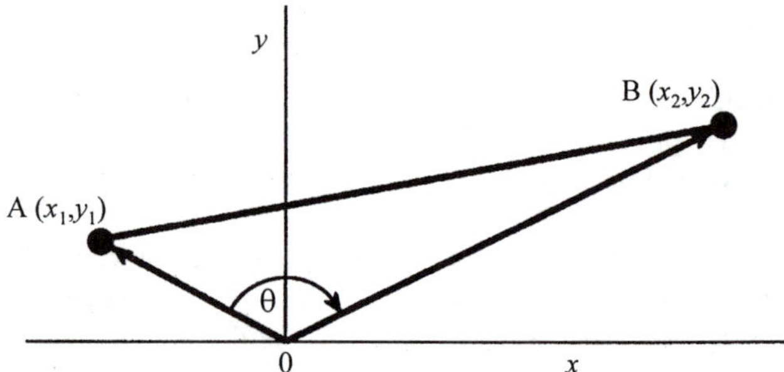

Figure 5.6 Two vectors (***OA*** and ***OB***) and inclusive angle θ.

+ length of side 2 squared
– 2 (length of side 1 × length of side 2 × cosine of their inclusive angle). (5.5)

2. Step two of the three-step general process is to express that concept in symbols. We can do that in terms of vectors or in terms of coordinates. In terms of vectors, the length of side one is $|\boldsymbol{OA}|$, where the vertical lines $|\,|$ indicate absolute magnitude. The length of side two is $|\boldsymbol{OB}|$ and that of side three is $|\mathrm{AB}|$. Our concept (Eq. 5.5) in terms of vector symbols therefore is

$$|\mathrm{AB}|^2 = |\boldsymbol{OA}|^2 + |\boldsymbol{OB}|^2 - 2|\boldsymbol{OA}|\,|\boldsymbol{OB}|\cos\theta. \tag{5.5a}$$

To express that definition in the alternative terms of known coordinates, we first get the length of each side from the distance formula (Eq. 4.3). For instance, the length of side 1, or, $|\boldsymbol{OA}|$ is $(x_1^2+y_1^2)^{0.5}$. Similarly, the length of side 2, or $|\boldsymbol{OB}|$, is $(x_2^2+y_2^2)^{0.5}$. The third side, namely from A to B, is $([x_2-x_1]^2+[y_2-y_1]^2)^{0.5}$. The cosine law (Eq. 5.5) squares those lengths. For that purpose, we just remove the exponent 0.5 from each of the definitions. Thus, $|\mathrm{AB}|^2$ in our example is $(x_2-x_1)^2+(y_2-y_1)^2$; $|\boldsymbol{OA}|^2$ is $x_1^2+y_1^2$; and $|\boldsymbol{OB}|^2$ is $x_2^2+y_2^2$. Substituting those symbol definitions into the word-definition of step one (Eq. 5.5) gives

$$(x_2-x_1)^2+(y_2-y_1)^2 = x_1^2+y_1^2+x_2^2+y_2^2-2([x_1^2+y_1^2]^{0.5}\,[x_2^2+y_2^2]^{0.5})\cos\theta. \tag{5.5b}$$

3. Step three of the three-step procedure is to simplify. We'll do that with Equation 5.5b. Squaring the left-hand side as indicated gives

$$x_2^2-2x_1x_2+x_1^2+y_2^2-2y_1y_2+y_1^2 = x_1^2+y_1^2+x_2^2+y_2^2-2([x_1^2+y_1^2]^{0.5}\,[x_2^2+y_2^2]^{0.5})\cos\theta.$$

Several terms (x_1^2, x_2^2, etc.) now occur on both sides of that equation. Cancelling those like terms leaves

$$-2x_1x_2-2y_1y_2 = -2([x_1^2+y_1^2]^{0.5}\,[x_2^2+y_2^2]^{0.5})\cos\theta.$$

The three main products in this last equation all include –2. We can eliminate that constant by dividing everything by –2, leaving

$$x_1x_2+y_1y_2 = ([x_1^2+y_1^2]^{0.5}[x_2^2+y_2^2]^{0.5})\cos\theta.$$

Rearranging to solve for $\cos\theta$:

$$\cos\theta = \frac{x_1x_2+y_1y_2}{(x_1^2+y_1^2)^{0.5}\,(x_2^2+y_2^2)^{0.5}}. \tag{5.5c}$$

That completes the three-step procedure and gives the required equation for determining θ. For three dimensions, the numerator also includes $+z_1z_2$, and the denominator is $(x_1^2 + y_1^2 + z_1^2)^{0.5}$ $(x_2^2 + y_2^2 + z_2^2)^{0.5}$.

Books on vectors typically write Equation 5.5c in shorter and more compact form by substituting vector symbols for coordinate values, where possible. The numerator in Equation 5.5c is none other than our new friend, the dot product, just defined (Eq. 5.4). In terms of Figure 5.6, that dot product is $\boldsymbol{OA} \cdot \boldsymbol{OB}$. In the denominator of Equation 5.5c, $(x_1^2 + y_1^2)^{0.5}$ is $|\boldsymbol{OA}|$ and $(x_2^2 + y_2^2)^{0.5}$ is $|\boldsymbol{OB}|$. Thus, in a vector book Equation 5.5c is apt to look more like

$$\cos\theta = \frac{\boldsymbol{OA} \cdot \boldsymbol{OB}}{|\boldsymbol{OA}| \cdot |\boldsymbol{OB}|} . \tag{5.5d}$$

To add one last comment on a related subject, cross-multiplying Equation 5.5d gives an alternate way of defining a dot product:

$$\boldsymbol{OA} \cdot \boldsymbol{OB} = |\boldsymbol{OA}|\ |\boldsymbol{OB}| \cos\theta. \tag{5.6}$$

One useful application of Equation 5.6 is the dot product of two **orthogonal** (mutually perpendicular) vectors. Being mutually perpendicular, the angle θ between two vectors is 90°. The cosine of that angle (90°) is zero. The entire right-hand side of Equation 5.6 for the special case of perpendicular vectors therefore becomes zero. In other words, the dot product for two mutually perpendicular vectors is zero. We can easily verify that important result graphically by drawing two orthogonal vectors and computing their dot product. For example, in Figure 5.7 vector $\boldsymbol{OA}$ is 3,3, and a perpendicular vector $\boldsymbol{OB}$ is 2,–2. The dot product $\boldsymbol{OA} \cdot \boldsymbol{OB}$ is, by definition (Eq. 5.4), $x_1 \cdot x_2 + y_1 \cdot y_2 = (3 \cdot 2) + (3 \cdot [-2]) = 6 - 6 = 0$.

Orthogonalization

Computing Lyapunov exponents can require deriving a set of orthogonal vectors $\boldsymbol{V}_i$ that are related to an observed set of nonorthogonal vectors $\boldsymbol{E}_i$, where i is 1, 2, and so on and here represents the ith vector. Let's see how to do that, beginning with the two-dimensional case.

The overall scheme is to choose any vector (any datum point) from among the nonorthogonal vectors (e.g. the values of the several variables measured at a given time) and build the new orthogonal set around it. In practice, choose the vector having the greatest magnitude. In other words, the first vector $\boldsymbol{V}_1$ in the new (orthogonal) set equals whichever measured vector $\boldsymbol{E}_1$ has the largest magnitude. Thus

$$\boldsymbol{V}_1 = \boldsymbol{E}_1. \tag{5.7}$$

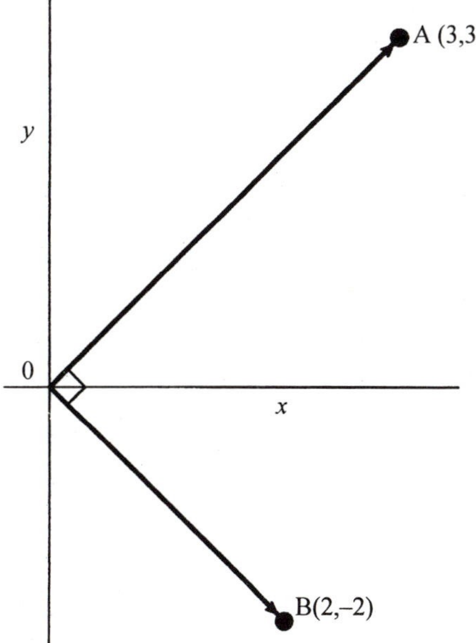

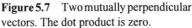

Figure 5.7 Two mutually perpendicular vectors. The dot product is zero.

The only remaining step in the two-dimensional case is to use old vector $\boldsymbol{E}_2$ and, from it, determine a related vector $\boldsymbol{V}_2$ that's perpendicular to $\boldsymbol{V}_1$. Graphically, we can begin by drawing two hypothetical vectors $\boldsymbol{E}_1$ (= $\boldsymbol{V}_1$) and $\boldsymbol{E}_2$ (Fig. 5.8a). (They might represent two measured values in the basic data or two points in lag space.) We then draw a line perpendicular to vector $\boldsymbol{E}_1$, starting at point B and terminating at point A (the terminal point of vector $\boldsymbol{E}_2$). The vector ***OB*** defined in this manner is the **projection** of $\boldsymbol{E}_2$ on $\boldsymbol{E}_1$. Vectors $\boldsymbol{E}_2$ (or ***OA***) and ***OB*** are two sides of a right triangle; the third side is vector ***BA***. Vector ***BA*** equals the new vector, $\boldsymbol{V}_2$, orthogonal to $\boldsymbol{V}_1$, which we're seeking.

From the triangle law of vector addition (Fig. 5.3b), the sum of two vectors ***OB*** and ***BA*** is vector $\boldsymbol{E}_2$, that is, $\boldsymbol{OB}+\boldsymbol{BA}=\boldsymbol{E}_2$. Rearranging that relation enables us to define new vector ***BA*** as

$$\boldsymbol{BA}=\boldsymbol{E}_2-\boldsymbol{OB}. \tag{5.8}$$

Following the laws of vectors, we can **translate** vector ***BA*** into a coordinate vector (i.e. one originating at 0,0) by maintaining its orientation and magnitude during the transfer (Fig. 5.8b). In its new position as a coordinate vector, orthogonal to $\boldsymbol{V}_1$, we'll rename it $\boldsymbol{V}_2$. That completes our task.

The process we've just gone through, namely determining a set of mutually perpendicular vectors from nonorthogonal vectors, goes by the name of **orthogonalization**. Some authors call it **Gram–Schmidt orthogonalization** or **Schmidt orthogonalization**.

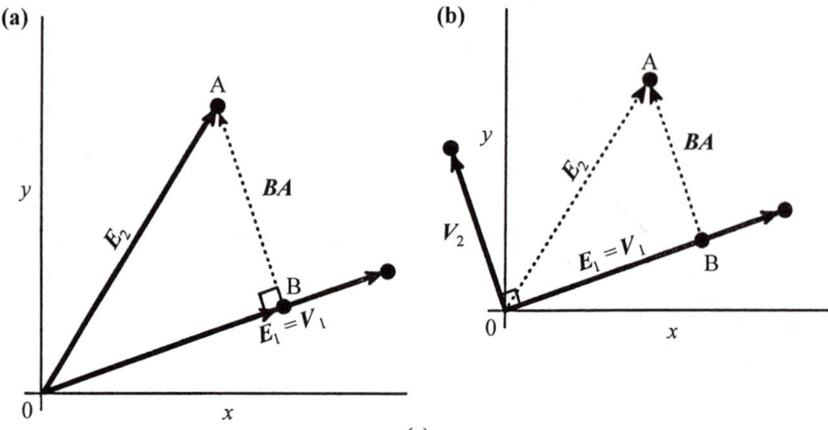

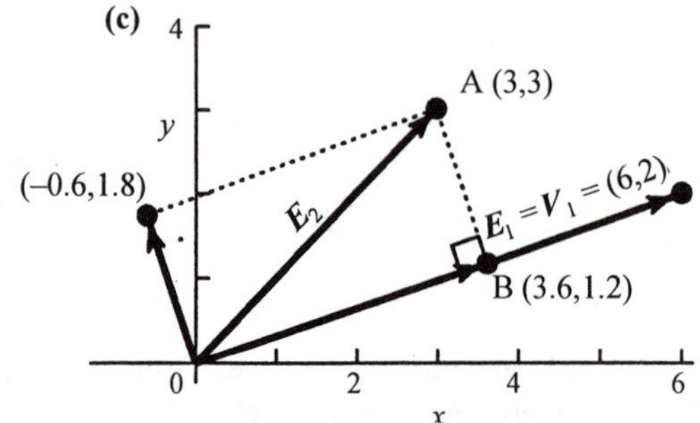

Figure 5.8 Orthogonalization of vectors: (a) graphical determination of a vector ***BA*** perpendicular to a given vector E_1; (b) formation of new orthogonal vectors V_1 and V_2 from the non-orthogonal (original) vectors E_1 and E_2 of part (a); (c) example using coordinate vectors.

We usually have to determine the new, orthogonal vector V_2 mathematically rather than graphically, for reasons of efficiency, time-saving, and computer-programming. The basic equation to use is Equation 5.8. We already know original vector E_2 (Fig. 5.8a). To get ***BA*** and hence V_2, we've yet to find vector ***OB***. That vector is simply vector V_1 multiplied by some scalar coefficient, b. In symbols,

$$\boldsymbol{OB} = \mathrm{b}V_1. \tag{5.9}$$

Substituting $\mathrm{b}V_1$ for ***OB*** and V_2 for ***BA*** in Equation 5.8:

$$V_2 = E_2 - \mathrm{b}V_1. \tag{5.10}$$

Having V_1 (= E_1) as well as original vector E_2, our only remaining job is to find b.

The key to getting b is the property that the dot product of two orthogonal vectors is zero, as discussed earlier (Fig. 5.7). Our two orthogonal vectors are V_2 and V_1. Setting their dot product equal to zero:

$$V_2 \cdot V_1 = 0.$$

In this last equation, we can substitute Equation 5.10 for V_2:

$$(\boldsymbol{E}_2 - \text{b}\boldsymbol{V}_1) \cdot \boldsymbol{V}_1 = 0.$$

A law of dot products, not derived here, says that we can rewrite this last equation in terms of two dot products as

$$(\boldsymbol{E}_2 \cdot \boldsymbol{V}_1) - \text{b}(\boldsymbol{V}_1 \cdot \boldsymbol{V}_1) = 0.$$

Rearranging that relation to solve for b:

$$\text{b} = \frac{\boldsymbol{E}_2 \cdot \boldsymbol{V}_1}{\boldsymbol{V}_1 \cdot \boldsymbol{V}_1}. \tag{5.11}$$

We can then insert that definition of b into Equation 5.10 to get $\boldsymbol{V}_2$:

$$\boldsymbol{V}_2 = \boldsymbol{E}_2 - \frac{\boldsymbol{E}_2 \cdot \boldsymbol{V}_1}{\boldsymbol{V}_1 \cdot \boldsymbol{V}_1} \boldsymbol{V}_1. \tag{5.12}$$

As an example, suppose our job is to take two nonorthogonal vectors $\boldsymbol{E}_1$ (6,2) and $\boldsymbol{E}_2$ (3,3) and to build from those vectors an orthogonal set (e.g. $\boldsymbol{V}_1$ and $\boldsymbol{V}_2$ in Fig. 5.8b, as reproduced in Fig. 5.8c). We take the first vector of the new set, $\boldsymbol{V}_1$, to be the largest vector of the old set, $\boldsymbol{E}_1$. Thus, $\boldsymbol{V}_1 = 6,2$. To get the new vector $\boldsymbol{V}_2$ that's perpendicular to $\boldsymbol{V}_1$ with Equation 5.12, we first compute the scalar b from Equation 5.11. Equation 5.11 is merely the ratio of two dot products. We define each dot product (the numerator in Eq. 5.11, and then the denominator) with Equation 5.4, plugging in the appropriate values of the coordinates of the vectors (3,3 for $\boldsymbol{E}_2$; 6,2 for $\boldsymbol{V}_1$). The numerator in Equation 5.11 then is the dot product $\boldsymbol{E}_2 \cdot \boldsymbol{V}_1 = (3 \cdot 6) + (3 \cdot 2) = 18 + 6 = 24$. Similarly, the denominator is the dot product $\boldsymbol{V}_1 \cdot \boldsymbol{V}_1 = (6 \cdot 6) + (2 \cdot 2) = 36 + 4 = 40$. The scalar b, per Equation 5.11, is the ratio of the two dot products, or 24/40, or 0.6.

Having b, we continue our example calculation of finding new vector $\boldsymbol{V}_2$ by using the rule of scalar multiplication (Eq. 5.3) to get the product b$\boldsymbol{V}_1$ (which is vector ***OB*** in Fig. 5.8c). That quantity is $\text{b}\boldsymbol{V}_1 = 0.6\ (\boldsymbol{V}_1) = 0.6\ (6,2) = 0.6\ (6), 0.6\ (2) = 3.6, 1.2$. Finally, according to Equations 5.10 and 5.2, vector $\boldsymbol{V}_2 = \boldsymbol{E}_2 - \text{b}\boldsymbol{V}_1 = (3,3) - (3.6,1.2) = (3-3.6), (3-1.2) = -0.6, 1.8$. So the new orthogonal coordinate vector $\boldsymbol{V}_2$ goes from the origin to the point located at –0.6,1.8 (Fig. 5.8c).

The same principles used in deriving orthogonal vectors for two dimensions apply to three or more dimensions. Without going through the derivation, I'll list here the equations for three dimensions. In that situation, we'd be given three nonorthogonal vectors ($\boldsymbol{E}_1$, $\boldsymbol{E}_2$, $\boldsymbol{E}_3$). Each is defined as a point in phase space. The task is to use those vectors to find three new vectors ($\boldsymbol{V}_1$, $\boldsymbol{V}_2$, $\boldsymbol{V}_3$) that are related to those data but that are orthogonal to one another. Each new vector has an equation. The first two equations (those for $\boldsymbol{V}_1$ and $\boldsymbol{V}_2$) are the same as those for the two-dimensional case (Eqs 5.7, 5.12). The only new equation is the one for the third vector. The equations are:

$$V_1 = E_1 \tag{5.7}$$

$$V_2 = E_2 - \frac{E_2 \cdot V_1}{V_1 \cdot V_1} V_1 \tag{5.12}$$

$$V_3 = E_3 - \frac{E_3 \cdot V_2}{V_2 \cdot V_2} V_2 - \frac{E_3 \cdot V_1}{V_1 \cdot V_1} V_1. \tag{5.13}$$

The entire process follows two simple principles:

- Define the main vector V_1 of the new (orthogonal) set as the largest observed vector, E_1.
- Get each of the other vectors of the new set by subtracting, from the corresponding observed vector, the projection(s) of that vector onto the newly created vector(s). For instance, get orthogonal vector V_2 by subtracting, from E_2, the projection of E_2 on V_1 (Eq. 5.12). In three dimensions, orthogonal vector V_3 equals observed vector E_3 minus the projection of E_3 on V_2 minus the projection of E_3 on V_1 (Eq. 5.13), and so on for higher dimensions.

Unit vectors and orthonormalization

In science, economics, and virtually all fields, it's often convenient to **normalize** a number or group of numbers. To "normalize" numbers means to put them into a standard form. Normalizing data doesn't change their relationship, but the values are easier to understand. Also, normalizing makes it easier for us to compare data measured in different units. The major step in normalizing data is to divide all values by some maximum reference quantity. The easiest case is a group of positive numbers. To normalize, just divide each value by the largest value of the group. For instance, if the dataset is 2, 12, and 27, divide each value by 27. The transformed (normalized) dataset then is 2/27, 12/27, and 27/27, or 0.074, 0.444, and 1. That transforms all values in such a way that they keep their same relationship between one another but now range from 0 to 1. A range of 0 to 1 is easier to understand and makes it easier for us to compare different datasets.

If the original set includes negative as well as positive numbers, the first step in normalizing is to increase all values by a constant amount to make them positive. Do that by adding the absolute value of the largest negative number to each value of the dataset. Then divide by the largest positive value, as before. For instance, say the basic data are 3, –2, –4, and 7. The largest negative value in that set of data is –4 (absolute value 4). Adding 4 to each value transforms the raw data to 7, 2, 0, and 11. That way, the separate values keep the same relation between one another, but now they're all positive. Dividing everything by 11 (the largest positive value) completes the **normalization**. The transformed values are 0.636, 0.182, 0, and 1.

For a vector, the most convenient reference quantity is its magnitude. Hence, to normalize a vector, just divide each coordinate by the vector's magnitude. That produces a so-called **unit vector**, with a magnitude of 1. For example, using the distance formula, the magnitude of the vector 2,3 (***OA*** in Fig. 5.9) is $(x^2+y^2)^{0.5} = (2^2+3^2)^{0.5} = (13)^{0.5}$. The corresponding unit vector in the same direction as ***OA*** then is each coordinate divided by $(13)^{0.5}$, or $2/(13)^{0.5}$ and $3/(13)^{0.5}$. We can verify that answer by computing the magnitude (length) of the vector $2/(13)^{0.5}$, $3/(13)^{0.5}$ (***OB*** in Fig. 5.9) to see if it indeed equals 1. That length, again using the distance formula, is $([2/(13)^{0.5}]^2+[3/(13)^{0.5}]^2)^{0.5} = [(4/13)+(9/13)]^{0.5} = (13/13)^{0.5} = 1^{0.5} = 1$.

For the two-dimensional case, in other words, the unit vector corresponding to the vector x,y is $x/[(x^2+y^2)^{0.5}]$, $y/[(x^2+y^2)^{0.5}]$. The same idea can be extended to three or more dimensions (e.g. by dividing each of three coordinates by $(x^2+y^2+z^2)^{0.5}$, and so on). Thus, in general a unit vector having the same direction as any vector ***OA*** is $\boldsymbol{OA}/|\boldsymbol{OA}|$. Unit vectors can go in any direction, just as regular vectors can.

Cross-multiplying the relation just mentioned (namely, that a unit vector = $\boldsymbol{OA}/|\boldsymbol{OA}|$) shows that the vector ***OA*** equals the unit vector times $|\boldsymbol{OA}|$. Therefore, since $|\boldsymbol{OA}|$ is a scalar quantity, any nonzero vector is a scalar multiple of a unit vector.

Let's designate normalized (unit) vectors by the global symbol $\boldsymbol{U}_i$. Then $\boldsymbol{U}_1 = \boldsymbol{V}_1/|\boldsymbol{V}_1|$, $\boldsymbol{U}_2 = \boldsymbol{V}_2/|\boldsymbol{V}_2|$, and so on. In general, for the ith vector,

$$\boldsymbol{U}_i = \boldsymbol{V}_i/|\boldsymbol{V}_i|. \tag{5.14}$$

In the chaos literature, you'll see unit vectors used in conjunction with the orthogonalization process just described. That is, once you've determined a new set of orthogonal vectors, you can normalize each of those new vectors by dividing

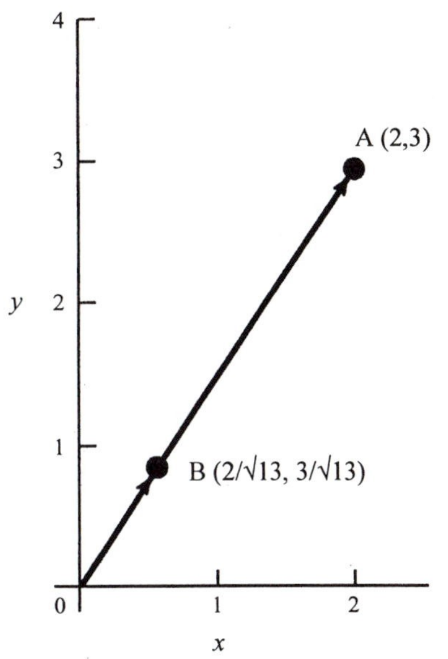

Figure 5.9 A vector ***OA*** and its associated unit vector ***OB***.

it by its magnitude. The process of reducing orthogonalized vectors to unit length is **orthonormalization**.

Summary

A vector is a quantity that has magnitude and direction. A vector whose starting point is at the origin of a graph is a coordinate vector, defined by the coordinates of its terminal point. A datum point in phase space therefore is a coordinate vector, often called simply a "vector." The vector's dimension is the number of components (coordinates) it has. A resultant is the sum of two vectors. Two ways to determine a resultant are the triangle law and the parallelogram law. Multiplying a vector by a scalar yields another vector. The dot product (or inner product or scalar product) of two vectors is the sum of the product of their x coordinates, the product of their y coordinates, the product of their z coordinates, and so on. Orthogonalization (sometimes called Gram–Schmidt orthogonalization) is a mathematical process for creating a set of orthogonal (mutually perpendicular) vectors that are related to an observed set of nonorthogonal vectors. A vector with a magnitude of 1 is a unit vector. Orthonormalization is the process of scaling orthogonalized vectors to unit length.

Chapter 6
Probability and information

Chaos theory relies heavily on several conventional concepts from statistics and probability theory. This chapter reviews the more relevant ones.

Variation

In statistics, a **population** is any well defined group of things. The group can be small or large. Examples are all the peanuts in a particular bag, students in your local school, and people in a country. For practical reasons, we usually can't or don't look at an entire population. Instead, we measure and evaluate just a sample of that population. Then we assume that the characteristics of the sample represent those of the population. One of the most important characteristics of a population or sample is its central tendency. Most aspects of chaos theory use the **arithmetic mean** (one of several possible measures) to indicate that central tendency. The arithmetic mean is just the sum of all the values divided by the number of observations. The chaos literature expresses that process and many related processes in symbols, as follows.

It's logical and conventional to represent the sum of anything with the capital letter S. In practice, the S gets modified slightly, depending on the application. For instance, when we add the areas of rectangles under a curve, the S gets stretched out vertically so that it looks like $\int$. For more general summations, people use the Greek translation or equivalent of S, which is Σ (capital sigma). In particular, Σ is used to symbolize the addition of many measurements of a variable. We'll use i to represent the various individual values and N for the total number of such values. The symbol x_i represents the entire list of values of variable x in a dataset; the list begins with $i=1$ and finishes with $i=N$. To indicate the addition of all such values, we write

$$\sum_{i=1}^{N} x_i .$$

Dividing that sum by N (to get the arithmetic mean) is the same as multiplying the sum by $1/N$. Hence, the symbol definition for the arithmetic mean $\bar{x}$ is:

$$\bar{x} = \frac{1}{N}\sum_{i=1}^{N} x_i . \tag{6.1}$$

Several important characteristics reflect variability of data. Here are three of them:

- *Range* The range is the largest value minus the smallest value.
- *Variance* **Variance** looks at variability in terms of deviations or differences of values from the **mean** (in symbols, $x_i - \bar{x}$). If those deviations are large, variability is great, and vice versa. Each such difference has a sign (+ or –) that's determined by whether observation x_i is larger or smaller than the mean. Right now it doesn't matter whether the individual observations are larger or smaller than the mean; our goal is just an absolute number representing the magnitude of the deviations as a group. One way of getting rid of the signs is to square each deviation. (Squaring it gives a positive number, regardless of whether a deviation is positive or negative.) The average of all such squared deviations reflects the general variability or spread of a group of values about their mean. That average is called the variance. The symbol for the variance of a sample is s^2. Thus

  variance = the average of the squared deviations from the mean

  $$= \frac{\textit{sum of squared deviations from the mean}}{\textit{number of observations}}$$

  or, in symbols,

  $$s^2 = \frac{1}{N}\sum_{i=1}^{N} (x_i - \bar{x})^2 . \tag{6.2}$$

  (And so we've just gone through the first two steps of that three-step procedure that we saw in the last chapter for defining a new quantity. In this case we can't do any simplifying, which is the third step.)

  One important feature about variance is that its magnitude reflects the units (miles, tons, etc.) of the variable being measured. Another is that those units are squared. For example, if we weigh a baby in kilograms on many occasions, the differences from the mean are in kilograms, and the average squared deviation (the variance) is in square kilograms. Other variances might be in "square dollars," "square degrees," and so on. Such numbers aren't easy to visualize. Plain old dollars, degrees, and the like, are much more understandable. The third measure of variability—the standard deviation—solves that problem.
- *Standard deviation* **Standard deviation** is merely the square root of the variance. Hence, in symbols, standard deviation s is:

$$s = \left[\frac{1}{N} \sum_{i=1}^{N} (x_i - \bar{x})^2 \right]^{0.5} . \tag{6.3}$$

Standard deviation, like variance, reflects the units of the variable involved.

Equations 6.2 and 6.3, when based on samples rather than on populations, give slightly biased or underestimated values. To correct for this bias, many statisticians compute variances and standard deviations by dividing by N–1 rather than by N, when working with samples. However, the difference is negligible except when N is small (say about ten or less). Consequently, I'll stick with N here for simplicity.

Frequency and probability

Suppose you and I flip a coin four times and get three heads and one tail. The **frequency** (number of observations in a class) of heads then is three, and the frequency of tails is one. A list showing the frequencies in each class (e.g. three heads, one tail) is a **frequency distribution**. For each possible outcome (head and tail), the **relative frequency** is:

$$\textit{relative frequency} = \frac{\textit{number of occurrences of an outcome}}{\textit{number of opportunities of occurrence}} . \tag{6.4}$$

Relative frequency for heads therefore is 3 heads divided by 4 tries or 0.75 (or 75%); for tails, it's ¼ or 0.25 (or 25%). The list of the various relative frequencies (the **relative frequency distribution**) is heads 0.75, tails 0.25.

By flipping the coin ten or twenty times instead of just four times, the relative frequency for heads probably would move from our original 0.75 toward 0.50. That for tails also would tend to move toward 0.50. In fact, out of a million flips we'd expect relative frequencies that are almost (but probably not exactly) 0.50. The limiting relative frequency approached as the number of events (flips) goes to infinity is called a **probability**. In practice, of course, it's impossible to get an infinite number of observations. The best we can do is estimate probabilities as our number of observations gets "large." Sometimes we have only a small number of observations, but we loosely call the result a probability anyway.

The variables measured in any probability experiment must be **mutually exclusive**. That means that only one possible outcome can occur for any individual observation. For example, the flip of a coin must be either heads or tails but can't be both; a newborn babe is either a boy or girl, not both; and a dynamical system can only be in one state at a time.

If an outcome never happens (no observations in a particular class), then the numerator of Equation 6.4 is zero. Hence, the entire ratio (the probability or relative frequency of that outcome) also is zero. At the other extreme, a particular outcome can be certain for every trial or experiment. Then the number of occurrences is the same as the number of opportunities, that is, the numerator equals the denominator in Equation 6.4. Their ratio in that case is 1.0. Possible values of relative frequency and of probability therefore range from 0 to 1; negative values of probability aren't possible.

A list of all classes and their respective probabilities is a **probability distribution**. A probability distribution shows the breakdown or apportionment of the total probability of 1 among the various categories (Kachigan 1986).

I mentioned earlier that two ways of measuring temporal phenomena are discretely or continuously. The same classification applies to a **random variable** (a variable that can take on different values on different trials or events and that can't be predicted accurately but only described probabilistically). A **discrete random variable** is a random variable that can only have certain specified outcomes or values, with no possible outcomes or values in between. Examples are the flip of a coin (heads or tails being the only possible outcomes) and the roll of a die (numbers one through six the only possibilities). A **continuous random variable**, in contrast, is one that can have any value over a prescribed continuous range. Examples are air temperatures and our body weights.

A popular graphical way to show a frequency distribution or probability distribution for one variable is to draw a **histogram**. (The suffix "gram" usually refers to something written or drawn; examples we've seen thus far include program, diagram, and parallelogram.) A histogram is a bar chart or graph showing the relative frequency (or probability) of various classes of something. Usually, the class boundaries are arbitrary and changeable. (Subdividing a range of values or a phase space into a grid of cells or bins is called **partitioning**. The collection of cells or bins is the **partition**.) The most common histograms consist of equal-width rectangles (bars) whose heights represent the relative frequencies of the classes. (If, instead, the variable is discrete, each possible outcome can have its own horizontal location on the abscissa. In that case, the graph has a series of spikes rather than bars.)

In chaos theory, we often deal with continuous random variables (even though measured discretely). As class width becomes smaller and sample size larger, a histogram for a continuous variable gradually blends into a continuous distribution (Fig. 6.1; Kachigan 1986: 35–7). The smooth curve obtained as sample size becomes very large is called a **probability density function**.

Having bar *heights* (relative frequencies) of a histogram sum to 1.0 is useful and straightforward. Another useful feature, mathematically and theoretically, is to have the *total area* of a histogram (including the area under the continuous curve of Fig. 6.1d) sum to 1.0 (Wonnacott & Wonnacott 1984: 101–102). To make the areas of the individual bars add up to 1.0, statisticians simply rescale the ordinate value of each bar to be a **relative frequency density**:

$$\textit{relative frequency density} = \frac{\textit{relative frequency}}{\textit{class width}}. \tag{6.5}$$

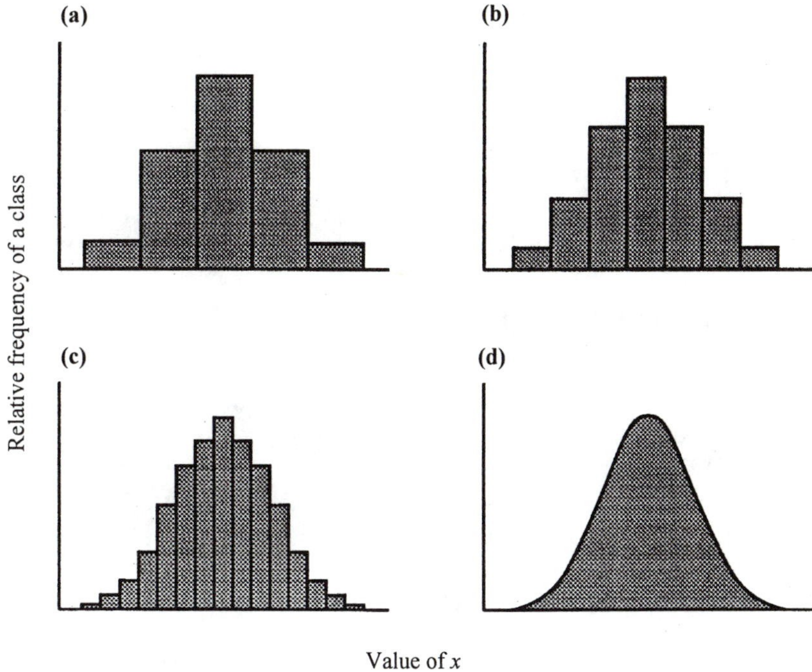

Figure 6.1 Progression of a histogram to a continuous distribution as sample size increases and bin width decreases (after Kachigan 1986: 35).

Figure 6.2a shows the rescaling idea. The figure is a histogram showing a fictitious relative-frequency distribution of the heights of women in Denver. I arbitrarily took the class width as 6 inches (0.5 foot). The height of each bar represents the relative frequency of that class. The sum of all bar *heights* is 1.0. Figure 6.2b shows the same basic data rescaled to relative frequency densities. Class width here is still 0.5 foot, so the ordinate value or height of each bar in Figure 6.2b is the relative frequency of a class divided by 0.5, per Equation 6.5. That height times the bar width (here 0.5) gives bar area. The sum of all such bar areas in Figure 6.2b is 1.0.

Statisticians don't use the word "density" in connection with a *discrete* random variable. Instead, "density" implies that the underlying distribution is continuous, whether measured on a discrete or continuous basis.

The probability density function is a basic characteristic of a continuous random variable. Some authors use other expressions for the probability density function but usually still include the word "density." Examples are **probability density**, **probability density distribution**, **probability density curve**, **density function**, and just **density**.

The related expression "probability distribution" can apply to either a discrete random variable or a continuous random variable. Also, people use "probability

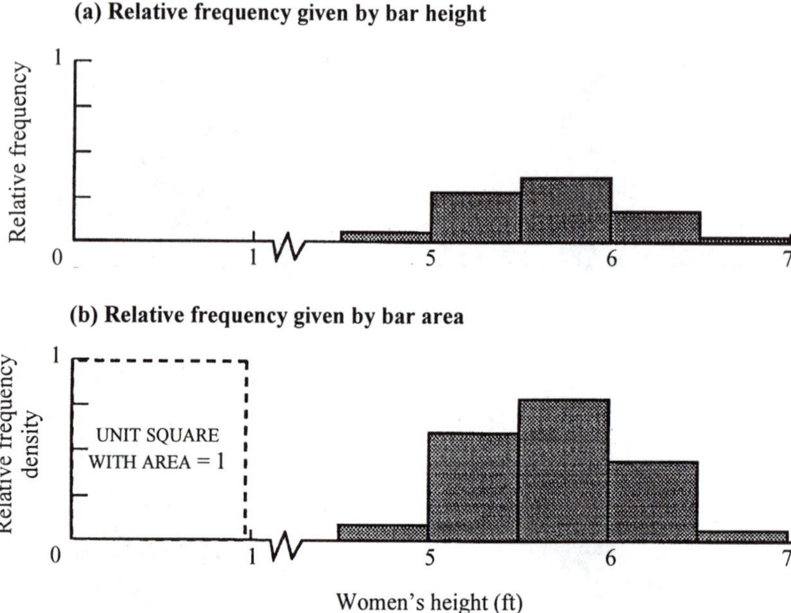

Figure 6.2 Histograms of hypothetical heights of women in Denver. (a) Relative frequency histogram with sum of bar *heights* equal to 1. (b) Relative frequency density histogram with sum of bar *areas* equal to 1 (after Wonnacott & Wonnacott 1984: 100).

distribution" in connection with the limiting relative frequency or with the limiting relative frequency *density*. For example, the "probability distribution" for a zillion flips of a coin is heads 0.50, tails 0.50. At the same time, the "probability distribution" for the heights of US women is a probability density function and very likely would look much like the curve of Figure 6.1d. Either way, the probability distribution shows how the total probability of 1.0 is distributed among the various classes (for a continuous variable) or among the individual outcomes (for a discrete variable).

Estimating probabilities

Several tools or measures in chaos theory require probabilities. We estimate such probabilities from our data. That's easy and straightforward for some purposes but not for others. In other words, depending on the tool, distribution of data, and other factors, the simple approach of setting up equal-width bins, counting observations in each bin, and so on, may not be particularly useful. Statisticians continually seek better ways. In fact, estimating probability distributions is a major research area in statistics. (The technical literature calls the general subject "**density estimation**."

In **parametric** density estimation, we assume a certain type of probability distribution. That means we know or assign values to certain parameters of that distribution [hence the name parametric]. In **nonparametric** density estimation, we make no assumptions about the nature of the probability distribution.)

Histograms

The easiest, oldest, and most popular probability estimator is the standard histogram. The procedure is just to divide up the possible range of values into adjoining but not overlapping intervals or bins of equal width, count the number of observations in each bin, and divide each count by the total number of observations. That gives relative frequencies. It's a very useful way to show the distribution of the data.

That said, histograms have four disadvantages:

- They carry various mathematical and statistical drawbacks (Silverman 1986: 10). For instance, the vertical boundaries of the bins are undesirable. They are what mathematicians call discontinuities in the distribution.
- Sample size affects the relative frequencies and the visual appearance of the histogram. For example, the more times we flip a coin, the closer a histogram becomes to two equal-size rectangles or spikes.
- The histogram for a given sample size or range of values varies with number of classes (i.e. with class width, for a given range of data). To demonstrate that, I made up a dataset of 87 values ranging from 10 to 30. Histograms based on those data look quite different, depending on how many bins we divide the data into (Fig. 6.3). For the ridiculous extreme case of just one bin, the histogram is one great big box.

  Mathematicians call a distribution having no discontinuities (no bin boundaries) a **smooth** distribution. The verb form, **smoothing**, means to reduce or remove discontinuities, irregularities, variability, or fluctuations in data. A histogram having relatively few bins has relatively few discontinuities. Hence, it's "smoother" than one with many discontinuities or bins. Thus, the first diagram in Figure 6.3 (two bins) is smoother than the fourth diagram (20 bins), in spite of visual appearance. In that sense, number of bins or bin width is called a **smoothing parameter**. (Other names for bin width are **window** width and bandwidth.)

  How do we choose bin width (or number of bins), for a given dataset? Briefly, there's no answer (no universally accepted way). Furthermore, the choice can be subjective or objective, depending on the use we'll make of the resulting density estimate (Silverman 1986: 43ff.). A subjective choice is appropriate if you want to explore the data for possible models or hypotheses. In such a case, try several different bin widths to see if different shapes appear for the estimated density distribution. An objective choice (automatic procedure), on the other hand, is appropriate for scientific research (where a standardized method for comparing results is desirable), inexperienced folk, or

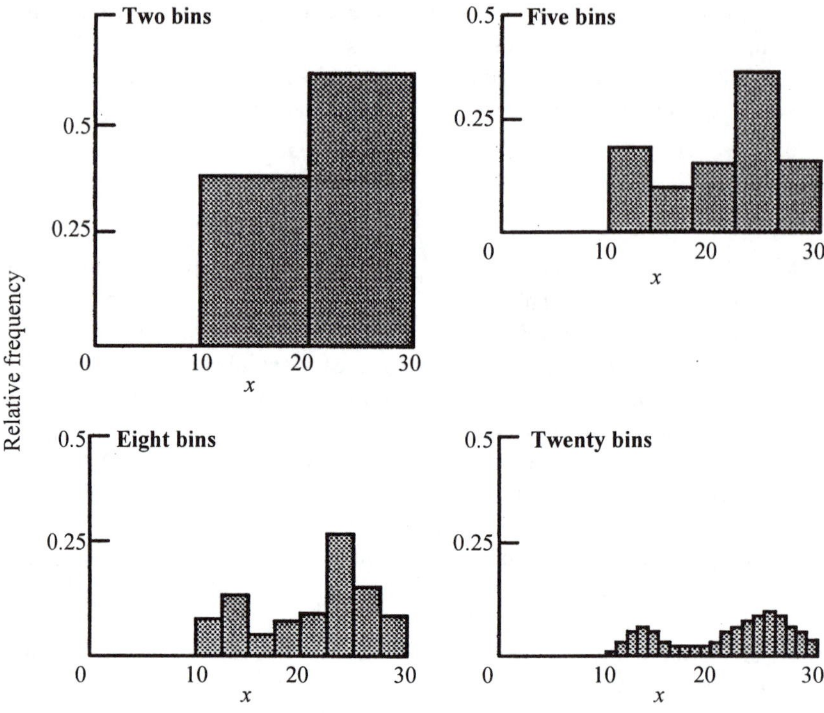

Figure 6.3 Effect of number of classes (or class width, for a given range of data) on a histogram, using the same 87 value dataset and the same axis scales.

routine usage where you have a large number of datasets.

There are a half dozen or more objective ways to select a bin width (Silverman 1986: 45–61). Even in the ideal case, where we somehow know in advance that the true distribution is bell-shaped into what statisticians call **normal**, there are several different proposed rules for the appropriate number of bins (Scott 1992: 55–6). Those proposals give number of bins according to sample size and in some cases also to other factors. Sturges' rule, rarely used nowadays, is an example. His rule says that the number of bins is $1+\log_2 N$, where N is the total number of observations. Thus, with 100 measurements, the rule says to use $1+\log_2(100) = 1+6.6 = 7.6$ or, rounding upward, 8 bins.

- Histograms vary according to the locations of the bin boundaries. For example, based on the same data of Figure 6.3, Figure 6.4 shows three histograms. They all have the same bin width (5, or actually 4.99) and the same number of bins (five). They differ only in the locations of the bin boundaries. Each histogram looks different. One approach that statisticians have explored to overcome this problem is to construct several histograms of slightly different (shifted) bin boundaries and then to average the lot.

With regard to chaos, histograms have a fifth disadvantage. A dynamical system might behave in such a way that vast regions of its phase space have only a few observations, and a relatively small region has a large majority of the observations (a clustering of data). A standard histogram for such data might have almost all observations in one class. The distribution within that class could have some important features in regard to possible chaos, but we'd lose that information with a standard histogram. For all of the above reasons, researchers are looking for more efficient ways (improvements over the standard histogram) to estimate probabilities.

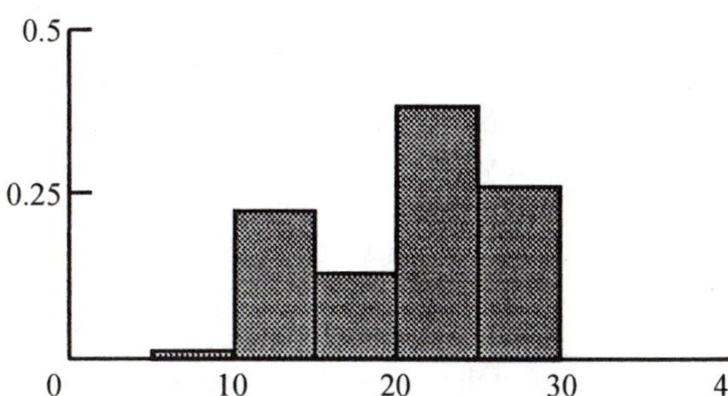

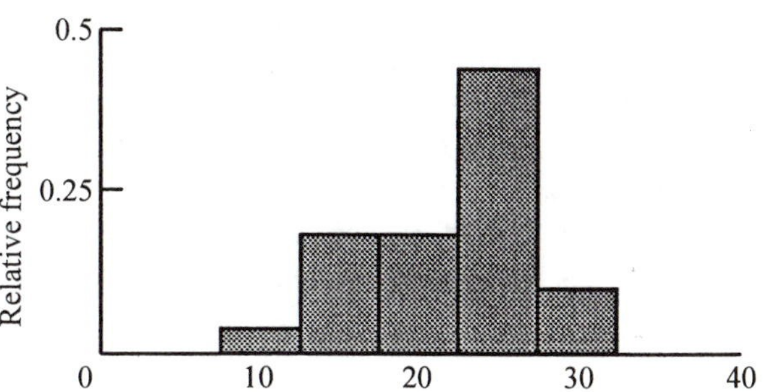

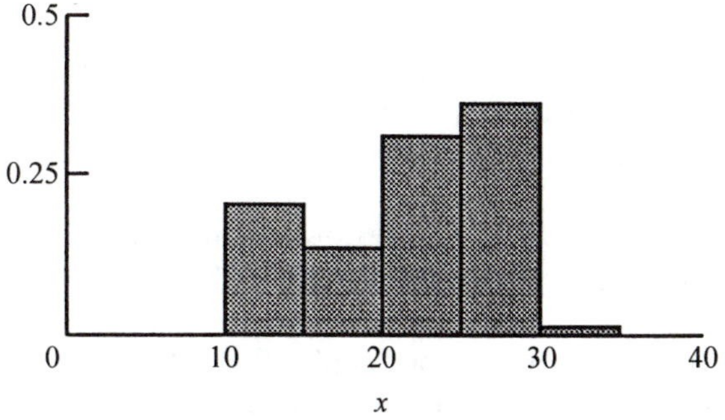

Figure 6.4 Effect of location of bin boundaries on visual appearance of histogram, for same hypothetical dataset and same bin width.

Other methods

In addition to standard histograms, there are a dozen or so other ways to estimate probabilities. Silverman (1986) and Izenman (1991) review those methods. Some of them might apply to chaos theory. Here are three that show promise.

ADAPTIVE HISTOGRAMMING (VARIABLE-PARTITION HISTOGRAMMING)

Data are often unevenly distributed over a frequency distribution. For example, they might be sparse in the tails (and sometimes elsewhere) and more numerous in the middle. A fixed bin width then might result in much of the information being swallowed up by one bin. Other regions might have isolated bins (isolated probability estimates), and still others might have zero probability. At least theoretically, such isolated estimates interrupt the continuity of the final probability density distribution (an undesirable result). To cope with that problem, statisticians developed a group of techniques that are data-sensitive or **data-adaptive**. That means they don't apply the same bin width to the entire distribution. Instead, the methods change bin (partition) width from one bin to another according to local peculiarities (especially number of observations) in the data. They adapt the amount of smoothing to the local density of data. They do so by using wide bins for ranges that have only a few data points (typically the tails of the frequency distribution) but narrow bins for ranges that contain many points. Equation 6.5 still applies, but the denominator (class width) varies from one class to another.

In the world of general statistics, the two most common approaches to **adaptive histogramming** are combining bins in the tails of the distribution (where observations tend to be fewer) or defining bin boundaries such that all bins contain the same number of data points (Scott 1992: 67). A prominent example of adaptive histogramming within chaos theory is that of Fraser & Swinney (1986). They cover the phase space with a grid in which the size of each bin depends on the number of points in the local neighborhood. Bins are smaller where there are more points, and vice versa.

KERNEL DENSITY ESTIMATORS

The general idea with this technique is to estimate a probability density for constant-width bins that are centered at each datum point. (Theoretically, we can center a bin wherever we want an estimate.) All points falling within a bin contribute to that bin's estimated density, but they do so in lesser amounts as their distance from the bin center point increases. In a bit more detail, the overall steps are:

1. Choose a bin width and center a bin at a datum point (such as point x in Fig. 6.5a).
2. Assume some sort of arbitrary local probability distribution. We'll apply that distribution not only to this datum point (really, this neighborhood) but, subsequently, to all other points as well. Figure 6.5b (after Härdle 1991) shows some of the more popular types of local probability distributions that people use. All such distributions have probability on the ordinate and standardized

(a)

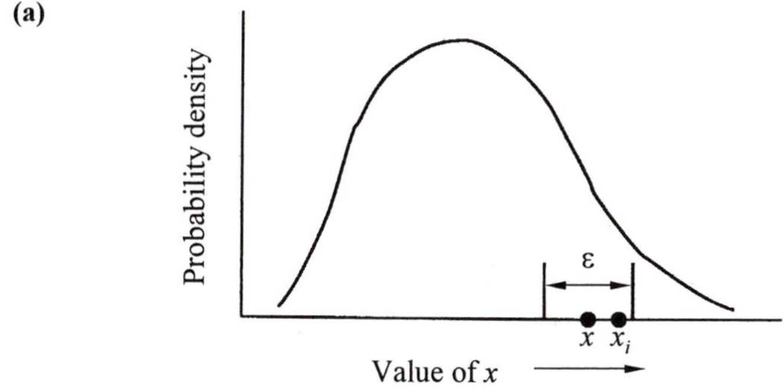

(b)

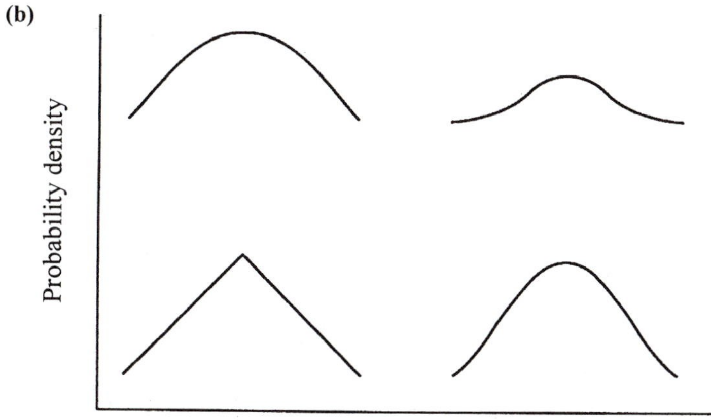

Figure 6.5 Basic concepts of kernel probability estimation. (a) Hypothetical probability density (the entire curve) for a group of measurements, showing bin width ε, data point x on which a bin is centered, and a neighboring point x_i. (b) Popular kernels (assumed local probability distributions).

distance to either side of the central datum point on the abscissa, as explained in step 3 below. They needn't be symmetrical but usually are. Each distribution looks like a bump on a flat surface (the abscissa). Hence, the distributions go by the general name "kernel," one everyday definition of which is "a hard swelling under the surface of the skin." (In fact, the local distributions commonly are called either "bumps" or "kernels" in the technical literature.)

3. For each datum point within the bin boundaries, estimate the probability density using the assumed distribution. For that purpose, first compute the standardized distance from the bin center as $(x-x_i)/\varepsilon$, where x is the central datum point, x_i is the ith neighboring point, and ε is bin width (Fig. 6.5a). (Ch.

9 describes standardization in more detail.) Then find that value on the abscissa of the assumed distribution, go straight up on the graph to where that value intersects the curve, and read the associated density value from the ordinate. (The kernel or assumed local probability distribution therefore is a **weighting** procedure. "Weighting" in general is any system of multiplying each item of data by some number that reflects the item's relative importance. With regard to kernels, the weighting assigns different probabilities depending on the distance of the datum point from the bin center.)

4. Average the estimated probabilities for that particular bin (add the probabilities and divide by the number of points), then translate that average into a density by dividing by class width.
5. Center a bin of the *same width* at each of the other data points and repeat steps 3–4 for each point, in turn. That gives an estimated density for the bin centered at each observation. The entire list of probabilities is the probability density distribution. It's common practice to plot those probabilities against x (the bin center-point) and connect all plotted points with a line, to simulate the continuous probability distribution.

Kernel density estimators solve two problems that plague the standard histogram: the vertical boundaries and arbitrary choice of bin locations. Arbitrary steps in the method are the choices of kernel and of bin width. Choice of kernel isn't too important. However, bin width has a major effect on the results (as with histograms). Too small a width brings a very irregular, wiggly probability distribution. On the other hand, making bin width too large smooths the distribution too much. (The extreme example of such "over-smoothing" is a bin width spanning the entire range of data, in which case the distribution consists of one large rectangle.) The best procedure is to try several bin widths and choose the one giving the results you think are most representative.

ADAPTIVE KERNEL DENSITY ESTIMATORS

This technique combines the data-adaptive philosophy (the philosophy of varying the bin width) with the kernel approach (Silverman 1986: 100–110). The intent, of course, is to gain the advantages of both. A necessary first step is to get some rough idea of the local density around each datum point. Almost any estimator works for such a pilot estimate; the standard kernel estimator with fixed bin width is a common choice. Next, assign a bin width to each datum point, tailoring that width to the local density as given by the pilot estimate. Then choose a kernel (e.g. Fig. 6.5b) and estimate the entire probability distribution in a way essentially like that described above for the kernel estimator. Authors have suggested ways to fine-tune one or more aspects of the procedure.

Probabilities involving two or more variables

The probability we've seen thus far has dealt with just *one* variable, such as height, weight, one coin, one die, or the air temperature. That probability represents the limiting relative frequency of each category or bin within that variable's range of possible outcomes. Probabilities can also be based on distributions that involve more than one variable ("multivariate" distributions). The multiple variables can be separate items (e.g. two dice; John, Sam, and Igor; temperature, air pressure, and humidity; etc.). For instance, if you and I each flip a coin at the same time, the combined distribution has four possible outcomes or categories (both of us get heads; both get tails; heads for you, tails for me; and tails for you, heads for me). Probabilities are 0.25 for each of the four categories. Furthermore, as explained in Chapter 3, chaologists also think of lagged values of one physical feature (e.g. x_t, x_{t+1}, etc.) as separate variables. Thus, they might categorize successive pairs (or triplets, etc.) of values of x and get a multivariate distribution in that sense. We'll see an example below.

Chaos theory uses three types of multivariate probabilities: joint, marginal, and conditional probabilities. (I'm using "type of probability" very loosely here. Strictly, there's no such thing as different types of probability. There are only different types of events on which those probabilities are based. "Event" can be defined on the basis of one, two, or many variables.)

Joint probability

"Joint" in statistics means "involving two or more variables." However, it has a more precise meaning in "joint probability": a **joint probability** is the likelihood of getting two or more specified outcomes.

A joint probability doesn't necessarily imply anything about time. Even when it does, the events can be simultaneous or sequential. For instance, one joint probability might be the probability that you presently subscribe to newspaper A and magazine B. Another joint probability might be the likelihood that anytime in July the daytime air temperature will exceed 95°F three days in a row.

Let's start with an example of joint probability for *simultaneous* events. Suppose I roll a pair of loaded dice (one die green, the other white) 100 times. The grid of Figure 6.6 shows hypothetical results. Values for the green die are x, those for the white die y. Each dot on the grid represents the x and y values that came up on a roll. For instance, the values 1,1 for x,y came up on three of the 100 rolls. Based on the 100 rolls, the numbers in each individual bin of the figure are estimated joint probabilities. For example, the estimated joint probability of getting 1,1 is 3 occurrences out of 100 rolls or 0.03, and so on. Statisticians symbolize that joint probability, namely that the green die (x) shows a 1 and the white die (y) also shows a 1, as $P(1,1)$ or $P(x_1,y_1)$, in which P stands for probability. The global symbol for any of the 36 bins is $P(x_i,y_j)$. Here the subscript i is an **integer**, specifically one

x \ y (Green die \ White die)	1	2	3	4	5	6	$P(x_i)$
1	••• 0.03	•• 0.02	• 0.01	• 0.01	•• 0.02	• 0.01	0.10
2	• 0.01	• 0.01	•• 0.02	••• 0.03	••• 0.03	•• 0.02	0.12
3	•• 0.02	• 0.01	•• 0.02	• 0.01	••• • 0.04	••• 0.03	0.13
4	•• 0.01	• 0.01	••• 0.03	•• 0.02	•••• • 0.05	••• ••• 0.06	0.18
5	• 0.01	•• 0.02	••• 0.03	••• • 0.04	•••• •• 0.06	•••• ••• 0.07	0.23
6	•• 0.02	•• 0.02	•• 0.02	••• • 0.04	••• ••• •0.07	••• ••• •0.07	0.24
$P(y_j)$	0.10	0.09	0.13	0.15	0.27	0.26	1.00

Figure 6.6 Hypothetical results of rolling a pair of loaded dice 100 times. Each dot represents the outcome of a roll. The grid of 36 numerical probabilities is the joint probability distribution, $P(x_i,y_j)$. The column on the far right, labeled $P(x_i)$, lists the marginal probabilities of each state x_i (each possible outcome for the green die). Likewise, the row along the bottom margin, labeled $P(y_j)$, is the marginal probability distribution of y_j.

through six, indicating the various possible values or bins of x (the green die); the subscript j does the same for y values (the white die). (An integer is a whole number, i.e. a number that doesn't include any fractional or decimal part.)

The entire array of joint-probability bins (36 of them in this example) with their respective values is the **joint probability distribution**. A joint probability distribution is a list, table, or function for two or more random variables, giving the probability of the joint occurrence of each possible combination of values of those variables.

Now let's move on to another example, this time for the joint probability of *successive* events. The common case in chaos theory involves a joint probability distribution based on two or more lagged values of a single feature (x). In that case, x in the grid of Figure 6.6 becomes x_t and y becomes x_{t+m}, where m is lag. The probability in each bin then is the probability that value x_t (as classed in the left-hand vertical column) was followed by value x_{t+m} (as classed in the column headings across the top).

Joint probability distributions based on lagged or successive values are very popular in chaos theory. We'll need some convenient terms or names to distinguish such probabilities from those based on other types of distributions. The statistics literature doesn't seem to have specific, standard terms for some of those probabilities. Therefore, I'll use two nonstandard terms, with apologies to statisticians. First, I'll call a joint probability based on lagged (sequential) measurements a **sequence probability**. A sequence probability in this book is the joint probability of getting a particular sequence of values for successive observations. In other words, it's the probability of getting a particular lagged series or pseudo phase space route. (Each possible combination on such lagged grids in a sense is a "bin." However, when we're talking about lagged measurements of a single variable and probabilities, I'll call it a "route," instead. So, for probabilities the terms "bin," "state," "class," and the like herein will refer only to one time; "route" will refer to a sequence of two or more successive times.) Secondly, I'll call the probability we dealt with earlier (involving just one variable) an **ordinary probability**.

Estimating sequence probabilities is easy. The routine first step is always to decide on the class limits (the partitioning) for the feature being measured. The next job is to choose the lag and number of events in the sequence. Then just have the computer count the number of occurrences of each sequence and divide each such total by the total of all sequences, for a given time.

A quick example shows the idea. Suppose we measure a variable 300 times and categorize each value into one of two possible states or bins, A or B. The first "sequence" by definition here involves just one event. That is, it uses ordinary probabilities. To get ordinary probabilities, we just (a) count how many observations fall into each of our two categories, and then (b) divide each tally by the total number of observations (here 300). Say that our dataset has 180 measurements in bin A and 120 in B. The ordinary probabilities then are 180/300 or 0.60 for A and 120/300 or 0.40 for B. So, at any arbitrary time ("time 1" in Fig. 6.7), the chances that the system is in A are 0.60 and in B 0.40.

The next sequence is the two-event sequence. In our example (Fig. 6.7), there are four possible sequences or time-routes over two events:

- into box A at time 1, then to A at time 2 (a pattern I'll call route AA or simply AA)
- into A at time 1, then to B (route AB)
- B at time 1, then A at time 2 (BA)
- B then to B again (BB).

We can't just assign probabilities of 0.25 to each of those four routes. That would imply that the system on average chooses each route about the same proportion of the time. Yet, for all we know, the system behaves in such a way that, once in A, it prefers to go next to B! The only way to get the best estimate is to count the number of occurrences of AA, of AB, of BA, and of BB. In the fictitious example of Figure 6.7, say those sequences happened on 126, 54, 96, and 23 occasions, respectively. (Those frequencies add up to 299 rather than 300 because the 300th observation doesn't have a new bin to go to.) Estimated sequence probabilities are number of occurrences for each route divided by the total. Thus, the probability of route AA (the sequence probability of AA) is 126/299 or 0.42; for route AB, 54/299 or 0.18; for BA, 96/299 or 0.32; and for BB, 23/299 or 0.08. (Fig. 6.7 doesn't include those values because I need the space on the figure for something else, to be explained.) Since those four routes are the only ones possible over two consecutive measurements, the four sequence probabilities sum to 1 ($0.42+0.18+0.32+0.08=1$). That is, sequence probabilities at any time add up to one.

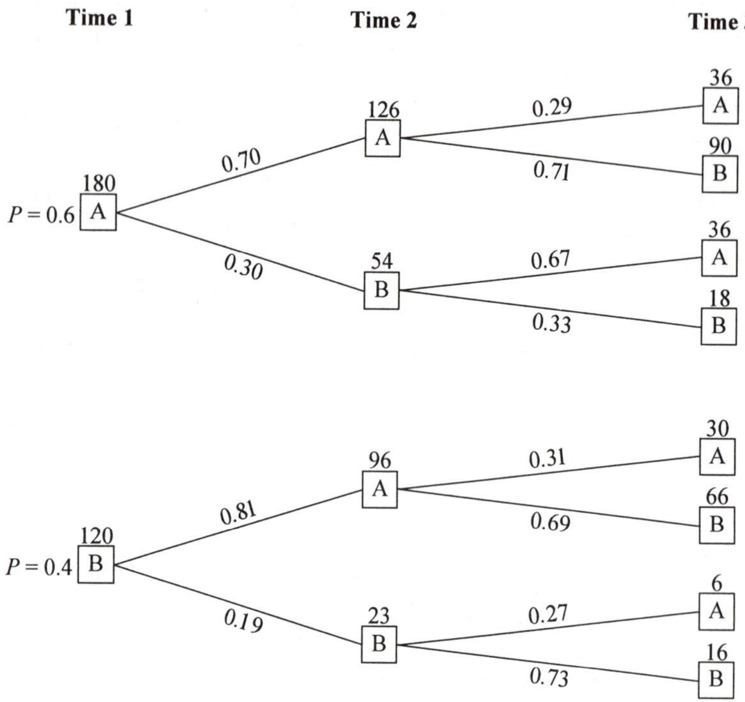

Figure 6.7 Branching pattern and hypothetical probabilities for two bins (A and B) over three time steps. Numbers along paths are conditional probabilities. Numbers on top of each box are number of observations.

For a three-event sequence, the possible routes in our example consist of AAA, AAB, ABA, and so on, or eight routes in all (Fig. 6.7, routes leading to "time 3"). For each of the eight routes, we count the number of occurrences and divide by the total number of all routes to get the estimated sequence probability for each. The total number of all routes for three sequential events in this example is only 298 (i.e. N–2), because the 299th and 300th values in the 300-member basic dataset don't have a third value to lead to.

When a system can go from any state at one time to any state at the next time (e.g. Fig. 6.7), then number of events and number of possible states greatly affect the required computer time. Each possible route has its own sequence probability, and the number of possible routes increases geometrically with number of events. (A **geometric progression** is a sequence of terms in which the ratio of each term to the preceding one is the same. An example is the sequence 1, 2, 4, 8, 16, and so on. In that series, the ratio of each term to the preceding term is 2.) Specifically, the number of possible routes, N_r, leading to any event within a sequence equals the number of possible states, N_s (our example has two—A and B) raised to a **power** that equals the sequential number (temporal position) of the event, n. That is,

$$routes = states^{events}$$

or, in symbols,

$$N_r = N_s^n. \tag{6.6}$$

Let's check that relation with our example of bins A and B. At time 1, the event number, n, by definition is 1. The number of possible routes N_r then is the number of states N_s (here 2) raised to the first power. That's 2^1 or 2 routes. (Again, those two possible routes are [1] into A or [2] into B.) Time 2 is the second event, so $n = 2$; N_s is still 2 (bins A and B), so there were 2^2 or 4 possible routes (AA, AB, BA, BB); at time 3 there were 2^3 or 8 routes, and so on. So, the relation is valid. Furthermore, as we increase the number of events n in our sequence, we have many more routes to monitor.

The number of possible states also has a huge effect. At time 1 we have only two probabilities to estimate, in our example of bins A and B. However, if we divide our unit interval or data range of 0 to 1 into ten compartments or possible states instead of two (0 to 0.1, 0.1 to 0.2, etc.), then we've got 10^1 or ten ordinary probabilities to estimate. At time 2, we have 2^2 or 4 sequence probabilities to estimate with two possible states (A and B) but 10^2 or 100 sequence probabilities if we use ten compartments. (Each of the ten possible states for time 1 can foster routes to any of the ten states for time 2.) At time 3, we have 2^3 or 8 probabilities for two bins but 10^3 or 1000 sequence probabilities if we use ten bins! And so on. So, we need powerful computers in this game.

Real-world systems may not use all of the mathematically possible routes, for a given number of possible states and events. Also, a dataset may be too small to have sampled some routes that the system takes. If either of those cases is true, one or more possible states or routes has a probability of zero. For instance, any class for which we have no observations gets an ordinary probability of zero. That, in turn, results in a sequence probability of zero for any route involving that class. Nonetheless, number of events and number of states still have a big influence on the computation requirements.

Marginal probability

Now back to our green and white die example (Fig. 6.6). The green die came up 1 (*irrespective* of what the white die showed) a total of 10 times. (Those ten consist of three dots for $x = 1, y = 1$; two dots for $x = 1, y = 2$; etc.) So, the probability of getting $x = 1$ is 0.10, since there were 100 total observations. We get that same answer by adding the separate joint probabilities across the row for $x = 1$. For instance, in Figure 6.6 the probability that $x = 1$ is 0.03+0.02+etc. = 0.10 (listed in the marginal column on the far right). In fact, we've just completed step one (conceptualization) of our three-step procedure for defining a new quantity. In this case we'll reword that concept as follows:

The probability of getting any one value of x by itself, regardless of the value of y (e.g. any row in Fig. 6.6), is just the sum of the joint probabilities of that x for each of the y's.

Step two of the general procedure is to express that probability in symbols. The probability that system X (here the green die) will be in bin x_i is $P(x_i)$. The symbol

$$\sum_{j=1}^{N_s}$$

represents the summation over the various j values (white-die values), for that value of x. (For instance, j here has six possible values, i.e. $N_s = 6$.) Finally, $P(x_i,y_j)$ is the global symbol for the pertinent joint probabilities (the probabilities of getting the pertinent bins in the joint probability distribution). (In our example, those pertinent joint probabilities are 0.03, 0.02, etc., and they sum to 1.0.) Hence, our rule in symbols is

$$P(x_i) = \sum_{j=1}^{N_s} P(x_i,y_j \; . \tag{6.7}$$

We can't simplify the expression (step three), in this case.

By convention, we'd evaluate Equation 6.7 by starting with the case of $x = 1$. In other words, what is the probability that $x = 1$, or, in symbols, what is $P(x_1)$? To get that probability, Equation 6.7 says to add $P(x_i,y_j)$ for all six values of y_j for $i = 1$. Next we want the probability that $x = 2$, that is, $P(x_2)$. That means adding $P(x_i,y_j)$ for all six values of y_j for $i = 2$. In our example (Fig. 6.6, second row), that's $0.01+0.01+0.02+0.03+0.03+0.02 = 0.12$. And so on.

The same idea also applies to system Y (here the white die). The symbols now have to reflect the changed roles of the two variables. The probability that system y will be in bin y_j is $P(y_j)$. To get the probability that $y = 1$, that is, $P(y_1)$, we take the first y bin and sum over the various observed values of x (denoted by i). There's a total of N_s such values of x. So, the summation goes from $i = 1$ to $i = N_s$, and the summation symbol becomes

$$\sum_{i=1}^{N_s} .$$

For instance, in Figure 6.6 the probability of getting $y = 1$ is the sum of the probabilities listed in the first column, or $0.03+0.01+\ldots = 0.10$ (written as the first value in the row for $P(y_j)$ along the bottom margin). We then add up the six x values for each of the other five classes of y, in turn. For $y = 2$, we get 0.09; for $y = 3$, we get 0.13; and so on. Finally, the symbol $P(x_i,y_j)$ remains valid for the pertinent joint probabilities. Thus, the rule in symbols for the probability of getting any chosen y value now is

$$P(y_j) = \sum_{i=1}^{N_s} P(x_i, y_j). \tag{6.8}$$

Because probabilities $P(x_i)$ and $P(y_j)$ are listed in the margins of the table (Fig. 6.6), they are called **marginal probabilities**. Thus, a marginal probability is the joint probability of getting a specified value of one variable, irrespective of the associated value of another variable. The entire ensemble of such probabilities for any one variable (such as the six probabilities in the example of the dice) is called the **marginal probability distribution** or simply the **marginal distribution** for that variable.

Conditional probability

Conditional probability is the probability of getting a certain outcome, given a specified restriction or condition. The specified condition on the one hand might be some known influential information. For example, what is the likelihood that you enjoy watching football, given that you are a male? Or the probability that Joe weighs more than 250 pounds, given that he is five feet tall? On the other hand, the specified condition might be the outcome of a preceding event in a sequence. For instance, what is the likelihood that Polly will get an A on the second mathematics exam, given that she got an A on the first exam? The known information enables us to adjust, refine, or reassess our estimate of the probability of the other outcome.

The general idea in estimating conditional probabilities is to focus only on those outcomes where the specified condition occurred. Thus, in the preceding paragraph the question isn't whether everybody (males and females) likes football. Rather, only males qualify. Similarly, to estimate the probability that Joe weighs more than 250 pounds, there's no sense estimating the likelihood that anyone selected by chance from the general population weighs more than 250 pounds; instead, we look only at males who are about five feet tall. In Polly's case, if we have no information about her or her record, we'd have to use some sort of general statistic based on students in general to guess her chances of an A on exam two. However, we'd rate her chances somewhat higher if we knew that she got an A on the first exam.

Figure 6.6 shows the difference between joint and conditional probabilities. For instance, the estimated joint probability that the green die (x) will come up five *and* the white die (y) on the same roll will come up six, that is, $P(x = 5, y = 6)$, is 0.07; that combination occurred 7 times out of the 100 rolls of the two dice. To get conditional probability for that same bin, we ask: given that $x = 5$, what is the likelihood that $y = 6$? Answering that question requires restricting our analysis to just the special subset of the data for which $x = 5$. In terms of numbers of occurrence, an x of 5 occurred 23 times out of our 100 rolls of the dice. Of those 23 times, a y value of 6 occurred 7 times. The conditional probability, here the probability that $y = 6$ given that $x = 5$, therefore is 7/23 or 0.30.

We can just as well represent the 23 occurrences of $x = 5$ by its ordinary probability within the dataset: 23/100 or 0.23. Similarly, we can represent the 7 occurrences of $x = 5$ and $y = 6$ in terms of its overall (joint) probability within the dataset (7/100 or 0.07). (Thus, just as 7/23 = 0.30, so too does 0.07/0.23 = 0.30.) Therefore, statisticians define conditional probability in terms of probabilities based on the entire dataset, rather than numbers of occurrences. Conditional probability with this approach is the joint probability of a particular bin (0.07 for our example) divided by the ordinary probability of getting a particular value of one variable (0.23 for our example): that's step one of the three-step derivational procedure!

Step two of the derivational procedure is to put that definition into symbols. Let's use just the case where the given information is an x value. (The case where y is given follows the same model.) We can use two of the same symbols as above, namely $P(x_i)$ for the probability of getting a particular x value and $P(x_i,y_j)$ for the joint probability of getting a particular x *and* a particular y. Conditional probability then is just $P(x_i,y_j)$ divided by $P(x_i)$ (such as 0.07/0.23 or 0.30 for our example). To complete the symbol statement, we need only a symbol for conditional probability. A common symbol for "given the value of" or "given" is a vertical slash (|). The symbol for the conditional probability of getting value y_j given that x_i has happened then is $P(y_j|x_i)$. Thus

$$P(y_j|x_i) = \frac{P(x_i, y_j)}{P(x_i)}. \tag{6.9}$$

Again, no simplification (step three) is possible.

In chaos theory, a typical "restricted condition" for conditional probability is that one or more specified outcomes have taken place over the preceding events (measuring occasions). In other words, we might come up with a better estimate of the state of our system for any specified time if we know where the system has been earlier. That extra knowledge might enable us to refine our estimate of the state of the system. Conditional probability is a tool that uses such extra knowledge. For example, conditional probability is an approximate answer to the question, what is the probability that the system will go to A, given that it's now in B? A more familiar example might be, "What is the probability that tomorrow will be warm and sunny, given that today is warm and sunny?"

Let's estimate conditional probabilities for the 300-observation dataset mentioned above. First we look at the 180 occasions when the system was in box A (time 1). For just those 180 selected observations, we count how many times it next went to A and how many times it went instead to B. We already tallied those frequencies in getting joint probability, above; sequence AA occurred on 126 occasions, AB on 54 occasions. Those are the data needed for conditional probabilities. Specifically, given that the system is in A at time 1, the probability that it will next go to A in our hypothetical example is 126/180 or 0.70 (Fig. 6.7, upper branch). The only other possibility is that the system will go from that initial A to B. The estimated conditional probability of that route is 54/180 or 0.30. The two conditional probabilities (0.70 and 0.30, listed on the branches in the figure) sum to 1.

In the same way, we can estimate where the system is apt to be at time 2, *given that it was in B (instead of A) at time 1*. The basic data showed 120 observations in bin B (time 1). We focus only on those 120 observations. (One of them happened to be the 300th measurement, so there are really only 119 two-event sequences.) In particular, we tally how many times the system next went to A and how many times it went instead to B. As mentioned above, those counts are 96 for route BA and 23 for the alternate route BB. The conditional probability that, given that the system is in B, it will next go to A is therefore 96/119 or 0.81 (Fig. 6.7). The conditional probability that it will instead go to B is 23/119 or 0.19.

To get conditional probabilities over three time periods, the question is of the same type but slightly more complicated. Now we first ask: given that the system has followed route AA for two successive measuring occasions, what is the probability that it will next go to A? To B? Then we repeat the question for the other three routes that led to time 2. Time 2 had 126 instances where the system had followed route AA. That's the first special subset to examine. The goal is to estimate the likelihood of going to a third A (overall route AAA) or instead to B (route AAB). The estimated probability of going to A at time 3, given that the system has followed AA over two measurements, is the number of occurrences of route AAA divided by 126. Similarly, the number of occurrences of AAB divided by 126 yields the conditional probability that the system will go to B at time 3, given that it has traversed AA for the previous two events. The two conditional probabilities sum to 1, since going to A or B are the only possibilities. Figure 6.7 shows the values for our hypothetical example. (For instance, out of the 126 occurrences of AA, the system went on to another A on 36 of those occurrences, so the conditional probability is 36/126 or 0.29.) Next, we proceed to the second of the four possible routes that led to time 2, namely route AB. The data showed 54 cases where the system followed that route. We look at only those 54 cases and estimate the conditional probabilities that the system will go to A and to B at time 3, given that it has followed AB over the preceding two values. Finally, we repeat the procedure for routes BA and BB.

Conditional probability, sequence probability, and ordinary probability are all interrelated. For instance, Equation 6.9 shows that conditional probability equals sequence probability divided by ordinary probability. Alternatively, our hypothetical data (Fig. 6.7) show that sequence probability is the product of the preceding ordinary and conditional probabilities along each route. First, let's check that with route AA. The ordinary probability of getting A at time 1 is 0.60. The conditional probability of getting an A at time 2 given that the system was in A at time 1 is 0.70. Multiplying the two gives $0.60 \times 0.70 = 0.42$, the same answer we got in our direct-count estimate of sequence probability. Similarly, ordinary probability times conditional probability for route AB is 0.60×0.30 or 0.18, and 0.18 also equals our answer by getting sequence probability directly (counting the number of routes AB in the data, etc.). We can calculate sequence probabilities over any number of observations in the same way. For example, the likelihood of AAA in Figure 6.7 is $0.60 \times 0.70 \times 0.29$ or 0.12, and so on. A sequence probability of 0.12 means that the route in question (here AAA) occurred only 12 per cent of the time out of all possible three-event routes.

The product of all legs except the last one in any sequence is actually a weighting factor. It weights the last leg according to the probability that the system had even traversed the prescribed route up to the last leg. Thus, in the sequence of $0.60 \times 0.70 \times 0.29$ (route AAA), there's only a 42 per cent chance (0.60×0.70 in decimals) that the system took route AA over the first two measurements. (The rest of the time, it took one of the other three possible routes, for two consecutive observations.) Hence, in evaluating all conceivable routes over three successive times, we *weight* our conditional probability of 0.29 (the chances of going to A, once AA has been traversed) by 0.42. That yields 0.12.

Probability as the core of information

Chaos theory uses probability not only by itself but also in getting a quantity called "**information**." Information has various meanings. In everyday usage, it usually refers to facts, data, learning, news, intelligence, and so on, gathered in any way (lectures, reading, radio, television, observation, hearsay, and so on). A stricter interpretation defines it as either the telling of something or that which is being told (Machlup 1983). In this stricter sense, it doesn't mean knowledge; the "something" being told may be nonsense rather than knowledge, old news rather than new news, facts that you already knew, and so on. But "information" also has a specialized, technical meaning. In this technical usage, it gets quantified into a numerical value, just like the amount of your paycheck or the price of a bottle of beer. The concept in this sense is the core of **information theory**—the formal, standard treatment of information as a mathematically definable and measurable quantity. Information theory turns up in chaos theory in several important ways, as we'll see.

To show how the specialized usage of "information" comes from probabilities, I'll first paraphrase Goldman's (1953: 1) example. Suppose Miami has a unique weather forecaster whose predictions are always right. On 3 July, that forecaster brazenly declares that "it won't snow in Miami tomorrow." Well, now, snow in Miami on 4 July? The chances are minimal. You'd say that the forecast didn't tell you much, or that the statement didn't give you much information (in the everyday usage sense). If a particular event or outcome is almost sure to happen, a declaration that it will indeed happen tells us very little or conveys little information.

Suppose, instead, the forecaster says "it *will* snow in Miami tomorrow." If you believe this and it does indeed snow, you'd be greatly impressed. In a sense, the forecast in that case carries much information. Thus, if an event is highly unlikely, a communique announcing that it will take place conveys a great deal of information. In other words, *the largest amount of information comes from the least likely or most unexpected messages or outcomes.*

The two alternatives regarding snow in Miami show *an inverse relation between probability and gain in information*: high probability of occurrence (no snow on 4 July) implies a small gain in information, and vice versa. That inverse relation makes it possible to define a specific expression for information in terms of just probabilities. Furthermore, because we can estimate numbers for probabilities, we can by the same token compute a number for information. The following several sections explain how.

Relating information to number of classes

In seeking an expression for information, early workers (dealing with something called entropy—the same thing as information for present purposes) began by considering a possible relation between information and number of possible outcomes, classes, or choices. If there's only one category or state for a system to be in, then the ordinary probability that the system is there is highest (1.0, the maximum). Accordingly, the information gained by learning the outcome of any move is lowest (zero), by the Miami snow analogy just explained. The next step up in complexity is two equally possible classes or outcomes. Here the ordinary probabilities are lower (0.5 for each class), and we gain some information by finding which class the system has entered. However, that gain isn't much, because we know in advance that the system will go into one of just two classes. If ten classes are equally likely and we're correctly told in advance that the system will go to class number seven, that announcement gives us a larger amount of information. Continuing the trend, information keeps increasing as number of classes (states or choices) increases. For the simplified case where all states are equally possible, people at first thought they could establish a relation between information (I) and number of possible states (N_s) simply by making the two directly proportional to one another:

$$I \propto N_s \tag{6.10}$$

where $\propto$ means "is proportional to." They made that proportionality into an equation by inserting a constant of proportionality, c_1:

$$I = c_1 N_s. \tag{6.11}$$

Researchers wanted the expression for information to satisfy two requirements. Both involved the idea of combining two separate systems into one composite system. (Merging two "systems" is very common in chaos theory. We'll see many more examples of it in later chapters.) The equation requirements were as follows: total information for the composite system must equal the sum of the information of one system plus that of the other; and the numbers of possible states in the two separate systems must be correctly related to those in the composite system. Figure 6.8 shows how the number of states of two systems combine into a composite

System X

x_1	x_2

System Y

y_1
y_2
y_3

Combined system

x_1,y_1	x_2,y_1
x_1,y_2	x_2,y_2
x_1,y_3	x_2,y_3

Figure 6.8 Ways to combine a system having two compartments (x) with a system of three compartments (y). The resulting combined system has $2\times3=6$ compartments.

system. System X has two possible outcomes or states (x_1 and x_2). System Y has three (y_1, y_2, and y_3). Joining systems X and Y into one composite system means that any x can occur with any y. There are six possible pairs in that combined system: x_1,y_1, x_1,y_2, x_1,y_3, x_2,y_1, x_2,y_2, and x_2,y_3.

Setting information directly equal to number of possible states (Eq. 6.11) implies that we can simply add the information values and also add number of possible states in each system, to get the totals for the combined system. Unfortunately, the number of possible states of a composite system isn't the sum of those in the individual systems (e.g. 2+3 or 5, in Fig. 6.8). Instead, it's the *product* (e.g. 2×3 or 6, per Fig. 6.8), as we've just seen. (In the same way, we had $6\times6=36$ possible combinations of the green die and white die in Fig. 6.6.) Consequently, Equation 6.11 was no good.

The problem was solved by changing Equation 6.11 to an **exponential equation**. An exponential equation is an equation that sets a **dependent variable** equal to a constant that's raised to a power, where the power (exponent) includes the independent variable. Examples are $y=\mathrm{c}^x$ and $y=\mathrm{ac}^{\mathrm{b}x}$, where a, b, and c are constants. (And so not just any equation that has an exponent is an exponential equation.) The constant c usually is some base of logarithms, such as 2, 10, or e (2.718). If the power (x or bx in these examples) is positive, the equation indicates an **exponential** growth; if negative, an exponential decay.

The chosen exponential equation doesn't directly relate number of possible states N_s to information I, as Equation 6.11 does. Instead, it relates N_s to a base number (a constant) raised to the I power:

$$N_s \propto \mathrm{base}^I. \tag{6.12}$$

Such a solution fulfils the two requirements for merging two separate systems into one composite system. That is, it adds information and multiplies the number of states. Here's how.

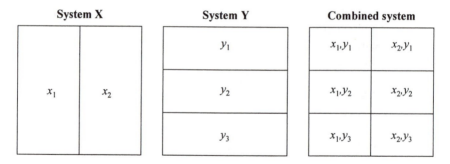

Proportionality 6.12 for system X is $N_X \propto \mathrm{base}^{I_X}$. For system Y, it's $N_Y \propto \mathrm{base}^{I_Y}$, where I_X and I_Y are the information values for systems X and Y, respectively. Multiplying those two expressions gives:

$$N_X N_Y \propto \text{base}^{I_X} \times \text{base}^{I_Y}.$$

One of the laws of mathematics (Appendix) is that $x^c(x^d) = x^{(c+d)}$. Hence

$$N_X N_Y \propto \text{base}^{(I_X + I_Y)}.$$

The correct number of possible states N_s in the composite system is the product of those in system X times those in Y, or $N_s = N_X N_Y$. Substituting N_s for $N_X N_Y$ gives

$$N_s \propto \text{base}^{(I_X + I_Y)}. \tag{6.13}$$

In that way, information moves into the position of an exponent. Proportionality 6.13 multiplies the number of possible states ($N_X N_Y$), thereby satisfying requirement two. It also adds the two information values, satisfying requirement one. Aside from the final step of reaching equation form, that arrangement solved the preliminary problem of finding a relation between information and number of possible classes or states.

Proportionality 6.13 leads to an expression for information alone. First we take logs of both sides:

$$\log N_s \propto (I_X + I_Y) \times \log(\text{base}).$$

Inserting a constant c_2 on the right-hand side makes the proportionality into an equation. Defining $I_X + I_Y$ as I then gives:

$$\log N_s = c_2 I \log(\text{base}).$$

To solve for I, we divide both sides by $[c_2 \log(\text{base})]$:

$$I = \log N_s / [c_2 \log(\text{base})].$$

Both c_2 and the log of the base are constant, regardless of which base (2, 10, e, etc.) we choose for the logs. Hence, $[c_2 \log(\text{base})]$, as well as its reciprocal $1/[c_2 \log(\text{base})]$, are constants. We'll designate the latter constant by the letter c. Our model equation for relating information to number of classes then becomes:

$$I = c \log N_s. \tag{6.14}$$

Equation 6.14 is a crude, preliminary example of quantifying information into a pure number. It's valid for both multiplying number of possible states and adding information. Other equations might also fulfil those two requirements but haven't been adopted. Shannon & Weaver (1949: 32) and Denbigh & Denbigh (1985: 102) mention several advantages to relating information to a logarithm, as in Equation 6.14.

Equation for information and equal probabilities

The development thus far relates information to number of possible states or choices, N_s. The next step is to estimate a probability associated with N_s and replace N_s with a probability. We'll still adhere to the simplified case where all possible choices are equally likely.

Rolling a standard six-sided die has an equal chance of producing any of the six sides. The number of possible outcomes or classes is 6. The probability of getting any particular side is one out of six or 1/6. So, when all possible outcomes are equally likely, the probability P for any particular class is simply the inverse of the number of possible choices N_s. That is, $P = 1/N_s$. Rearranging $P = 1/N_s$ gives $N_s = 1/P$. Replacing N_s with $1/P$ in Equation 6.14 gives:

$$I = \text{c}\log(1/P). \qquad (6.15\text{a})$$

Log($1/P$) is the same as $-\log P$ (Appendix). Hence, Equation 6.15a sometimes appears in the alternative form

$$I = -\text{c}\log P. \qquad (6.15\text{b})$$

Following Equation 6.14, Equation 6.15 is a slightly more advanced example of quantifying information into a number. More importantly, Equation 6.15 begins to relate information to probability in a specific, numerical way.

The constant c in Equation 6.15 is only a choice of a measuring unit. It's convenient to assign c = 1, because then it doesn't occur as a separate item in the equation. That gives:

$$I = -\log P. \qquad (6.15\text{c})$$

The base to use for the logarithms is arbitrary. Logarithms to the base two are common in information theory. (The logarithm to the base two of a number is the power (exponent) to which we raise two to give the number.) Logs to base e are less common in information theory, and base 10 is rare. (Logs to different bases differ only by constant factors, anyway. For example, the $\log_2$ of x is just 1.443 times $\log_e$ of x. Hence, to get $\log_2$ of any number, take your calculator and get $\log_e$ of that number, then multiply by 1.443.) I'll use base two or base e, depending on the situation. If Equations 6.15 involve logs to base two and if c = 1, Equation 6.15a becomes:

$$I = \log_2(1/P) \tag{6.16a}$$

and Equation 6.15c becomes:

$$I = -\log_2 P. \tag{6.16b}$$

According to Equation 6.15a with c = 1, the occurrence of an event whose probability is P communicates an amount of information equal to $\log(1/P)$. Let's digress briefly here to talk about the names (units) given to that information. The name for the units depends on the base of the logarithms. In chaos theory, logs usually are taken to base two. The name of the units derives from the special case of two equally likely, mutually exclusive possibilities. The technical term for such a two-possibility situation is **binary**. Common examples of binary systems are on/off switches and the dots and dashes of Morse code. The two possibilities in computer programming are **binary digits**—either of the two digits, conventionally 0 and 1, used in a binary code system. The usual abbreviation or contraction of "binary digit" is "**bit**." Hence, when logs are taken to the base two, a "bit" is the unit of information. Furthermore, although it derives from the case of just two possible outcomes, "bit" applies to any number of possible outcomes, as long as logs are taken to base two.

Here are two examples of information in bits:

- Suppose you declare that you and your spouse have just had a new baby girl. Probabilities for the two possible outcomes (boy or girl) were equal, namely 0.5 for each. Hence, Equation 6.14 (with $N_s = 2$ and c = 1) or its counterpart, Equation 6.16 (with $P = 0.5$), apply. From Equation 6.14, your announcement gives an amount of information equal to $\log_2 N_s = \log_2 2 = 1.443 \log_e 2 = 1.443 (0.693) = 1$ bit of information. (Eq. 6.16 gives the same answer: $I = \log_2 (1/P) = \log_2 (1/[0.5]) = \log_2 2 = 1$ bit.)
- As an honest gambling shark, you roll your unloaded die and have an equal chance of getting any of the six sides. No matter which of the six equal possibilities turns up, you get $\log_2 (1/P) = \log_2 N_s = \log_2 6 = 2.59$ bits of information. The 2.59 bits are somewhat more information than the 1 bit of the two-possibility case (boy or girl), because here there are six possible outcomes.

When logarithms are taken to base e, the unit of information by convention is a *nat*ural digit or nat, nit, or natural bel; when to the base 10, a Hartley, bel, digit, decimal digit, or decit. So much for units.

Table 6.1 gives another view of how the three ingredients of number of possible states, probability, and information interact. The first column of the table shows geometric grids, the second column their number of possible states or compartments N_s, and the third column the corresponding probabilities P. The probabilities are the equal probabilities of landing in a particular square of the grid by any impartial method. Column 4 of the table is simply the numerical value of information (in bits), computed from Equation 6.16. The bit values happen to be integers only because the number of compartments in each grid in the table equals two raised to an integer power.

Table 6.1 Probabilities and information for equally likely outcomes (after Rasband 1990: 196).

1 Grid	2 Number of possible states N_s	3 Probability P $=1/N_s$	4 Information I $= \log_2(1/P)$ $= \log_2 N_s$ (in bits)
	1	1	0
	2	0.5	1
	4	0.25	2
	8	0.125	3
	16	0.062	4

The top row of data in Table 6.1 is the one-compartment case. In that somewhat trivial case, the result of any experiment is certain to fall within that all-inclusive box, numerically. Any such "sure thing" has a probability P of 1. The computed information (Eq. 6.16a) is $\log_2 (1/P)$, here equal to $\log_2 (1/1) = \log_2 (1) = 0$. In other words, Equation 6.16a tells us that, from the information point of view, we don't learn anything new when an outcome is preordained or known in advance. Finding out what we already knew is an information gain of zero. And that, by the way, is another reason why the equation for information (Eq. 6.16) involves a logarithm. In other words, we'd like the computed information gain to be zero when the predicted outcome is certain ($P = 1$); using the log (Eq. 6.16) does that.

The numerical value of information (col. 4) increases downward in the table. In general, the table shows that, when the possible outcomes are equal, the amount of information gained upon learning a result increases with the number of possible outcomes, N_s (col. 2). It also increases as the equal probabilities of the outcomes decrease (col. 3).

Generalized equation relating information to probabilities

Thus far, we've related information to probability when the possible choices or outcomes are all equally likely. In many situations, they aren't all equally likely.

Instead, some bins or choices are more popular than others. A more general version of Equation 6.16 covers cases where the possible choices may or may not be equally probable.

Let's begin with the one-compartment case and use an analogy. Suppose you earn \$10000 per year but pay 20 per cent of that for taxes, health insurance, and so on. Your annual take-home pay therefore is only 80 per cent of \$10000 or 0.80 (\$10000) or \$8000. (You need a pay-raise!) The potential amount (\$10000) is weighted or reduced by a certain factor, in this case a factor that represents your withholdings. In the same way, compartment i in phase space contributes information I_i to the total information for the system. Over time, however, a trajectory visits that bin only a proportion P_i of the time. That reduces its information contribution over the time interval to P_iI_i. In other words, we've got to weight a compartment's overall information by the compartment's relative frequency or popularity. The weighting factor P_i is the estimated probability that the dynamical system will be in bin i at a given time.

Now let's say that more than one person (or compartment) contributes. Besides your \$10000 salary (x) from which your employer withholds 20 per cent, your spouse makes \$18000 per year (y) but has 30 per cent withheld. Spouse's after-tax income therefore is $0.70y$ or 0.70 (\$18000) or \$12600. What is the after-tax income for your entire household or system? It's simply $0.80x+0.70y$. Here that's \$8000+\$12600, or \$20600. Each person's salary has been weighted by the different appropriate factors (i.e. taxed), and the two resulting values (the after-tax incomes) have been added. The total obtained (\$20600) is a **weighted sum**.

In the same way, two phase space bins might have information I_1 and I_2, respectively. However, the system only visits bin 1 part of the time, namely with a relative frequency or probability of P_1. It visits bin 2 with a relative frequency or probability of P_2. Compartment 1's weighted contribution of information over some time interval therefore is P_1I_1. That of compartment two is P_2I_2. If the entire system only has those two compartments, the information (the weighted sum) for the system is $P_1I_1+P_2I_2$, just as we added the two taxed salaries (take-home pays) to get the \$20600 annual income for you and your spouse. Other family members might also contribute their various earnings to the household's annual income. Similarly, if a trajectory visits N_s bins (each having its own frequency or estimated probability), the information or weighted sum for the trajectory is $P_1I_1+P_2I_2+\ldots+P_{N_s}I_{N_s}$.

Equation 6.15a, $I = c \log(1/P)$, with $c = 1$ says that $\log(1/P)$ equals information I. The information for compartment one, I_1, therefore is $\log(1/P_1)$. The weighted contribution of that compartment over the time interval is P_1 times that information or P_1I_1 or $P_1\log(1/P_1)$. Similarly, the weighted information that compartment two contributes is $P_2\log(1/P_2)$, and so on. Just as with the household's after-tax income, the information that all N_s compartments contribute together (i.e. the weighted sum I_w) then is:

$$I_w = P_1\log(1/P_1)+P_2\log(1/P_2)+\ldots P_{N_s}\log(1/P_{N_s}). \tag{6.17a}$$

Equation 6.17a, a straightforward and simple summing of terms, is the goal toward which we've been striving—a general relation between information and probability. However, you'll almost always see it written in a shorter, more concise way, using the mathematician's language and symbols. We've already done that to some extent with other quantities, using Σ for summation. Also, authors commonly invoke a general symbol P_i to represent the entire ensemble of individual probabilities $P_1, P_2, \ldots P_{N_s}$. Using those symbols, Equation 6.17a takes on a shorthand disguise:

$$I_w = \sum_{i=1}^{N_s} P_i \log (1/P_i) \tag{6.17b}$$

or, without reciprocals,

$$I_w = -\sum_{i=1}^{N_s} P_i \log P_i . \tag{6.17c}$$

The summation over N_s bins includes only bins that have at least one observation; probabilities for all other bins are zero. If P is zero, then $P \log P$ also is zero for any i, so such bins don't play any role in calculating I_w.

In deriving information I_w (Eq. 6.17b or c), what we've done, of course, is taken the three steps in deriving a new quantity. The concept (step 1) is that total information is the sum of the individual contributions of information. We then wrote the general concept in terms of symbols (Eq. 6.17a, step 2). Finally (step 3), we simplified the cumbersome symbol definition into Equations 6.17b and c.

A weighted sum (Eq. 6.17) is a sum in one sense, but it's also an average value. Let's see why. There's a long way and a short way to get an average (an arithmetic mean). The long way—the way I was taught in school—is to add up all N values and divide by N. The short way is to compute a weighted sum. For example, say someone loads a die so that it tends to come to rest with a five or six showing. We toss the die 100 times and get side one 10 times, side two 5 times, side three 12 times, four 10 times, five 28 times and side six 35 times. Now, what is the average value for our 100 tosses? Calculating it the long way, we first add 10 ones, 5 twos, 12 threes, and so on. That sum is 446. Then we divide the sum by the total number of observations, 100. The average therefore is 4.46. Calculating it the short way (the weighted sum), we first use Equation 6.4 and compute the frequency or probability for each of the six sides as number of occurrences divided by total number of tosses. Side one's frequency or probability therefore is 10/100 = 0.10; side two has 5/100 or 0.05; and so on, through 0.35 for side six. The weighted contribution of each side then is its probability times its value. For instance, side one has probability 0.10 and a value of one, so its weighted contribution is 0.10 (1) or 0.10. The weighted sum for all six sides therefore is

$$0.10(1)+0.05(2)+0.12(3)+0.10(4)+0.28(5)+0.35(6) = 0.10+0.10+0.36+0.40+1.40+2.10=4.46,$$

the same answer we got the long way. (It's immaterial that 4.46 doesn't correspond to any particular side of a die and that we'd never get it on any one toss; the average has meaning nonetheless, just as the average family in the USA has, say, 2.37 children.) Hence, Equation 6.17 (version a, b, or c) gives an average value of information of the set of probabilities P_i. Any similar equation with a Σ sign followed by weighted values also produces an *average*, whether or not the author mentions it.

Equation 6.17 was introduced into information theory by the electrical engineer Claude Shannon in the late 1940s (e.g. Shannon & Weaver 1949). Subsequently, it's been found to be very general and to have applications in many other fields as well. (For instance, here's something important that I've only alluded to earlier: the probabilities to use in Equation 6.17 can be any of several kinds, including ordinary, sequential, and conditional probabilities.) The equation is one of the most important and common relations in all of chaos theory. Three examples of its use in that field are the information dimension, Kolmogorov–Sinai entropy, and mutual information, all discussed in later chapters.

Summary

Probability is the limiting relative frequency. A common estimate of probability is the ratio of observed or theoretical number of occurrences to total possible occurrences. A probability or frequency distribution is a list or graph showing how the various classes or values of a variable share the total probability of 1. Estimating the probability distribution for a continuous random variable is a monumental problem in mathematical statistics. The usual histogramming is often inadequate for various reasons. Statisticians have proposed many alternate procedures. Three of those that may be useful in chaos theory are adaptive histogramming, kernel density estimators, and adaptive kernel density estimators.

Probability is closely and inversely related to information. Therefore, by estimating a number for probability, its inverse is the basis for a numerical estimate of information. If the probabilities of the possible classes or outcomes of an experiment are equal, information I is $\log(1/P)$, which also equals the log of the number of states or classes. Usually, the probabilities of the various classes or outcomes aren't equal, so we have to weight the information of each class. That weighting involves multiplying $\log(1/P_i)$ (which is the information of each class) by the probability P_i of that class, yielding $P_i\log(1/P_i)$. Information then is the sum of the weighted information values of the separate classes, or

$$\sum_{i=1}^{N_s} P_i \log(1/P_i).$$

In that relation, the probability can be any of several types. For instance, it can be ordinary probability (the theoretical or empirical chance of getting a given state or outcome at a given time). It can also be joint probability (the likelihood of the joint occurrence of two or more particular outcomes, whether for simultaneous or sequential events). Two common types of joint probability are marginal probability (the joint probability of getting a particular value of one variable, regardless of the value of another variable for a simultaneous event) and sequence probability (a nonstandard term I use to represent the joint probability of getting particular values of one variable over successive events in time). The probability used in computing information can also be conditional probability (the likelihood that a particular outcome will happen, given one or more specified outcomes for the preceding measurements). The number of possible sequences or routes over time increases drastically (geometrically) with increase in either or both of number of possible states and number of events (time): $routes = states^{events}$.

Chapter 7
Autocorrelation

A time series sometimes repeats patterns or has other properties whereby earlier values have some relation to later values. **Autocorrelation** (sometimes called **serial correlation**) is a statistic that measures the degree of this affiliation. As an example, a time series of daily measurements might repeat itself every two weeks (i.e. show a lag of 14 days). That means the value for day 15 is similar to that for day 1, day 16 is similar to day 2, and so on. The autocorrelation statistic (explained below) is a number that, by its magnitude, shows that similarity. (In practice, using measurements of just one feature, we first compute autocorrelation using a lag of one. That is, we first compare day 1 to day 2, day 2 to day 3, and so on. Then we do repeat computations using lags of two, three, and so on. Methodically evaluating the computed statistic for successively larger lags reveals any dependence or correspondence among segments of the time series.)

Autocorrelation in a time series can cause problems in certain kinds of analyses. For example, many statistical measures are designed for unrelated or independent data (as opposed to correlated data). Applying measures built for independent data to correlated data can produce false or meaningless results. Secondly, autocorrelation might influence our choice of prediction methods, as least for short-term predictions.

In chaos tests, autocorrelated data carry at least two additional problems. The first deals with a lag-space graph (x_{t+m} versus x_t, m being the lag). Plotting positively autocorrelated data (explained below) amounts to plotting pairs of similar (nearly equal) values and hence gives a 45° straight-line relation. (One way to resolve such a problem is to limit the plot to measurements that are spaced far enough apart in time to be uncorrelated. An autocorrelation analysis indicates this minimum required time interval.) The second problem is that autocorrelated data introduce several complications in determining Lyapunov exponents and the correlation dimension (important measures explained in later chapters).

Because of those potential problems, a routine early step in analyzing any time series is to find out whether the data are autocorrelated. If they are, it's often a good idea to transform them into a related set of numbers that aren't autocorrelated, before going ahead with the analysis. Chapter 9 explains several methods for doing that.

The formula for autocorrelation has two ingredients—autocovariance and variance. (Such intimidating names!) We'll now construct those ingredients and put them together. Both of them are quite easy and straightforward.

Autocovariance

The first ingredient, **autocovariance,** literally means "how something varies with itself." In our case, a time series gets compared to itself. Lag is the main tool. Figure 7.1a is an example. It shows a simple sine curve (variation of sin y with time) and a lagged curve that has a lag of two time units. Autocovariance is a way of jointly evaluating the vertical differences or deviations between each of the two time series (the one unaltered, the other lagged) and a straight horizontal reference line drawn at the value of the arithmetic mean. In Figure 7.1a the mean is zero, for simplicity. At each value of time, the product of the two deviations reflects their joint magnitude, in a manner analogous to computing variance (Eq. 6.2). Also, as with variance, the average of those products then represents all of them as a group. That's the general idea of autocovariance (step 1 of the derivation).

Step 2 in defining this new quantity is to symbolize it. At any time t, the original time series has value x_t. The mean of the entire series (a constant for a given dataset) is $\bar{x}$. The deviation or difference for each observation is $x_t - \bar{x}$. The associated deviation for the lagged time series at that same value of t is $x_{t+m} - \bar{x}$. The product of the two deviations, representing their joint magnitude, is $(x_t - \bar{x})(x_{t+m} - \bar{x})$. Thus, at time 1 the product is $(x_1 - \bar{x})(x_{1+m} - \bar{x})$; at time 2, $(x_2 - \bar{x})(x_{2+m} - \bar{x})$; and so on. The average of all such products (their representative value) is their sum divided by the number of products involved. Rather than write out a string of products in symbols the long way, we'll go directly to the third and final step—simplification of the symbol definition. As with variance, that's just a matter of using the global symbols x_t and x_{t+m} for individual values, the summation sign Σ to show that all the deviation-products have to be added, and $1/N$ as a multiplier to get their average. (In fact, there are only N–m products to add, as explained in Ch. 3, so the multiplier actually should be $1/[N-m]$ rather than $1/N$. For most datasets, however, the difference is negligible.) The formula for autocovariance therefore becomes:

$$autocovariance = \frac{1}{N}\sum_{t=1}^{N-m} (x_t - \bar{x})\,(x_{t+m} - \bar{x}). \tag{7.1}$$

Let's go through some calculations with a hypothetical time series, using a lag of one (Table 7.1). Columns 3 and 4 list the "measured" data—the basic time series and the lagged series, respectively. (For a lag of one, observation number 1 of the original time series gets paired with observation number 2 of the lagged series, and so on.) The general idea with all equations involving summations (and with many other equations as well) is to break the entire equation into its component steps,

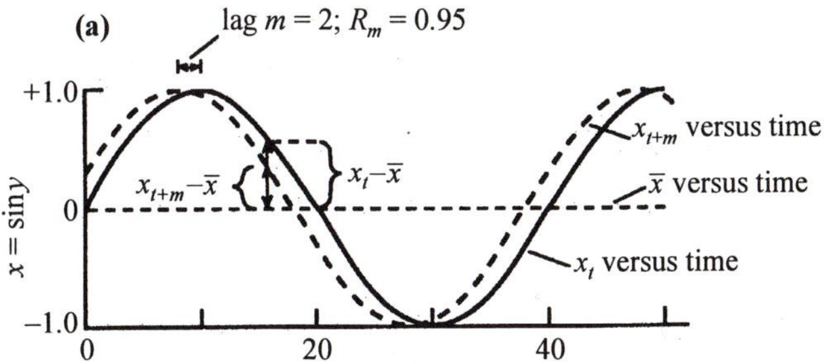

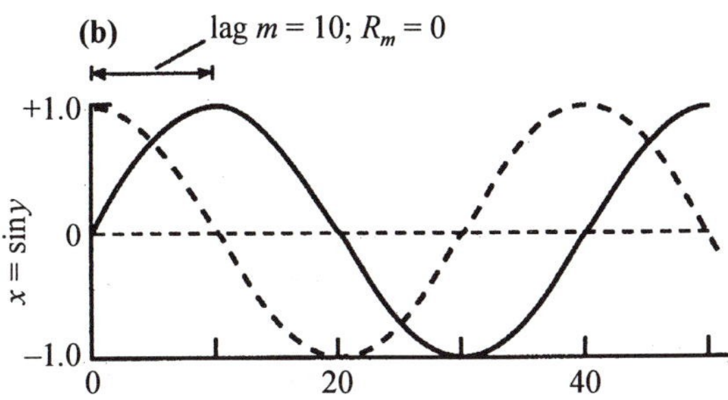

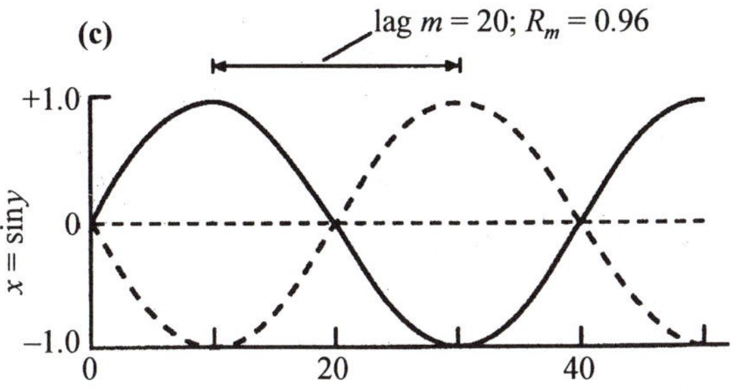

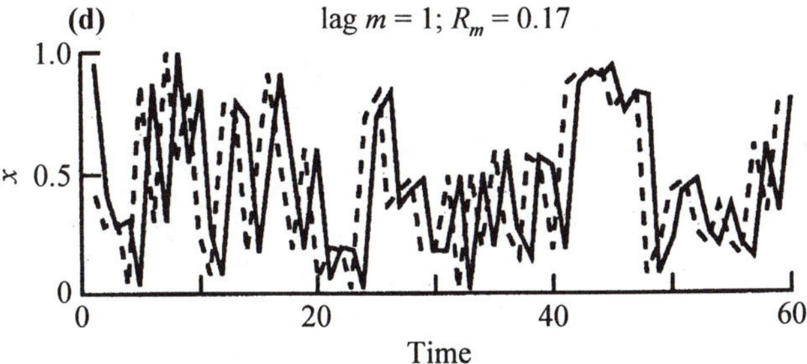

Figure 7.1 Time series (solid lines) and various lags (dashed lines): (a), (b), (c) sine curve with lags of 2, 10, and 20, respectively; (d) uncorrelated noise with lag of 1.

then do the steps starting at the right side of the equation and proceeding leftward. The first component step usually is to compute the individual values of whatever is to be added up. Equation 7.1 tells us to add up the individual products of $(x_t - \bar{x})(x_{t+m} - \bar{x})$. To get those deviations and then their products, we need the arithmetic mean ($\bar{x}$) of the time series. For the data of Table 7.1, $\bar{x}$ is 10.1.

Table 7.1 Hypothetical time-series data showing computation of autocovariance, variance, and autocorrelation, for a lag of one.

1	2	3	4	5	6	7	8
Obs.	Time t	Observed time series x_t	Lag-of-one time series x_{t+1}	$x_t - \bar{x}$	$x_{t+1} - \bar{x}$	Product $(x_t - \bar{x})(x_{t+1} - \bar{x})$	$(x_t - \bar{x})^2$
1	1	15	18	4.9	7.9	38.7	24.0
2	2	18	16	7.9	5.9	46.6	62.4
3	3	16	14	5.9	3.9	23.0	34.8
4	4	14	15	3.9	4.9	19.1	15.2
5	5	15	10	4.9	–0.1	–0.5	24.0
6	6	10	15	–0.1	4.9	–0.5	0.0
7	7	15	5	4.9	–5.1	–25.0	24.0
8	8	5	9	–5.1	–1.1	5.6	26.0
9	9	9	6	–1.1	–4.1	4.5	1.2
10	10	6	1	–4.1	–9.1	37.3	16.8
11	11	1	4	–9.1	–6.1	55.5	82.8
12	12	4	3	–6.1	–7.1	43.3	37.2
13	13	3	–	–7.1	–	–	50.4
	Sums	131				247.6	398.8

Mean $\bar{x}$ = 131/13 = 10.1
Autocovariance = 247.6/13 = 19.0
Variance = 398.8/13 = 30.7
Autocorrelation = 247.6/398.8 = 19.0/30.7 = 0.62

Column 5 of the table shows the difference or deviation between the value of the observed time series (x_t) and the mean, at each time; column 6 does the same for the lagged series. At each time, the joint magnitude of those two deviations is the one difference multiplied by the other: $(x_t - \bar{x})(x_{t+1} - \bar{x})$, as in column 7 of the table. That completes the computations for the first component step.

Proceeding leftward in Equation 7.1, the next task is to sum all those products. There's a total of N–m (here 13–1 or 12) of them to be added up. For the data of Table 7.1, the sum of the products is 247.6.

Again moving leftward in the equation, the final step is to multiply that sum by $1/N$. So, autocovariance for this particular lag is simply the average product (actually the approximate average product). That is, it's the average of the various values of $(x_t - \bar{x})(x_{t+1} - \bar{x})$. The autocovariance or average product for our data is 247.6 times $1/N$, or 247.6 (1/13), or 19.0.

Table 7.1 and the explanation up to this point only give the autocovariance for a lag of one. In an analysis, we next do similar calculations to get an autocovariance for a lag of two, then for a lag of three, and so on. To get meaningful results, there are usually two rules of thumb. One is that the dataset (time series) should have at least 50 values. The other is that the largest lag for which to compute an autocovariance (and hence an autocorrelation) is one-fourth the number of values in the time series (Davis 1986: 223).

Variance

Now for the second of the two major ingredients in autocorrelation. Autocovariance (Eq. 7.1) reflects the units of the physical feature we measured. Moreover, those units are squared (resulting in "square degrees," "square grams," and so on). Autocovariance by itself therefore can be difficult to interpret. Standardizing the autocovariance makes it dimensionless and changes it into a form that can be directly compared to other standardized autocovariances. The quantity we use to do that standardization is the variance (Eq. 6.2). Equation 6.2 as applied to a time series is:

$$variance = \frac{1}{N}\sum_{t=1}^{N}(x_t - \bar{x})^2. \tag{7.2}$$

To compute variance, we again start at the far right in the equation and proceed leftward. Hence, the first job is to get $(x_t - \bar{x})$ for each observation, then square that value. Column 8 of Table 7.1 shows $(x_t - \bar{x})^2$ for each observation in the original time series. Moving leftward in Equation 7.2, we next sum those squared deviations, beginning with the one labeled $t = 1$ and including all N values. The sum of those squared deviations in Table 7.1 is 398.8. Finally, dividing that sum by N, per Equation 7.2, gives the variance (the average squared deviation) as 30.7.

Autocorrelation for a given lag

Autocorrelation *for a given lag* is the autocovariance for that lag as standardized by the variance of the observations. That is, it's the autocovariance divided by the variance (Salas et al. 1980: 38; Makridakis et al. 1983: 366). (And there again is step one—conceptualization.) So

$$autocorrelation = \frac{autocovariance}{variance}.$$

To symbolize that concept (step two):

$$autocorrelation\ R_m = \frac{\text{Equation 7.1}}{\text{Equation 7.2}}$$

$$= \frac{\frac{1}{N}\sum_{t=1}^{N-m} (x_t - \bar{x})\ (x_{t+m} - \bar{x})}{\frac{1}{N}\sum_{t=1}^{N} (x_t - \bar{x})^2}.$$

Step three is to simplify, if possible. Both the numerator and denominator start with $1/N$. Hence, that common factor cancels out, leaving

$$autocorrelation = \frac{\sum_{t=1}^{N-m} (x_t - \bar{x})\ (x_{t+m} - \bar{x})}{\sum_{t=1}^{N} (x_t - \bar{x})^2}. \tag{7.3}$$

Equation 7.3 or its approximation appears in many forms or variations. In any case, it's called either the **lag-*m* autocorrelation coefficient** or the **lag-*m* serial correlation coefficient**. The entire spectrum or series of computed coefficients (those computed for a series of successive lags) is the **autocorrelation function**.

The variance is well suited to be the standard. First, it has the same units as autocovariance (deviations squared, with deviations in x units). Secondly, it's larger than the absolute value of the autocovariance for all lags except those that bring the two time-series segments into exact superimposition, and at exact superimposition the autocovariance and variance are equal. That means the absolute magnitude of their ratio (Eq. 7.3) can never be more than 1.0. Thirdly, its value is always positive.

Autocovariance (and hence also autocorrelation, since variance is always positive) carries a sign that reflects positive or negative autocorrelation. Let's see why. Figure 7.1a shows a simple sine curve (variation of sin y with time) and a lagged curve that has a lag m of two time units. We plot a lagged curve by pretending that what was measured at time $t+m$ was measured at time t. So, at time $t = 1$ we plot what was measured at time $t+m$ or $1+m$. For this example, lag is 2, so at time 1 the value we plot for our lagged relation is the measurement of time 3; at time 2 the value of our lagged relation is the measurement of time $2+m$ or time 4; and so on. The plot of the original series and the lagged series gives a graphical comparison of the two series.

A positively correlated time series is one in which one high value tends to follow another high value, or one low value tends to follow another low value. Examples are the data of Table 7.1 and the sine curve of Figure 7.1a. In Figure 7.1a the lag (2) is small (there's only a minor time offset between the two curves). The two deviations to be multiplied at each time in Equation 7.1 are therefore somewhat similar in magnitude. Also, most such products have the same sign. (If the curve is above the mean [the horizontal line at $x = 0$ in Fig. 7.1a], the sign for that deviation is positive; if below the line, negative.) As a result, almost all products of the paired deviations also are positive. Hence, so are the autocovariance and correlation coefficient.

Figure 7.1b shows the same sine curve as Figure 7.1a but with a larger lag, namely $m = 10$. For the sine curve and number of observations of this example, that lag happens to correspond to an offset whereby, over many measurements, the deviation products balance each other in both magnitude and sign. As the curves indicate, the two deviations (and hence their products) are both positive over some small range of time. They have opposite signs for the next small range of time (giving negative products); and they are both negative (again yielding positive products) for the final small range. The sum of the many positive and negative products, for this particular lag, turns out to be close to zero. The autocovariance (Eq. 7.1) therefore also is nearly zero.

Figure 7.1c shows the same curve at a still larger lag of $m = 20$. For the sine curve and number of observations used in this diagram, that lag puts the two curves exactly out of phase with one another. Such a situation yields large-magnitude products and hence also an autocorrelation of large absolute magnitude. However, in this case the two deviations for any time always have opposite signs, so the products are all negative. Therefore, the autocovariance and autocorrelation are also negative.

Autocorrelation coefficients R_m, therefore, range from +1 to −1. A +1 arises when the two time segments being compared are exact duplicates. In other words, they're perfectly and positively correlated with one another. Regardless of how complex or erratic the time series, this condition ($R_m = 1$) always pertains to lag 0. That's because there's no offset at all between the two segments; they plot on top of one another. So, when $m = 0$, $x_{t+m} = x_t$. The numerator in Equation 7.3 then becomes identical to the denominator, and their ratio (R_m) is 1. Other lags can produce an R_m almost equal to +1, especially with periodic curves such as the sine curve. (Eq. 7.3 gives an R_m of exactly 1.0 only at lag 0. At all subsequent lags there are only $N–m$ values being summed in the numerator.)

An R_m of −1 (or approximately −1) means that the two segments are perfect mirror images of one another (e.g. a positive value of x_t corresponds to a negative but equal-in-magnitude value of x_{t+m}, as in Fig. 7.1c). R_m is zero when the chosen lag results in a sum of products that is close (or equal) to zero (Fig. 7.1b) or when there's absolutely no correlation at all between the two segments, as with uncorrelated data (Fig. 7.1d).

Autocorrelation is a statistic that measures how well the paired variables (x_t and x_{t+m}) plot on a *straight* line. That is, we would take each pair of measurements and plot x_{t+m} on the ordinate and the associated x_t on the abscissa. (The straight line can have any slope.) If, instead, data fall on a *curve* (even with no scatter), such as part of a **parabola**, then the correlation coefficient is not 1.0. Instead, it's less than 1.0. (The method fits a straight line to the curve.) In fact, data that plot exactly as a complete parabola (horseshoe) have a correlation coefficient of zero. Thus, autocorrelation measures *linear* dependence between x_t and x_{t+m}. If the data don't have a linear relation, the correlation coefficient may not be a relevant statistic.

The correlogram

After computing the autocorrelation coefficient for about $N/4$ lags, the final step in analyzing actual data is to plot autocorrelation coefficient versus lag. Such a plot is a **correlogram**. Figure 7.2 shows some idealized time series (left column) and their corresponding correlograms. Figure 7.2a is the time series and correlogram for a sine curve. Figure 7.2b is a time series and correlogram for data that have no mutual relation. Figures 7.2c–e show other examples. The first important feature of these diagrams is this: for uncorrelated data (Fig. 7.2b), the autocorrelation coefficient is close to zero for all lags (except, of course, a lag of zero). In contrast, correlated data yield a correlogram that shows a nonzero pattern. The second important feature is that, to a large extent, the correlogram indicates or summarizes the type of regularity in the basic data.

For data with no trend and no autocorrelation, 95 per cent of the computed autocorrelation coefficients theoretically fall within about $\pm 2/(N^{0.5})$, where N is the total number of values in the basic data (Makridakis et al. 1983: 368). Thus, about 5 per cent of the coefficients could exceed those bounds and the data could still be "uncorrelated."

Summary

Autocorrelation or serial correlation is a dimensionless statistic that reflects the extent to which a variable correlates (positively or negatively) with itself, over time. It measures the degree of correlation between values x_t and x_{t+m}, within a time series. For reasons of statistical theory, prediction, and chaos analysis, it's important to find out whether raw data are autocorrelated. Mathematically, autocorrelation for a given lag (technically called the lag m autocorrelation coefficient or the lag m serial correlation coefficient) is simply the ratio of autocovariance to variance. Autocovariance is a dimensional measure (in squared units) of the extent to which a measured feature (temperature, heart rate, etc.) correlates with itself. Dividing it by the variance (which is also in those same squared units) makes the autocovariance dimensionless and standardizes it. A typical analysis involves computing the autocorrelation coefficient for each successive lag ($m = 1, 2$, etc. out to about $m = N/4$). A plot of autocorrelation coefficient versus lag is a correlogram. The pattern on the correlogram indicates whether or not the raw data are autocorrelated. If the data are autocorrelated, the pattern summarizes the type of regularity in the data.

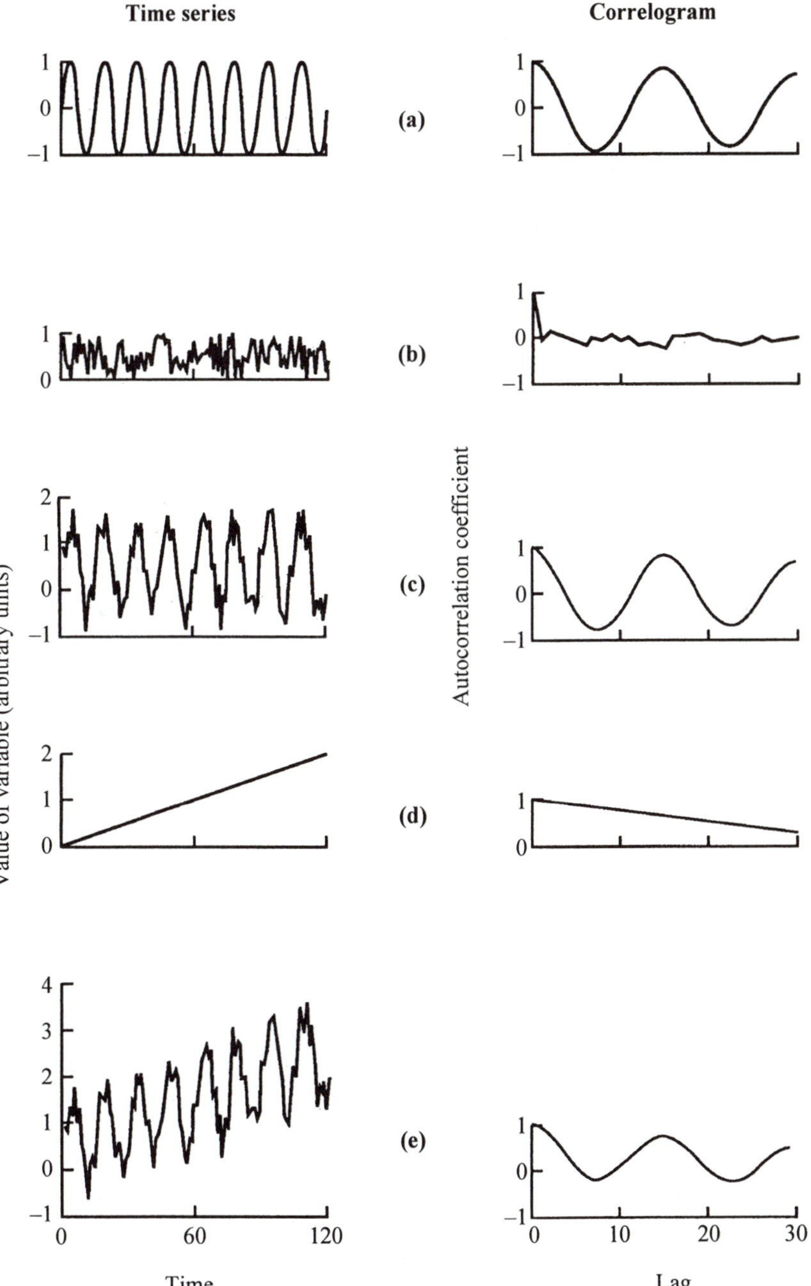

Figure 7.2 Time series and associated correlograms (after Davis 1986): (a) sine curve; (b) uncorrelated data (noise); (c) sine curve with superimposed noise (a+b); (d) trend; (e) sine curve with superimposed noise and trend (a+b+d).

Chapter 8
Fourier analysis

Fourier analysis is a standard step in time-series analysis, regardless of whether the data are chaotic or not. It's a basic analytical technique in all branches of the physical sciences, mathematics, and engineering.

Introduction

This chapter is about **periodicity** (the quality of recurring at a regular interval). A random time series doesn't have any periodicity. A chaotic time series may or may not have periodicity. Consequently, any test for periodicity can't, by itself, indicate chaos. However, knowing whether or not a time series is periodic is important information that suggests how to proceed in a chaos analysis. In addition, a test for periodicity can reveal autocorrelation.

One way to detect periodicity is with an autocorrelation analysis. A more common way of finding it is by the related technique of **Fourier analysis** (also known as **spectral analysis**, **frequency analysis**, or **harmonic analysis**). The name stems from Jean Joseph Fourier (1768–1830), the French analyst and mathematical physicist who developed the method.[1]

Fourier analysis is a mathematical technique for uniquely describing a time series in terms of periodic constituents. Phenomena made up of periodic constituents are common. A beam of sunlight, for example, can be split up into many constituent colors, each characterized by its own unique wavelength. Similarly, a musical tone consists of various constituent tones. A Fourier analysis gives special attention to the relative strengths of the periodic constituents. The general idea is somewhat analogous to looking at a stack of books in terms of the thicknesses of the individual books in the stack. The goal is to find the relative thickness of each book. The method can be used with any time series, not just a series that has periodicity.

Periodicity is a type of structure in a time series. A time series can also include structure that is not periodic. A Fourier analysis won't identify nonperiodic structure.

The time series in a Fourier analysis usually involves just one measured feature. If you measured more than one feature, you can run a separate Fourier analysis on each.

Wave relations

Fourier analysis looks at periodicity in the context of waves that repeat themselves. That's because periodic motion makes approximately regular rise-and-fall (wave-like) patterns on a time chart. All waves that we'll discuss will be periodic. A good model is a simple rotating disk and recording chart that transform rotary motion into linear motion (Fig. 8.1; Davis 1986). The disk spins at constant speed and is periodic because it regularly makes identical, complete revolutions. A stiff rod with one end fixed to the edge of the disk moves back and forth in a straight slot as the disk rotates. Under the slot, a sheet of paper or chart moves horizontally at a constant speed. As the disk revolves, a pen attached to the tip of the rod traces a wave on the moving paper.

The **wavelength** of this trace is the horizontal distance from any point on one wave to the equivalent point on the next wave. Hence, wavelength is in distance

---

1. **Fourier, Jean Joseph (1768–1830)** Jean Joseph Fourier was a Frenchman who combined scientific achievements with a successful career in public administration and other activities. At age 13 he developed a passionate interest in mathematics, although he also excelled in literature and singing. When he finished school at age 17, he tried to enter the French military (artillery or engineers). However, he was rejected because, as the son of a tailor, he wasn't of noble birth. He then began teaching mathematics and preparing to become a monk. That in turn changed drastically during his early twenties (around the time of the French revolution), when he became interested in politics. As a member of a local revolutionary committee during 1793–4 (age 25–26) he was imprisoned twice and came close to being sentenced to death. Surviving that crisis, he resumed teaching in 1795 but left in 1798 to accompany Napoleon Bonaparte to Egypt, in the role of a savant. While in Egypt, and also for another 15 years after his return to France in 1801, Fourier was appointed to various high positions in local civilian government, where he distinguished himself as an administrator.

   Throughout his administrative years, Fourier kept up his interest in mathematics and physics. During about 1804–11 (around age 40 or so), as a full-time administrator and part-time physicist–mathematician, he created and embellished his most important scientific contribution—his "theorie analytique de la chaleur." A mixture of theoretical physics and pure and applied mathematics, it was one of the most important science books of the nineteenth century. In it Fourier not only made revolutionary advances in the theory of heat, he also introduced his now-famous Fourier series (the idea of a mathematical function as consisting of a series of sine and cosine terms) and other mathematical advances. Several leading French scientists and mathematicians of the time (Laplace, Lagrange, Biot, and Poisson) publicly disagreed with various aspects of Fourier's treatise (Lagrange over the trigonometric series idea), but Fourier vigorously and diplomatically repulsed their attacks. Today his work is generally accepted as a great scientific achievement (Herivel 1975: 149–59).

   Fourier was elected to the French Academy of Sciences in 1817, to co-secretary of that group in 1822 and to various foreign scientific societies. A lifelong bachelor and often in poor health, he died of a heart attack in 1830 at age 62.

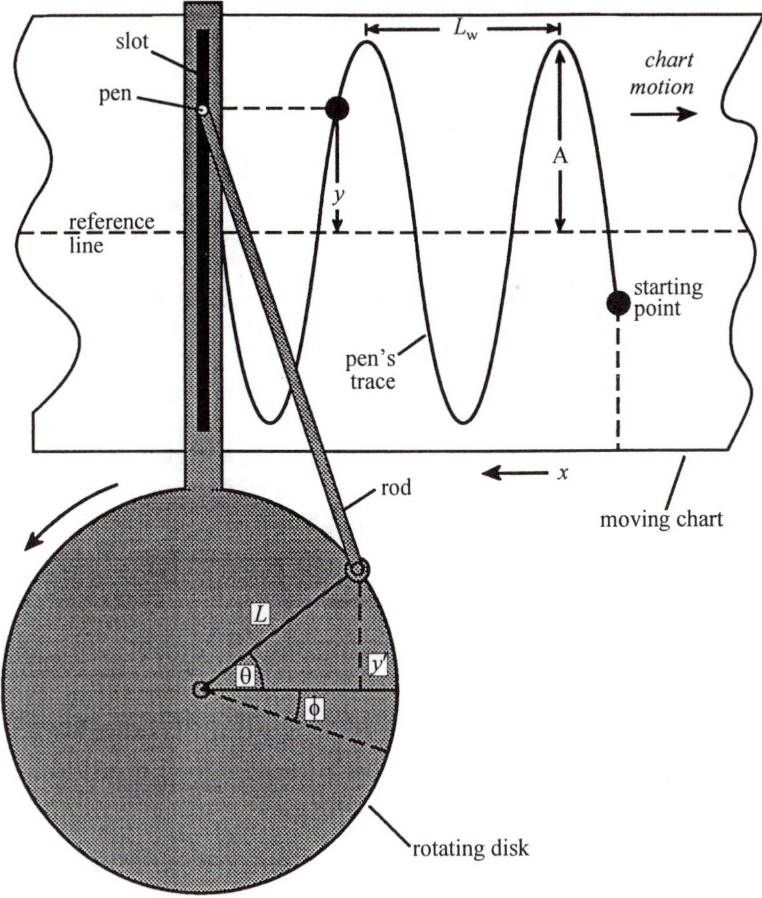

Figure 8.1 Disk-and-chart apparatus for transforming rotary motion into linear motion (after Davis 1986). Chart (above) indicates that disk has undergone two full revolutions and part of a third. L_w = wavelength, A = wave amplitude, y = present height of latest wave.

units. If the horizontal axis is time, "**wave period**" or "**cycle**" is used instead of wavelength. The wave period is the time needed for one full wavelength or cycle. A cycle is a complete wave and is a number that has no units. The **wave frequency** is the oscillation rate; that is, frequency is the number of wavelengths (waves), periods, cycles, or radians (depending on our framework) in a unit of time. (And so this frequency is similar to, but not exactly the same as, the frequency of ordinary statistics.) **Wave amplitude** is half the vertical height from the trough to the crest of the wave (Fig. 8.1), or half the extreme range of motion. On the chart, we measure it from a horizontal reference line drawn at the midpoint.

The long (horizontal) axis on the chart in Figure 8.1 is parallel to the 0° radius of the disk. That means there's a direct association between the point where the rod is attached to the disk and the pen's location on the chart and wave. For example, when the rod attachment rotates up to the very top (90°) on the disk, the pen is at the very top (90°) on the chart (the peak of the wave). As a result, several disk features are numerically equal to, or directly proportional to, wave features. For instance, a given disk radius produces the greatest vertical displacement of the pen from the reference line when the disk goes through a quarter of a revolution, from 0° to 90°. Hence, disk radius equals wave amplitude. Also, one complete revolution of the disk traces out one wavelength on the chart. Because of such equalities or proportions, trigonometric relations involving the *disk's* radius and rotational position also apply to the *wave's* geometry.

We define such trigonometric relations by drawing a right triangle from the center of the disk, with the disk radius as hypotenuse (Fig. 8.1). The most important such relation is easy and straightforward: the sine of the angle θ on Figure 8.1 is the opposite side of the triangle (height y') divided by the hypotenuse L. In symbols, $\sin\theta = y'/L$. Cross-multiplying gives height $y' = L\sin\theta$. Height y' also equals the vertical location or height y of the pen on a wave form on the chart. That is, $y' = y = L\sin\theta$. Finally, since disk radius L equals wave amplitude A, we have

$$y = A\sin\theta. \tag{8.1}$$

Many experiments might start when the rod's attachment to the disk is at a location other than 0°. The total angle covered during a new experiment then equals the angle at the start, ϕ (Fig. 8.1), plus the additional angle subtended, θ. The height y of a wave at any time then becomes

$$y = A\sin(\theta+\phi). \tag{8.2}$$

The original angle ϕ is the **phase angle** or simply the phase. "Phase" in this context is more specific than when used with "phase space." Phase here means the fraction of a waveform's cycle which has been completed at a particular reference time. Phase angle, wavelength, and amplitude fully characterize any wave.

Besides expressing height y in terms of angles, we can also express it in terms of distance along the chart or time during which the disk has been spinning. Let's look at distance first. Suppose successive experiments start the disk and chart at the same point (a constant location on the disk perimeter or wave form) but stop the device at a different point each time. The angle at the start, ϕ, is a constant for the several experiments. The additional angle θ is proportional to the horizontal distance, x, traversed along the chart during each experiment. Within any one wave the maximum possible values of θ and x, respectively, are one disk revolution (usually expressed as 2π, in radians) and one wavelength, L_w. Hence, $\theta/x = 2\pi/L_w$. Rearranging gives $\theta = 2\pi x/L_w$. Substituting that expression into Equation 8.2 gives the vertical coordinate y of any point on a wave form in terms of distance x:

$$y = A\sin([2\pi x/L_w]+\phi). \quad (8.3)$$

A comparable transformation lets us express y in terms of time. That's just a matter of substituting the time-counterparts, namely time (t) for distance x and wave period (T) in place of wavelength. That gives

$$y = A\sin([2\pi t/T]+\phi). \quad (8.4)$$

We can also develop expressions for height y in terms of cosines rather than sines. Derivation using cosines is exactly as for sines, except that the reference line on the rotating disk is vertical (at 90°) rather than horizontal (at 0°). That puts a cosine wave 90° out of phase with a **sine wave**. Except for this difference in phase angle, the sine wave and cosine wave are identical. Textbook discussions of Fourier analysis might deal with sine terms only, cosine terms only, or both.

Adding waves

Fourier analysis uses the principle of algebraically "adding" the heights of two or more different but simultaneous (superposed) waves. Their sum is a third wave, also periodic, that represents their combined effect. It's essentially like stacking one water wave on top of another and seeing what elevation the new water surface comes to.

Let's look at an example of how to add waves. Figure 8.2a is a sine wave with a wavelength of 2 "units." Figure 8.2b is a sine wave having a different wavelength, namely 1 unit. To draw both waves, I used Equation 8.3 ($y = A\sin([2\pi x/L_w]+\phi)$, with the phase angle $\phi = 0$. Now pretend that both those waves occur simultaneously. That means we don't see them individually; instead, we only see a composite wave that represents the sum of their action. In other words, the composite wave is the sum of the two component waves. An observed time series can be looked upon as a composite wave.

Adding the two component waves is just a matter of computing their heights at a particular distance or time, adding the two to get the height of the composite wave at that time, then repeating for successive times. We'll compute wave heights in this example with the distance version (Eq. 8.3). That equation requires A, x, and L_w (we'll take $\phi = 0$, for simplicity). To begin, we arbitrarily choose a horizontal location x, say $x = 0.3$, and compute the heights of the two waves. Wave A has $A = 1$ and $L_w = 2$ (Fig. 8.2a). Hence, at $x = 0.3$ the height y of wave A is $y = A\sin([2\pi x/L_w] = 1\sin[(2)(3.14)(0.3)/2] = 0.81$, in our arbitrary height units. At that same x value of 0.3, the height of wave B (which has $A = 1$ and $L_w = 1$, as shown in Fig. 8.2b) is $y = 1\sin[(2)(3.14)(0.3)/1] = 0.95$ height unit. Next, just add those two wave heights to get the height of composite wave C at $x = 0.3$. Thus, the composite wave height y at that location is $0.81+0.95 = 1.76$ units. That's all there is

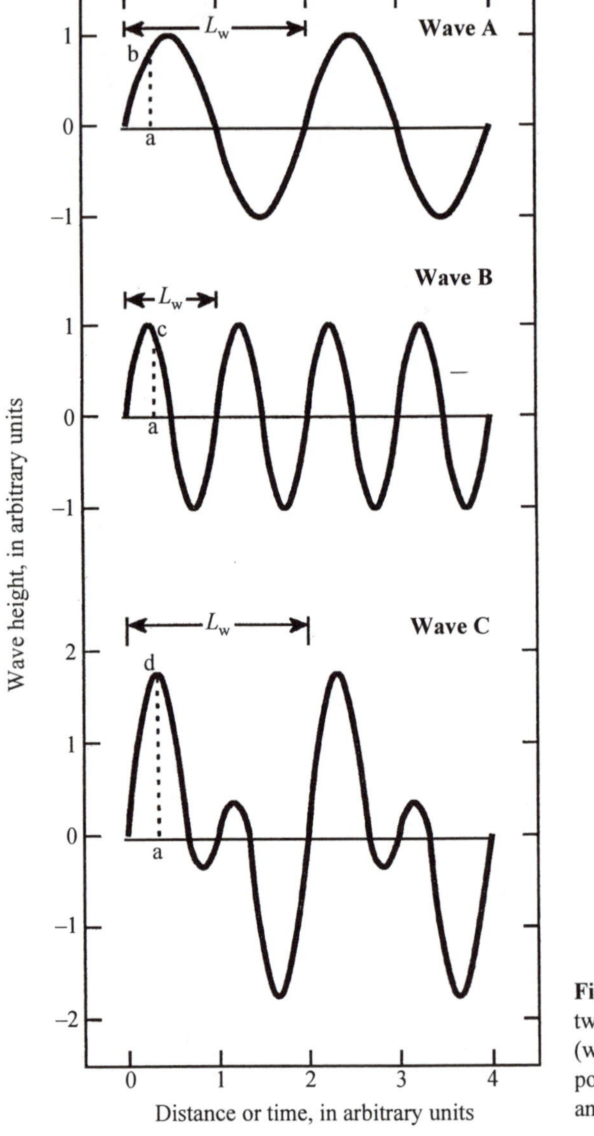

Figure 8.2 Combining two periodic (sine) waves (waves A and B) into a composite wave C. Heights ab and ac add up to height ad.

to the calculating. We then go to another x and repeat the procedure. For instance, at $x = 0.75$, Equation 8.3 gives heights y of 0.71 for wave A and –1.00 for wave B. Adding those two height values gives $y = -0.29$ for wave C, at $x = 0.75$. Repeating at various other values of x gives the entire trace of the composite wave (Fig. 8.2c). So, if waves A and B act simultaneously, what shows up is composite wave C. Using cosine waves rather than sine waves verifies the same principle.

The constituent waves (A and B) in Figure 8.2 are **sinusoids** (curves describable by the relation $y = \sin x$). However, composite wave C isn't **sinusoidal**. Hence, it's possible to compose a nonsinusoidal waveform out of sinusoids. In other words, a composite wave doesn't necessarily look like any of its constituent waves. Figure 8.3 shows that same idea with other waves. In Figure 8.3 I added five sine waves of different wavelengths, amplitudes and frequencies and got the composite wave shown at the bottom of the figure. The composite wave looks very strange and unlike any constituent, but it's still periodic (repeats itself at regular intervals). (Fig. 8.3 shows only one wavelength of the composite wave.) Also, the wavelength of the composite wave in Figures 8.2 and 8.3 is the same as that of the longest wavelength in the constituents. In summary, a composite wave:

- doesn't look like any of its constituents

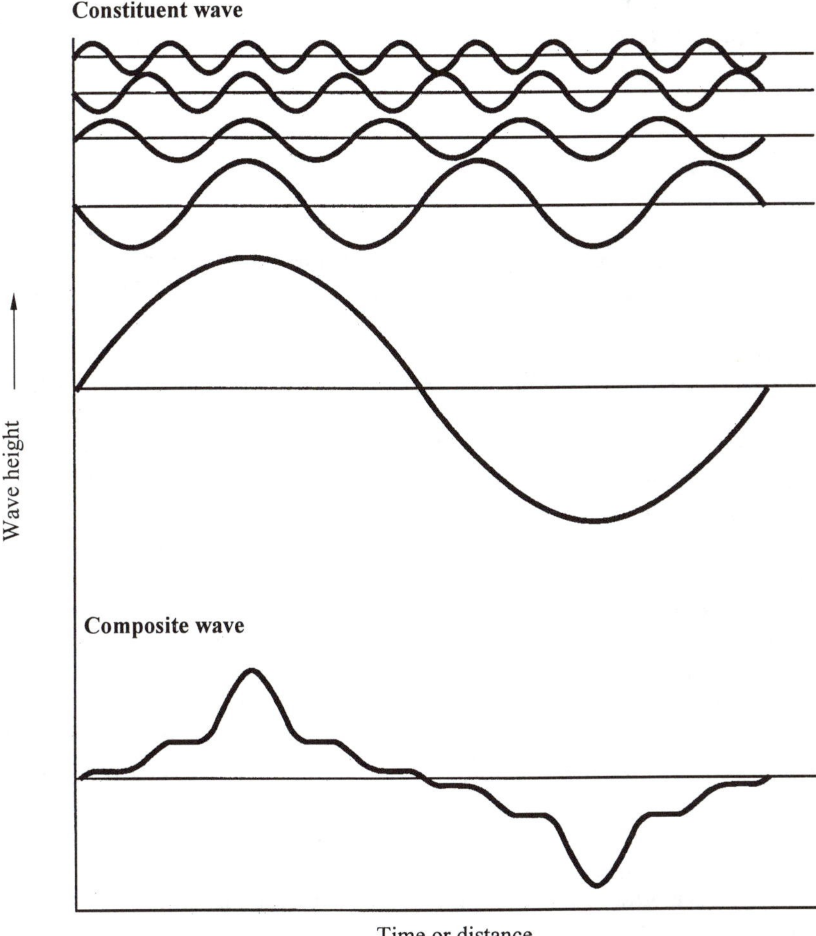

Figure 8.3 Addition of five waves of different properties (upper) to get a composite wave (lower).

- isn't sinusoidal, even though its constituents may be
- is periodic if its constituents are
- has a wavelength equal to the longest wavelength represented among the constituents.

Frequency, as mentioned, is the number of wavelengths or cycles completed in one time unit. Let's see what the various waves of Figure 8.2 have done at that basic reference time of one time unit. Wave A after one time unit has completed half a wavelength or cycle, so its frequency by definition is 1/2. Also, its wavelength L_w is 2 units, so its frequency is the reciprocal of its wavelength. Wave B at a time of one unit has completed one full wavelength and so has a frequency of 1. Here wavelength L_w again is 1 unit; frequency again is the reciprocal of wavelength. Composite wave C at time 1 has gone through half a cycle (a frequency of 1/2). With a wavelength of 2, frequency again is the reciprocal of wavelength. The reciprocal relation between frequency and wavelength is true in general.

Harmonics

Waves added in a Fourier analysis usually differ from one another in wavelength (and frequency), amplitude, and phase. A **fundamental** or **basic wave** serves as a standard or reference. That wave usually is the longest wave available; in practice, it's usually the length of the record. In Figure 8.2 it's wave C, the composite wave (which has a length of 2 units). The wavelength and frequency corresponding to the basic wave are the **fundamental wavelength** and **fundamental frequency**.

Even at constant wave amplitude and phase, there can be an infinite number of sinusoidal waves of differing wavelengths or frequencies. For any designated fundamental frequency, Fourier analysis deals only with a special subclass of that infinite number of waves. Specifically, it uses *only* those waves whose frequencies are an *integer multiple* of the fundamental frequency (two times the fundamental frequency, three times the fundamental frequency, and so on). Such a constituent wave is a "**harmonic**." In Figure 8.2 the basic wave is C, and it has a frequency of 1/2. Wave *A* also has a frequency of 1/2, or 1 times that of the basic wave, so wave A is a harmonic. Wave B's frequency is 1, or twice that of the basic wave, so it's also a harmonic.

Harmonics carry number labels based on their frequency relative to the fundamental frequency. A constituent having the same frequency as the basic wave is the **first harmonic** (wave A in Fig. 8.2). The wave having twice the frequency of the basic wave is the **second harmonic** (wave B in Fig. 8.2). The third harmonic has three times the frequency of the basic wave, and so on. The first harmonic is said to have a **harmonic number** of 1, the second harmonic has a harmonic number of 2, and so on

Dealing only with harmonics leads to a generic symbol statement for adding constituent waves. It can be written for wave heights in terms of angles, distances, or time (Eqs 8.1, 8.3, and 8.4, respectively). I'll show the idea with angles, that is, with $y = \sin\theta$ (Eq. 8.1). Everything is based on a unit time. That's arbitrarily defined as the time needed for the composite wave to complete one full wavelength. Referring again to Figure 8.1, the central angle θ then becomes the angle associated with the fundamental wavelength. Harmonics deal only with integer multiples of the fundamental wavelength and hence only with integer multiples of θ (2θ, 3θ, etc.). Thus, using $y = \sin\theta$ as the model, the wave height of the fundamental wavelength or frequency (the first harmonic) is $A_1 \sin\theta$, in which the subscript of A indicates the harmonic number. The second harmonic, having twice the frequency of the first, involves completing exactly two full wavelengths during the unit time. The angle subtended is 2θ. The height is $A_2 \sin 2\theta$. The third harmonic has a height of $A_3 \sin 3\theta$, and so on. Summing the heights of all such constituent harmonics gives the height y of the composite wave as

$$y = A_1 \sin\theta + A_2 \sin 2\theta + A_3 \sin 3\theta \ldots \tag{8.5}$$

We'll discuss how to get the coefficients (wave amplitudes, A_i) below.

One of the key principles of Fourier analysis is that we can approximate virtually any wave (or time series) by selecting certain harmonics and adding enough of them together. The more constituent waves we include in the summation, the closer the sum comes to duplicating the composite wave. For example, Figure 8.4a (after Borowitz & Bornstein 1968: 480–82) shows a "sawtooth" wave. Suppose that's our composite wave. Its frequency by definition is the fundamental frequency (or wavelength). Arbitrarily, wave height ranges from 0 to π. The unit time (the time needed for one full wavelength or cycle) is 2π radians. For use in Equation 8.5, the coefficients for the successive terms, not derived here, turn out to be $A_1 = 2$, $A_2 = -1$, $A_3 = 0.67$, and $A_4 = -0.50$. For this example, that's as many as we'll use. Thus, at any time or distance the height of the wave called the first harmonic is $A_1 \sin\theta$ or $2\sin\theta$; that of the second harmonic is $A_2 \sin 2\theta$ or $-\sin 2\theta$; and so on (Fig. 8.4b). Figure 8.4c shows that the first harmonic (here $y = 2\sin\theta$) by itself only crudely approximates the sawtooth relation. The sum of the first and second harmonics ($y = 2\sin\theta - \sin 2\theta$) gives a better fit. Adding more and more constituents approximates the composite (sawtooth) wave closer and closer.

The coefficients A_i act as weighting factors. Their absolute values (2, 1, 0.67 and 0.50 in Fig. 8.4) become smaller as wave frequency or harmonic number increases. The resulting effect on computed height y is that a harmonic contributes less and less to the composite wave as the frequency or number of the harmonic increases. Figure 8.4c shows that tendency. In general, the constituents having the lower frequencies (longer wavelengths, i.e. the first few harmonics) contribute most to the overall *shape* of the composite wave. Those of higher frequency have more influence on the finer details.

Just as summing waves of different wavelengths produces a composite wave, the reverse process also works: a composite wave (such as a time series, no matter how strange it looks) can be decomposed into a series of sinusoidal constituent

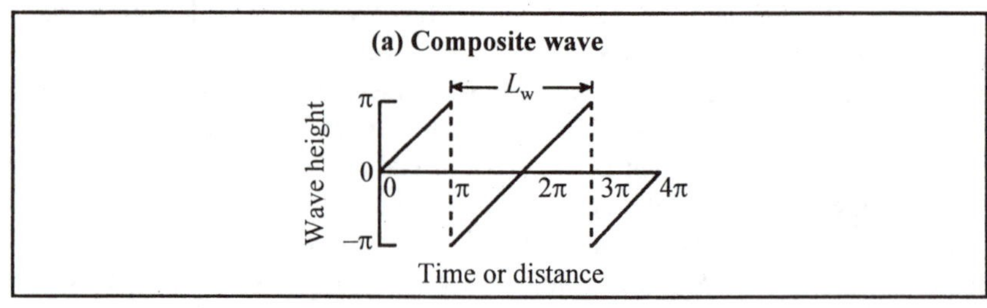

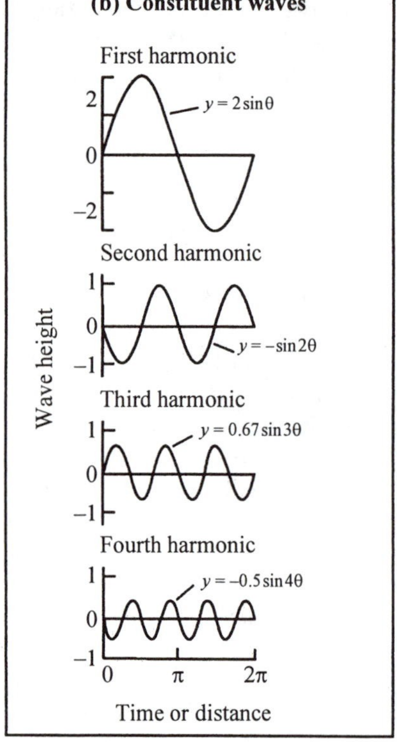

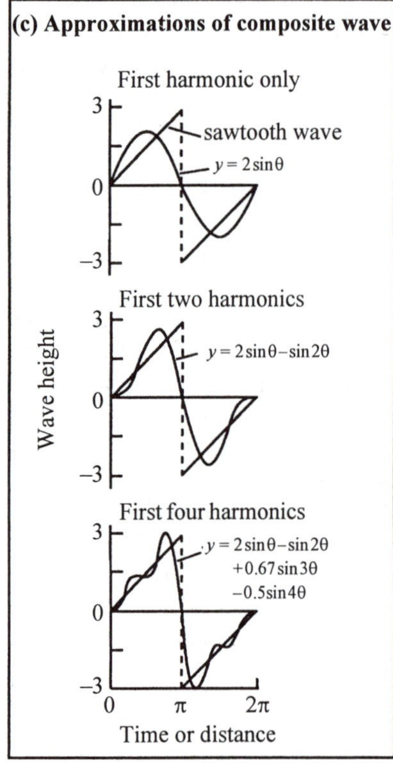

Figure 8.4 Approximations of the sawtooth wave by adding more and more harmonics (after Borowitz & Bornstein 1968).

waves. In fact, that's just what a Fourier analysis does. In practice, some harmonics are missing or are insignificant. Such harmonics have a coefficient of zero or nearly zero. Our eventual goal is to determine which frequencies are present and the relative importance of each, as reflected in their coefficients (wave amplitudes).

The Fourier series

Equation 8.5 involves only sines and is therefore a **Fourier sine series**. In the more general case, phase angle ϕ also is involved. The equation for each individual harmonic then has the model form

$$y_h = A_h \sin([h\theta]+\phi_h) \tag{8.6}$$

where h is the number of the harmonic. Taking a 90° phase shift into account leads to the same sort of expression with cosines (a **Fourier cosine series**).

Let's now get a more general and inclusive version of the Fourier sine and cosine series. Equation 8.6 derives from Equation 8.2, $y = A\sin(\theta+\phi)$. The right-hand side of Equation 8.2 involves the sine of the sum of two angles, namely the sine of $(\theta+\phi)$. A fundamental trigonometric identity for the sine of the sum of two angles is:

$$\sin(\theta+\phi) = \cos\theta\sin\phi+\sin\theta\cos\phi.$$

Inserting that identity into Equation 8.6 gives:

$$y_h = A_h(\cos[h\theta]\sin\phi_h+\sin[h\theta]\cos\phi_h).$$

Multiplying each term by A_h as indicated,

$$y_h = A_h\cos[h\theta]\sin\phi_h + A_h\sin[h\theta]\cos\phi_h. \tag{8.7}$$

To simplify that relation, we define two coefficients, α_h and β_h: $\alpha_h = A_h\sin\phi_h$ and $\beta_h = A_h\cos\phi_h$. Those coefficients now largely represent the wave amplitude, A_h, of any given harmonic. Substituting them into Equation 8.7 gives

$$y_h = \alpha_h\cos[h\theta]+\beta_h\sin[h\theta]. \tag{8.8}$$

Summing wave heights such as those defined by Equation 8.8 for all constituent waves (all harmonics) gives a composite wave. One goal of Fourier analysis is to find the chief contributors (the relatively large constituent amplitudes, if any). As mentioned, the *coefficients* (here α_h and β_h) for each wave reflect the magnitude of those amplitudes. Identifying the relatively large amplitudes in turn identifies major periodicities in the time series. For instance, the haphazard-looking time series may hide a regular periodicity of every month or every year. That's useful information.

Writing Equation 8.8 for the first harmonic or first constituent gives $y_1 = \alpha_1\cos\theta+\beta_1\sin\theta$. Doing the same for the second harmonic gives $y_2 = \alpha_2\cos[2\theta]+\beta_2\sin[2\theta]$. Similarly, the third harmonic is $y_3 = \alpha_3\cos[3\theta]+\beta 3\sin[3\theta]$, and so on. Now we'll write a statement (equation) expressing the addition of many such constituent waves. In so doing, we'll put the cosine terms in one group and the sine terms in another. The sum gives the composite wave in the general expression known as the **Fourier series**:

$$y = \alpha_1 \cos\theta + \alpha_2 \cos 2\theta + \alpha_3 \cos 3\theta + \ldots + \alpha_n \cos h\theta$$
$$+ \beta_1 \sin\theta + \beta_2 \sin 2\theta + \beta_3 \sin 3\theta \ldots + \beta_n \sin h\theta. \quad (8.9)$$

How many harmonics should we include in that addition? In practice, the summation of harmonics by convention starts with the first harmonic. At the other extreme, the largest harmonic number to include in the summation is usually the one numerically equal to half the number of observations in the dataset. That's $N/2$ (rounded downward if N is odd). Thus, with a dataset of 1001 values, stop at $h = 500$. (Including more than $N/2$ harmonics doesn't contribute any additional information. That is, look upon the coefficients as periodic, repeating themselves every $N/2$ harmonics.) The lengthy Fourier series statement just written therefore condenses into a generalized and more compact form:

$$y = \sum_{h=0}^{N/2} (\alpha_h \cos[h\theta] + \beta_h \sin[h\theta]). \quad (8.10)$$

(And again we've gone through the three steps for formula derivation. The concept [step one] was to add constituent waves to get a composite wave. Eqs 8.5 and 8.9 expressed that idea in symbols [step two]. And various simplifications [step three] led to Eq. 8.10.)

As Equations 8.8 and 8.10 show, each harmonic has two coefficients, α_h and β_h. The many coefficients generally designated as α_h and β_h are **Fourier coefficients**. From the association indicated in Equation 8.10, α_h is a *cosine coefficient* and β_h a *sine coefficient*.

Equation 8.10 (the Fourier series) is in a sine–cosine form. We also saw an alternate or second form, namely in terms of just the sine or just the cosine. A third form, known as the complex or exponential form, involves an imaginary number (see, for example, Ramirez 1985: 28). Different authors discuss the Fourier series equation in any of these various forms.

The Fourier series as just derived is in terms of the angle θ, so the series is in polar (angular) terms. Space or position (linear distances x and L_w) can replace the polar terms (per Eq. 8.3):

$$y = \sum_{h=0}^{N/2} (\alpha_h \cos h[2\pi x/L_w] + \beta_h \sin h[2\pi x/L_w]). \quad (8.11)$$

The same approach leads to the more commonly used Fourier series in terms of time (per Eq. 8.4):

$$y = \sum_{h=0}^{N/2} (\alpha_h \cos h[2\pi t/T] + \beta_h \sin h[2\pi t/T]). \quad (8.12)$$

Discrete Fourier coefficients

Because Fourier coefficients (the various α's and β's) reflect the relative contributions of the constituent waves, the coefficients are a major goal of Fourier analysis. Specific equations for Fourier coefficients come from multiplying both sides of the Fourier series equation by sine and cosine terms, integrating over the length of the record, and then rearranging (see, for instance, Kreyszig 1962: 467–80; or Rayner 1971: 21ff.). There's no need to go through that derivation here. The resulting definitions can take on slightly different forms, depending on the form of the Fourier series equation and on other factors. Also, wave features (wavelength, frequency, and period) usually aren't known. The standard assumption in that case is that they correspond to the length of the record (the number of observations, N). We'll restrict our attention, as usual, to data taken at equally spaced, discrete intervals (once every minute, once every year, etc.). Definitions for the discrete Fourier coefficients (hereafter called simply the Fourier coefficients) are:

$$\alpha_h = \frac{2}{N}\sum_{t=0}^{N-1} y_n \cos\frac{2\pi h t_n}{N} \tag{8.13}$$

and

$$\beta_h = \frac{2}{N}\sum_{t=0}^{N-1} y_n \sin\frac{2\pi h t_n}{N} \tag{8.14}$$

where α_h and β_h are the Fourier coefficients for harmonic number h, and y_n is the data value observed at time t_n. (Time here conventionally begins at an arbitrary origin designated as $t = 0$. Assigning the first data value to $t = 0$ instead of to $t = 1$ means that the last observation corresponds to $t = N–1$ rather than to $t = N$. That's why the summation in Equations 8.13 and 8.14 goes from $t = 0$ to $t = N–1$.) For the special case where N is even and $h = N/2$, the definition for the coefficient α involves $1/N$ rather than $2/N$:

$$\alpha_h = \frac{1}{N}\sum_{t=0}^{N-1} y_n \cos\frac{2\pi h t_n}{N}\,. \tag{8.13a}$$

Mathematically, Fourier coefficients are determined in such a way that, at each value of t (or x), the average squared difference between the value of the composite wave (time series) and the value of the sum of the components is a minimum. In other words, the method produces least-squares estimates, in exactly the same way that an ordinary least-squares regression gives the parameters (coefficients) of an equation (see, for example, Bloomfield 1976: 9ff.).

To see the computation method in a Fourier analysis, we'll now run through an abbreviated example of fictitious time-series data (Table 8.1). We'll use just $N = 6$ observations (far fewer than we'll need in practice) of the variable, y. The discrete time intervals can be in seconds, hours, etc.

Table 8.1 Hypothetical time-series data.

Observation number, N	1	2	3	4	5	6
Time, t_n	0	1	2	3	4	5
Value of y_n	8	20	32	13	10	26

Table 8.2 (first four columns) presents the results of a Fourier analysis of these data. The general procedure is as follows.

Table 8.2 Fourier analysis of data of Table 8.1.

	Harmonic number h	Fourier cosine coefficient α_h	Fourier sine coefficient β_h	Variance (power) of harmonic number h s_h^2	% contribution
Fundamental harmonic	1	−1.00	4.62	11.17	14.8
Second harmonic	2	−7.70	−8.08	62.06	82.2
Third harmonic	3	−1.50	0.0	2.25	3.0
Sums				75.48	100.0

Our immediate goal is to find the Fourier coefficients (α and β). Equations 8.13 and 8.14 are the ones to use. We break each equation into its component steps and then systematically take those steps, starting at the right side and proceeding leftward. We have to repeat the computations for each harmonic. Advancing leftward in Equations 8.13 and 8.14, the three steps in the computation (for a given harmonic h) are:

1. For every value of y and t, compute the value of $\cos(2\pi h t_n/N)$ (for Eq. 8.13) or $\sin(2\pi h t_n/N)$ (for Eq. 8.14).
2. Sum all such values.
3. Multiply the sum by $2/N$ (or, for Eq. 8.13a, by $1/N$).

We begin with the first harmonic ($h = 1$). That involves evaluating Equations 8.13 and 8.14 using all successive pairs of y_n and t_n (the full dataset). Let's start with the cosine coefficient, α_1. Our dataset has six observations. Step one is to get the value of $y_n\cos(2\pi h t_n/N)$ for each observation. For a given harmonic h, three ingredients within that cosine expression are constant, namely π (= 3.14 . . .), h, and N (here 6). Hence, the only items that change from one computation to the next are the successive pairs of y_n and t_n.

1. The first pair of values (Table 8.1) is $y_n = 8$ at $t_n = 0$. Plugging in those values, $y_n\cos(2\pi h t_n/N) = 8\cos[2(3.14)(1)(0)/6] = 8\cos(0) = 8(1) = 8.00$. (Before taking a trigonometric function—here the cosine—be sure to set your calculator or computer for radians.)
2. The second pair of values in the basic data is $y_n = 20$ and $t_n = 1$. Here $y_n\cos(2\pi h t_n/N) = 20\cos[2(3.14)(1)(1)/6] = 20\cos(1.0472) = 20\,(0.5) = 10.00$.
3. The third pair of values ($y_n = 32$, $t_n = 2$) gives $32\cos[2(3.14)(1)(2)/6] = 32\cos(2.0944) = -16.00$.

4. The fourth pair gives $13\cos[2(3.14)(1)(3)/6] = -13.00$.
5. The fifth pair gives $10\cos[2(3.14)(1)(4)/6] = -5.00$.
6. The sixth and final pair of the dataset gives $26\cos[2(3.14)(1)(5)/6] = 13.00$.

With those basic calculations (step one) done, the next task (step two) that Equation 8.13 says to do is to sum the terms just computed for all pairs of the dataset. That means starting with the value for the pair at $t_n = 0$ and adding all values through the final pair at $t_n = N-1 = 5$. For our example, that sum is $8.00+10.00-16.00-13.00-5.00+13.00 = -3.00$. Finally (step three), we multiply that sum by $2/N$ (here 2/6). That yields $\alpha_1 = (2/6)\,(-3.00) = -1.00$. (This final step is not yet standardized in regard to the use of N. Some people choose to multiply the sum by some other function of N, rather than by $2/N$.)

To get $ß_1$, we repeat the same steps but use $y_n \sin(2\pi h t_n/N)$, per Equation 8.14. For the data of Table 8.1, that yields $ß_1 = 4.62$ (Table 8.2).

That completes computing the two Fourier coefficients that correspond to the first harmonic. We would next have to go through the entire routine for the second harmonic. That means repeating the three basic steps using the same six pairs of data for each of the two equations (8.13 and 8.14), the only difference now being that $h = 2$. As mentioned, the maximum number of harmonics for which to compute Fourier coefficients is up through the largest integer less than or equal to $N/2$. In this example $N = 6$, so we compute only 6/2 or three pairs of Fourier coefficients, namely those for the first, second and third harmonics.

Variance (power)

Fourier coefficients by tradition usually aren't evaluated in the form just computed. Instead, we translate them into variances, the basic statistic explained earlier (Eq. 6.2). Using y for our variable, the variance s^2 is

$$s^2 = \frac{1}{N}\sum_{n=1}^{N} (y_n - \bar{y})^2 \qquad (8.15)$$

where n indicates the successive y values and $\bar{y}$ is the arithmetic average of the y values over one period (a constant). For Fourier coefficients, the next steps (not shown here) in deriving variances are to (1) algebraically multiply out the expression $(y_n - \bar{y})^2$, (2) substitute the square of the Fourier series for y_n^2, and (3) collect nonzero terms (e.g. Rayner 1971: 24). That translates the common variance s^2 into the variance s_h^2 of harmonic number h. Also, it expresses s_h^2 in terms of the Fourier coefficients of that harmonic:

$$s_h^2 = (\alpha_h^2 + ß_h^2)/2. \qquad (8.16)$$

The variance of harmonic number h, therefore, is merely the average of the two squared Fourier coefficients. In other words, that variance is a single number that reflects the magnitude of the two Fourier coefficients, for a given harmonic. It therefore indicates the relative importance of that particular harmonic. It shows the relative contribution of that harmonic in making up the composite wave (time series).

Equation 8.16 applies to all situations except the special case of the highest harmonic (namely at $h=N/2$) with an even number of y values. For that special case (i.e. for the highest harmonic when N is even), $\beta_{h=N/2}=0$ (e.g. Table 8.2), and the variance is

$$s_h^2 = \alpha_{h=N/2}^2. \tag{8.17}$$

Table 8.2 includes variances for the first three harmonics, computed according to these last two equations. The sum of all the component harmonic variances ought to equal the variance of the group of y values as given by Equation 8.15.

A final worthwhile calculation is the percentage of variance contribution that each harmonic makes to the total variance (Table 8.2, last column). That's simply the variance of the harmonic divided by the total variance, with the resulting proportion then multiplied by 100 to put it into percentage form.

Because of terminology used in electrical engineering, the variance is often called **power** instead of variance. I'll use the two terms interchangeably.

Graphical description (domains) of sinusoidal data

There are three graphical ways to show a set of sequential measurements (a time series or composite wave) of a feature. They are:

- *The time series (or spatial distribution plot, if distance takes the place of time)* The abscissa is time or distance, and the ordinate is the value of y (Figs 1.1, 8.2–8.4). (Always make such a plot of the basic data as a standard step, regardless of subsequent analysis.) In this raw or unaltered form, the data are in the **time domain** (or spatial domain, if distance takes the place of time).
- *The pseudo phase space plot (Ch. 3) of the selected variable against itself, in two or three dimensions* Several variations—difference plots, first-return maps, and stroboscopic maps (all explained in Ch. 18)—are common.
- *The wave-characteristic plot* Here the abscissa usually is wave frequency, although some authors use harmonic number or wave period. The ordinate usually is some parameter that reflects wave amplitude. Examples are the amplitude itself (producing a so-called amplitude–frequency plot), the Fourier coefficients (on separate graphs), the variance (power), or the variance per interval of frequency (producing a so-called spectral-density plot). Data characterized by these wave features (frequencies, amplitudes, Fourier

coefficients, variances, etc.) are in the so-called **frequency domain**. Such plots show the relative importance of the various constituent waves of a temporal or spatial record. That's useful information not readily apparent in the raw data. One graph conveniently includes all the constituents, since each wave has a different frequency, period or harmonic number.

Because wave-characteristic graphs show a parameter's array or **spectrum** of values (whether those values are on the abscissa or ordinate), people often attach the word "spectrum," as in "**power spectrum**" or "**frequency spectrum**." The terminology isn't yet fully standardized. Commonly, a "variance spectrum" or "power spectrum" is a plot with variance on the ordinate and with wave frequency on the abscissa. It shows the distribution of the total variance among the various intervals or bands of frequency. In the same manner, there can be a spectrum of amplitudes, a spectrum of frequencies, and so forth. Spectral analysis is the calculation and interpretation of a spectrum.

Some authors reserve the term "spectrum" for the case of a continuous (as opposed to discrete) distribution of powers, amplitudes, and so on. More commonly, however, a spectrum can be continuous or discrete. A discrete spectrum has values only at certain frequencies. The graph of a discrete spectrum is a **periodogram** or a **line spectrum**, and it consists of separate vertical lines at those discrete frequencies. The height of each line equals the value of the ordinate (for example, the variance) at each frequency. A continuous spectrum, on the other hand, has values at all frequencies. In that case, just think of the successive ordinate values as being connected to one another by a continuous line.

Statisticians just love frequency-domain transformations of a time series. That's because, in the frequency domain, adjacent variances have the highly desirable qualities of being nearly independent and uncorrelated, adding up to the power spectrum or total variance, and being approximately normally distributed (Shumway 1988: 47).

Fourier analysis is the tool for determining the frequency-domain description of time-domain data. The whole purpose of Fourier analysis is to search for possible periodicity in the basic data by looking at their frequency-domain characterization.

Records of infinite period and the Fourier integral

Except for Tables 8.1 and 8.2, our examples thus far have dealt with time series (composite waves) that have finite periods. Finite wave periods in fact are the basis of the whole idea of the Fourier series. Unfortunately, real-world datasets in the time domain often don't show finite periods (noticeable periodicity). Consequently, we wouldn't know (on a theoretical basis) whether Fourier analysis could be applied. Mathematicians had to figure out some way around that problem. It turned out that the Fourier-related idea of decomposing any observed record into constituent waves and into a frequency spectrum is still viable. The conceptual solution was to let the period become infinitely long. An "infinitely long period" merely means that the period doesn't repeat itself within the range of the data. That relaxation led to an extension or more general expression of the Fourier series, called the **Fourier integral**. In practice, people use the Fourier integral instead of the Fourier series (Jenkins & Watts 1968: 24–5).

An **integral** in general is just the sum obtained by adding together a group of separate elements. A Fourier integral is a mathematical operation that decomposes a time series into sinusoidal components at all frequencies and combines the variances thus obtained into a continuous distribution of variances. To imagine building such a continuous distribution, think of the horizontal spacing between discrete frequencies on a periodogram becoming smaller and smaller. In the limit, frequency spacing becomes infinitesimal. As that happens, wave period (the inverse of frequency) goes to infinity. Also, the distribution of frequencies becomes continuous rather than discrete. When those two events occur—wave period reaching infinity and spacing between successive harmonics or frequencies becoming infinitesimal—the Fourier series turns into an integration of variances over the full range of frequencies. In other words, the Fourier series becomes a Fourier integral. Thus, both the Fourier series and Fourier integral are ways of representing a time series in terms of linear combinations (sums) of sines and cosines. The Fourier series theoretically applies to finite-period time series, whereas the Fourier integral applies to any time series. Hence, the Fourier series is really a special case of the Fourier integral.

Using the Fourier integral requires a mathematical equation describing the observed time series. Real-world data can have many patterns or functions. Therefore, many variations of the general Fourier integral are possible. Engineering textbooks tabulate the more common variations, but it's well nigh impossible to write an equation describing many time series (especially for the kind of data I usually seem to get). In such cases, there's no way at all to use the Fourier integral. Another practical disadvantage is that the integral applies to continuous data, whereas chaos can also deal with discrete data. For these reasons, the Fourier integral in its pure form is rarely used in chaos analysis.

In practice, people determine the Fourier coefficients and power spectrum by either of two practical variations of the Fourier integral. One variation or method is the **Fourier transform** of the autocorrelation function, the other the **discrete Fourier transform**. (Don't be impressed by the fancy-sounding names.) Both techniques break a time series into a sum of sinusoidal components. Both of them achieve their mutual and singular purpose of transforming time-domain data to a frequency-domain counterpart. The frequency-domain spectrum they produce is continuous rather than discrete. Here's a brief rundown of those two techniques.

Fourier transform of the autocorrelation function

Strictly, the Fourier transform is the end result or finished product of the Fourier integral. The transform is a continuous frequency-domain characterization of the strength of the wave feature (variance, amplitude, phase, etc.) over the continuum of frequencies. However, as applied to the autocorrelation function and in the discrete Fourier transform discussed below, it doesn't directly involve the Fourier integral. As a result, it's useful in many practical applications.

Stripped to its essentials, the method called the Fourier transform of the autocorrelation function consists of three steps:

1. Compute the autocorrelation of the time-series data, as described in the preceding chapter.
2. Transform the autocorrelation coefficients into the frequency domain. (There's an easy equation for this, e.g. Davis 1986: 260, eq. 4.102.)
3. "Smooth" the results to reduce noise and bring out any important frequencies.

Some people do the smoothing, discussed below, as step two rather than step three. That is, they smooth the autocorrelation coefficients rather than the variances or powers.

Historically, this method of getting frequency-domain information from a time series was the primary technique of Fourier analysis, especially in the several decades beginning in the 1950s. Davis (1986: 260) says that many people still use it. I'm not treating it in more detail here because in chaos analysis the more popular technique seems to be the discrete Fourier transform, discussed next.

Discrete Fourier transform

The discrete Fourier transform (DFT) is a discrete-time system for getting the Fourier coefficients. It's a mathematical operation that transforms a series of discrete, equally spaced observations measured over a finite range of time into a discrete frequency-domain spectrum. Each of the discrete variances stands for a narrow band of frequencies. The bands join one another and are of equal width. A final step in the procedure averages the variances across the tops of the bands to produce a smoothed continuous frequency spectrum (a spectrum that includes all frequencies).

Most commonly, people visualize the DFT as a discrete approximation of the Fourier integral. However, it has enormous practical advantages over the integral, especially since it uses discrete data and doesn't require an equation describing the continuous time series.

The DFT is a sequence of **complex numbers**. A complex number is any number of the form a+jb, where a and b are real numbers and j here is a so-called **imaginary number** (a number equal to the square root of –1). In our case, a and b are Fourier coefficients α and ß, respectively. Hence, we compute a value of the DFT for each harmonic, h. The general expression for the DFT is:

$$\text{DFT} = \alpha_h + \text{j}\beta_h. \tag{8.18}$$

Since the term $\text{j}\beta_h$ includes the imaginary number j, β_h is called the "imaginary part" of the DFT for harmonic h. The other coefficient, α_h, is the "real part" of the DFT for harmonic h. Both coefficients in practice are still computed with Equations 8.13 and 8.14, for each of the various harmonics. Also, Equations 8.16 and 8.17 still give the variance of harmonic number h.

As the sample calculations associated with Tables 8.1 and 8.2 showed, Fourier-type computations largely consist of the same types of calculations over and over. Also, many calculations are needed. Our particular procedure (e.g. Tables 8.1, 8.2) involved N calculations for each of $N/2$ harmonics, for each coefficient. That amounts to $N\times(N/2)$ or $N^2/2$ calculations for each of the two coefficients, or N^2 total computations for the entire job. The use of discrete data, along with the many calculations and their repetitive nature, make the job ideal for computers.

Mathematicians have tried various types of computer programs to compute Fourier coefficients. One family of programs today has emerged as the best. All members of this family have the common characteristic that they reduce the N^2 operations to about $N\log_2 N$ operations. For large datasets, this is a significant saving in computer time. For example, with $N = 1000$, there'd be 1000^2 or 1000000 operations the long way but only $1000\log_2 1000$ or about 10000 operations (one one-hundredth as many) the shorter way. For larger datasets the savings in computer time can be much greater; jobs that would take weeks can be done in seconds or minutes.

Computer programs for calculating the DFT in about $N\log_2 N$ operations are known in general as **fast Fourier transforms**. A fast Fourier transform (FFT), in other words, is any member of a family of computer programs for quickly and efficiently calculating the real and imaginary parts (Fourier coefficients) of the DFT.

The mathematics behind the DFT are vast and complicated. Many books, including Rayner (1971), Bloomfield (1976), Bracewell (1978), Kanasewich (1981), Priestley (1981), Elliott & Rao (1982), Ramirez (1985), and Brigham (1988), cover the subject in more detail.

Sample periodograms

The periodogram or line spectrum is the most common way of displaying the results of a Fourier analysis. Periodicities (or lack thereof) can appear in different ways on such a plot. Figure 8.5a (left side) shows a time series that has a constant value with time. There are no deviations from the mean, so the variance or power spectrum (Fig. 8.5a, right side) is zero at all frequencies.

Figure 8.5b is a sinusoidal wave. All of its variance (dispersion about the mean

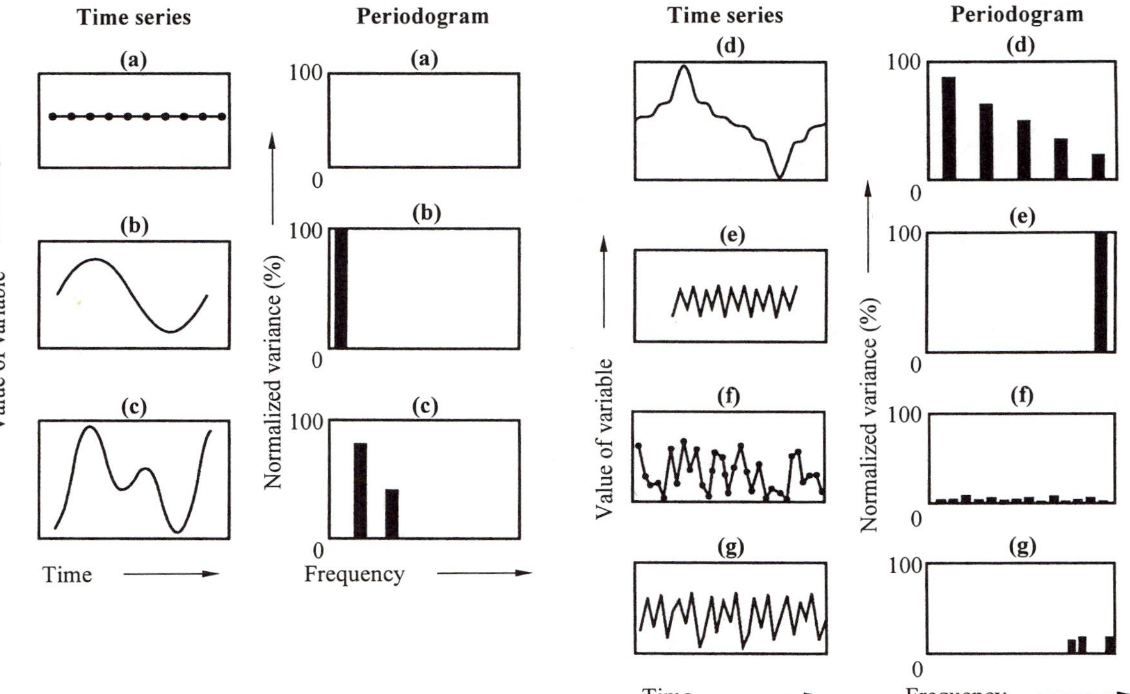

Figure 8.5 Idealized examples of time series and associated power spectra. (a) Constant value of variable with time. (b) Sine wave (one cycle). (c) Composite of two sine waves. (d) Composite of five sine waves. (e) Periodic nonsinusoidal data. (f) Random numbers. (g) Chaotic data.

value of the variable) is the result of the periodic and systematic fluctuation above or below the mean. Variance of this sort is concentrated at the frequency corresponding to the associated wave period. We can compute the particular wave period for such special wavelike patterns because wave relations apply. Wave frequency F is related to a frequency counter or index i and number of observations N according to $F=2\pi i/N$. Rearranging gives $2\pi/F=N/i$, in which $2\pi/F$ or N/i is the period length, in time units. Fourier analysis indicates the frequency F and associated i value at which the maximum variance is concentrated. Using either F or i to compute $2\pi/F$ or N/i then indicates the period length of the wave.

Figure 8.5c shows the same time-series curve as Figure 8.2c. That curve consists of two constituent sine waves (harmonics). Those components spring up in the power spectrum (Fig. 8.5c, right side). Figure 8.5d shows the same time-series curve as the foot of Figure 8.3. The five constituents appear in the frequency-domain analysis (plot of variance versus frequency).

Figure 8.5e is a time series consisting of a cycle of four successive values (0.38, 0.83, 0.51, 0.89), repeated five times. Such a cycle is periodic but not sinusoidal. Wave relations such as N/i therefore don't indicate the period length. However, Fourier analysis does show that the data are periodic.

Figure 8.5f shows random (uncorrelated or disassociated) numbers and their corresponding periodogram. No strong periodicity appears. (The "random" numbers came from a random-number table and hence are really only "pseudo-random," since a computer generated them. However, any one digit in such tables has an equal chance of being followed by any other digit. Similarly, any pair of digits, any triplet of digits, and so on, is equally likely. For practical purposes, therefore, the random numbers are close enough to being truly random.) Finally, Figure 8.5g shows chaotic data. The periodogram for the chaotic data isn't distinctive enough to justify any generalizations other than that there isn't any strong periodicity. Bergé et al. (1984: 48–62) give further remarks on the periodograms of various types of time series.

Like the correlogram, therefore, to a considerable extent the periodogram summarizes regularity in the basic data.

Figure 8.5 shows the magnitudes of the variances by bars for easier viewing. In fact, however, each variance applies to just a finite, single value of frequency rather than to a small range of frequencies. Thus, each bar really ought to be a thin-as-possible spike of appropriate height, because it represents just one discrete frequency.

A statistical test, called Fisher's test, reveals whether the dominant peak is significant or not (see, for example, Davis 1986: 257–8). Often, no particular frequency turns out to have a significant peak. In that case, the spectrum is a "**broadband spectrum**." Figure 8.5f is an example.

Size of dataset

There isn't any firm rule to indicate the desirable number of data points for a Fourier analysis. The reason is that both the underlying periodicity (usually unknown in advance) and noise affect the results. To take a ridiculous and extreme example, only a few data points are enough to reveal a perfect noiseless sine wave. On the other hand, identifying periodicity in noisy data that have a long wavelength might require millions of points (covering several cycles). It's a dilemma. People commonly run Fourier analyses with 50–100 points, but more than that certainly is desirable. In any case, you can always run the analysis on a dataset and use Fisher's test to see if any resulting peaks are significant.

Sampling intervals in Fourier analysis

Fourier analysis requires data taken at equally spaced intervals. If you didn't measure the data that way (e.g. if you measured them continuously or at uneven intervals), interpolate to get values at equally spaced points, for the Fourier analysis. (I'll call the interval at which basic data were originally measured the *sampling* interval. Some authors don't use the term in exactly that way.)

The sampling interval influences the amount of detail attainable in the analysis. The shortest resolvable period is twice the sampling interval. In other words, defining any given frequency component requires at least two samples (more than two are better) per cycle. (If the sampling interval isn't small enough to identify the shortest periods that contribute significant variance to the record, the method won't lose this short-period variance but reports it as harmonics of the shortest computed period [Wastler 1963: 18].) A small sampling interval also makes possible a greater number of samples for a given length of record and gives you more confidence in the estimates of the power spectrum.

Concluding remarks

Fourier analyses can produce different results, depending on arbitrary choices made in the procedure. Sampling interval is one such arbitrary choice. As another example, increasing the number of lags while keeping the sampling interval constant enables us to make estimates for a greater number of spectral bands. It also increases the **precision** of each estimate. This could reveal information that we miss if we use fewer lags in the computations. Therefore, do several Fourier analyses on a dataset, rather than just one analysis (Rayner 1971).

This chapter has dealt mainly with basic concepts and the "plain vanilla" (classical) Fourier analysis. However, many datasets have special characteristics (excessive noise, few observations, and other problems). Furthermore, classical Fourier analysis has many inherent weaknesses. Sophisticated variants of classical Fourier analysis have been developed to cope with such difficulties. Examples are windowed periodograms, maximum-entropy spectral analysis, **singular spectral analysis**, and **multitaper spectral analysis** (e.g. Lall & Mann 1995).

Once you identify periodic components, look for physical explanations. Only then does the tool of Fourier analysis fully justify its use.

Summary

Fourier analysis is a mathematical procedure that reveals periodicity (or lack thereof) in data. It's based on the idea that any time series can be decomposed (approximately) into constituent sine waves. Consequently, the method uses wave concepts and definitions, such as wavelength, period, frequency, phase, and harmonics. Fourier coefficients indicate the relative importance of the various constituent waves that make up a given time series. Fourier coefficients have to be computed for each harmonic, except that the highest harmonic worth computing is about $N/2$.

People usually transform Fourier coefficients into "variances" ("powers"). A variance spectrum or power spectrum is a plot with variance on the ordinate and with frequency on the abscissa. For the discrete spectrum, people call that plot a periodogram or line spectrum. Characterizing a continuous or discrete time series in that way puts it into the so-called frequency domain. Thus, there are three methods or "domains" of mathematically describing a time series: the domains of time, pseudo phase space, and frequency. Fourier analysis produces the frequency-domain description of time-domain data. The purpose of Fourier analysis is to search for possible periodicity in the basic data by looking at their frequency-domain characterization. The two main practical ways of determining the Fourier coefficients and power spectrum are the Fourier transform of the autocorrelation function and the discrete Fourier transform (DFT). The DFT is becoming the more popular of the two. It's a mathematical operation that transforms a series of discrete, equally spaced observations measured over a finite range of time into a discrete frequency-domain spectrum. Computer programs for calculating the DFT in about $N\log_2 N$ operations are known in general as fast Fourier transforms. There's no simple rule to indicate the desirable number of data points to have for a Fourier analysis; that's because both noise and underlying periodicity affect the size of the dataset you'll need.

Chapter 9
Preliminary analysis of time-series data

Regularity of time intervals between measurements

Most techniques and concepts in a chaos analysis assume that raw data were measured at equally spaced time intervals. Only a few techniques have sophisticated variants to account for irregularly spaced data. Hence, the best approach by far is to measure data at equal time intervals. In fact, some assumptions or modifications usually have to be applied to any data that weren't measured at a fixed rate, so that they represent a constant time interval. There are three approaches or methods for doing this:

- Estimate new values by interpolation (Ch. 4). This is the most popular method.
- Adopt the value of the nearest neighbor, at each equal-time interval.
- Ignore the irregularity and assume equal intervals.

Our discussions will be based on equal time intervals, unless specified otherwise.

First examination of data

Any time series should be examined carefully before being tested for chaos. For one thing, familiarizing yourself with the data is simply a basic obligation and necessity in research. For another, the data in raw form may have features that render them unsuitable for chaos tests (or, indeed, for many standard statistical tests). Minor details of any recommended preliminary examination differ from one author to another. Here's a reasonable procedure.

1. Plot the data (each variable against time) and scrutinize the plot. Although this step might seem self-evident, many people these days by-pass it completely and send the data directly into a computerized statistical analysis. By doing that they lose the vital familiarity that a graph provides, especially a graph they plotted by hand. A time series can show any of a wide variety of

patterns. Long-term trends and periodicity are common and important ones. In forecasting, for example, a straight trend line could give a very unreliable prediction if the system really is periodic.

2. Do a Fourier analysis and inspect the resulting power spectrum.
3. Test the raw data for autocorrelation.

Sometimes we can analyze data in their raw form, but on other occasions we might transform them. Some reasons for transforming data are to:

- reduce them to a standard scale
- reduce or eliminate autocorrelation
- get a straight-line relation as opposed to a curved relation on a graph
- bring about a less skewed distribution of the data (strongly preferred for many statistical analyses).

Opinions differ as to how best to handle autocorrelation in chaos analyses. The three possible routes, in no particular order of preference, seem to be:

(a) Ignore the autocorrelation on the philosophy that it's an inherent part of the entire dynamics and that the dynamics ought to be analyzed as a whole. In other words, analyze the raw data for chaos, even though they are autocorrelated. (This alternative means that the autocorrelation might affect the results.)

(b) Remove the features responsible for the autocorrelation (typically trend and periodicity), as discussed below. (In principle, decorrelated data still retain the nonlinear and possibly chaotic structure of the original time series.)

(c) Declare immediately (whether rightly or wrongly) that the dynamical system isn't chaotic (and is instead cyclic or whatever).

Major concerns or problems with autocorrelated data, as mentioned in Chapter 7, are:

- false pseudo phase space plots (e.g. attractor reconstructions that are spurious)
- errors or complications in determining the correlation dimension, Lyapunov exponents, and other "invariants"
- ineligibility for many statistical analyses
- possible influence on prediction methods.

For those reasons, many people choose to transform raw, autocorrelated data by removing the features responsible for the autocorrelation (option (b) above). Such a transformation can help us understand the time series better and, depending on the method we choose, identify the relative importance of the reasons for the autocorrelation.

Some research, on the other hand, raises concern about removing autocorrelation. Theiler et al. (1992) and Prichard (1994) suggest that applying such transformations can affect estimates of the correlation dimension. Theiler & Eubank (1993) report that, under certain circumstances, it can degrade evidence for nonlinearity, obscure underlying deterministic structure in chaotic data, and amplify noise. Those authors recommend the use of surrogate data (Ch. 19) as a possible way around the problems.

Since periodicity and trend probably are the most common causes of autocorrelation, we'll spend the rest of this chapter discussing ways to remove those two features.

Removing periodicity and trend

Filters

Many aspects of time-series analysis (not just the removal of periodicity and trend) involve the use of **linear filters**. A linear filter (herein simply called a **filter**) is any mathematical technique that systematically changes an input series (e.g. raw data of a time series, power spectrum, or autocovariance) into a new (output) series having certain desired and predictable qualities. Filters are linear when output varies linearly with input. Examples of "predictable qualities" might be the elimination or reduction of periodicity or trend, the elimination or reduction of noise, or the suppression of certain component frequencies (as in Fourier analysis or decomposition, discussed below). The term "filter" comes from electrical engineering. A radio tuner, for example, might block or suppress all unwanted frequencies except for a narrow band centered on a particular desired frequency.

Low or slow frequencies correspond to "long waves" (waves having frequencies of a few cycles per unit time). High or fast frequencies, in contrast, are associated with short waves (waves having frequencies of many cycles per unit time). Under certain circumstances, we might want to remove low frequencies from a series and let high frequencies pass relatively unaffected. The name for a mathematical technique or filter that suppresses or removes low frequencies and that lets high frequencies pass is, naturally, a **high-pass filter**. Differencing (discussed below) is an example. In other cases, long-term trends (low frequencies) might be our main interest. That means removing the distracting effects of small aberrations or noise, including high frequencies. We then need a **low-pass filter**.

Because a low-pass filter smooths out high frequencies and noise, and gives a clearer picture of overall trends, applying a low-pass filter is called smoothing. (We met smoothing in Ch. 6, where it was defined as reducing or removing discontinuities, irregularities, variability, or fluctuations in data.) Smoothing, therefore, is a special type of filtering. An example of a common low-pass filter or smoothing procedure is a **moving average** (discussed below).

A third type of filter, the **band-pass filter**, screens out both high and low frequencies and retains only those frequencies within a particular range or band. There are various other types of filters. Many books (e.g. Bloomfield 1976, Shumway 1988) give considerable attention to them.

Standardization

Where a time series is *periodic* (even if noisy) but has no trend or other regularity, the usual technique for isolating and removing that periodicity is **standardization**. Standardization is a transformation that converts data of whatever units to a common or standard scale. The mean on that scale is 0, and the standard deviation is 1. The new scale measures the transformed variable in terms of the deviation from its mean and is dimensionless. Therefore, besides removing periodicity, standardization offers the big advantage of enabling us to directly compare data measured in different units.

To standardize any value, just subtract the mean and then divide by the standard deviation. That is,

$$\textit{standardized value} = \frac{\textit{observed value minus the mean value}}{\textit{standard deviation of the values}}$$

$$= \frac{y - \bar{y}}{s} \tag{9.1}$$

where y is any observation, $\bar{y}$ is the arithmetic mean of the observations, and s is the standard deviation. (Dividing by the standard deviation puts each deviation in terms of standard deviations. It's like dividing a large number of eggs by 12 to express that quantity in units of dozens of eggs.)

For taking out periodicity from a time series, the particular "observations" to use in Equation 9.1 take on special definitions. The general rule to keep in mind is to think in terms of "seasons" or periods. That is, the y, $\bar{y}$ and s apply just to a particular season, not to the entire dataset as a unit. To explain that concept, let's first talk about what a "season" is. The definition of a season or period is strictly arbitrary but usually falls out naturally from the data. The idea is to identify the major periodicity in the data and then to divide that period into standard subperiods (seasons), for analysis. For example, a common periodicity in nature is one year. That periodicity shows up either from a glance at the time series or by inspecting the correlogram or power spectrum. We'd then declare appropriate subperiods, such as the four annual seasons (summer, etc.) or calendar months. As another example, our work routine tends to have a periodicity of one week, in which case a logical subdivision is seven one-day "seasons." The first step, therefore, is to determine the most prominent periodicity in the data and then to decide how to divide that repeated interval into seasons.

Once we've defined the seasons, we rearrange the measured time series on the basis of those seasons. That is, the rearranged data are simply in little subgroups in which each subgroup pertains to a particular season. Suppose, for instance, that we have ten years of daily measurements and that those data show a cycle that repeats itself every year. Now we need to subdivide that one-year cycle into "seasons." Let's arbitrarily decide to divide that one-year periodicity into twelve seasons of one calendar month each. We therefore rearrange the ten-year time series into twelve subgroups, each having only the data for a particular season (month). That is, we collect all data for season 1 (January) into one subgroup. January has 31 days and there are 10 years' worth of January values, so our first subgroup (January data) has 310 values. (While rearranging the data in this way, we retain the separate Januaries as separate subunits.) The 10 batches of observations for season 2 (February) make up a second subgroup (February data), and so on.

As mentioned, the values y, $\bar{y}$ and s to use in the standardization apply only to a particular subgroup or season. Let's take January. The y value should be a representative y for January of each year, in turn. To get the representative y value for January of each year, let's accept the arithmetic average of the 31 daily values for the year. Thus, for our ten years of data there will be ten representative y values to compute. The first is the average y for the 31 January observations of year one, the second is the average y for the 31 observations of January of year two, and so on. Those ten seasonal y values (one for January of each year) become the "basic data" for computing the standardized variables for January data. We then calculate the mean ($\bar{y}$) and standard deviation (s) of those ten January values. Finally, for each year, we determine the value of $(y-\bar{y})/s$. That means first taking the representative y value for January of year one, subtracting the average y of the ten Januaries, then dividing by the standard deviation of the ten January y values. Then we repeat the procedure for the representative y for each subsequent January, in turn. That produces ten standardized data values for the first season (January) and thereby completes the transformations for the January data. Then we repeat the entire procedure for each of the other seasons (February, March, etc.), in turn. Each season (here month) has its own unique set of ten y values and hence its own unique $\bar{y}$ and s.

Although computed on a seasonal basis, the transformed data then are put back in correct chronological order to make up the standardized time series. The first value in the new time series is the transformed value for January of year one, the next value is that for February of year one, and so on. Now let's go through all of this with some real data.

Figure 9.1a (upper part) shows the average monthly maximum air temperatures at Durango, Colorado, for the eight-year period of 1904–1911. That raw time series suggests a regular annual seasonality (high in summer, low in winter), along with ever-present noise. (The plot shows no indication of any significant trend. We'll assume for simplicity that there isn't any.) The lower part of Figure 9.1a shows the autocorrelation coefficients at successive lags. The coefficients show a wavelike variation with a wavelength of 12 months. That supports the idea of an annual periodicity. Many coefficients are far from zero, so that periodicity definitely causes some significant autocorrelation.

To begin the standardization, we subdivide the annual cycle into 12 monthly "seasons" (January, February, etc.). We then rearrange the data according to season. Table 9.1 shows one way to do that. The table includes only the first two months (January and February) because two months are enough to show the method of computation.

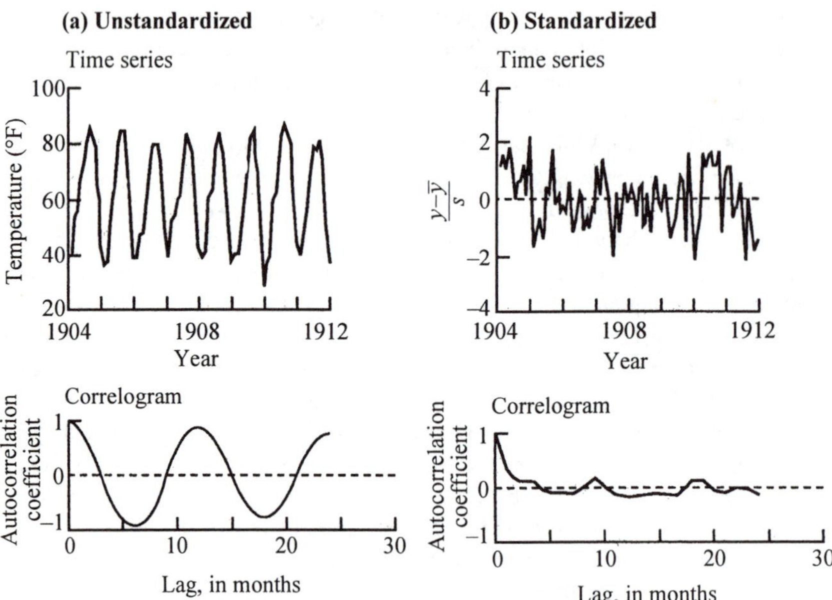

Figure 9.1 Effects of standardization on a time series whose only pattern is noisy annual seasonality (average monthly maximum air temperatures at Durango, Colorado, 1904–1911).

Table 9.1 Sample calculations for standardizing average monthly maximum air temperatures (Fahrenheit) at Durango, Colorado, 1904–1911.

	Raw data (°F)		Standardized values	
	January	February	January	February
1904	41	53	1.2	1.6
1905	36	37	–1.7	–1.1
1906	38	47	–0.5	0.6
1907	39	52	0.1	1.4
1908	39	40	0.1	–0.6
1909	40	40	0.6	–0.6
1910	37	40	–1.1	–0.6
1911	41	40	1.2	–0.6
Mean	38.9	43.6		
Standard deviation	1.7	5.8		

Let's go through the routine for January. The eight January values (each of which is the average of 31 daily measurements) have an arithmetic mean of 38.9°F and a standard deviation of 1.7°F. Those values are the $\bar{y}$ and s, respectively, to use in standardizing the data for the eight Januaries. For example, the representative maximum air temperature for January 1904 (the first of our eight January values), was 41°F. The standardized variable for January 1904 therefore is $(y-\bar{y})/s$ =

(41–38.9)/1.7 or 1.2 (Table 9.1). January of 1905 (the second of the eight January values) had a representative maximum temperature of 36°F, so the transformed value is (36–38.9)/1.7 or –1.7. After transforming the eight January values in that manner, we go through a similar routine for the second season, February. For instance, the mean and standard deviation for the eight Februaries are 43.6°F and 5.8°F, respectively. We therefore standardize February of 1904, which has a representative temperature of 53°F, as $(y-\bar{y})/s = (53-43.6)/5.8 = 1.6$. We then do February of 1905, February of 1906, and so on, in the same way. Finally, we follow the same procedure to do the eight values of each of the other months (March–December), in turn.

Figure 9.1b shows the time series consisting of the 96 standardized values (12 values per year times eight years). That series seems to have no significant autocorrelation. However, we can't simply assume that the standardized values have been decorrelated. Rather, we've got to do another autocorrelation analysis, this time on the standardized data. Figure 9.1b (lower plot) shows the correlogram for the standardized data. It verifies that our standardization effectively removed the seasonality of the raw data and that the transformed data are decorrelated.

Let's take a quick look at what standardization does and doesn't do. In so doing I'll use the word "periodicity" in a general way and arbitrarily break it into two subtypes: **seasonality** (periodicity at frequencies of less than or equal to one year) and **cyclicity** (longer-term periodicities). With that subdivision, the basic components of a time series are **trend**, *seasonality*, *cyclicity*, and *noise*. (Noise usually shows up as random-like deviations and is sometimes called error, disturbance, or randomness.) Some people don't use the two subtypes of periodicity, in which case the components are simply trend, periodicity, and noise.

What does standardization do to a time series that has not only a pronounced annual seasonality but also more frequent seasonalities? Using the Durango temperatures as a base, I constructed an artificial four-year time series having not only the fundamental one-year periodicity (peaking in July) but also two lesser six-month seasons (peaking in March and November, respectively) (Fig. 9.2a). To get more than four years of record for correlogram purposes, I repeated the four-year cycle two more times, thereby producing a 12-year time series to analyze for autocorrelation. Standardizing that 12-year time series (Fig. 9.2b) took out not only the one-year periodicity but also the two lesser seasonalities (March and November). So, standardization eliminates not only a basic seasonality but also shorter (more frequent) seasonalities, at least if those shorter ones are harmonics of the basic seasonality.

A third issue is whether standardization also gets rid of less frequent periodicities (in our example, cycles of more than one year). Figure 9.3a shows an artificial time series, again based on Durango temperatures. That concocted series has the basic annual seasonality. It also has a three-year cyclicity whereby the temperatures generally increase over a three-year period and then return to the approximate values of the first year to repeat the pattern. Standardizing those data on a basis of one year still left a periodicity of 36 months, as the correlogram of Figure 9.3b indi-

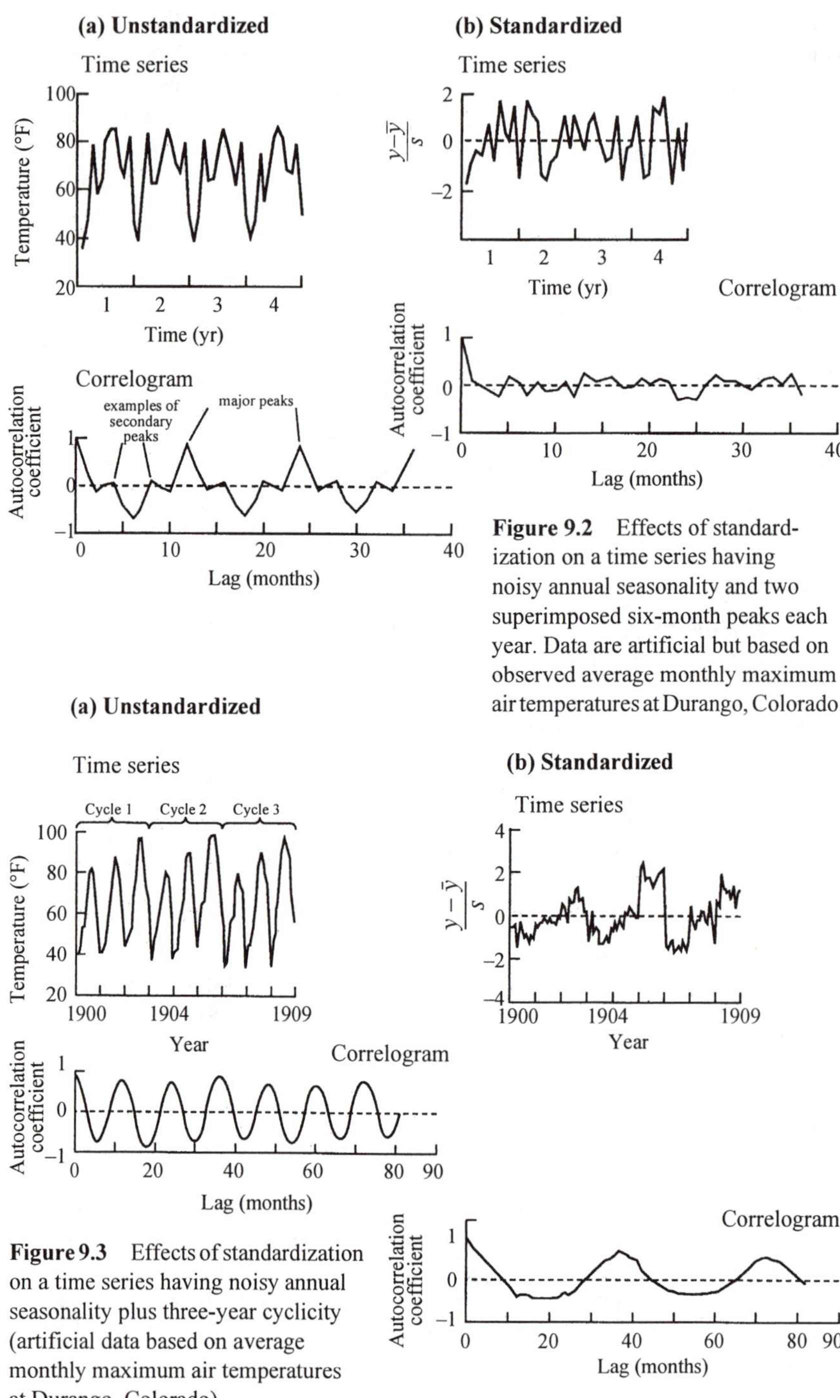

Figure 9.2 Effects of standardization on a time series having noisy annual seasonality and two superimposed six-month peaks each year. Data are artificial but based on observed average monthly maximum air temperatures at Durango, Colorado.

Figure 9.3 Effects of standardization on a time series having noisy annual seasonality plus three-year cyclicity (artificial data based on average monthly maximum air temperatures at Durango, Colorado).

cates. In other words, the standardization didn't affect the three-year cycle. Thus, standardization doesn't remove periodicities that are less frequent than the one on which the standardization is based.

A final question is whether standardization eliminates trend. Figure 9.4a shows regular annual temperature fluctuations but with an artificially introduced trend. The correlogram for the standardized data (Fig. 9.4b, lower plot) does not fall to within the noise level (autocorrelation coefficients near zero) within the first few lags, as desired. Instead, it decreases very gradually. Hence, standardization doesn't remove trend, and autocorrelation remains.

In conclusion, standardization takes out seasonality but not longer-term periodicity (cyclicity) or trend. Other types of transformations, such as classical decomposition and differencing (discussed next), can isolate or eliminate all of those features.

Classical decomposition

Decomposition is the process of numerically breaking down or transforming each observation into its various components. Those components usually are considered to be trend, seasonality, cyclicity, and noise (random deviations). The resulting filtered (transformed) time series consisting of just the noise presumably still has the nonlinear structure of the original raw data.

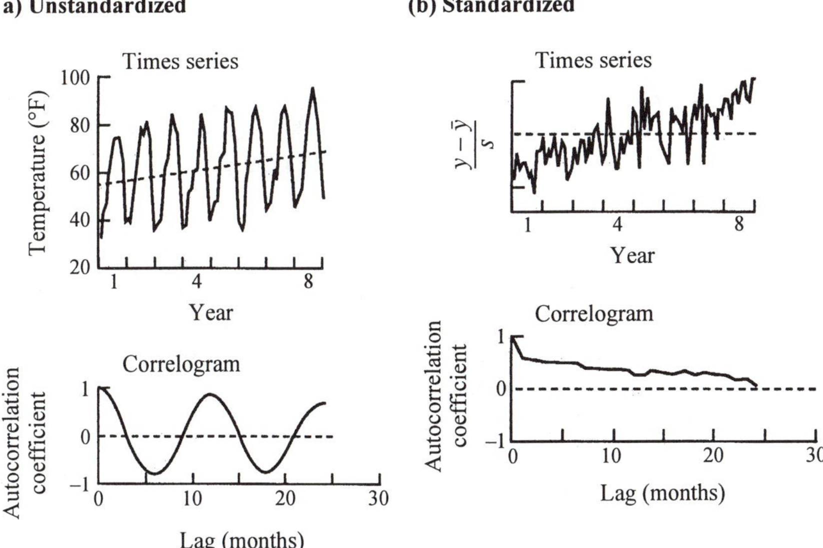

Figure 9.4 Effects of standardization on a time series having annual seasonality plus a steady long-term trend (artificial data based on average monthly maximum air temperatures at Durango, Colorado).

MODELS FOR DECOMPOSITION

Classical or traditional **decomposition** (e.g. Makridakis et al. 1983: ch. 4) is so called because it's the old, established, standard method for breaking down a time series into its four main components (trend, seasonality, cyclicity, and noise). It takes each value of a variable *y* and numerically identifies the proportions of that value that each of those four components contributes. There are two ways to describe those proportions.

- On an additive basis, that is, in terms of values that are added to yield the observed *y* (an "additive model"):

$$\textit{value of time-series variable} = \textit{trend} + \textit{seasonality} + \textit{cyclicity} + \textit{noise}. \quad (9.2)$$

- On a multiplicative basis, that is, in terms of values that are multiplied to yield the observed *y* (a "multiplicative model"):

$$\textit{value of time-series variable} = \textit{trend} \times \textit{seasonality} \times \textit{cyclicity} \times \textit{noise}.$$

As with standardization, it's common to first prepare raw time-series data such that each observation represents a "season" (if the basic measurements weren't made that way). So, for example, if we measure a variable each day but want to define a season as three months, we'd add up the appropriate three months' worth of daily measurements and average them to get a single value that embodies that particular season. Then we'd repeat for subsequent three-month seasons.

THE MOVING AVERAGE

A basic tool in classical decomposition is the moving average. (Many other types of data analysis, such as Fourier analysis [Ch. 8], also use the moving average.) A moving average is a transformed time series—actually a subseries—in which each value in the subseries is replaced by an average. The average is based on the measured value and several adjacent observations that come before and after it. That is, each new (transformed) value is the arithmetic mean of a subseries or constant number of successive observations. We assign the computed mean to the middle observation of the subseries. (Hence, computations are slightly easier if we use an odd number of data values to figure each average. I'll explain in a moment what to do if the number of observations is even.)

As an example of how to compute a moving average, suppose five successive measurements of a variable yield values of 7, 3, 15, 10, and 1. For a moving average based on three values (a "three-point moving average"), the first average is (7+3+15)/3 or 8.3. We assign that average to the middle value of the set, namely to the 3. To "move" the average, just advance the subseries by one observation (by discarding the oldest piece of data and incorporating the newer one) and compute a new average. Then repeat for each successive measurement. Thus, for our data the next step is to move one observation later in time and compute a new three-value average. To do that, we first drop the oldest of the original three values (7)

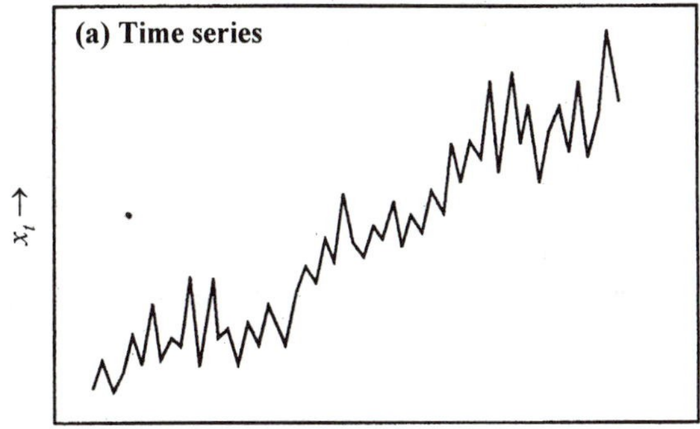

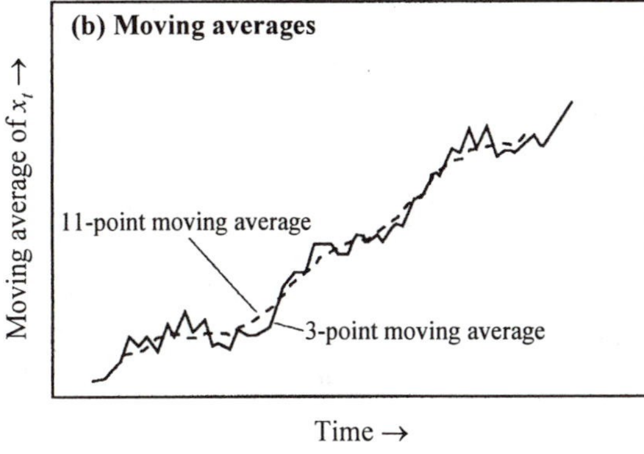

Figure 9.5 (a) Hypothetical time series with trend and noise. (b) Low-pass filters (moving averages) of (a), retaining the overall trend of the time series and showing greater smoothing as more and more data values are included in the moving average.

and incorporate the fourth observation (here 10) into the subseries. The new three-value subset then becomes 3, 15, and 10. The average of that subset is (3+15+10)/3 or 9.3. That average gets assigned to the midpoint of that series (15). The third average is (15+10+1)/3 or 8.7 and is assigned to the 10. The three averages—8.3, 9.3, and 8.7—reflect or represent the basic data, but they greatly dampen or smooth the relatively strong short-term fluctuations of those data.

The more data values included in the moving average, the greater the smoothing of the short-term fluctuations. Figure 9.5a is a hypothetical noisy time series. Figure 9.5b shows two moving averages of that series. Both moving averages show smaller fluctuations than the original time series of Figure 9.5a. Furthermore, the 11-point moving average (dashed line) smooths the original series much more than does the three-point moving average.

For convenience in the analysis, there must be a perfect midpoint to the group of observations involved in a moving average. Such a perfect midpoint works out naturally if the number of observations is odd, as we've just seen. However, when the number is even, no observation is the middle one. People traditionally remedy such a problem with a weighting technique or second smoothing. The second smoothing includes the next sequential observation (to make the number of observations odd) while giving the first and last observations only half **weight**. For example, to center the average of four observations, use five observations but give only half weight to the first and fifth. The general relation for the moving average $\bar{y}_t$ of four sequential observations then is:

$$\bar{y}_t = (0.5y_{t-2}+y_{t-1}+y_t+y_{t+1}+0.5y_{t+2})/4. \quad (9.3)$$

Here we've centered the average of four observations (divided the sum in Eq. 9.3 by four) because the total number of observations within the parentheses is four. (The middle three members each have weights of 1, and the two end members each have a weight of 0.5, so the total is four.) At the same time, such a moving average is "centered on" the group's middle observation, y_t. In our example here, y_t is observation number three, so in a table we'd list that moving average opposite the third observation of the subseries.

Equation 9.3 in a form applicable to any even number of observations is:

$$\bar{y}_t = (y_t+y_{t+1}+y_{t-1}+\ldots+0.5y_{t+i}+0.5y_{t-i})/N \quad (9.4)$$

where N (an even number) is the total weight (sum of the coefficients) of the observations and $i = N/2$. It's actually a compact, shorthand expression for two smoothing operations merged into one calculation. In other words, the value $\bar{y}_t$ in Equation 9.3 really comes from averaging two moving-average values that are separated by a lag of one. Let's take a paragraph here to show why that's true. Using the example of four seasons ($N = 4$), we'll add two successive moving averages and divide that sum by two:

$$\begin{aligned} \bar{y}_t = {} & [(y_1+y_2+y_3+y_4)/4 && \text{(first average)} \\ & + (y_2+y_3+y_4+y_5)/4] && \text{plus (second average)} \\ & \text{divided by two.} && \text{(average of the two)} \end{aligned}$$

Adding the two averages and collecting like terms, the entire equality in symbols is:

$$\bar{y}_t = [(y_1+2y_2+2y_3+2y_4+y_5)/4]/2.$$

Dividing as indicated:

$$\bar{y}_t = (y_1+2y_2+2y_3+2y_4+y_5)/8.$$

Multiplying everything by 0.5 to simplify:

$$\bar{y}_t = (0.5y_1+y_2+y_3+y_4+0.5y_5)/4 \quad (9.3)$$

which is the same as Equation 9.3: a "double smoothing" (averaging two successive moving averages) produces the weighted moving average of Equation 9.3.

The little algebra we've just gone through reveals three different ways of computing a moving average when N is even:

- Compute the two sums (the first sum in our example being y_1 through y_4, the second being y_2 through y_5). Then average those sums (here by adding the two sums and dividing by two). Finally, divide by N, the number of observations (here 4) (Firth 1977: 67).
- Compute the two moving-average values, then average the two (Wonnacott & Wonnacott 1984: 640).
- Use Equation 9.4.

Those three ways all involve weights of 0.5 for the two end members and 1.0 for all other terms. All three methods give the same answer.

INITIAL DATA ANALYSIS OF A CASE STUDY

With our tools in place, let's now go through the classical decomposition procedure with some real data: the flow of water (discharge, in technical parlance) in the Colorado River near Grand Lake, Colorado, for 1954–68. (The record actually continues beyond 1968; I used 1954–68 only because that period shows, to some extent, the four components of a time series.) We'll arbitrarily define a "season" as one calendar month. Water discharge was measured daily. Since classical decomposition uses seasonal values, we average the 30 or so daily values for each month to get representative seasonal (monthly) values (Table 9.2).

Table 9.2 Monthly mean water discharge, in cubic feet per second, for Colorado River near Grand Lake, Colorado, 1954–68.

Year	Jan	Feb	Mar	Apr	May	June	July	Aug	Sept	Oct	Nov	Dec
1954	5.97	5.74	5.84	38.5	100.6	69.8	27.3	11.1	22.9	27.7	12.4	8.45
1955	7.25	5.65	5.09	38.0	108.0	143.6	51.0	24.8	15.4	12.8	11.4	8.63
1956	5.35	4.85	5.53	31.7	248.1	193.4	34.5	19.6	11.8	9.25	7.43	4.56
1957	3.91	4.17	5.65	14.3	116.8	462.3	312.1	44.3	29.2	27.1	15.0	9.35
1958	8.41	7.26	6.63	13.1	315.2	267.8	36.5	21.9	15.9	10.9	9.1	5.22
1959	4.68	4.64	4.89	14.4	127.9	211.6	49.2	29.3	25.5	37.6	26.9	12.6
1960	8.58	8.28	10.1	56.7	143.3	242.0	57.1	20.6	25.5	19.1	13.3	10.0
1961	8.0	7.00	7.52	15.5	159.6	359.0	56.3	28.6	75.5	83.7	37.2	17.4
1962	10.3	8.54	8.88	74.5	216.0	306.8	142.4	28.5	17.2	18.0	9.54	6.57
1963	4.37	4.78	5.57	34.0	169.4	140.4	41.5	39.4	29.1	18.9	12.6	8.55
1964	8.11	6.83	7.23	13.1	204.3	230.2	89.9	37.2	17.8	12.4	8.76	6.84
1965	5.41	5.17	5.66	21.7	157.7	506.7	177.9	50.2	37.7	39.3	21.1	13.9
1966	10.2	8.84	9.32	35.1	146.4	126.8	41.9	17.4	16.9	22.6	12.9	7.58
1967	6.05	6.23	8.17	35.7	159.3	398.3	232.3	31.3	39.2	31.3	16.3	9.76
1968	7.39	8.17	8.57	12.3	97.7	445.4	110.1	34.3	25.2	19.4	10.9	9.3

The standard beginning is the general preliminary steps outlined under "First Examination of Data," above. Step one is to plot the data. An arithmetic scale for water discharge (Fig. 9.6a) produces a very skewed (uneven) distribution of discharges in that many values are small and few are large. Using logs of discharges (a "log transformation") (Fig. 9.6b) brings about a much more even distribution.

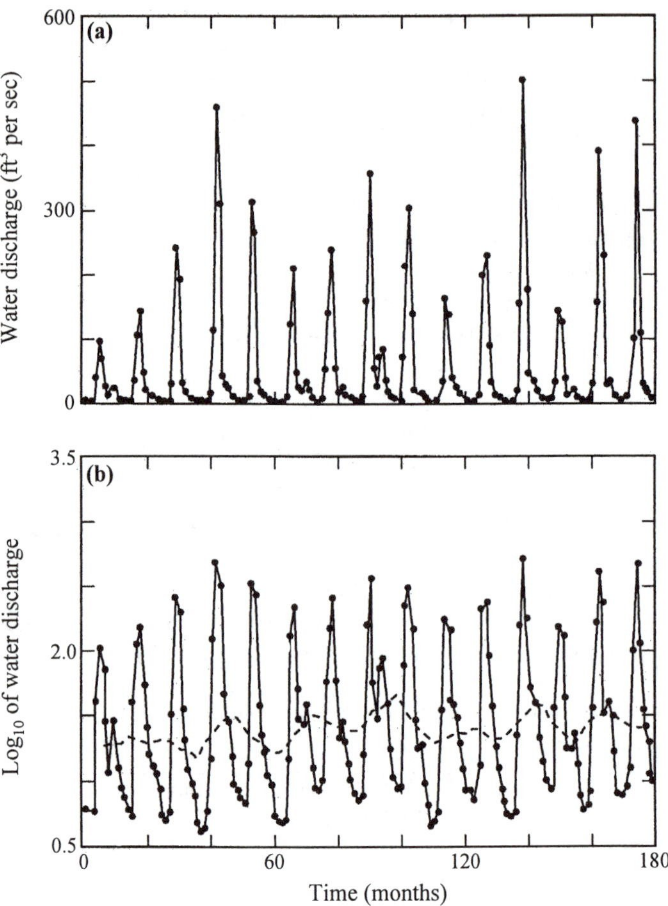

Figure 9.6 Time series of average monthly streamflow, Colorado River near Grand Lake, Colorado, 1954–68. (a) Raw data for water discharge. (b) Logarithms of water discharge. Dashed line in (b) is a centered 12-point moving average.

That's a highly desirable feature for statistical tests. Hence, we'll analyze logs of discharges rather than discharges themselves.

Table 9.2 and Figure 9.6b show that water discharge has a pronounced periodicity, as expected. In particular, it tends to be relatively high around May and June of each year. The next steps in the analysis should confirm this. Any trend over the 15-year period isn't very apparent.

Step two of the preliminary analysis is a Fourier analysis (Fig. 9.7). It shows spikes at three frequencies—about 0.5236, 1.0472, and 2.0944 radians. Dividing each frequency into 4π gives the associated periodicities in time units. The respective frequencies just mentioned therefore translate into regular oscillations of 24 months (a relatively minor one, judging from the low variance), 12 months (very strong), and 6 months (intermediate).

The third required preliminary step is an autocorrelation test (Fig. 9.8). The sinusoidal trace on the correlogram confirms our visual conclusion of a regular periodicity (cf. Figs 7.2a and 7.2c). The wavelength of the sinusoids is 12 months, as expected (one oscillation occurs each year). The pattern and coefficients on the correlogram confirm that there's significant seasonality. (As an aside, both the correlogram [Fig. 9.8] and Fourier analysis [Fig. 9.7] show the annual seasonality [12-month periodicity] that was obvious in the basic data [Fig. 9.6]. However, the Fourier analysis shows the secondary oscillations more clearly than does the autocorrelation test.)

The correlogram of Figure 9.8 shows a slight tendency for the sinusoids to be

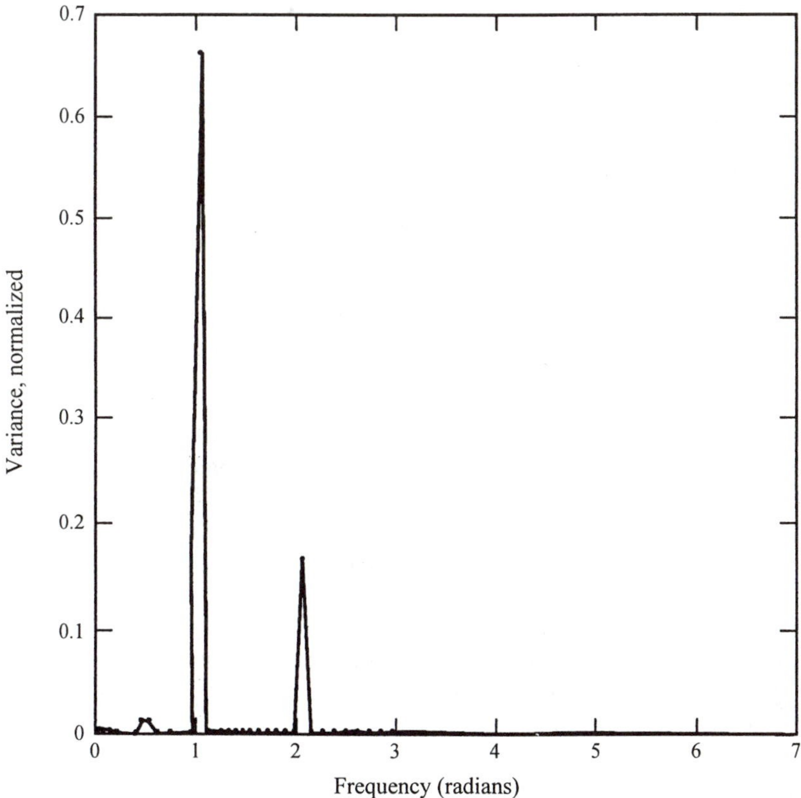

Figure 9.7 Periodogram (Fourier analysis) of monthly average streamflow, Colorado River near Grand Lake, Colorado, 1954–68. Total number of observations, N, is 180 (15 years with 12 monthly values per year). A Fourier analysis deals with a range of $N/2$ (here 90) observations and assigns this range to 2π (= 6.28) radians.

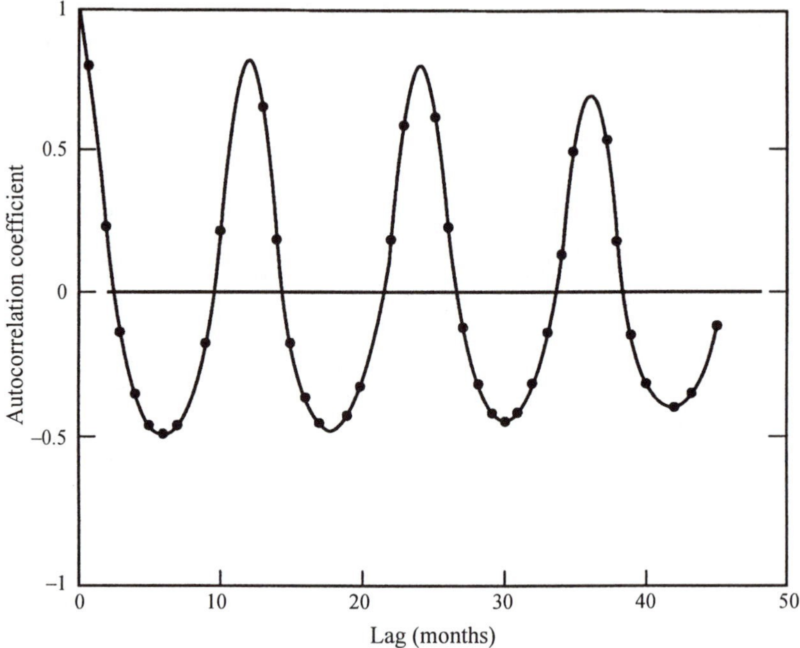

Figure 9.8 Correlogram of logs of monthly average streamflow, Colorado River near Grand Lake, Colorado, 1954–68.

offset above the horizontal reference line of zero autocorrelation. That might or might not indicate a minor trend of increasing streamflow with time (cf. Fig. 7.2e). For the time being, we won't rule out the possibility of trend.

In summary, the preliminary analysis reveals that the Colorado River time series has seasonality (6- and 12-month oscillations), cyclicity (24-month oscillations), possibly a minor trend, and of course noise.

STEPS IN CLASSICAL DECOMPOSITION

Having finished that standard preliminary routine, we'll now go through the steps in classical decomposition, using the Colorado data. Following that, I'll summarize the steps.

Decomposition's general purpose is to transform the raw data (in this case the logs of the monthly mean discharges) into a form in which they lack autocorrelation. Since the data are now in log units, we'll use the additive model (Eq. 9.2). Adding logs is tantamount to multiplying the unlogged values.

- *Step 1* The first step is to smooth the raw time series with a moving average. The purpose of the moving average is to remove seasonality and noise. Our example will show how that works.

The particular moving average to use must include two features. First, since the purpose of the moving average is to smooth out seasonality, each of our transformed (averaged) values should include any given season only once. Why is that? Say we have a hypothetical time series consisting of alternating high and low values (a seasonality of two). Averaging any successive two values would remove those two high and low points and result in a single intermediate value. Hence, we set the "length" of each subseries in the moving average to be the length of the seasonality, where "length of seasonality" means the interval between repetitions of the cycle (in other words, the wavelength). Doing that will average out (remove) the seasonality.

Colorado River discharges show seasonality on a one-year basis (large flows in the springtime, small in late summer). Therefore, the length of the subseries for our moving average should be 12 months.

The second required feature of the moving average is that each average be "centered" right on a particular season. That is, there has to be a perfect midpoint to the group of seasons involved in the computation. Since 12 is an even number, we have to use a weighting technique to center each average on a particular observation. I used Equation 9.4.

Figure 9.6b plots that moving average as a fluctuating dashed line. The moving average shows four critical features:

- The 6- and 12-month periodicities (the seasonality) are gone.
- Most of the irregularities (the noise), including the very high and very low points, also are gone. That is, the moving-average curve is much smoother.
- A roughly sinusoidal pattern remains, with a wavelength of about two years. That periodicity represents the cyclicity.
- The moving average dashed line shows a slight tendency to move upward on the graph with time. That represents the possibly minor trend suggested by the correlogram. (Neither the cyclicity nor trend were particularly apparent in the basic data.)

Summarizing, the moving average is a modified time series that effectively removes seasonality and noise, still contains cyclicity and trend, and by itself doesn't define or numerically isolate any of the four components individually.

Table 9.3 lists the first two years' worth of data and associated calculations for the classical decomposition method, to illustrate the computation procedure. Column 3 is the values of the centered moving average (Eq. 9.4 with $N = 12$). Run your eye down column 3 and notice the relatively small variability of those smoothed values of $\log Q$, compared to the raw data of column 2. (The first six and last six observations of the entire 180-row dataset don't have a 12-point centered moving-average because the range of values needed in the calculation extends beyond the measured data.)

Computing the moving average completes step one of the classical decomposition technique. The order of the remaining steps (except for the last one) can vary slightly from one author to another.

- *Step 2* Step two is to use the moving average to define trend, if there is one. The moving average (replotted on Fig. 9.9) has a sinusoidal-like pattern whose overall direction can be approximated by a straight line. Because our purpose here in fitting a straight line will be to estimate the dependent variable (the moving-average of log$Q$), ordinary least squares is the line-fitting technique to use. The equation of that straight line for the 168 observations is

$$\textit{estimated moving average} = 1.295 + 0.001t \tag{9.5}$$

where t is time, in months.

The coefficient of time, 0.001, is close to zero. If it were zero, the estimated moving average would be constant (1.295) and wouldn't depend on time at all. Thus, any trend is very minor. In any case, we ought to compute confidence intervals to insure that any trend is significant (e.g. Wonnacott & Wonnacott 1984: 343–4). And it almost goes without saying that we can't assume that the trend continues beyond the range of our data, namely beyond 1968. For our tutorial purposes here, it would be nice to have some trend, so I'll by-pass the statistical test and pretend the coefficient of 0.001 for 1954–68 is significantly different from zero.

The fitted straight line represents trend; its value at any time gives the trend component of log$Q$, at that time. Thus, to identify trend numerically at each successive time, we plug in the appropriate t and solve the least-squares equation (here Eq. 9.5). Based on that empirical least-squares equation, column 4 of Table 9.3 gives the "trend factor" of log$Q$ (the straight-line estimate of the moving average) for each successive time, t.

- *Step 3* Step three is to compute the value of "cyclicity" at each time. We

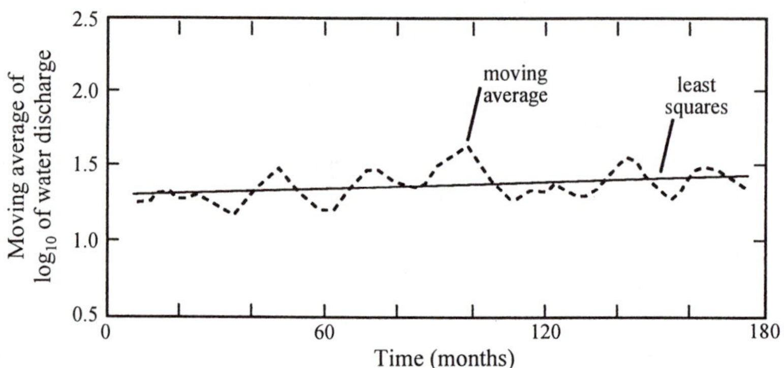

Figure 9.9 Cyclicity and trend, shown by the moving-average dashed line, and isolation of trend by the least-squares straight line, for logs of monthly average streamflow of the Colorado River near Grand Lake, Colorado, 1954–68.

Table 9.3 Classical decomposition technique used on logarithms of monthly mean water discharges, Colorado River near Grand Lake, Colorado, 1954–68. Only years 1954 and 1955 are listed here, for space reasons.

1 Year/month	2 Observed y = log of water discharge	3 12-point centered moving average	4 Trend = least-squares value of moving average	5 Cyclicity = moving average minus trend	6 Seasonality + noise = log Q minus moving average	7 Seasonality = seasonal mean of (log Q minus moving average)	8 Noise = time (seasonality + noise) minus seasonality
1954							
1	0.776						
2	0.759						
3	0.766						
4	1.585						
5	2.003						
6	1.844						
7	1.436	1.257	1.302	–0.045	0.179	0.474	–0.295
8	1.045	1.260	1.303	–0.043	–0.215	0.053	–0.268
9	1.360	1.257	1.304	–0.047	0.103	0.003	0.100
10	1.442	1.254	1.305	–0.051	0.188	–0.032	0.220
11	1.093	1.255	1.306	–0.051	–0.162	–0.236	0.074
12	0.927	1.270	1.307	–0.037	–0.343	–0.439	0.096
1955							
1	0.860	1.294	1.308	–0.014	–0.434	–0.556	0.122
2	0.752	1.320	1.309	0.011	–0.568	–0.588	0.020
3	0.707	1.327	1.310	0.017	–0.620	–0.550	–0.070
4	1.580	1.306	1.311	–0.005	0.274	0.006	0.268
5	2.033	1.291	1.312	–0.021	0.742	0.821	–0.079
6	2.157	1.290	1.313	–0.023	0.867	1.032	–0.165
7	1.708	1.284	1.314	–0.030	0.424	0.474	–0.050
8	1.394	1.276	1.315	–0.039	0.118	0.053	0.065
9	1.188	1.275	1.316	–0.041	–0.087	0.003	–0.090
10	1.107	1.273	1.317	–0.044	–0.166	–0.032	–0.134
11	1.057	1.285	1.318	–0.033	–0.228	–0.236	0.008
12	0.936	1.305	1.319	–0.014	–0.369	–0.439	0.070

found above that the moving average at any time does not contain seasonality and noise but does represent trend and cyclicity. Using our additive model, that means that:

$$\textit{moving-average value of } \log Q = \textit{trend} + \textit{cyclicity}. \qquad (9.6)$$

According to that equation, cyclicity at a particular time is simply the moving-average value minus the trend value (col. 5 of Table 9.3).

After completing that step, we've isolated (defined numerically) each of trend and cyclicity, at each time (season). That leaves just seasonality and noise to determine.

- *Step 4* Step four is to calculate, for each season (time), a single value representing the combined total of seasonality + noise. That's just a matter of another easy subtraction. Log$Q$ represents the sum of all four components (Eq. 9.2); the moving average represents trend + cyclicity (Eq. 9.6). Thus, subtracting the moving average from the observed log$Q$ at each time gives the sum of the two remaining components, namely seasonality + noise (col. 6 of Table 9.3). Once done with that step, all we need to do to find each of seasonality and noise individually is to get one of these and then subtract it from the combined factor of column 6 to get the other. Seasonality is the one we'll determine first, as follows.
- *Step 5* Step five uses the same "combined value" relation of the previous step to determine (estimate) seasonality. The underlying assumption is that seasonality for any given season is equal to the long-term arithmetic mean of the sum of seasonality + noise for that season. In other words, we assume that, since some noise is positive and some is negative, many noise values offset one another or "average out."

  The first season is January. We want the average value of seasonality + noise for that month. Although we have a 15-year dataset, there are only 14 values of seasonality + noise for any season. (January–June of the first year have no 12-point centered moving average because the required observations begin prior to the first observation. Similarly, July–December of the last year have no centered moving average because the necessary data extend beyond the period of record.) The arithmetic mean of the 14 seasonality + noise values for January will be our "seasonality" factor for January, the first season. That mean value is 0.474 (col. 7 of Table 9.3). Then we do the same for the 14 Februaries to get the seasonality factor for February (the second season), and so on for the other seasons. Such calculations give the seasonality factor for each month. We've now gotten numerical values for three of the four time-series components, namely trend, cyclicity, and seasonality.
- *Step 6* Step six is to isolate the fourth and final component—noise—for each month. We can do so in either of two ways. One is to add trend + cyclicity + seasonality and subtract that sum from log$Q$, for each successive time, per Equation 9.2. The other is merely to subtract seasonality (col. 7 of Table 9.3) from (seasonality + noise) (col. 6). The sequential monthly noise values (col. 8 and Fig. 9.10a) make up a transformed series of Colorado River discharges. In particular, they are a time series that supposedly has the trend, seasonality, and cyclicity removed.
- *Step 7* The final step is always to run an autocorrelation analysis on the noise, to see if the noise has any remaining autocorrelation. If it does, then in the subsequent chaos analysis use a lag large enough to provide values that are uncorrelated. For this example, we wouldn't need to use any such lag because the

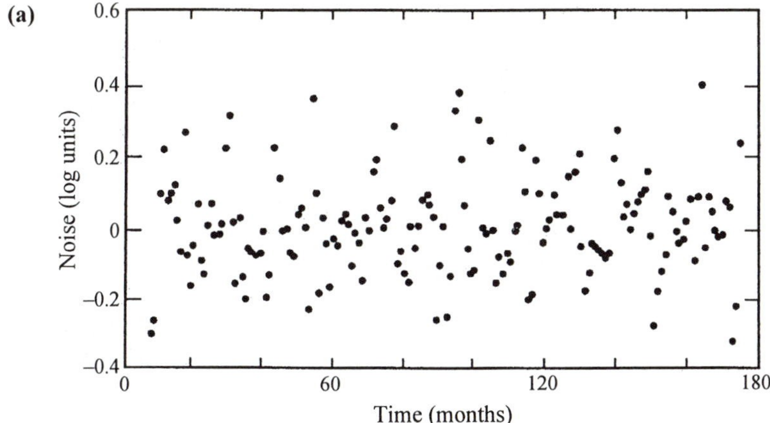

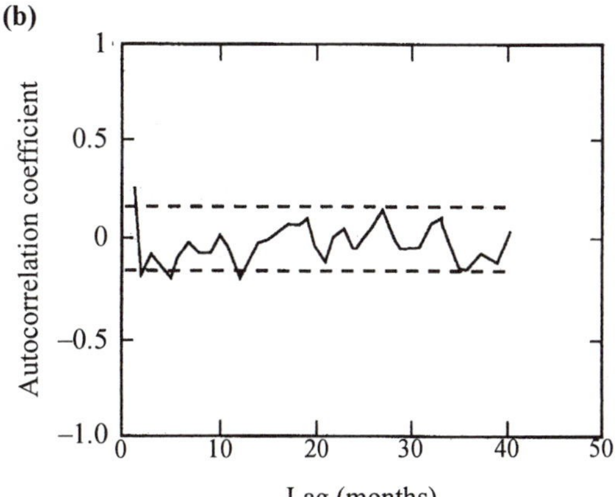

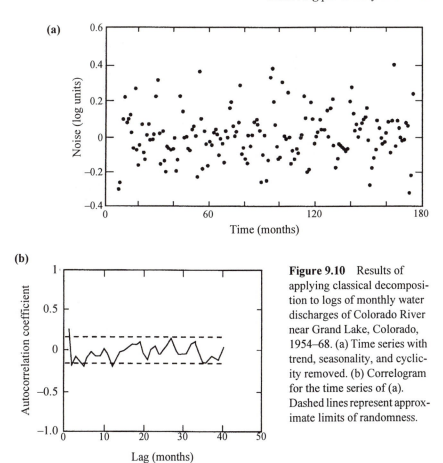

Figure 9.10 Results of applying classical decomposition to logs of monthly water discharges of Colorado River near Grand Lake, Colorado, 1954–68. (a) Time series with trend, seasonality, and cyclicity removed. (b) Correlogram for the time series of (a). Dashed lines represent approximate limits of randomness.

autocorrelation coefficients generally fall within the approximate limits for randomness (Fig. 9.10b). In other words, for our Colorado River data the values in the transformed series (the noise) have no mutual relation. This unconnected time series (col. 8 of Table 9.3), rather than the raw data (col. 2), are the values to analyze for chaos if we choose first to decorrelate the data.

Summarizing the steps in the classical decomposition technique:

1. Smooth the raw time series with a moving average. Choose the length of the moving average specifically to eliminate seasonality. The smoothing also takes out or at least reduces the noise. However, it does not remove any trend or cyclicity. In other words, the moving-average value at any time equals trend, if any, plus cyclicity, if any.
2. Identify the type of trend (linear, exponential, etc.) and calculate a best-fit line to the moving average. The equation for this line then gives the value of "trend" at a given time.
3. Determine cyclicity for each successive time by subtracting the trend value

("**detrending**") from the moving-average value.

4. Calculate, for each season, a single value that stands for seasonality + noise. Such a value equals y minus the moving-average value.
5. Compute the average value of (y minus moving average) for each season. Assume that this average value smooths out the noise and therefore reflects just the seasonality factor.
6. Estimate the final component, noise, for each successive time as (seasonality + noise) minus seasonality. (Each such value is the same as y–[trend + seasonality + cyclicity].)
7. Do an autocorrelation analysis on the noise, to insure that it's mostly free of trend, cyclicity, and seasonality.

What we've gone over is just the basic bare-bones "vanilla" type of classical decomposition. Modern versions include many embellishments.

Differencing

Classical decomposition usually involves the assumption that trend, seasonality, and cyclicity are constant over time. For some practical situations, there may be no basis for that assumption. The computationally easier method of differencing is an alternate technique that removes both trend and periodicity.

Differencing is a transformation made by systematically subtracting one value from another throughout a dataset, with each pair of values separated by a designated lag. That creates a new, transformed time series in which the bits of data are differences or intervals. Thus, the new data are arithmetic differences between pairs of observations that are separated by a constant, selected time.

To illustrate, suppose a time series of some variable is 1, 6, 3, 17, and 8. The simplest differencing merely involves a lag of one—subtracting the first observation from the second, the second from the third, and so on. (The observations might be iterations of an equation or actual measurements.) To difference such data using a lag of one, we just take $y_{t+1}-y_t$. So, the first value in the transformed (differenced) dataset is $y_2-y_1 = 6-1 = 5$; the second item is $y_3-y_2 = 3-6 = -3$; the third is $17-3 = 14$; etc. Our new, transformed time series is a sequence of *intervals*: 5, –3, 14, –9. Because we've differenced the original data just once, the sequence of intervals makes up a **first-difference transformation** or, less formally, a **first-differencing**.

A first-differencing can be done using any lag, not just a lag of one. For instance, declaring a lag of two defines each difference as $y_{t+2}-y_t$. For the data of the above example, the first such difference is $y_3-y_1 = 3-1 = 2$; the next difference or value in the transformed series is $y_4-y_2 = 17-6 = 11$; and the third value is $8-3 = 5$.

As the above examples show, y_t always moves up by one value in the raw data, regardless of the lag. That is, we always advance by one observation to compute the next difference. Then we apply the lag to find the measured value from which to subtract y_t.

First-differencing with a lag of one usually removes trend. For example, suppose a series of exact (noise-free) measurements consists of the values 2, 4, 6, 8, 10, and so on (Makridakis et al. 1983: 381). Those data indicate a linear trend—each new value is 2 greater than the preceding value. Suppose we want to transform the data such that they have no trend. We decide to eliminate the trend (detrend) by the method of first differences using a lag of one. Our first item of new data therefore is $y_2-y_1=4-2=2$. Our second value is $y_3-y_2=6-4=2$. The third interval is $y_4-y_3=8-6=2$, and so on. Thus, the new (transformed, or first-differenced) dataset consists of 2, 2, 2, etc. The trend is gone from the original data.

First-differencing of a time series is an example of a high-pass filter. That transformation or filter isolates and focuses attention on just the relatively short intervals between successive observations (the high or fast frequencies). Those short-term intervals generally don't show long-term trends (low frequencies). That is, the new (output) series masks or removes any long-term trends. For example, Figure 9.11a is a hypothetical time series showing trend and noise. First-differencing the data (Fig. 9.11b) eliminates the trend. (The noise, however, remains.)

In just as simple a fashion, first-differencing also can eliminate periodicity, whether seasonal or cyclical. The only change is the choice of lag. To eliminate any periodicity, the appropriate lag isn't one but rather the wavelength or time between peaks in the series. For example, suppose we measure values of 1, 2, 7, 1, 2, 7, 1, 2, 7, so on. That series peaks every third observation and so has a periodicity of three. We therefore take differences using a lag of three. The first item of data in the new

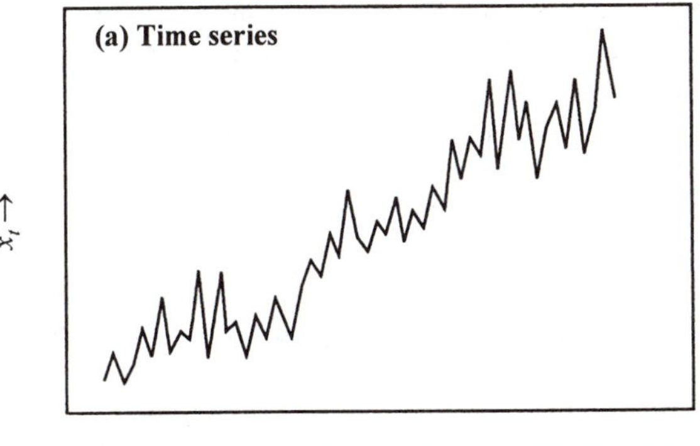

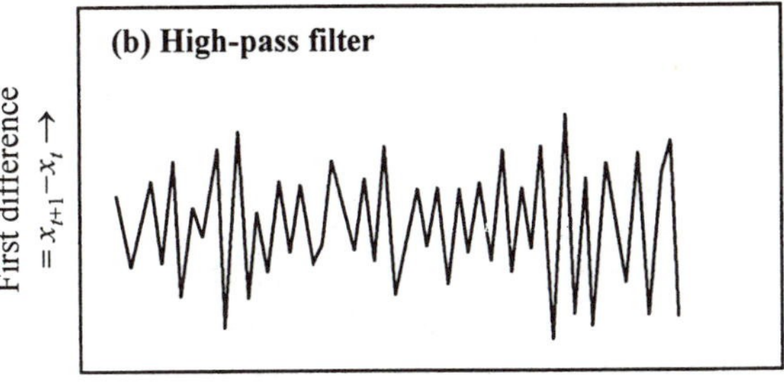

Figure 9.11 (a) Hypothetical time series with trend and noise. (b) High-pass filter (here the first difference between successive values) of (a). Such a filter removes the trend but not the noise.

(differenced) series based on a lag of three is y_4-y_1. For our data (1, 2, 7; 1, 2, 7; etc.), that's $1-1=0$. The second item is $y_5-y_2=2-2=0$. The third interval is $y_6-y_3=7-7=0$, and so on. Our filtered dataset therefore consists of the values 0, 0, 0, etc. This differenced set of data doesn't have the periodicity of the original data.

Eliminating periodicity as just described removes any trend at the same time. To show that, let's make up a simple, hypothetical dataset that has both trend and periodicity. To do so, we'll combine the data having trend (2, 4, 6, etc.) with those having periodicity (1, 2, 7, 1, 2, 7, etc.). We do that by adding the two individual sets:

Description	Values								
Observation number	1	2	3	4	5	6	7	8	9
Data of first set (having trend)	2	4	6	8	10	12	14	16	18
Data of second set (periodicity)	1	2	7	1	2	7	1	2	7
Combined set (trend + periodicity)	3	6	13	9	12	19	15	18	25

The new dataset grows with time (has a trend). It also still peaks every third observation (has periodicity). We therefore base the differencing on a lag of three. The first item in a transformed dataset then is y_4-y_1. Here $y_4-y_1=9-3=6$. The second interval is $y_5-y_2=12-6=6$. The third value is $y_6-y_3=19-13=6$. Next is $15-9=6$, and so on. The differenced set consists of 6, 6, 6, and so on. We've removed both the trend and the periodicity of the original data.

Real-world data of course include noise. In contrast to the previous examples, differenced values of a noisy time series won't be exactly constant. As with the classical decomposition method, always do an autocorrelation test on differenced data to see whether any significant autocorrelation remains. Let's use the monthly mean water discharges of Table 9.2 to see why that step is important.

The water discharges (and hence also their logs) showed what I called a minor trend. They also showed periodicities of 6, 12, and 24 months. The 12-month periodicity was by far the strongest feature. It might therefore seem that a lag of 12 seasons (12 months) will banish the periodicity. Applying a lag-12 differencing to the logarithms for the 15-year record means computing the differences from January to January, February to February, and so on for successive years. Hence, the first value in the transformed dataset is the log of discharge for January 1955 minus the log of discharge for January 1954. The second value is the log of discharge for February 1955 minus that for February 1954, and so on. Such a lag-12 differencing produces a filtered time series, not shown here. Next, we need to know whether any autocorrelation still remains in that filtered time series.

Figure 9.12a is the result of the autocorrelation test. The autocorrelation coefficients don't show the desired random-like fluctuations around a value of zero with time (lag). Instead, they show distinct periodicity, here with a period of 24 months. Many values exceed the 95 per cent confidence bands (the approximate limits for randomness). Hence, there's still some periodicity in the data. Without doing the autocorrelation test on the differenced data, we wouldn't recognize that important fact.

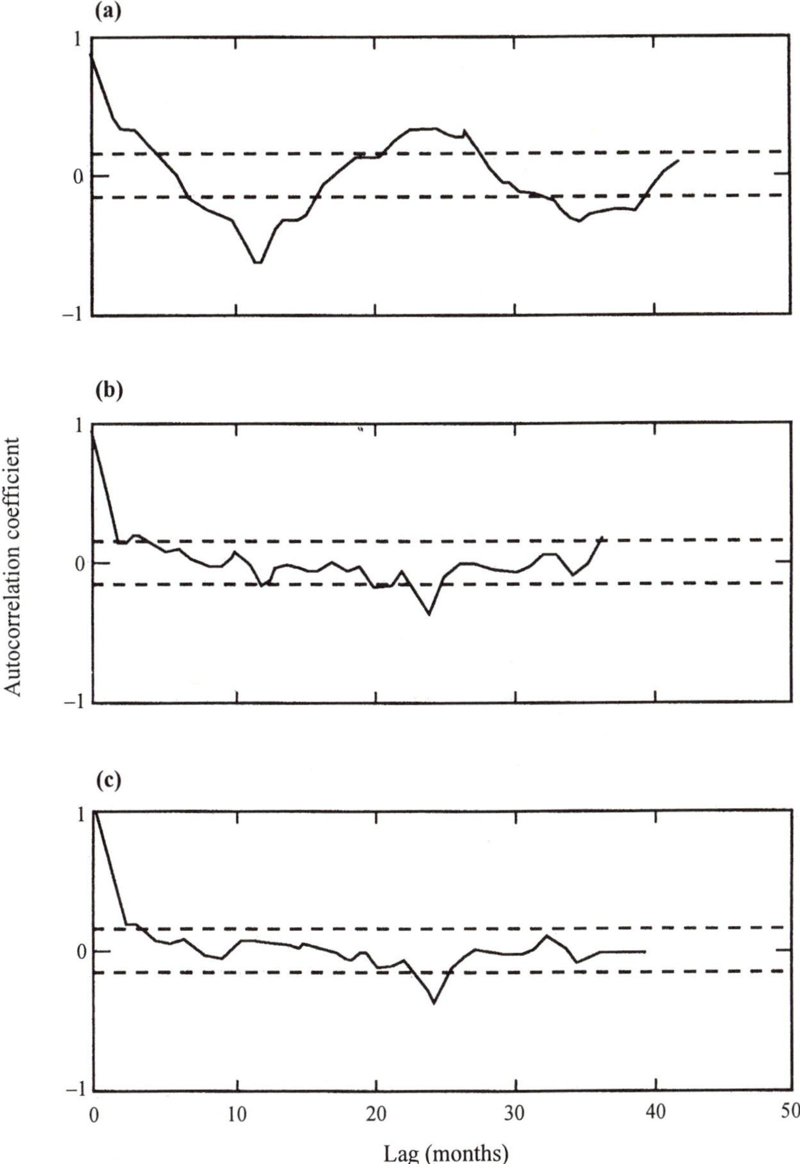

Figure 9.12 Correlograms for various differencings of logs of mean monthly water discharges, Colorado River near Grand Lake, Colorado, 1954–68. (a) Lag-12 differencing. (b) Second-order differencing, here a lag-24 differencing of the lag-12-differenced values. (c) Lag-24 differencing. Dashed lines are 95 per cent confidence limits.

When a first-differencing doesn't remove all the periodicity from a dataset, we can select for our chaos analysis only values that are far enough apart in time as to be unrelated. However, time-series specialists usually opt for a **second-differencing** on the filtered data. That is, they difference the differenced time series. In most cases, a second-differencing accomplishes the goal. As with any differencing designed to remove periodicity, the first step in the second-differencing is to choose the lag. Figure 9.12a suggests that the filtered time series has a periodicity of 24 months, so 24 months is the lag to choose. In other words, we're going to apply a lag-24 differencing to the lag-12-differenced data. The first item in the new time series therefore is [the difference from January 1957 to January 1956] minus [the difference from January 1955 to January 1954]. (There's exactly a two-year or 24-month gap or lag between those two differences.) The second item is [the difference from February 1957 to February 1956] minus [the difference from February 1955 to February 1954], and so on.

Figure 9.12b shows the results of the autocorrelation test done on the second-differenced time series. Autocorrelation coefficients fall to within the random-like zone after the first lag and, for the most part, remain in that zone thereafter. The second-differenced time series, in other words, consists of unrelated values.

For these particular data, the periodicities of 6, 12, and 24 months happen to be harmonics of 24. When the several periodicities are all harmonics of a single value, such as 24 in this case, just one differencing using that highest lag can remove all periodicities. That means we'd use a lag of 24 instead of 12 in our first-differencing. Differencing the logs of monthly mean water discharges using a lag of 24 (January 1956 minus January 1954, etc.) and running an autocorrelation test on the resulting time series yields the correlogram of Figure 9.12c. Sure enough, the values in that new series generally are largely unrelated to one another.

Summary

The first steps in examining time-series data are to plot the data, do an autocorrelation test, and run a Fourier analysis. If data are autocorrelated, opinions differ as to the advisability of decorrelating them before testing for chaos. Some reasons for decorrelating them are that autocorrelated data:

- can give false pseudo phase space plots
- can lead to erroneous estimates of the correlation dimension and Lyapunov exponents
- are unsuitable for many statistical analyses
- might influence our choice of prediction methods.

On the other hand, decorrelated data can cause problems in detecting deterministic structures in chaotic data and in identifying nonlinearity.

Removing periodicity and trend are the two chief ways of accounting for excessive variability in the autocorrelation coefficient. Removing periodicity or trend is done by transforming the raw data with a filter. A filter (linear filter) is any mathematical technique that systematically changes an input series, such as a time series or power spectrum, into a new (output) series that has certain desired qualities. There are many types of filters. Standardization is a filter or transformation that removes periodicity; it converts data of whatever units to a common or standard scale. To standardize any value, simply subtract the mean value from the observed value, then divide by the standard deviation. All three ingredients of that formula pertain just to a "season" rather than to the entire dataset. After standardizing data, run another autocorrelation test to verify that those transformed data are not autocorrelated.

Classical decomposition is a filter that removes both periodicity and trend. It's the traditional way of breaking down a time series into trend, seasonality, cyclicity, and noise. As with standardization, calculations are done with seasonal data. A major tool in the method is the moving average—a transformed time series in which each value in the original series is replaced by the arithmetic mean of that value and a specified, constant number of adjacent values that come before and after it. The general steps in classical decomposition are:

1. smooth the raw time series with a moving average
2. fit a representative (best-fit) line to any trend and use this line to get a value for "trend" for each season
3. remove the trend component for each season to get "cyclicity"
4. calculate the combined total of seasonality + noise, for each season
5. compute seasonality as the average value of (y – moving average) for each season
6. from (seasonality + noise), subtract seasonality to get noise
7. run an autocorrelation test on the noise to make sure that the noise is essentially free of autocorrelation.

Differencing is an alternate and easy way to remove both periodicity and trend. Differencing is a transformation of raw time-series data into successive (lagged) differences or intervals. First-differencing with a lag of one eliminates trend. First-differencing in which the lag equals the duration of each period (whatever its length) eliminates periodicity (seasonality or cyclicity). It also eliminates trend, in the process. Always run an autocorrelation test on differenced data to make sure no significant autocorrelation remains. If some does remain, a chaos analysis can be done on values separated by a lag large enough to ensure no autocorrelation between the values. However, alternative procedures are either to first difference the raw data using a lag equal to the longest periodicity present in the data or do an additional differencing (a second-differencing) of the filtered (first-differenced) data.

PART III

HOW TO GET THERE FROM HERE

Chapter 10
The parameter as king

Now that our toolkit is in good order, we're ready to tackle some of the more important aspects of chaos. A simple approach using a stripped-down mathematical equation shows many principles of chaos. One of those important principles is this: even with the basic ingredients (a nonlinear, dynamical system that evolves in time), chaos (at least when developed from a mathematical equation) can come about only with the consent of a key player—the controlling parameter. Let's see why.

The logistic equation

The most popular equation for studying chaos is the so-called **logistic equation**. Devised in 1845, it's a one-dimensional feedback system designed to model the long-term change in a species population (May 1976, Briggs & Peat 1989). The population is assumed to change at discrete time intervals, rather than continuously. Typical time intervals are a year or the time from one breeding season to the next.

A crude model of a change in population size might say that the population during any one year (x_{t+1}) is some fixed proportion of the previous year's population, x_t. In symbols, that's $x_{t+1} = kx_t$, where k is a constant that reflects the net birth or death rate. (In general, k is an environmental or **control parameter**.) Here k can take on any realistic value, such as 0.5, 1 or 1.87.

The relation is more useful with normalized data, ranging from 0 (the minimum possible population) to 1 (the normalized maximum possible population). To normalize logistic-equation data, divide all original values by some theoretical maximum population (say, the largest population that the region could ever support). The quotients then are decimal proportions of that maximum population.

As long as k is greater than 1, the equation $x_{t+1} = kx_t$ unfortunately predicts that, in due course, the species would inherit the Earth, since the creatures would go on multiplying forever. A more realistic variation, for reasons explained below, comes from multiplying the right-hand side by the term $1-x_t$. That produces what is now known as the logistic equation:

$$x_{t+1} = kx_t(1-x_t). \tag{10.1}$$

The word **logistic** has many meanings. One is "provision of personnel and material" (as in logistics, the military meaning). Another is "skilled in computation." In our case (Eq. 10.1), "logistic" has a mathematical meaning and refers to a particular type of so-called growth curve (an expression that specifies how the size of a population varies with time). The equation is ideal as an example in chaos because (a) it's simple (is short and deals with just one quantity, x, and (b) it demonstrates several key concepts of chaos, as we'll see.

Equation 10.1 often appears with a left-pointing arrow (←) in place of the equals sign, to indicate that the entire right-hand side of the equation becomes, goes to, or produces x_{t+1} (the new x).

Taking a small value for x_t (say slightly above zero) leaves the quantity $1-x_t$ close to 1. The entire right-hand side of Equation 10.1 (equal to the computed x_{t+1}) then becomes almost equal to kx_t. Thus, the population increases as x_t increases, although not in direct proportion to x_t. In other words, population is apt to grow when x_t is close to zero. Conversely, at relatively large values of x_t (say a little less than the maximum possible value of 1), the quantity $1-x_t$ becomes small (approaching 0). The right-hand side of the equation then becomes small. In other words, growth is small. Including $1-x_t$ in the equation therefore makes predictions much more realistic. Assuming environmental conditions are suitable, growth increases when populations are low. However, the equation dictates an optimum population beyond which densities become so high that they inhibit further successful reproduction and in fact even promote a decrease in population.

The logistic equation plots as a parabola or humpbacked curve (Fig. 10.1). With

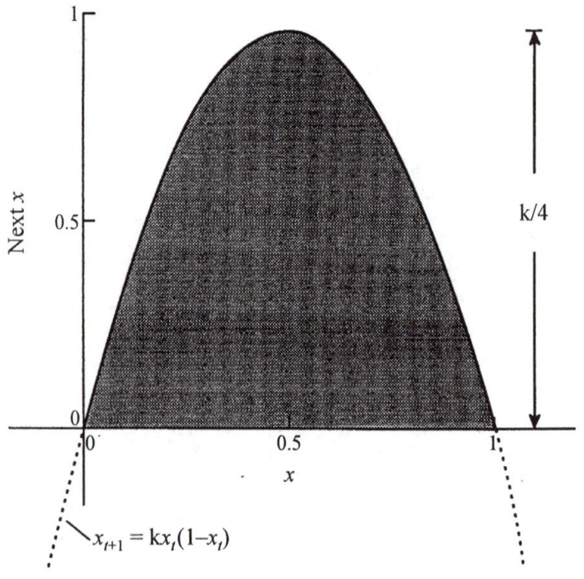

Figure 10.1 One-dimensional map of logistic equation.

x (or x_{t+1}) on the ordinate and the previous x (or x_t) on the abscissa (a one-dimensional map), a computed point on the parabola falls on or above the abscissa only when x_t has values between 0 and 1. For x_t values beyond the range of 0 to 1 the parabola plots below the abscissa, implying that the iterated population (x_{t+1}) is less than zero (becomes extinct). Thus, x_t (and hence also x_{t+1}) for this model range only from 0 to 1. That's another justification for normalizing the values of x_t and x_{t+1}.

The constant k governs the steepness and height of the parabola above the abscissa. The larger k becomes, the higher the parabola's peak (i.e. the larger the population). Mathematically, the peak's height equals k/4, in units of x_{t+1}. Therefore, because x_{t+1} only ranges from 0 to 1, k (for all of our applications of the logistic equation) can only range from 0 to 4.0. If k exceeds 4, iterations of Equation 10.1 produce values of x_{t+1} that aren't possible (either <0 or >1). The limitations on x_t, x_{t+1} and k mean that, for our population model, the plotted parabola rises smoothly from the origin (at 0,0) to a peak or maximum population at $x_t = 0.5$ and $x_{t+1} = k/4$. As x_t increases beyond 0.5, the curve falls back (population recedes) to $x_{t+1} = 0$ at $x_t = 1$ (Fig. 10.1).

Now we're going to assign some trial values to k and iterate the equation. We'll say that those iterations, with k constant, will simulate the evolution of a dynamical system over time. For instance, it might show how a population might change with time. Secondly, and here is the critical part, we'll repeat that procedure for different values of k so as to examine the effect of k on the predicted variability of population with time.

Low parameter values

First, let's set k = 0.95 (keeping in mind that $0<k<4$). Equation 10.1 then becomes $x_{t+1} = 0.95x_t(1-x_t)$. Let's also assume that the population is already at a starting value of $x_0 = 0.4$ (meaning 40 per cent of its maximum possible population, since $0<x<1$). Inserting the arbitrary starting value of $x_t = 0.4$ into the logistic equation, the first solution of the equation yields $x_{t+1} = 0.95(0.4)(1-0.4) = 0.228$. For the second iteration (perhaps the population the following year), we feed in 0.228 (the value computed during the first iteration) as the new x_t. The new computed x_{t+1} (with k remaining at 0.95) then is 0.167. Subsequent values for x_{t+1} (iterates) turn out to be 0.132, 0.109, 0.092, 0.080, and so on. In fact, the values **converge**[1] to zero, suggesting that the population becomes extinct. Figure 10.2a shows the time series and the associated phase space plot (or "one-dimensional map," in chaos jargon). (By far the best way to gain an appreciation for such a curious evolution is to do the iterations yourself, right now. A hand-held calculator, especially a programmable one, does the job.)

Similar calculations show that with any other x_0 and k = 0.95, the computed values still converge to zero. In a sense, that point (zero) "attracts" all computed points

1. "Converge" means to approach a finite limiting value.

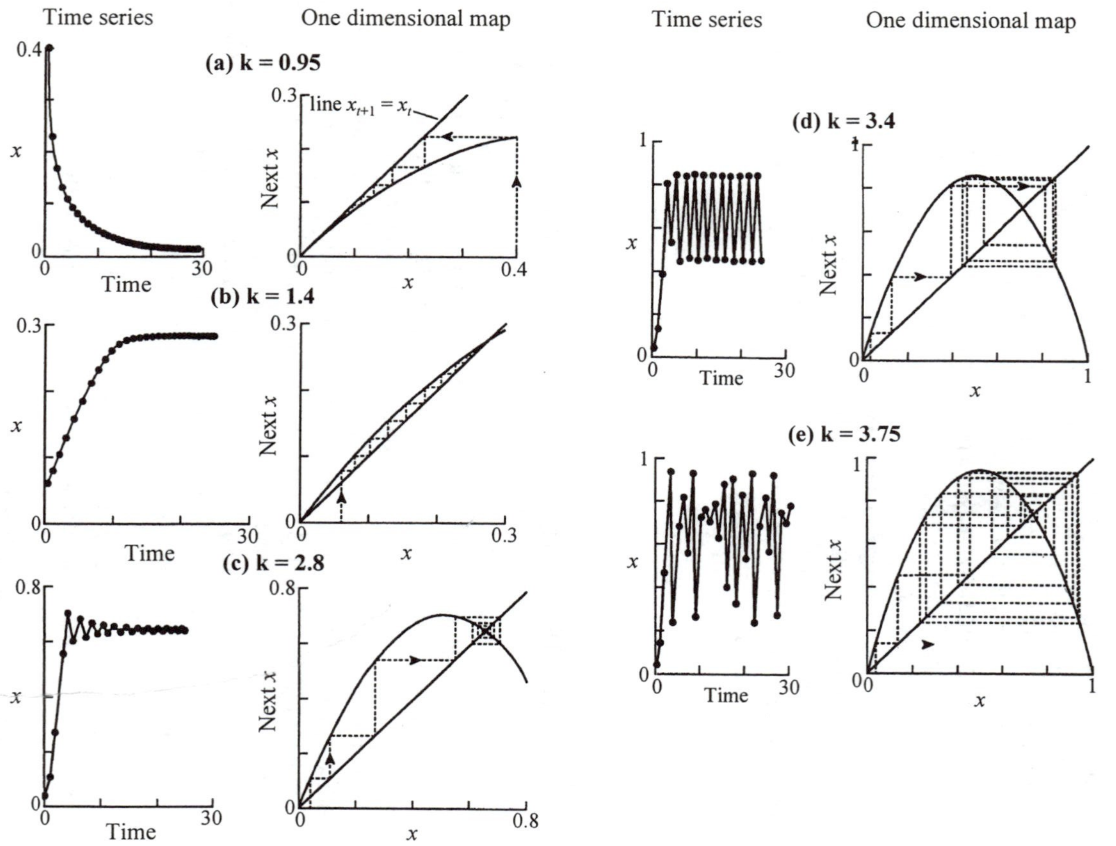

Figure 10.2 Time series and one-dimensional maps of logistic equation, for five values of control parameter k.

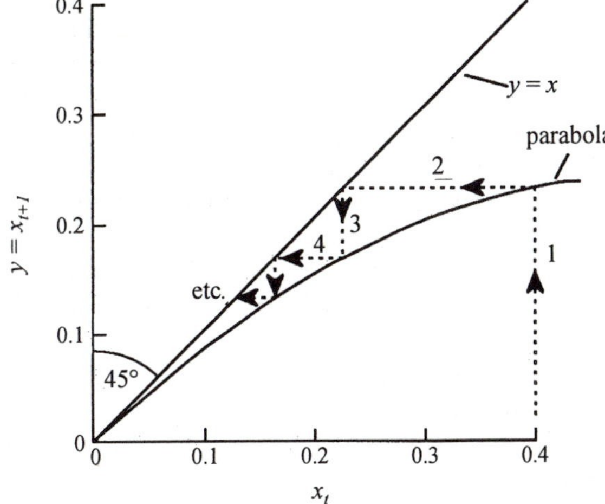

Figure 10.3
Graphical iteration. Numbers on graph are order of steps.

(Hofstadter 1981, 1985), when k = 0.95. It's therefore called an **attractor**, hereafter symbolized by x^*. An attractor is the phase space point or points that, over the course of time (iterations), attract all trajectories emanating from some range of starting conditions (the "basin" or "**basin of attraction**"). In this particular example, the attractor (zero) consists of just one point. That point is therefore called a **fixed point, fixed-point attractor**, or **final state**. Inserting $x_t = 0$ and computing x_{t+1} gives $x_{t+1} = 0$. In other words, a fixed point is a value that is its own iterate.

Trials show that all iterations at constant k for k < 1 decay to zero, regardless of x_0. On a practical basis, a fixed point of zero could suggest that disease, unsuitable temperature, scarcity of food, or some other barrier prevented the population from surviving, as reflected in the value of k.

Graphical iteration

For a given value of k, any solution to Equation 10.1 falls somewhere on the general curve (parabola) that the equation describes. That is, the plotted curve is a graph of all possible iterates. Therefore, knowing the curve and the starting value x_0, we can do all the iterations graphically (although a bit crudely), without using a calculator at all. Figure 10.3 shows how. First draw two lines on the graph, namely the parabola and a straight 45° line[2] representing the relation $y = x$ (or $x_{t+1} = x$). Now we'll go through the iteration procedure with the same values just used (k = 0.95, $x_0 = 0.4$).

Beginning at $x_t = x_0 = 0.4$ on Figure 10.3, the parabola indicates the value of x_{t+1} produced by solving the logistic equation. So, at $x_t = 0.4$, simply go upward on the

2. The diagonal 45° line is sometimes called the **identity line**.

graph to the intersection with the parabola (labeled as step 1 or simply "1" in Fig. 10.3). That intersection is at $x_{t+1} = 0.228$ (the first iterate). The next iteration begins by setting 0.228 equal to the new x_t. Therefore, go horizontally to the line $y = x$ (= 0.228) (step 2 in Fig. 10.3). At that location, $x_t = 0.228$. For $x_t = 0.228$, the parabolic curve again gives the new x_{t+1}. Hence, keep x_t at 0.228 by going vertically (downward in this case) to meet the parabola (step 3). That meeting is at $x_{t+1} = 0.167$ (the second iterate). Then plug in 0.167 as the new x_t. That means moving the graphical route horizontally to the line $y = x$ again, and so on.

Summarizing, the general procedure for drawing the iteration route is first to go along the abscissa to the starting (initial) x_t (population at a given time). Then repeat the following two steps:

1. Move vertically (up or down) to the parabola. (That gives x_{t+1}, the population predicted for the new time, for that particular x_t.)
2. Move horizontally (right or left) to the diagonal line $y = x$ (the starting population or x_t for the next iteration).

In other words, it's simply a matter of parabola to diagonal to parabola to diagonal, and so on.

Intermediate parameter values

Now back to our calculator to explore further the influence of k. We'll repeat the iteration procedure with a different k value, say k = 1.4. Let's arbitrarily take x_0 = 0.06 and do a series of iterations. The values converge by successively smaller increments to a single point (a fixed point or final state) located at $x^* \approx 0.286$ (Fig. 10.2b). (Again, the best way to get a feel for this is to work through some calculations on a desk calculator.) It also turns out that, for our chosen value of k, the single point to which all computed iterations converge is the same (0.286), regardless of x_0, as long as x_0 isn't zero. So we have another fixed-point attractor.

Continuing our experiment of varying the k value, let's take k = 2.8. Again choosing any value between 0 and 1 for x_0 and iterating the equation, the sequentially computed points seek a new fixed-point attractor, now at $x^* = x_{t+1} \approx 0.643$ (Fig. 10.2c).

Figure 10.2a–c, along with similar experiments for k values in the range of 0 to almost 3, show several interesting features:

- Each different x_0 gives birth to its own trajectory, but all trajectories *for a given* k *value* lead to the same point attractor. (The practical implication is that, for any k value within the range 0<k<3, the same eventual population materializes, no matter what the starting population was.)
- The attractor is zero for any k<1. As k exceeds 1 and increases to almost 3, the attractor increases from zero to about 0.667.
- Either of two shortcuts immediately gives the value of the point attractor x^*, without iterating the equation.
  - The attractor can be computed directly. It equals (k–1)/k or, dividing that

expression by k, 1 − (1/k). (Negative answers mean that the attractor is zero, as with the example of k = 0.95 above.) For instance, at k = 1.4 we found after many iterations that the fixed-point attractor equals 0.286; that same value results from 1 − (1/k) = 1 − (1/1.4) = 0.286.

- The attractor can be found graphically as the value of x_t corresponding to the intersection of the parabola and the diagonal line of $y=x$. Again, for the logistic equation these two quick methods of determining the attractor are valid only for k values less than 3.

- For $0<k<3$ the fixed-point attractor resides in successively different and predictable phase space regions relative to the parabola's peak. Specifically, on one-dimensional maps as in Figure 10.2, the attractor is left of the parabola's peak when $k<2$ (diagrams a and b). It's at the peak when k = 2 (not shown). It's to the right of the peak when $k>2$ (diagrams c–e).
- When $k<2$, the trajectory converges to the attractor by drawing nearer from one side only (Fig. 10.2a,b). In contrast, when $k>2$, the trajectory converges to the attractor by first getting into the general neighborhood and then bouncing back and forth on opposite sides of the attractor, while drawing ever nearer (Fig. 10.2c–e).

Periodic points

Onward now to higher k values. When k is 3.0 or greater, a surprising thing happens: the trajectory no longer converges to a single x^* value. The trajectory's behavior now becomes more and more sensitive to the value of k. For example, if k = 3.4, then regardless of our chosen x_0, the iterations converge to not one but two x^* values alternately, namely at $x_1^* \approx 0.452$ and $x_2^* \approx 0.842$. Computed values systematically approach the two x^* values and then alternate back and forth between just those two points, in a regular oscillation (Fig. 10.2d). This implies that one year the population has the relatively low value of 0.452, the next year it has the relatively high value of 0.842, and that it keeps fluctuating between those two values from year to year.

The two x^* values together now are said to constitute the attractor or final state. So, the control parameter has now brought about a very different and slightly more complex type of attractor—one that's no longer just a single point. At k = 3.4, for instance, it produces a "two-point attractor," that is, an attractor or final state consisting of two **periodic points**.

The critical k value that marks the transition from one x^* value to two, as mentioned, is k = 3.0. As k increases beyond 3.0, the two x^* values that make up the attractor (the two steady population sizes) change. That is, they are different for each k. Moreover, they change such that they get further apart as k increases. Considering the two values as a cycle, the cycle gets bigger as k gets progressively larger than 3.

Continuing the experiment with the same simple equation (Eq. 10.1), let's increase k to 3.5 and again start with any x_0. Another surprise: four periodic points or fixed x^* values (a four-point attractor) now come forth from the iterations, namely at $x^* \approx 0.875$, 0.383, 0.827, and 0.501. Weird! Fascinating! Those particular values mean that the bug population has a four-year cycle, attaining 87.5 per cent of its maximum possible population one year, 38.3 per cent the next, and so on. Trial and error shows that the transition from two to four x^* values takes place at k ≈ 3.4495...

Methodically increasing the value of k still further, using any arbitrarily selected value for x_0, shows that the attractor continues to have four periodic points until k reaches approximately 3.5441, at which point eight x^* values suddenly pop up. When k reaches about 3.5644, 16 values abruptly appear. Further computations with slightly larger k values reveal sudden jumps to 32, 64, 128 periodic points, and so on.

Once the attractor consists of two or more values, there's no longer any shorthand method (such as [k–1]/k) for computing those values. Instead, we've got to plow through the iterations.

The number of x^* values or iterates in a cycle is the **period** of the attractor. (A period is the amount of time or number of iterations needed for a system to return to its original state.) For example, an attractor consisting of two repeating x^* values (a two-point attractor) is an attractor of period two. We've calculated attractors of period one, two, four, eight, and so on. The x^* values in a new period come about by a splitting of each attractor value of the previous period into two new attractor values. The result is that each new domain of k has twice as many x^* values as the preceding domain. Such a bifurcation is known as **period-doubling** (sometimes also called **flip bifurcation** or **subharmonic bifurcation**). The entire series has formed a cascade—an infinite sequence of period-doubled periodic orbits.

A **bifurcation** is a sudden qualitative change in a system's behavior, occurring at a fixed (critical) value of a control parameter. The change can be subtle (small change in the system's behavior) or catastrophic (large change in behavior). Similarly, a **bifurcation point** is a parameter value at which a bifurcation occurs. In the particular cases just described, bifurcation involved a division into two parts or branches. (Owing to certain mathematical characteristics, it's a type of bifurcation known as **pitchfork bifurcation**.) In a more general mathematical sense, the "bi" in bifurcation may not necessarily imply two. That is, more than two branches of new behavior may appear. Some authors stick to the "two branches" interpretation; others prefer the more general interpretation. Thus, you might see "bifurcation point" defined as a parameter value from which precisely two branches emerge (Seydel 1988: 51) or several branches emerge (Bergé et al. 1984: 271). Bifurcation theory goes back at least to Poincaré and is a separate and important class of mathematics.

Moving into chaos

At k values of around 3.55 and 3.56 the periods begin doubling after particularly small increases of k, such as increases in the fifth or sixth or more decimal place. That is, the range of stability for any periodicity becomes drastically shorter.[3] At a k value of about 3.57 the number of periods (the number of points that make up the attractor) becomes infinitely large. The range from k ≈ 3.57 to 4.0 has chaos. It's a region where a trajectory's route for some k values looks erratic, with no apparent order (Fig. 10.2e). Other k values within that range produce windows of apparent stability, as discussed later.

In summary, for the logistic equation and its period-doubling route to chaos, the parameter k has four distinct ranges or cases, as follows.

- At k≤1, the attractor is a fixed point that has a value of 0.
- At 1<k<3, the attractor is a fixed point that has a value greater than 0 but less than about 0.667.
- At k≥3 but less than about 3.57, period-doubling occurs. The attractor consists of 2, 4, 8, etc. periodic points, as k increases within that range.
- At k greater than about 3.57 but ≤4, we're in the region generally known as chaos. An attractor here can be erratic (chaotic, with infinitely many points) or **stable**.

The control parameter therefore plays a dictatorial or decisive role in a system's evolution and in the onset of chaos (given that the system is deterministic, nonlinear, and dynamic). A variable can undergo an unbelievable variety of possible evolutions, depending on the value of the control parameter. For certain values within a particular range of the control parameter, chaos is *unavoidable* when iterating the logistic equation and certain other equations (Ch. 13). We don't have a choice, at those values. Furthermore, the overall range of the parameter has certain critical levels. At each critical level, the system changes its behavior, suddenly and drastically. With increase in control parameter, the trajectory at each new level is more complex and less orderly.

Let's take a quick look at a second example, again involving period-doubling. A more general version of Equation 1.1 is:

$$x_{t+1} = \mathrm{k} - x_t^2. \qquad (10.2)$$

Some authors call this equation the **quadratic map**—map for reasons already explained, and **quadratic** because it squares the value of x_t (and doesn't contain any terms raised to a power greater than two). It's really a special case of a general quadratic equation. As such, it's a cousin of the logistic equation. Values of x_t for Equation 10.2 as applied to chaos generally range from about −2 to +2 (Grebogi

3. "Stable" means that, with any perturbation, the system eventually returns toward the attractor. A "**perturbation**" in chaos jargon is any displacement in a trajectory, or any difference between two neighboring trajectories or observations, at any given time.

et al. 1983, 1987). For k less than –0.25 and greater than 2.0, all trajectories become asymptotic to $-\infty$, where ∞ is infinity. Between those two k values, however, iterations of the equation produce attractors, period-doubling, and chaos. For k ranging from –0.25 to 0.75, the attractor consists of just one value. At k = 0.75, the attractor becomes period two; at k = 1.25, it becomes period four; and so on. Chaos sets in at about k = 1.4. The chaotic domain therefore ranges from k = 1.4 to k = 2. Again, to really get a feel for the route to chaos and for chaos itself, take out your pocket calculator and go through those quick and easy computations yourself.

The entire transition and ensuing chaos in both of the above examples were entirely self-generated from within the system, with no external forces. Researchers want to find out whether that happens in nature. For instance, Lorenz (1976) studied how climate might change, even when environmental effects remain constant. Many avenues still need to be explored.

The two equations we've looked at (Eqs 10.1, 10.2) and corresponding chaos reveal several interesting generalities.

- Systems governed by physical laws or by deterministic equations can produce regular results under some conditions but irregular or disorderly results under others.
- A random-like or even chaotic evolution doesn't have to be the result of a random operation. Instead, it can arise by design.
- Chaos can result from a simple equation (involving, for example, just one variable); it doesn't require a complex origin or many variables.
- Deterministically based random-like patterns generated with an equation can look identical to those measured for natural phenomena.
- Period-doubling followed by irregular fluctuations in some cases indicates that those fluctuations are chaotic (Olsen & Degn 1985: 194; Glass & Mackey 1988: 49).

Descriptions of chaos as "random-like behavior" are mostly justified. Chaotic time series not only look uncorrelated or unsystematic, they often pass every statistical test for randomness (Stewart 1984). To that extent, therefore, chaotic data are both random and deterministic.

Like random data, we can describe chaotic data by their average or statistical properties. Therefore, where reliable long-term predictions are impossible, a statistical approach may be the only viable alternative.

Terminology

Let's take a brief timeout here for further definitions or amplifications of some common terms or expressions:

- A **two-dimensional map** is a pair of iterated equations involving two physical features (e.g. x and y), such that x_{t+1} is some function of both x_t and y_t, and y_{t+1} is some other function of x_t and y_t.

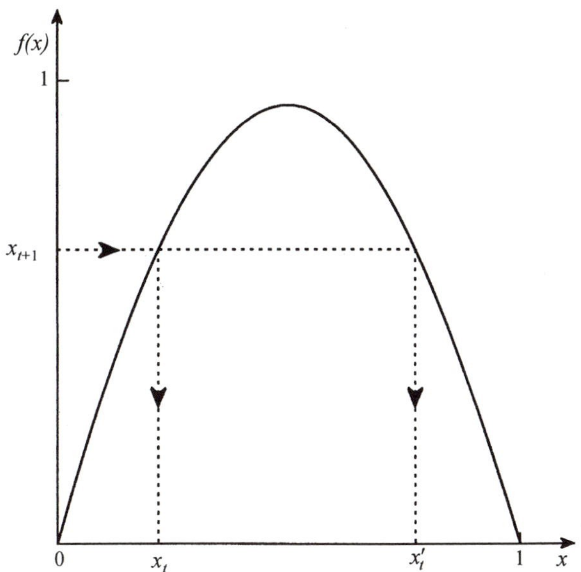

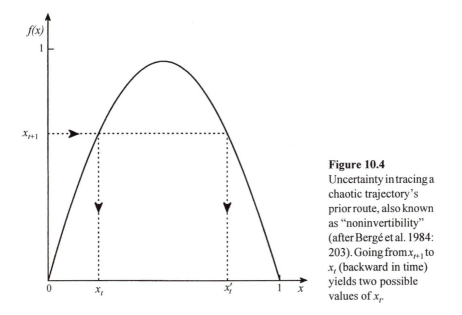

Figure 10.4
Uncertainty in tracing a chaotic trajectory's prior route, also known as "noninvertibility" (after Bergé et al. 1984: 203). Going from x_{t+1} to x_t (backward in time) yields two possible values of x_t.

- An **interval** is a set of real numbers that fall between two end-member real numbers.
- "On the interval" in chaos literature refers to the interval between any two values. Usually that range or interval is from $x_t = 0$ to $x_t = 1$, sometimes called the "unit interval."
- "Map of the interval onto itself" is an expression for the output of any equation that generates x_{t+1} from x_t for a certain interval. That is, the equation or map is solved by iterating x_t, and the results apply to a certain specified range or interval of x_t.
- **Invertible** is a term applied to a one-dimensional equation that projects forward or backward in time, uniquely. In other words, we can solve an invertible equation uniquely for either of x_t or x_{t+1}, given the other.
- **Noninvertible** is a term describing an iterated equation that goes forward in time but not uniquely backward in time. In other words, the equation yields just one value of x_{t+1} from any x_t (i.e. going forward in time) but yields more than one value of x_t from a given value of x_{t+1} (i.e. going backward in time) (Fig. 10.4).
- A *trajectory* is a mapping—a sequence of numbers calculated from a map. The trajectory of any point is the list of successive iterates (the sequence of computed values of x_{t+1}). The trajectory therefore consists of certain points that are located "on the map." In a sense, a trajectory is that selected range of the map that we choose to calculate.
- "**Stretching**" has slightly different meanings, depending on whether we're looking at one trajectory or comparing two neighboring trajectories:

- For one trajectory, stretching is the process whereby a given range of input values emerges from iteration as a larger range of output values. For example, iterating the logistic equation with k = 2.8 produces a parabola whose peak is at $x_t = 0.5$ and $x_{t+1} = 0.7$ (Fig. 10.5a). Look at just the left half of that bilaterally symmetrical parabola (where x_t goes from 0 to 0.5). Input values (along the horizontal or x_t axis) from 0 to 0.5 (a range of 0.5 units) emerge as output values (along the vertical or x_{t+1} axis) of 0 to 0.7 (a range of 0.7 units). Thus, the original range of 0.5 units expanded or stretched to an output range of 0.7 units (Fig. 10.5b). In the same way, the right half of the parabola shows that input (abscissa) values of 0.5 to 1 (which again is a range of 0.5 units) get iterated to an output (ordinate) range of 0.7 to 0 (i.e. 0.7 units). As before, the input range is 0.5, and the output range is 0.7 (again, Fig. 10.5b). Hence, iteration stretches the original (input) range.
- For two neighboring trajectories, stretching is the process whereby the trajectories separate over time, in phase space. This is the more common notion of stretching and is a local phenomenon (usually lasting over just a small zone on the attractor).

- "**Folding**" is a radical change in a trajectory's direction. It often takes place

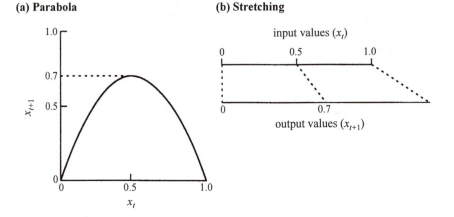

(a) Parabola

1.0
0.7
x_{t+1} 0.5
0
0 0.5 1.0
x_t

(b) Stretching

input values (x_t)
0 0.5 1.0
0 0.7
output values (x_{t+1})

(c) Folding

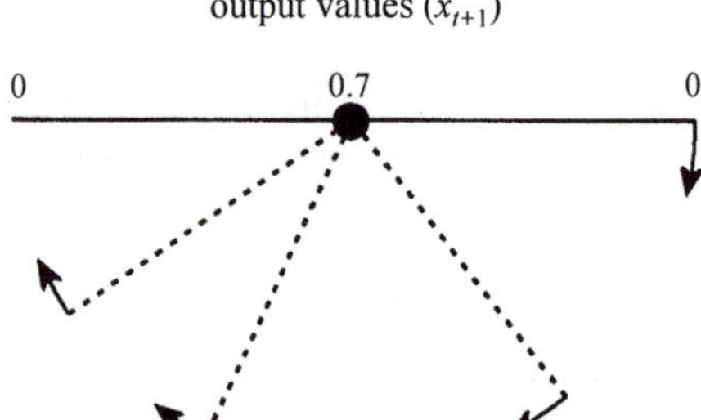

Figure 10.5 Stretching and folding with the logistic equation (k = 2.8).

> at the edge of the attractor as viewed in phase space, that is, around the limiting values of the phase space variables. In more pronounced cases, the trajectory nearly reverses itself and comes back very near the route it just traversed. (In chaos jargon, the trajectory in that case gets "mapped onto itself." Or, the points, after getting "mapped" from one phase space region to another, end up getting mapped back to their original region.) A finite space (not part of the attractor) always remains between the nearby paths. The result is a layered structure to the attractor. The effects of such a fold can last over much of the attractor, so folding can be global as well as local. It's often likened to the mixing of dough or other foods (hence a kind of mathematical food mixer).

Figure 10.5c shows one way to visualize folding, again using the logistic equation with k = 2.8. As we just saw (Fig. 10.5a), the same range of output values (between 0 and 0.7) applies to two input ranges—one where x_t is within 0 to 0.5 and another where x_t is within 0.5 to 1.0. Think of the stretched-out iterates (values of x_{t+1}) as a straight-line continuum of numbers that start at 0, go through 0.7 at the midpoint and then decrease back to 0 at the far end (Fig. 10.5c). The second half of the string of numbers therefore is a mirror-image of the first half. Then imagine a hinge at the midpoint (0.7), so that the second half of the continuum can figuratively swing back or fold back to merge with the first half (Fig. 10.5c). Peitgen et al. (1992: 536–48) give more extensive discussions of stretching and folding.

Summary

Partly because of its simplicity, the logistic equation is the most popular way to show many key features of chaos. It's especially useful for illustrating the importance of the control parameter. Depending on that parameter's value, iterations can lead to a fixed point (zero or some positive number), period-doubling, chaos, and to many other features we haven't yet explored. The control parameter therefore acts like a sort of tuning knob, governing the system's type of behavior. As the parameter increases from 0, the entire road to chaos has critical levels of the control parameter. (We'll analyze those critical levels in more detail in Ch. 12.) Common terms used in connection with the processes we've discussed are mapping, interval, map of the interval onto itself, invertible, noninvertible, trajectory (orbit), stretching, and folding.

Chapter 11
Nonchaotic attractors

About attractors

Possibly the most fundamental concept in chaos analysis is that of an attractor. An attractor is a dynamical system's set of stable conditions. A good way to see a concise summary of those conditions is with a standard phase space plot. There, an attractor shows a system's long-term behavior. It's a compact, global picture of all of a system's various possible steady states. (A **steady state** is a state that doesn't change with time, or the state toward which the system's behavior becomes asymptotic as time goes to infinity.) In short, the attractor is the system's identification (ID) card.

The finite volume of phase space that the attractor occupies usually is quite small relative to the volume of the phase space itself. You might liken an attractor to the relatively small zone that your cat gets to as it romps around in your living room. Much of the room, especially the higher part in this example, is a zone the cat never visits. Out of the many numerically possible combinations of the variables of the phase space, only relatively few combinations ever really happen.

Another example of an attractor is Pool's (1989e) analogy of temperature and wind-velocity variability at any place, such as the control tower of Los Angeles International Airport. A standard phase space graph shows how temperature and wind velocity change with time at that location. Each plotted point is the temperature and velocity value at a particular time. After enough samplings (perhaps over some years), a pattern emerges, showing the various possible combinations of values that occur. Those possible values as a group make up the attractor. On the two-dimensional phase space graph, they occupy only a certain select zone (a subspace). Some pairs of values (e.g. the point for 70°F and 10 miles per hour) are on the attractor, whereas others (say 135°F and 200 miles per hour) aren't. (At least, the local residents hope not.) A more complicated phase space consists of three, four, or more variables (dimensions), such as temperature, wind velocity, humidity, and air pressure.

Trajectories might originate anywhere within the attractor's basin of attraction. They gravitate toward the attractor over time. In other words, the conditions we might specify initially (e.g. the value x_0 at which we might start iterating an equation) don't influence the attractor; it exists independently of initial conditions. Furthermore, because an attractor is a steady-state or equilibrium condition, it's stable.

The attractor or model in one sense is just an abstract mathematical object. However, we can draw it on paper, and it often makes a fascinating picture. A phase space can harbor many attractors, rather than just one. Their respective basins of attraction can overlap or intermingle, in the phase space.

Some attractors (namely fixed-point attractors) are as simple as a single dot. Others are complex geometrical objects. To interpret the curious shapes of attractors, mathematicians often look upon them as distorted versions of a basic line, surface, or volume. The idea is to imagine that the object is flexible or malleable. That means we can stretch, squeeze, twist, bend, or fold (but not tear) it into a phase space figure (attractor). It's like pulling taffy (toffee) or playing with bread dough. For example, we can deform a circle into an oval or into the highly irregular shape of an island coastline. We can mash a tennis ball into the shape of a watermelon, or fashion a blob of dough into a cupcake, cinnamon roll, or loaf of bread. In a special sense, mathematicians consider the original line, surface, or volume (such as the blob of dough) to be unchanged or to retain its identity, even though it's evolved to a different form and maybe also a different size. This general field of study is called **topology**—the study of geometrical figures that are unaltered by imposed, continuous deformations. Topology's basic premise is that any two geometrical shapes are equivalent (topologically similar) if we can smoothly deform one into the other.

Here are a couple of curious features about attractors (Schaffer & Kot 1985). First, if the chosen starting point x_0 already lies on the attractor, then the trajectory stays there forever (never leaves the attractor). Some of the above examples of Equation 10.1, $x_{t+1} = \mathrm{k}x_t(1-x_t)$, show that feature. For instance, where the attractor was zero and $x_0 = 0$, then, regardless of k, the computed $x_{t+1} = 0$, and all successive iterations yield 0. Another example was an attractor of $x^* \approx 0.643$ for $\mathrm{k} = 2.8$. Using 0.643 as the starting point of the iterations yields 0.643 for the second iteration, 0.643 for the third, and so on. All iterations keep producing $x_t = x^* = 0.643$; the trajectory never leaves the attractor.

The second strange feature is that a trajectory never gets completely and exactly all the way onto an attractor. Instead, it only approaches it asymptotically.[1] This asymptotic feature perhaps is more common with mathematical models. It means that iterations get closer and closer, by smaller and smaller amounts, but they never get all the way there. For example, the logistic equation with $\mathrm{k} = 2.8$ has an attractor of *about* 0.643. Actually, on my calculator the 15th iteration is 0.638, the 25th is 0.6424, the 35th is 0.64281, the 45th is 0.642851, the 55th is 0.642856, the 65th is 0.64285708, and the 90th is 0.64285714. The successive iterates keep changing by smaller and smaller amounts, but they keep changing nonetheless. Of course, for practical purposes (maybe two or three decimal places) the trajectory has

1. There are two exceptions. One is systems that come to rest with time. The other is when the trajectory starts out right on the attractor.

reached the attractor. However, the decision as to when the iterates are changing by small enough increments to enable us to say that the trajectory is "on the attractor" is strictly arbitrary on our part. Because of that asymptotic approach, Bergé et al. (1984), speaking as mathematicians, define an attractor as an asymptotic limit (as the number of iterations goes to infinity) of the solutions to an iterated equation.

Experimental data always carry some error or background noise. Once a trajectory gets close enough to an attractor to be within the characteristic scale of the noise, we can't track the trajectory any longer. Thus, for deterministic equations as well as physical systems, the expression "on the attractor" is a bit loose.

Here's an exciting feature about attractors: looking at all disciplines (physics, economics, biology, etc.) as a group, the attractors fall into just a few basic types. That is, systems from many different fields often behave in the same way, insofar as their attractors are concerned. In other words, as Schaffer & Kot (1985) put it, there's a "taxonomy of motion" within dynamical systems.

All attractors are either nonchaotic or chaotic. Nonchaotic attractors are of three types: **point**, **periodic**, and **toroidal attractors** (in order of increasing complexity and number of dimensions involved). Nonchaotic attractors consist of regular, predictable trajectories. That is, predictions of long-term evolution or recurrent behavior can be quite accurate, even far into the future. Two neighboring trajectories, once "on" their attractor, stay a calculably close distance from one another (or are identical, in the case of point attractors) as they evolve.

Chaotic or strange attractors, discussed in Chapter 15, arise only after the onset of chaos. They take on many interesting and complex shapes in phase space. There's no way to predict long-term evolution on those attractors with any reliability.

Curiously, periodic attractors turn up not only in the nonchaotic domain but also under chaos, as we'll see. There aren't any point or toroidal attractors in the chaotic domain. Now let's look more closely at nonchaotic attractors.

Point attractors

The simplest type of attractor, the point attractor, is a single fixed point in phase space. It represents all systems that come to rest with the passage of time (Crutchfield et al. 1986) or that progress to a state where they no longer vary with time. Examples are a bouncing ball that comes to rest, a pendulum (that either finally stops because of friction or that forever swings at the same amplitude), a marble rolling in a bowl and coming to rest at the bottom, and an airplane cruising at a steady velocity. Once that steady-state condition arrives, the point attractor is independent of time. It "stays fixed," and the system no longer evolves. In phase space, the system is static.

The fixed state of a point attractor isn't as interesting as how its phase space trajectory gets there. One way in which a trajectory might approach the point is directly. Examples mentioned above in discussing the logistic equation are the value of $x^* = 0$ for k = 0.95 and $x^* = 0.286$ when k = 1.4. Values of the iterates converge directly to the attractor, from one direction only.

A second way in which a trajectory can approach a point attractor in phase space is in an alternating fashion. The trajectory goes to its attractor by first getting to one side of it and then to the other. The successive iterates alternate in being larger and smaller than the attractor while systematically getting closer and closer. An example from the logistic equation is the x^* of 0.643 for k = 2.8 (Fig. 10.2c).

A third way of approaching a point attractor is spirally. An example is a pendulum with friction. Two variables—position and velocity—fully describe the pendulum's behavior (Fig. 11.1a). Just before the pendulum starts its swing at the far left end in Figure 11.1a, velocity is zero and position or displacement is a maximum. In the corresponding phase space, taking leftward displacement as negative, the point plots at a far left location on the horizontal axis. Midway through its first swing the pendulum is in a vertical orientation. At that location its position by definition is zero. Its velocity there is highest. With velocity during rightward swings plotted upward on the graph, that point plots at the very top and middle of the phase space diagram. The pendulum finishes its first swing at the far right end (bottom diagram in Fig. 11.1a). At that location it again has maximum displacement (now positive) and lowest (zero) velocity. The corresponding location on the phase space plot is at the far right end of the horizontal axis.

Now for the pendulum's return trip to the left. Here we plot the associated velocities downward instead of upward on the phase space's vertical axis, since the pendulum now is traveling in the opposite direction. Because of friction, the pendulum doesn't make it all the way back to its original starting point. The phase space plot now describes a path similar to—but shorter than—the path made for the first swing. The pendulum gradually swings through shorter and shorter arcs or damped oscillations, finally coming to rest in a vertical orientation (zero position and zero velocity) (Fig. 11.1b). The successively plotted points on the phase space graph of velocity and position describe a spiralling route that ends at the point 0,0. That point is the attractor.

We reach the same attractor (0,0 in this example) no matter where the pendulum began swinging, that is, irrespective of the starting point and of the route over which the phase space trajectory travels. That's true of point attractors in general.

Point attractors involve two variables, so the phase space plot has two dimensions (Fig. 11.1a,b). However, the attractor itself has a dimension of zero, because it's just a point. (In this sense, a point has no length, breadth, or height. Alternatively, it's a line that has zero length.)

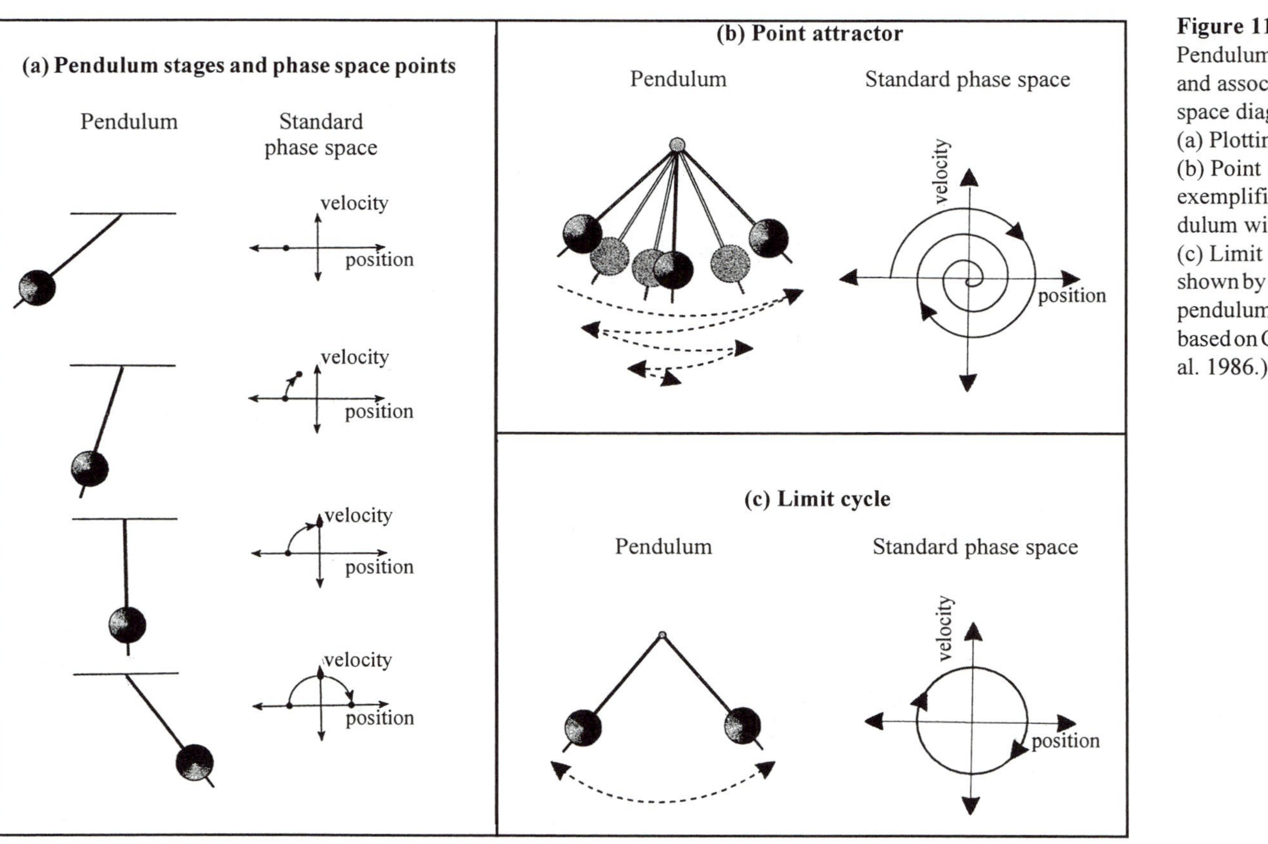

Figure 11.1 Pendulum behavior and associated phase space diagrams. (a) Plotting scheme. (b) Point attractor, as exemplified by a pendulum with friction. (c) Limit cycle, as shown by a frictionless pendulum. (b and c based on Crutchfield et al. 1986.)

Periodic attractors (limit cycles)

The next simplest attractor is the periodic attractor. It consists of two or more values that keep repeating in the same order. A good example is any attractor within the period-doubling regime of the logistic equation. With period two, for instance, the system cycles back and forth sequentially between two attractor values. The duration of a cycle or period, in terms of number of iterations or measuring occasions, is its periodicity.

Values of a periodic attractor act as a limit or goal for any trajectories that originate within the basin. For that reason, people often call the periodic attractor a **limit cycle**. A limit cycle describes a system of regular periodicity.

In the conceptual world of mathematics, the repetition or oscillation can go on forever. In the real world, the repetition usually slows down and stops after some time, unless the system receives energy from some source. Examples where a system gets a boost from such an outside source are mechanical devices that must be wound, such as clocks and metronomes.

The basic data for many limit cycles plot as a loop in standard phase space. Drawing the graph requires at least two dimensions or coordinates. An example is a mechanically boosted pendulum that swings repeatedly back and forth between the same two end points (Fig. 11.1c). The phase space again consists of velocity versus position. The pendulum now traverses the exact same route over and over again. Also, it always has the same velocity at any given location along its arc. The trajectory on the phase space diagram therefore describes a cycle that's repeated over and over.

Something else is important: even if we nudge or otherwise perturb the pendulum, it tends to return to its standard cycle. Thus, a periodic attractor is stable—it resists change or perturbations.

Stewart (1989b) cites the simple, idealized analogy of the shark–shrimp predator–prey relation (see also Abraham & Shaw 1982: §2.4). In setting the stage, Stewart says (and I hope you're ready for this) "the shrimp population is limited solely (shrimply?) by the number of shrimp . . ." Starting with lots of shrimp, the sharks thrive and increase (Fig. 11.2, far right side of graph). Under all that predation, the shrimp then begin to decline (Fig. 11.2, top part of loop). That in turn causes the number of sharks to decrease because food dwindles (Fig. 11.2, left side of loop). Finally, under the reduced predation, the number of shrimp increases (bottom part of loop). The cycle then begins again.

The normal rhythmic beating of the heart is a periodic attractor or limit cycle. The cycle is an attractor because the heart usually keeps about the same rhythm. If perturbed (changed), it returns to that rhythm as soon as it can.

Most examples mentioned above dealt with two variables or dimensions, for simplicity. Limit cycles also occur in three or more dimensions. An example might be predator–prey data for three interrelated creatures, such as the populations of herons, frogs and flies in a lake region. The phase portrait then shows a closed loop in three dimensions (Fig. 11.3).

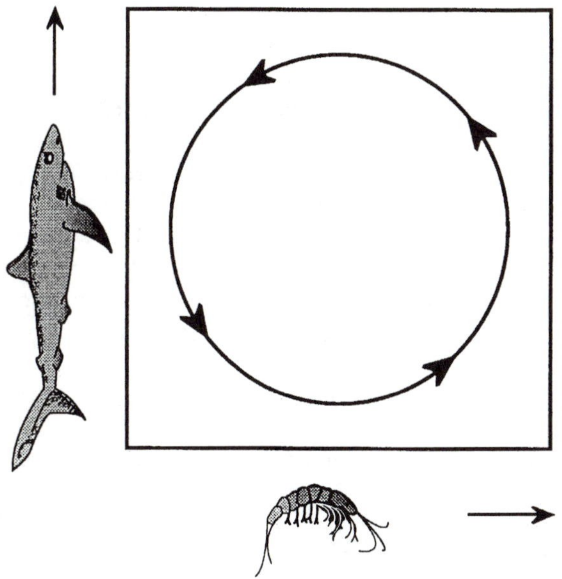

Figure 11.2 Phase space diagram showing cyclical relation between number of sharks and relative number of shrimp, over time (after Stewart 1989b).

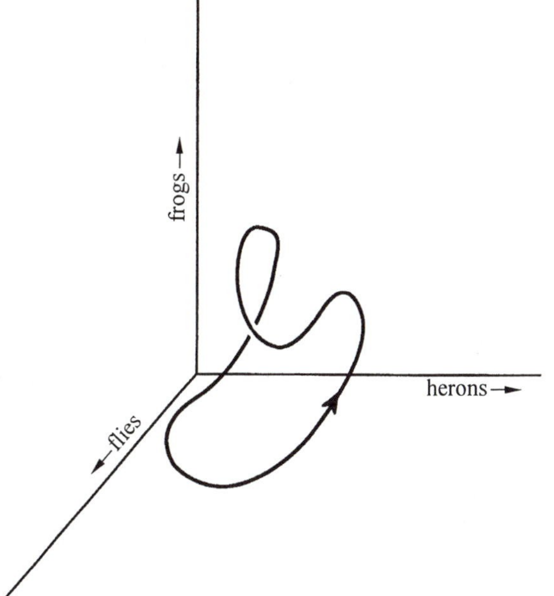

Figure 11.3 Hypothetical three-dimensional phase space plot of heron, frog, and fly populations, showing periodic attractor (after Abraham & Shaw 1982: 33).

The torus

The next step up in complexity of attractor types is an attractor that combines the action of two or more limit cycles. Each cycle stands for a separate dynamical system. I'll start the explanation by considering just two systems. They may or may not be independent of one another; however, plotting them on the same graph implies that we're considering them as one composite system. The attractor here is a phase space shape that accommodates a trajectory for each cycle or system.

If the simplest attractor is a point and the next simplest a loop, what form does this new attractor have? Let's answer that by building such an attractor from scratch, borrowing extensively from Abraham & Shaw (1982). It will consist of two limit cycles.

Each of the two interrelated limit cycles is a loop in a two-variable standard phase space. The values of a system's two coordinates at any time define the state of its system. Our goal is to figure out what the attractor looks like. In other words, we want to find a phase space geometric object or pattern that accommodates cycles or repetitions of trajectories, represents an attractor for both dynamical systems together, and is as simple as possible.

Let's start with one of our two systems, say the shark-and-shrimp cycle. The phase space plot (Fig. 11.2) reflects evolution over some unspecified time. Figure 11.4 depicts, on one composite graph, a series of individual phase space plots showing sequential measurements in the numbers of shark and shrimp. Each separate phase space graph, here oriented vertically and normal to the time axis, shows a change in the relative numbers. Taken as a group, the graphs show a history (much-abbreviated) of these numbers as they complete one cycle. Interspersing more such individual graphs would fill in the history in more detail but would still give the same loop.

The plotted points for the successive times describe a path on the surface of an imaginary cylinder. The cylinder consists of the hoops of many imaginary but complete limit cycles. In Figure 11.4 the long axis of the cylinder is parallel to the time axis and extends from the first graph ($t = 0$) to the last ($t = 4$). In this example, the fifth measurement ($t = 4$) coincides with the first, so the trajectory has come full circle and has exactly completed one shark–shrimp cycle.

The frequency of a periodic motion is the number of periods or return trips per unit time, such as cycles per second. Low frequency means few periods per unit time; high frequency means many periods per unit time. As frequency increases, the trajectory covers more and more of the surface of the imaginary cylinder, for a given duration of observation. In the limit of infinitely high frequency, trajectories cover the tube's entire surface.

The key aspect of the conceptual framework is the series of graphs (loops) oriented perpendicular (normal) to the time axis. At any time, a graph or ring that circles the imaginary cylinder includes all the possible states of the system. Each ring for that particular time has one plotted point—the values of the two coordinates (here shark and shrimp).

Now let's consider how to portray the second dynamical system, say the temperature and salinity of the water in which the shark and shrimp live. Like the first system, we base this second dynamical system on the idea of a phase space oriented normally to the time axis. However, to distinguish the new plotting concept from that used for the first system, we adopt a different structural framework. Instead of a series of rings oriented normally to the time axis, the plotting scheme now consists of many straight longitudinal lines parallel to the time axis (Fig. 11.5). Each longitudinal line is a streamline that's a certain unique combination of values of the two variables (here temperature and salinity). At each successive time, we simply locate the streamline that corresponds to the measured values of the variables, move along that streamline to the corresponding time, and plot the point. The path of successive points again falls on the surface of a cylinder, because with increase

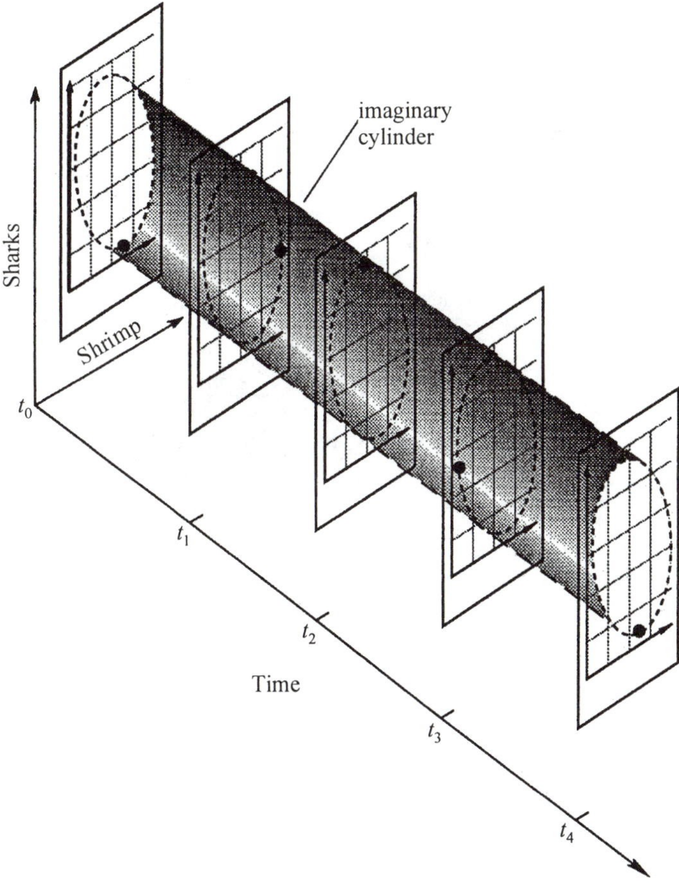

Figure 11.4 Sequential phase space plots of shark–shrimp limit cycle, in framework normal to time axis.

of time the trajectory of the limit cycle closes on (rejoins) itself to form a spirally loop. As with the first system, the trajectory eventually covers the surface of the cylinder if the frequency becomes infinitely high.

To combine the two systems, imagine that the same cylinder can serve both. It will act as a sort of all-purpose community cylinder. We'll just sketch the two conceptual frameworks, namely the circumferential rings (system 1) and the longitudinal streamlines (system 2), on the same cylinder. The result is a skeletal framework that looks like a section of fuselage from an old airplane (Fig. 11.6).

Since time goes along the longitudinal axis of the cylinder, the cylinder might need to be unrealistically long to show a long history. Therefore, we'll play topologist and change it into a compact and efficient form. To do that, we pretend it's made of rubber. We then chop it off at some arbitrary length (i.e. time), grab the two ends, bend the cylinder uniformly around in a circular direction and glue the two ends together (Fig. 11.7). That way we can keep going round and round

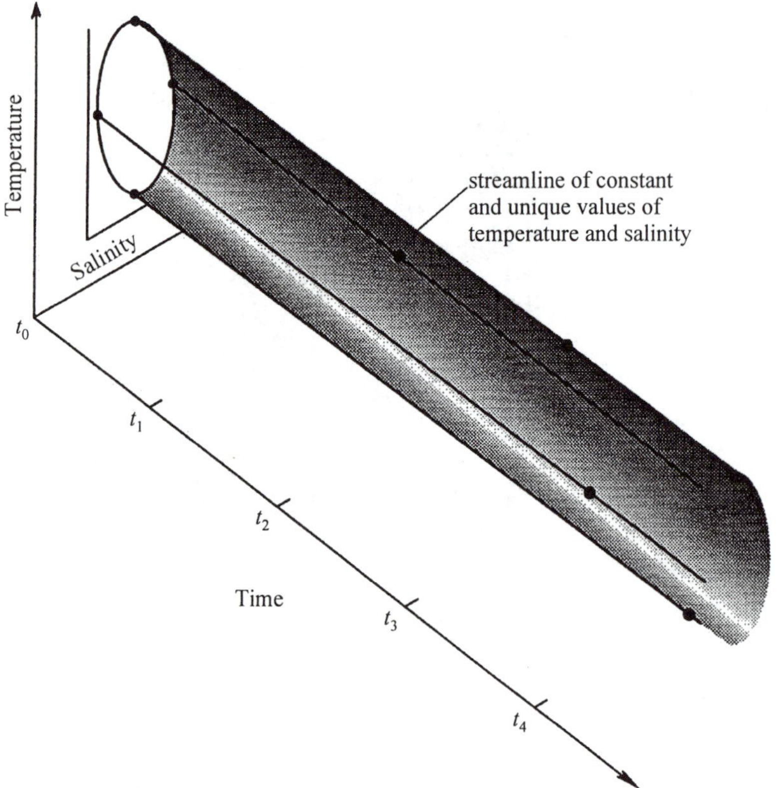

Figure 11.5 Sequential phase space plots of temperature–salinity limit cycle, in framework parallel to time axis.

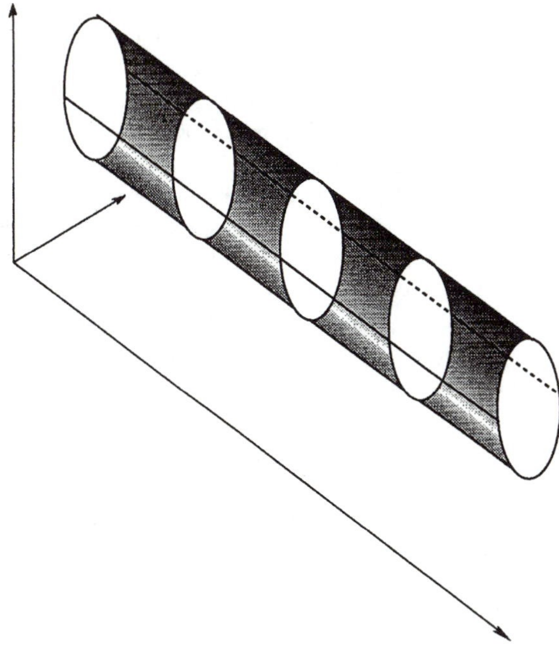

Figure 11.6 Combined plotting frameworks for two interrelated limit cycles.

forever, without being limited by time as far as plotting capability is concerned. The result of our handiwork is an object that, with a covering on its surface, looks like the inner tube of a tire, or a doughnut. All the action—all the dynamics—take place on the surface or shell of the doughnut. Mathematicians call such an object a **torus**.

Plotting two different but interrelated systems on the same graph brings them together into one compound system of four variables. That is, any single point on the torus is now the unique combination of four variables—in this example, number of shark, number of shrimp, water temperature, and salinity.

A torus can consist of more than one—in fact, many—trajectories. A point can go from one spot to another (values at one time to those at another) only by moving

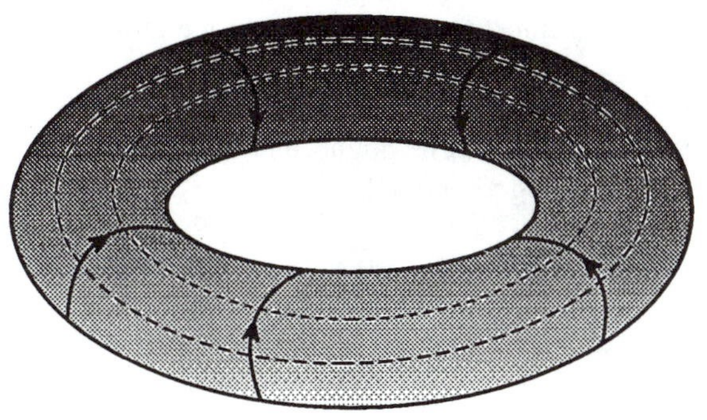

Figure 11.7 Torus with plotting frameworks.

on the surface of the torus. Since time (the long axis of our original cylinder) continues to increase, any trajectory travels around on the surface or shell of the doughnut in a way that results in a net forward travel. The trajectory doesn't zigzag or reverse its direction; rather, it follows a curved path. Generally, it winds around the doughnut or torus like the stripes on a candy cane or barber's pole, or like the wire of a coiled spring. Think of the path as consisting of two components, namely our systems 1 and 2 as described above. The two components or systems differ in terms of their relation to the doughnut's central hole. One component is the path as it winds *through* that central hole (system 1 above). The other component is the path as it winds *around* the central hole (system 2 above).

Mathematicians think of the two basic components or paths as independent frequencies or oscillations. They determine the frequency associated with one oscillation—system 1 above—by how fast the trajectory orbits the doughnut in the short direction (i.e. through the hole). The other frequency—system 2 above—depends on how fast the trajectory circles in the long direction (i.e. around the hole) (Crutchfield et al. 1986).

So far, we've considered the simplest case, namely the case with just two interacting limit cycles. Other cases might involve any number of interacting periodic cycles. Each cycle or system is an additional independent frequency on the torus. Attaching the number of independent frequencies to the name "torus" labels the torus more precisely. Thus, with two frequencies the torus is a "2-torus". Additional frequencies bring 3-tori, 4-tori, and so forth.

There are two general categories of toroidal attractors: periodic and quasiperiodic (Schaffer & Kot 1986). In the **periodic toroidal attractor**, the composite trajectory's motion is periodic—it winds around the torus an integer number of times and then comes back on itself exactly. That is, the trajectory takes a trip consisting of a certain number of axial rotations, finally returning to its starting point. Then it repeats that trip, over and over. It therefore never visits some regions of the torus.

In the **quasiperiodic toroidal attractor**, the trajectory almost but not quite repeats the same route periodically. Therefore, the frequency of the motion around the hole of the doughnut isn't exactly an integer multiple of the frequency through the hole. The behavior recurs as far as each separate frequency is concerned but not as far as the composite trajectory is concerned. As a result, the trajectory never repeats itself exactly. Given enough time, it eventually covers the torus's entire surface.

Two trajectories that start close to one another on a torus stay separated by that same gap forever (Tsonis 1989). Such trajectories might come from two measurements (beginning states) that differ by some amount.

Summary

An attractor is the set of phase space conditions that represent a system's various possible steady states. Trajectories generally approach attractors asymptotically and, at least with most mathematical examples, never get all the way "onto" an attractor. Nonchaotic attractors are of three types:

- the point attractor (a fixed point, therefore occupying zero dimensions in phase space)
- the periodic attractor or "limit cycle" (two or more values that recur in order, occupying two dimensions in phase space)
- the torus (a combined representation of two or more limit cycles, generally taking the shape of a three-dimensional doughnut in phase space).

A torus can be periodic (meaning that the combined trajectory of the various limit cycles exactly repeats itself) or quasiperiodic (in which the combined trajectory almost but not exactly repeats itself).

Chapter 12
Routes to chaos

With the logistic equation, the transition from steady, regular behavior to chaos came about by period-doubling. For other dynamical systems, the transition from orderly to chaotic motion might take place in any of several other ways. As a group, these types of transition make up a "rich and intricate landscape between order and chaos" (Percival 1989). (Actually, at least some types can also occur *within* the chaotic regime.) The transition may well be more recognizable than the chaos itself. In other cases, clear transition between order and chaos might not occur.

Routes to chaos are important for greater understanding of chaos and for practical purposes. For example, identifying pre-chaotic patterns or behavior might help us anticipate the occurrence of chaos. It's not yet possible to identify, in advance, the particular path that a dynamical process will follow in going to chaos. Worse yet, a system might follow one route on one occasion and another on the next. Where a transition stems from orderly behavior, that behavior usually is periodic. In such case, the roads to chaos differ from one another by the way in which the periodic regime loses its stability (Bergé et al. 1984).

All the possible routes to chaos probably haven't yet been discovered. The three methods discussed below—period-doubling, **intermittency**, and **quasiperiodicity**—have received most of the attention in the literature thus far. Grebogi et al. (1983) and Swinney (1986) mention other types of transition. Some of these seem to be variations on the three themes described here. All of them need further study. One of these other types—crises—shows up only after the onset of chaos (Ch. 16).

Period-doubling

Period-doubling, introduced in Chapter 10 and related sections, is the most extensively studied type of transition. It's a systematic cascading progression to chaos. Period-doubling occurs in fluid convection, water waves, biology, electricity, acoustics, chemistry, and optics, to name a few (Swinney 1986). In more familiar examples, it shows up in electrical circuits, whale populations, the transition to turbulence in a pot of water being heated, a dripping faucet (as in my kitchen!), and in the electrocardiograms of dogs (Fisher 1985). No one knows whether it's more common than other routes to chaos (Kadanoff 1983).

We saw earlier that period-doubling for the logistic equation occurred only in the range of $3 < k < 3.57$, with k restricted to values from 0 to 4. Within that critical range of the controlling parameter k (i.e. the period-doubling range), a small change in that parameter produces quite a radical change in the system's behavior.

The successive iterates during period-doubling conceal a fascinating fact: all (or nearly all) period-doubling sequences, whether observed in electromagnetism, chemical interactions, fluids, and so on (or simply in an equation), have in common certain underlying numerical characteristics or "scaling laws."[1] Those common characteristics make period-doubling look virtually the same for all physical processes that experience it; in other words, if we see period-doubling in data, we probably couldn't guess the physical process. There is, then, a **universality** or common behavior to systems that undergo period-doubling. The behavior is common both qualitatively and quantitatively (Feigenbaum 1993). Discovering that universality was a major advance in knowledge of the onset of chaos and of nonlinear science in general. Let's look at two of those scaling laws.

The bifurcation-rate scaling law

One **universal** feature of period-doubling deals with the intervals Δk between the critical k values at which new periods appear along the road to chaos. This law says that the critical k values increase at an approximately exponential rate. To see the law, let's use the k values associated with the logistic equation. I mentioned some of those values (rounded to four decimal places) in Chapter 10. Specific k values for any given bifurcation (period change) as listed in the literature differ slightly from one author to another, beyond three or four decimal places. (Computers are the cause; Ch. 14 discusses this problem.) I'll use Lauwerier's (1986a: 44) values here.

The first three columns of Table 12.1 list the bifurcation events, new periodicities, and k values at which logistic-equation bifurcations happen, respectively. Values of k from zero to just less than 3 produce one value of x^* (a periodicity of one). The first bifurcation event or period-doubling is at $k = 3.0$, where the system goes from one x^* value to two. Further increases in k continue to yield two x^* values until $k = 3.449499$, at which point the second bifurcation (from two to four x^* values) occurs. So, the range of k values over which period two reigns is from $k = 3$ to $k = 3.449499$. That's an interval of 0.449499 k units. The third bifurcation (from period four to period eight) happens at about $k = 3.544090$. Hence, period four lasts from $k = 3.449499$ to $k = 3.544090$, or a range of only 0.094591 k units. In general, the k values at which new periods appear show a smaller and smaller range of applicable k values for each periodicity. That feature shows up quite

1. To **scale** something is to reduce or enlarge it according to a fixed ratio, that is, by a constant factor.

Table 12.1 Evolution of universal constants in period-doubling, based on 5000 iterations in each case.

1	2	3	4	5	6	7	8
		Parameter convergence		Fork width			
Bifurcation event	New periodicity of attractor = 2^n	k value at which bifurcation occurs	$\Delta k_i/\Delta k_{i+1}$	Approx. k value at intersection of line $x^* = 0.5$	Approx. x^* value of associated fork	Fork width, in x^* units	$\Delta x_i/\Delta x_{i+1}$
0	$2^0 = 1$	–	–	–	–	–	–
1	$2^1 = 2$	3.00	–	3.236060	0.809015	0.309015	2.6547
2	$2^2 = 4$	3.449499	4.7520	3.498562	0.383598	0.116402	2.5319
3	$2^3 = 8$	3.544090	4.6558	3.554641	0.545975	0.045975	2.5085
4	$2^4 = 16$	3.564407	4.6684	3.566668	0.481672	0.018328	2.5049
5	$2^5 = 32$	3.568759	–	3.569243	0.507317	0.007317	–
...	...						
∞	$2^\infty = \infty$	3.569946	4.6692	–	–	–	2.5029

strikingly on a plot of x^* (the attractor or final state) versus parameter k, a so-called **bifurcation diagram** or "final state diagram" (Fig. 12.1).

We're now going to see something important: the critical k values for each new periodicity come forth at a definable, known, approximately constant, and asymptotic rate. That rate is our first scaling law: the **bifurcation-rate scaling law**. It takes either of two forms.

One form requires k_∞ (the k value at which chaos begins) and two other constants (Grossman & Thomae 1977). That form of the law directly estimates the critical k value (k_n) at which any given bifurcation event n takes place within the sequence of bifurcations:

$$k_n = k_\infty - c_1 e^{-cn} \tag{12.1}$$

where c and c_1 are constants that vary from one period-doubling system to another, e = 2.718 (the base of natural logarithms), and n = 1, 2, 3, . . . to infinity (∞). Equation 12.1 is an exponential equation, defined in Chapter 6. Now, as n increases, the quantity e^{-cn} gets smaller. Hence, as the bifurcation event n and associated periodicity go to infinity, e^{-cn} approaches zero. That means c_1e^{-cn} also approaches zero. According to Equation 12.1, the computed k_n then converges to k_∞, the threshold of chaos.

The constants carry a certain amount of associated error, so the equation isn't exact. It becomes more accurate as the periodicity becomes larger. Also, the constants vary from one model equation (logistic, etc.) to another. For the logistic equation, $k_\infty = 3.569946 \pm 1.3 \cdot 10^{-7}$, $c = 1.543 \pm 0.02$, and $c_1 = 2.628 \pm 0.13$ (Grossman & Thomae 1977). As a quick example of a predicted k value using those constants,

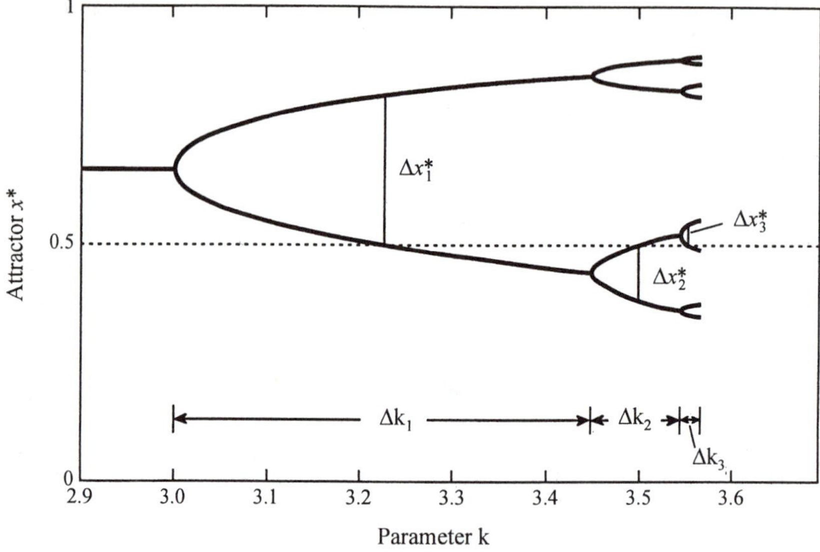

Figure 12.1 Period-doubling with the logistic equation and attributes for scaling laws.

let's take the fifth bifurcation along the series, where periodicity switches from 16 to 32 (Table 12.1). Inserting $n = 5$ and the constants into Equation 12.1 yields an estimated k value of 3.568774. Table 12.1 shows 3.568759 (which is close).

The second form of the bifurcation-rate scaling law requires three successive critical k values. This approach gives the approximate and asymptotic rate at which the critical k values converge to chaos. The procedure is to compute the range of applicable k values for two successive periodicities and then to take the ratio of those ranges. In symbols, the range for one stable periodicity is $k_n - k_{n-1}$, where k_n is the critical k value for bifurcation to period n and k_{n-1} is the critical k value for bifurcation to the preceding period. The range for the next stable periodicity then is $k_{n+1} - k_n$, where k_{n+1} is the critical k value for bifurcation to the periodicity next above n. The ratio of the two ranges is: (range of stable periodicity for any chosen periodicity) / (the next higher range of stable periodicity), or $(k_n - k_{n-1}) / (k_{n+1} - k_n)$. For example, let's take the series shown in Table 12.1 (periods 1, 2, 4, 8, etc.) and choose k_n to be 3.449499 (the k value for bifurcation from period 2 to period 4). The calculated ratio then is (k range for period 2) / (k range for period 4) = (3.449499–3.0) / (3.544090–3.449499) = 0.449499/0.094591 = 4.7520 . . .

That ratio in generalized form is $\Delta k_i / \Delta k_{i+1}$, where the Δk's are the ranges or lengths (in k units) of any two successive branches on the bifurcation tree. Ratios computed in that fashion are in column four of Table 12.1. The table lists three such ratios, not counting the one at a periodicity of infinity. (Beyond 32 periods, differences in k values become so small that the k values must have many decimal places for us to calculate meaningful ratios.) Collet & Eckmann (1980: 36) give a more detailed example using another equation.

The computed ratios (4.7520, 4.6558, etc.) are surprisingly close to one another. That means the critical k values converge toward chaos at about a constant rate, as shown earlier. Let's examine that a bit further. More detailed k values and their associated ratios would reveal a curious trend: as we compute more and more ratios, their values get closer and closer to 4.6692 . . . That number, then, is a constant that the ratios approach asymptotically as the number of the period becomes larger and larger. Mitchell Feigenbaum discovered that number in the mid-1970s, while at the Los Alamos National Laboratory, New Mexico (Feigenbaum 1978).[2] Therefore, it's appropriately known as the **Feigenbaum number** or **Feigenbaum constant**.

The Feigenbaum number plays a key role in dynamical systems theory. However, to avoid misinterpretations, here are three features to keep in mind:

- The ratio or Feigenbaum number computed for any given pair of periods, although close to the true constant, is only an approximation.
- Computed ratios get closer and closer (converge) to 4.6692 . . . as periodicity gets larger.
- The constant is unattainable exactly, no matter how large the periodicity becomes.

A succinct mathematical statement summarizes those features and defines the Feigenbaum number:

$$\lim_{n \to \infty} \frac{\Delta k_i}{\Delta k_{i+1}} = 4.6692\ldots \qquad (12.2)$$

where n is the period number.

The limit term ("$\lim_{n \to \infty}$" in Eq. 12.2) shows up frequently in chaos literature and needs clarification. "Lim" is a kind of goal or magnet, but it doesn't mean that the indicated goal or limit is reachable. For instance, a limit as something goes to infinity is unreachable, since a quantity never gets to infinity. Thus, an equation with $\lim_{n \to \infty}$ tells what the solution approaches as n gets bigger and bigger, converging on

2. **Feigenbaum, Mitchell J. (1944–)** Mitchell Feigenbaum grew up in New York City (Brooklyn), attending high school and college there (with a major in electrical engineering). As a teenager, he was mathematically inclined and liked to calculate sines and logarithms in his head. After getting a PhD in theoretical high-energy physics at the Massachusetts Institute of Technology in 1970, Feigenbaum spent two years at Cornell University and another two at Virginia Polytechnic Institute. He did his work on the universality theory at the Los Alamos National Laboratory, New Mexico, during 1974–80, when he was in his early thirties. His life during that period apparently consisted largely of (a) a diet of coffee, cigarettes, red wine, and "the reddest possible meat" (Gleick 1987: 179) and (b) eternal battles with the local mainframe computer. His findings on universality first became known only through his lectures and preprints, because technical journals rejected his papers for at least two years before eventually accepting some. (Journals do make mistakes!) Today he is a Professor of Mathematics and Physics at Rockefeller University in New York City.

infinity. It says to evaluate the equation while getting closer and closer to the indicated end point. Similarly, $\lim_{n\to 0}$ means "the limiting value as n gets smaller and smaller, approaching zero." That's why the limits are written with an arrow rather than with an equals sign. Thus, any constant designated as the limit as something goes to infinity (e.g. Eq. 12.2) can only be approximated, and at values other than at the limit.

Just as Equation 12.1 can estimate successive critical k values, so too can the Feigenbaum number (4.6692 . . .). Estimating critical k values with the help of the Feigenbaum number requires the bounding critical k values that define one periodicity. Knowing the beginning and ending k values for any periodicity, subtract one from the other to get the range. Then just increase that range of k by the reciprocal of the Feigenbaum number, that is, by 1/4.6692016 or 0.214 17 (a little more than one fifth). That gives the approximate range of k over which the next periodicity lasts. (Stated another way, the range of k values for any given periodicity is about 0.214 or only a little more than one fifth of the range for the preceding periodicity.)

Let's do a quick example. Suppose period 4 starts at k = 3.449499 and period 8 sets in at k = 3.544090. What is the k value at which period 16 first appears? First, subtract 3.446499 from 3.544090, thus getting a range of 0.094591 k units for period 4. Next, multiply that range by the reciprocal of the Feigenbaum number to get the range for period 8:

$$\text{range of one periodicity} \times 0.21417 = \text{range of next periodicity},$$

or, for period 4,

$$0.094591 \times 0.21417 = 0.020259.$$

That calculation shows that period 8 lasts for 0.020259 k units. Since it started at k = 3.544090, period 8 ends (and period 16 sets in) approximately at k = 3.544090+0.020259 = 3.564349, according to our approximate rule. (Table 12.1 indicates 3.564407.) Incidentally, since such a procedure requires the inverse of the Feigenbaum number, you might occasionally see the general relation that defines the Feigenbaum number (Eq. 12.2) written in inverse fashion, namely as $\lim_{n\to\infty} (\Delta k_{i+1}/\Delta k_i) = 0.21417$.

Equation 12.1 and the Feigenbaum number, then, are two different ways of characterizing the essentially constant rate of convergence of critical k values during period-doubling. In other words, the bifurcation-rate scaling law shows the life or longevity of any given periodicity (in k units) relative to the longevity of the preceding periodicity.

Having a reliable scaling law, even if only approximate, can be useful from a practical or experimental viewpoint. If the period-doubling and parabolic-curve requirements are fulfilled, and if we know the necessary values for either of the two forms of the law, then we can estimate the approximate critical parameter values at which each new period will arrive. (However, there's no known way to predict whether a new time series will show period-doubling. Direct observation is the only way.)

The fork-width scaling law

A second universal constant in period-doubling was also found by Feigenbaum (1978). The pitchforks at the bifurcation points of Figure 12.1 become smaller and smaller with increase in k. The second universal constant gives the rate at which they shrink.

The standard approach is to describe a fork's size by its width. That means measuring the span or width between the two tines (prongs) of a fork. Since the forks in Figure 12.1 are horizontal, the span between tines will be in units of whatever is on the ordinate. In this case, that's x^*.

A close look at the figure reveals a couple of complications. One is that the width of any given pitchfork isn't constant. That is, the two prongs aren't parallel. Instead, the "width" of any fork increases with k. So, at what horizontal location on the figure, that is, at what k value for a particular fork, do we measure that fork's width? Another problem is that, once there are two or more pitchforks for any chosen k (i.e. beyond k = 3.449499), no two pitchforks within a given periodicity are the same size. In other words, even if a genie gave us a k value at which to measure fork width, there'd be two or more different-sized pitchforks at that k value or periodicity. How do we decide which fork to measure?

The same two steps solve both problems. Step one is to adopt a standard or reference value of x^*, to be used as explained below. For the logistic equation, that reference value is $x^* = 0.5$. The reason for choosing that value is that it's the critical point (in mathematical language, the point at which the derivative is zero) on the one-humped parabolic curve of x_{t+1} versus x_t. Graphically, it's the highest point on the curve (the peak of the hump). For some other model equation, that point might not be at $x^* = 0.5$.

The second step in the solution is to adopt a simple rule: within each successive periodicity, the only fork eligible for measurement is the one for which one of the prongs has a value of $x^* = 0.5$. Measure the fork width at precisely that particular k value. All other forks for that periodicity are ineligible (unrepresentative). That rule pinpoints which branches to measure and disqualifies most pitchforks from the analysis. It also specifies where to measure the fork width.

The eligible forks are easy to spot graphically because they intersect a horizontal line drawn at $x^* = 0.5$ (Fig. 12.1). Measure the fork width right at that k value where the reference line and prong intersect. Measuring fork width or vertical gap requires the value of x^* for each prong, at that k value. For one prong, x^* is 0.5 by definition. For the counterpart prong, either compute x^* or read it from the graph. The fork width for that periodicity then is the absolute difference between the two x^* values (one of which is 0.5).

Let's go through the procedure. First, go up the vertical axis on the left edge of

Figure 12.1 to $x^* = 0.5$. That gets you to the reference line. Next, follow the dashed horizontal reference line rightward to where it first intersects a prong. That prong identifies the first eligible fork. At the intersection, measure the vertical distance (in x^* units) between the two prongs (labeled Δx_1^* in the figure). The lower prong is at $x^* = 0.5$. Although it might be possible to read the x^* value of the other prong graphically, let's do it a little more accurately by iterating the logistic equation. The graph shows that we're in the region of a period-two attractor and that k is about 3.236. We have to make several trial-and-error runs, until we find the k value that produces $x^* = 0.5$ as one of the two values of the attractor. For this first case, the more exact k value at which this takes place turns out to be k = 3.236060 (Table 12.1, col. 5). The other value of the attractor at that k value turns out to be at $x^* = 0.809015$ (Table 12.1, col. 6; Fig. 12.1). The fork width (vertical distance between prongs, on the graph) therefore is 0.809015–0.5 = 0.309015, in x^* units (Table 12.1, col. 7).

Now for the next eligible fork. Following the dashed horizontal base line of $x^* = 0.5$ further rightward, the next intersection with a prong is at k ≈ 3.498. The periodicity at that k value is four, making up two forks. However, the upper fork isn't eligible, because neither of its branches has a value at $x^* = 0.5$ (i.e. neither of its branches intersects with the horizontal reference line). The lower fork is the one that qualifies; its upper prong has $x^* = 0.5$. To get the x^* value of the lower prong, I iterated the logistic equation until finding, by trial-and-error, the more exact k value for which one of the four attractor values is 0.5. That turned out to be at k = 3.498 562. The lower prong at that k has $x^* = 0.383\,598$. The fork width therefore is 0.5–0.383 598 = 0.116402 (Table 12.1, cols 5–7). A similar procedure for larger k values yields the other fork widths listed in Table 12.1.

The first two measured fork widths were 0.309 115 and 0.116402. Forming the ratio of the first to the second gives 0.309 115/0.116402, or 2.6547 (Table 12.1, col. 8). As with the first law, we generalize the ratio into the form $\Delta x_i^*/\Delta x_{i+1}^*$ for any two successive periodicities. Measuring more fork widths and computing the ratio of successive widths gives ratios of 2.5319, 2.5085, and so on (Table 12.1). As with parameter (k value) convergence, those ratios are close to one another. Furthermore, they get closer and closer to a constant as periodicity increases. The constant has a value of 2.5029 . . . and is a limit in the sense of being a target number as the periodicity approaches infinity. Thus, the second scaling law—the **fork-width scaling law**—is:

$$\lim_{n\to\infty} \frac{\Delta x_i^*}{\Delta x_{i+1}^*} = 2.5029\ldots. \quad (12.3)$$

The constant reflects the approximate amount or factor by which a fork is wider than its immediate eligible offspring. The law means that, using just the special subset of admissible forks (namely those for which one prong has $x^* = 0.5$) and measuring fork widths at just those k values, each parent fork is about 2.5 times as wide as its eligible child. Alternatively, any fork is only about 1/2.5 or 0.4 as wide as its eligible predecessor.

Mathematicians actually knew of period-doubling (but not the scaling laws) many decades ago. As with most subjects, many individuals (including some not mentioned here) contributed. The main advance and stimulus for chaos was Feigenbaum's (e.g. 1978, 1979, 1980) discovery of the two scaling laws and his emphasis on their universality—the idea that his constants 4.6692 . . . and 2.5029 . . . apparently apply to all members of a certain class. That class consists of equations that both lead to period-doubling and show a one-humped curve like a parabola. Bifurcation diagrams (Fig. 12.1) for all such equations look alike, as long as all diagrams use the same distance and type of scale for the ordinate (x^*) and also for the abscissa (k). Different constants apply to other types of curves (Stewart 1989a: 206). Most (but not all) chaologists agree with Feigenbaum's (1978, 1980) ideas about universal application. Of course, "universal" in the chaos sense means "pertaining to all of something specified," such as period-doubling.

Other characteristic numbers

Data in Table 12.1 show a third virtually constant feature within period-doubling. The successive k values at which we measure fork widths (col. 5) scale like the k values that mark the bifurcations (col. 3) (Schuster 1988: 41). That is, the fork-width k values of column 5 are themselves in ratios that, in the limit, approach the parameter-convergence constant 4.6692. For example, the three ratios computable from the data in Table 12.1 are 4.6809, 4.6628, and 4.6707.

Period-doubling has still other characteristic numbers. Collet & Eckmann (1980: 36–55) and Lichtenberg & Lieberman (1992: 485ff.) give examples.

Characteristic numbers as a group make up a fascinating order or structure to the period-doubling road to chaos. Says Farmer (in Fisher 1985): "It's the universality that's exciting. It says that all nonlinear equations that do this particular thing do it in the same way." The group of numbers applies to such diverse topics as electrical circuitry, optical systems, solid-state devices, business cycles, populations, and learning (Briggs & Peat 1989: 64).

Intermittency

Intermittency, a second type of transition, consists of orderly periodic motion (regular oscillations, with no period-doubling) interrupted by occasional bursts of chaos or noise at irregular intervals (Fig. 12.2). In mathematical modeling, the periodic motion (limit cycle) typically shows up under relatively low values of the control parameter. Gradually increasing the control parameter brings infrequent chaotic bursts in the time series. These bursts set in abruptly, rather than gradually. With further increase of the control parameter, chaotic bursts are more frequent and last longer, until the pattern eventually becomes completely chaotic.

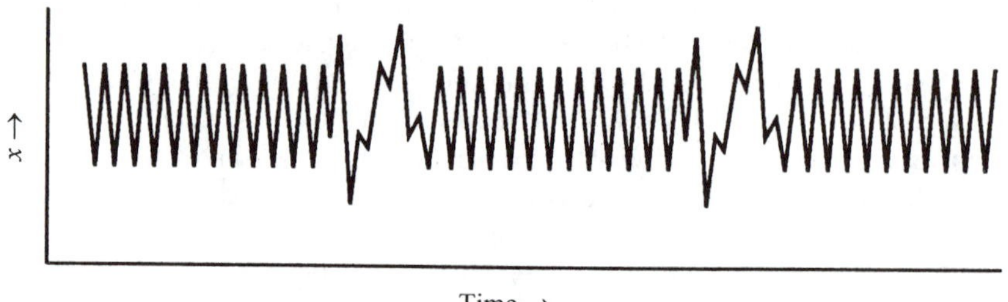

Figure 12.2 Hypothetical sketch of intermittency.

For a graphical explanation of intermittency, let's first define a few terms. Figure 12.3 graphs the parabola $x_{t+1}=2.8x_t(1-x_t)$, an equation you recognize by now. The 45° straight diagonal line (the "identity" line) is the relation $x_{t+1}=x_t$. Wherever that identity line intersects the parabola, the indicated value of x_t immediately iterates to its same value, since $x_{t+1}=x_t$ everywhere along the diagonal. The graph

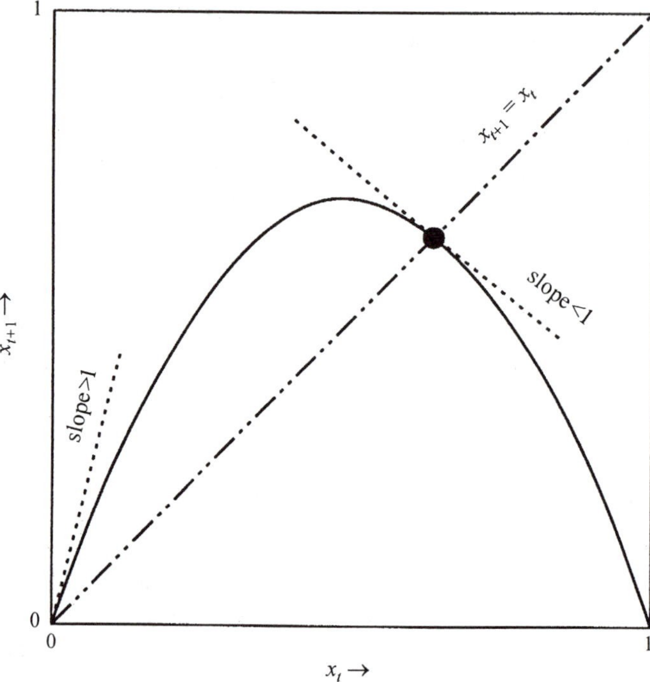

Figure 12.3 Parabola of the logistic equation, diagonal identity line ($x_{t+1}=x_t$), and two fixed points where identity line intersects parabola (unstable fixed point at origin; stable fixed point at $x_t \approx 0.643$). Slope of dashed line tangent to fixed point indicates whether fixed point is unstable (slope > 1) or stable (slope < 1).

shows two such fixed points—one at the origin of the graph (0,0) and one at the attractor ($x^* = x_{t+1} = 0.643 \ldots$). For example, plugging $x_t = 0$ into our equation yields $x_{t+1} = 0$. Then the next iteration gives $x_{t+2} = 0$, and so on. Similarly, starting with $x_t = 0.643$ gives $x_{t+1} = 0.643$, $x_{t+2} = 0.643$, and so on.

There's an important difference between the two fixed points in this example. Iterations starting from a point close to the origin (say, $x_t = 0.001$) move away from the origin (and in fact are attracted to $x_t = 0.643$). The origin therefore is an **unstable fixed point**—a fixed point from which all nearby points diverge upon iteration. In contrast, the fixed point at $x_t = x_{t+1} = 0.643$ is a **stable fixed point**, because iterates of all nearby points (in fact, all points except 0 and 1 in this case) converge to it. (In other words, when used for fixed points, **unstable** in chaos jargon means "tending to amplify initial differences or perturbations." "Stable" means "tending to reduce perturbations.")

Instead of iteration, a simple graphical rule also can reveal whether a fixed point is stable or unstable. If the slope of a line drawn tangent to the fixed point has an absolute value greater than 1, the point is unstable; if less than 1, the point is stable (Fig. 12.3).

With the above definitions in mind, let's now see how intermittency occurs and develops into chaos. Following Bergé et al. (1984) and Schuster (1988), I'll use the Lorenz equations (not written out here, for brevity). Intermittency begins with the simple case of regular oscillations (a limit cycle). Figure 12.4a shows part of the relation of x_{t+1} versus x_t and the diagonal identity line. The identity line intersects the curve at two fixed points. Applying to each point the slope-of-tangent rule just described reveals that the lower point is stable whereas the upper one is unstable. (Iterations shown on the graph therefore move in the directions indicated by the arrows.) When a one-dimensional map contains just one stable fixed point, the dynamical system is a limit cycle, discussed in Chapter 11.

For the equations involved in this example, increasing the control parameter results in translating the curve's plotted position upward. This brings the two fixed points closer together. Starting from the conditions of Figure 12.4a, suppose we increase the parameter just enough that the curve slides upward and plots exactly tangent to the identity line (Fig. 12.4b). The two fixed points now have coalesced into one. At that single point of tangency, the function (i.e. the curve), as far as its geometry is concerned, shows a sort of bifurcation in that two branches of the curve emanate from the point of tangency. This special situation whereby the curve is just tangent to the identity line therefore is known as **tangent bifurcation**. It's a threshold or critical intermediate condition along the path to chaos.

A further minor increase in the control parameter moves the curve slightly above the identity line, so that a narrow gap or channel appears between the two lines (Fig. 12.4c). Although there's no longer any fixed point, there's almost a fixed point. It's at the narrowest point of the channel, at the former point of tangency. Now that the control parameter exceeds the critical value for tangent bifurcation, the dynamical system is in the realm of possible intermittency. The intermittency, as mentioned, consists of periods of regular or oscillatory behavior interrupted by occasional bursts of chaos. A seemingly regular or nonchaotic stage corresponds to the time interval during which the trajectory passes through the narrow gap. That's because, as the graphical iteration route indicates, successive iterates during the interval differ only by tiny amounts and therefore can look almost the same ($x_{t+1} = x_t$). In other words, the trajectory thinks it is approaching (or is essentially at) a fixed point. The fixed point implies a limit cycle or regular periodicity in the dynamical system, as explained in a later chapter.

Once the trajectory successfully negotiates the narrow gap, it no longer has to move in tiny increments. Instead, it's now free to make large and relatively uninhibited jumps between the curve and the identity line. Such unrestricted motion corresponds to a chaotic stage. The chaotic stage lasts until the trajectory either re-enters the same narrow channel or comes to another comparably narrow channel, at which time another almost periodic stage sets in.

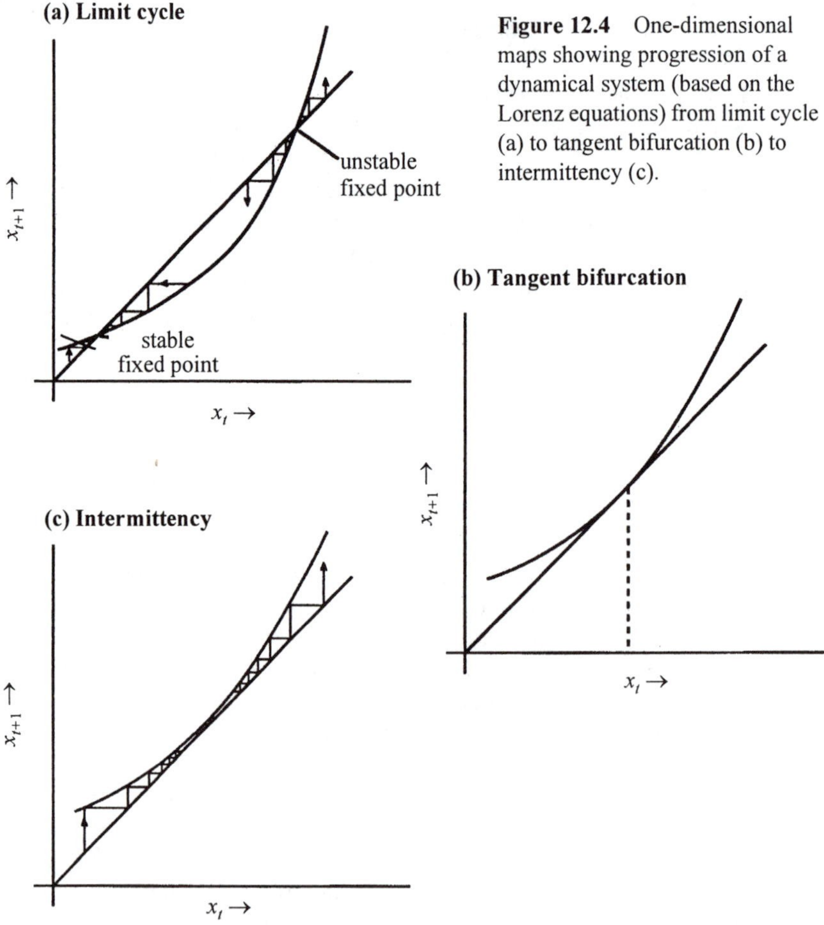

Figure 12.4 One-dimensional maps showing progression of a dynamical system (based on the Lorenz equations) from limit cycle (a) to tangent bifurcation (b) to intermittency (c).

When the control parameter exceeds its critical value only slightly, the trajectory needs many iterations to work its way through the gap, because its incremental steps here are so small. That corresponds to a long, seemingly periodic time segment within the time series. Chaotic stages therefore last for relatively short durations. Further increases in the control parameter move the curve upward on the graph, widening the gap between the curve and identity line; the trajectory then can travel much more quickly through the gap. Durations of apparent regularity thereby become shorter and those of chaotic behavior relatively longer. As the control parameter becomes still larger, the entire time series eventually becomes completely chaotic. A universal scaling law relates the average duration of regular motion (i.e. the time needed to go through the narrow channel) to the amount by which the parameter differs from its critical (tangent-bifurcation) value (Pomeau & Manneville 1980; Bergé et al. 1984: 230; Cvitanovic 1984: 31).

There are three types of intermittency. The description just given pertains to Type I, the best known of the three. Bergé et al. (1984) and Schuster (1988) give more complete details.

Intermittency appears in electronics, chemistry, and hydrodynamics, including boundary layers and pipe flow (Pomeau & Manneville 1980, Cvitanovic 1984).

Quasiperiodicity

Chapter 11 defined a quasiperiodic toroidal attractor as one having two unequal frequencies. In keeping with that idea, quasiperiodicity (a third possible route to chaos) is motion or behavior caused by two or more simultaneous periodicities whose different frequencies are out of phase (not commensurate) with one another. Since the frequencies are independent and lack a common denominator, the motion never repeats itself exactly. However, it can almost repeat itself, or seem at first glance to repeat itself. Hence, the name "quasiperiodic." It doesn't show up readily on a time-series graph and usually requires more sophisticated mathematical techniques to be seen. The periodicities characterize trajectories that trace out a torus or doughnut in phase space.

The quasiperiodic road to chaos can pass through at least three critical thresholds, with increase in control parameter (Robinson 1982; Bergé et al. 1984: 162ff.; Schuster 1988: 145ff.). At a low value of the parameter, the system is in a steady state. In phase space, such a steady state is a fixed point and has a frequency of zero. At a larger value of control parameter, the system passes the first threshold and behaves periodically (limit cycle). As such, it has one frequency. With further increase in control parameter, the system passes the second threshold. The trajectory now doesn't repeat itself, and it has two incommensurate frequencies. This stage, therefore, is quasiperiodic.

Further increase of the parameter brings the system to a third threshold. Beyond that threshold, there are at least two possible scenarios. The simplest scenario involves a breakdown of the 2-torus, so that the system goes from its quasiperiodic regime with two frequencies directly into chaos. In the alternative scenario, the quasiperiodicity changes to three frequencies before chaos arrives. There is in that case some uncertainty, because of noise, as to how long the quasiperiodicity with three frequencies lasts before the onset of chaos. Where such a quasiperiodic three-frequency regime looks stable, the system might go through a fourth threshold to arrive at chaos. In other instances, chaos might set in quickly after the three-frequency quasiperiodicity appears.

In any event, the entire transition ideally goes from steady to periodic to quasiperiodic to chaotic motion. Each of the various thresholds prior to chaotic motion is known as a **Hopf bifurcation**. As with period-doubling, the quasiperiodicity path to chaos has certain universal properties, involving parameter scaling and scaling of distances between cycle elements (Schuster 1988: 155ff.).

This route to chaos has been seen in fluid dynamics (flow between rotating cylinders), electrical conductivity in crystals, and heart cells of chickens (Schuster 1988: 175ff.).

Summary

As a control parameter increases, chaos can emerge from various routes or scenarios. The period-doubling route is a systematic doubling of periodicity (1, 2, 4, 8, etc.) by exponentially smaller increases of the control parameter, until chaos sets in. Several "universal" laws of period-doubling describe critical values, rates of change of attractor values, and rates of change of control parameter. The universality means that all systems that undergo period-doubling do so according to the same rules. Intermittency, another way in which a system can go to chaos, consists of periodic behavior with periods of chaos or noise interspersed at irregular intervals. A third route, quasiperiodicity, is almost-periodic behavior caused by two different periodicities that aren't quite in phase with one another.

Chapter 13
Chaotic equations

This chapter looks at various nonlinear equations that can give rise to chaos. Books (including this one) and technical papers on chaos tend to dwell to a disproportionate extent on the logistic equation (Eq. 10.1). What, you may ask, is the big deal about the logistic equation? Also, aren't there any other equations that produce chaos when iterated? The logistic equation is popular because it's simple and shows several key concepts. For example, it shows fixed-point attractors, period-doubling (limit cycles), and chaos (Fig. 13.1). Now it's time to see some other equations that lead to chaos and their distinctive features.

Some publications seem to imply that **power laws** (explained below), exponential equations, sine and cosine expressions, and many other nonlinear relations lead to chaos. Also, Bergé (in Bergé et al. 1984: 267) states that period-doubling (and, by implication, subsequent chaos) can come from almost any nonlinear function of x. Those implications or claims may be premature or subject to misinterpretation, as discussed next.

Equation requirements

A few quick and easy calculations prove that not just any nonlinear equation leads to chaos. Simple power laws, for instance, don't necessarily produce chaos. A **power law** is any equation having the general form $y=\mathrm{a}x^{\mathrm{b}}$, where a and b are constants. It's called a power law because y changes according to a certain power of x. Except where b = 1 or 0, a power law is nonlinear. An easy experiment is to take a simplified power law, such as $y=x^2$, and iterate it to see what happens. For x_0 less than 1 but greater than zero, iterates keep getting smaller and smaller, eventually becoming infinitely small. No chaos develops. If, instead, x_0 is larger than 1, then iteration produces larger and larger numbers, ad infinitum. Again, no chaos.

Exponential equations (e.g. $y=\mathrm{ac}^{\mathrm{b}x}$, in which a, b, and c are constants) also don't necessarily lead to chaos when iterated. Appearance of the independent variable x as an exponent in an exponential equation distinguishes an exponential equation from a power law. Even with a simple exponential equation, such as $y=x_{t+1}=\mathrm{e}^{x}$,

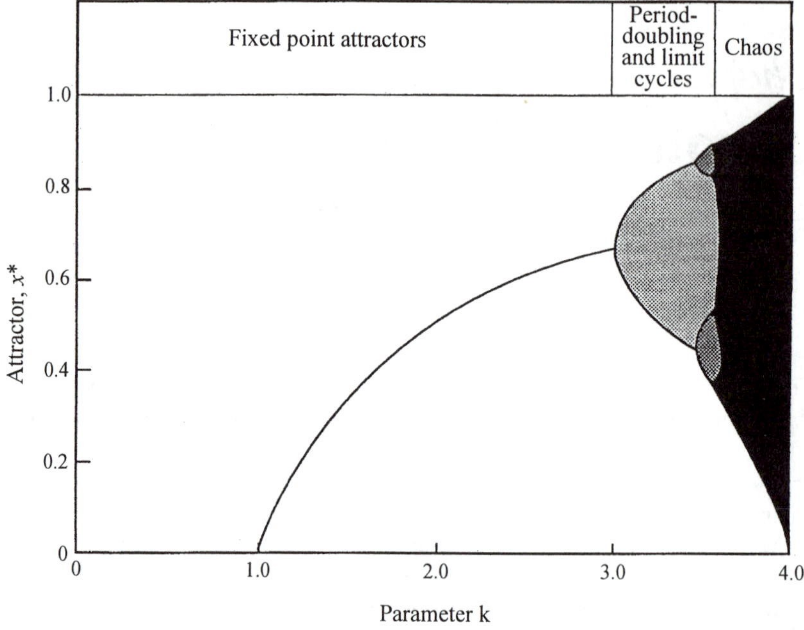

Figure 13.1 Variation in the attractor x^* with control parameter k, for the logistic equation. Depending on the value of k, the equation can show fixed-point attractors (for k<3), limit cycles (for k between 3 and about 3.57), or chaos (for k greater than about 3.57). Regardless of the value of k, possible values of x^* are extremely limited. The computed x^* falls right on the plotted line (for fixed-point and limit-cycle attractors) or within the blackened zone (for chaos). No computed values of x^* occur in the white zones that occupy most of the graph. The chaotic zone drawn here necessarily omits many details because of the scale.

iterates eventually end up getting infinitely large. No chaos develops.

Hofstadter (1981: 24) and Peterson (1988: 150) point out the problem. They state that, to lead to cyclic or erratic behavior, the curve must have some sort of switchback or hump. By that, they mean that x first has to plot in one direction and then bend back in the other direction—a feature mathematicians call **nonmonotonic**, meaning the slope of the line changes sign at some point. Even that criterion, however, doesn't necessarily mean that chaos can develop. For instance, the power law $y = -ax^2$ is nonmonotonic (a parabola), but iterating it doesn't lead to chaos. May (1987) states that period-doublings and dynamics that look chaotic aren't peculiar to the logistic equation but are general to most or all maps with one hump (one maximum in x). Perhaps further research will show that to be the answer.

Other requirements that people have proposed for a discrete, nonlinear, one-dimensional equation to be capable of chaos are:

- Noninvertibility: as explained in Chapter 10, noninvertibility means that inserting a value for x_{t+1} into an iterative equation won't lead to a unique value of its antecedent, x_t (Fig. 10.4).

- At least one unstable fixed point (a point from which a neighboring point diverges upon iteration).

The consensus among mathematicians seems to be that most discrete, nonlinear equations can lead to chaos, *given an appropriate choice of the parameters*. The two keys to accepting this thesis are to think in terms of general equations and to choose the right values of the parameters. A good example is the general quadratic equation $y = \mathrm{a}x^2+\mathrm{b}x+\mathrm{c}$, where a, b, and c are parameters. Setting b and c equal to 0 gives a simple power law ($y = \mathrm{a}x^2$), from which chaos doesn't develop. On the other hand, setting c = 0 and a = –b gives the logistic equation, from which chaos does develop. Again, the trick is to start with a general equation and to specify appropriate values of the parameters. For certain values of the parameters, no chaos arises, whereas for other values it does.

Sample equations

Published examples of nonlinear equations that can lead to chaos are becoming more and more common. There are too many to list here. (May & Oster 1976, and May 1987, give some examples from biological literature.) Following is a sampling of equations associated with chaos. Some (perhaps many) are special cases or variants of a more general equation. The listed authors are only examples of the equations' use. They didn't necessarily propose the equation originally.

Lorenz (1963) (the "tent" map):

$$\begin{aligned} x_{t+1} &= 2x \quad \text{If } 0 \leq x \leq 0.5 \\ x_{t+1} &= 2-2x \quad \text{If } 0.5 < x \leq 1. \end{aligned} \tag{13.1}$$

Devaney (1990) (a quadratic map):

$$x_{t+1} = \mathrm{c}+x_t^2 \tag{13.2}$$

where c is a constant (a parameter). This is a special case of the general quadratic equation $y = \mathrm{a}x^2+\mathrm{b}x+\mathrm{c}$, with a = 1 and b = 0.

Grebogi et al. (1983) (a quadratic map):

$$x_{t+1} = \mathrm{c}-x_t^2. \tag{13.3}$$

This is both a more general version of Equation 1.1 and another special case of the quadratic equation $y = \mathrm{a}x^2+\mathrm{b}x+\mathrm{c}$, here with a = –1 and b = 0.

May (1976) (the logistic equation):

$$x_{t+1} = kx_t(1-x_t). \tag{13.4}$$

As mentioned, this is yet another special case of $y = ax^2+bx+c$, with c = 0 and a = –b.

Lorenz (1987) (a quadratic map):

$$x_{t+1} = (x_t-2)^2. \tag{13.5}$$

Expanding the right-hand side gives $x_t^2-4x_t+4$. Hence, this too is a special case of the general quadratic equation $y = ax^2+bx+c$, this time with a = 1, b = –4, and c = 4.

Lorenz (1976) (climatology):

$$x_{t+1} = a(3x_t-4x_t^3). \tag{13.6}$$

May (1976) (biology):

$$x_{t+1} = x_t e^{a(1-x_t)} \tag{13.7}$$

where e is the base of natural logarithms.

Thompson & Stewart (1986) (various physical systems):

$$\theta_{t-1} = \theta_t+(a/2\pi)\sin 2\pi\, \theta_t+b \tag{13.8}$$

where θ is an angular variable, $\pi = 3.14$, and b is a constant.

Jensen (1987) (various physical systems):

$$x_{t+1} = x_t+y_{t+1}$$

and

$$y_{t+1} = y_t+a\sin x_t. \tag{13.9}$$

Devaney (1988) (the “gingerbread man”):

$$x_{t+1} = 1-y_t+|x_t|$$

and

$$y_{t+1} = x_t. \tag{13.10}$$

Hénon (1976) (the **Hénon map**):

$$x_{t+1} = y_t + 1 - ax_t^2$$

and

$$y_{t+1} = bx_t. \quad (13.11)$$

Lauwerier (1986b):

$$x_{t+1} = y_t$$

and

$$y_{t+1} = ay_t(1 - bx_t - [1-b]y_t) \quad (13.12)$$

Much chaos literature refers to one or more equations of the continuous category (differential equations). It's common to solve them (and others like them) for discrete times. Prominent examples of this group are:

Rössler (1976):

$$\begin{aligned} dx/dt &= -(y+z) \\ dy/dt &= x + ay \\ dz/dt &= b + z(x-c) \end{aligned} \quad (13.13)$$

where dx/dt is the first derivative of x with respect to time t and dy/dt and dz/dt are similarly defined.

Lorenz (1963) (fluid flow, climatology):

$$\begin{aligned} dx/dt &= -v_1 + v_1 y \\ dy/dt &= -xz + v_2 x - y \\ dz/dt &= xy - az \end{aligned} \quad (13.14)$$

where v_1 and v_2 are special variables.

Rössler (1979):

$$\begin{aligned} dx/dt &= -y - z \\ dy/dt &= x + ay + w \\ dz/dt &= b + xz \\ dw/dt &= c_2 z + c_1 w. \end{aligned} \quad (13.15)$$

I've included this chapter and a listing of sample equations because, in learning the principles of chaos, model equations play a key role. They are good and necessary teaching tools. However, in practice (with real-world problems) we usually don't have an equation and may not discover one. That's the reason for the tools discussed later.

Summary

Not just any nonlinear equation can lead to chaos. For example, iterating a simple power law ($y = ax^b$) or exponential law ($y = ac^{bx}$) won't necessarily produce chaos. For a nonlinear equation to produce chaos, proposed requirements in the equation are a switchback or hump, noninvertibility, and at least one unstable fixed point. A suitable choice of parameter values in a general nonlinear equation can often fulfil one or more such requirements and lead to chaos. The literature contains many examples of nonlinear equations that can be iterated to yield chaos.

PART IV

CHARACTERISTICS OF CHAOS

I've mentioned some characteristics of chaos in earlier chapters; others come in later chapters. To get a more complete picture, here's a list of chaos's main characteristics:

1. Chaos results from a deterministic process.
2. It happens only in nonlinear systems.
3. The motion or pattern for the most part looks disorganized and erratic, although sustained. In fact, it can usually pass all statistical tests for randomness.
4. It happens in feedback systems—systems in which past events affect today's events, and today's events affect the future.
5. It can result from relatively simple systems. With discrete time, chaos can take place in a system that has only one variable. With continuous time, it can happen in systems with as few as three variables.
6. For given conditions or control parameters, it's entirely self-generated. In other words, changes in other (i.e. external) variables or parameters aren't necessary.
7. It isn't the result of data inaccuracies, such as sampling error or measurement error. Any particular value of x_t (right or wrong), as long as the control parameter is within an appropriate range, can lead to chaos.
8. In spite of its disjointed appearance, it includes one or more types of order or structure.
9. The ranges of the variables have finite bounds. The bounds restrict the attractor to a certain finite region in phase space.
10. Details of the chaotic behavior are hypersensitive to changes in initial conditions (minor changes in the starting values of the variables).
11. Forecasts of long-term behavior are meaningless. The reasons are sensitivity to initial conditions and the impossibility of measuring a variable to infinite accuracy.
12. Short-term predictions, however, can be relatively accurate.
13. Information about initial conditions is irretrievably lost. In the mathematician's jargon, the equation is "noninvertible." In other words, we can't determine a chaotic system's prior history.

14. The Fourier spectrum is "broad" (mostly uncorrelated noise) but with some periodicities sticking up here and there.
15. The phase space trajectory may have **fractal** properties. (We'll discuss fractals in Ch. 17.)
16. As a control parameter increases systematically, an initially nonchaotic system follows one of a select few typical scenarios, called routes, to chaos.

The next few chapters give a closer look at some of the more important of these characteristics. We'll discuss other items in the list later in the book.

Chapter 14

Sensitive dependence on initial conditions

This chapter looks more closely at an important feature that can help identify chaos.

Trajectories in both the nonchaotic and chaotic regimes of nonlinear, dissipative systems have at least one thing in common: once we choose the control parameter, all trajectories (no matter what value we use for x_0 to start iterating) go to an attractor. There's also an important difference in trajectories of the two regimes. For nonchaotic circumstances, trajectories for two different values of starting conditions (x_0) get closer together (in most cases) or remain equidistant (on tori), for the same value of the control parameter. In the chaotic domain, however, it's just the opposite: trajectories diverge.

The trajectory divergence in the chaotic regime means that two neighboring values of x_0, differing only by the minutest of magnitudes, such as in the fourth or fifth decimal place, can evolve to different trajectories under the same value of the control parameter. A tiny difference or error, compounded over many iterations (a long time), grows into an enormous difference or error. Eventually, there's no relation at all between the two trajectories. A loose analogy (not necessarily a chaotic process) is the shuffling of a deck of cards. Two cards that start next to one another may end up quite far apart after many shuffles. Another analogy is a pair of water molecules in a river or a pair of air molecules in the breeze. The molecules may be close at one time and far apart some time later. In general, the slightest change in a variable's first value or the system's state at one time leads ultimately to very different evolutionary paths. Chaologists refer to such a trait as **sensitive dependence on initial conditions**, or simply sensitivity to initial conditions. ("Initial" in this sense means any time at which we begin comparing the pair of neighboring trajectories.)

Some chaologists use "sensitivity to initial conditions" in a more restrictive sense. They use the expression only when the rate of divergence of two nearby trajectories is *exponential* with time. (We'll discuss exponential divergence in Ch. 25.)

From tiny to huge differences

Let's look at a quick example of long-term effects of minor differences in initial conditions. We'll use the logistic equation, take a k value (3.75) that leads to chaos, and see what happens to two trajectories whose starting points differ by only a tiny amount, say 0.0001. Specifically, we'll compare the trajectory (list of iterates) for $x_0 = 0.4100$ to the one for $x_0 = 0.4101$. Here are the first five iterations, with those for $x_0 = 0.4100$ listed first and those for $x_0 = 0.4101$ in parentheses: 0.9071 (0.9072), 0.3159 (0.3157), 0.8104 (0.8102), 0.5761 (0.5767), and 0.9158 (0.9154). Short-term predictions (the early iterates) don't vary significantly despite the minor difference in x_0. This implies that short-term predictions in a chaotic system can be reasonably reliable. (In contrast, we can't make reliable predictions for uncorrelated noise, even for one time-step into the future.)

At iterations 20–24, on the other hand, the values are 0.6135 (0.6366), 0.8892 (0.8675), 0.3694 (0.4309), 0.8736 (0.9196), and 0.4142 (0.2773). Even after only these few iterations, some computed values diverge by quite a bit (e.g. 0.4142 versus 0.2773). With other equations or examples—that is, depending on noise, the system, and other features—sensitivity to initial conditions might become manifest over fewer iterations or less time. In fact, within the noise range, large differences might be possible after just one iteration. At the other extreme, differences might not become significant until many more iterations or measuring events than those associated with the example I gave here. The important point is that differences sooner or later do become significant (even though I didn't rigorously prove that here). Any original separation at all, even infinitesimal, sooner or later can lead to important separations.

The difference between the two starting values of 0.4100 and 0.4101 is only 0.024 per cent. Differences or errors somewhat larger than that are typical of most data (and certainly of mine). Figure 14.1 shows the effects of errors of 0.1 per cent and 2 per cent in the starting input values, again using the logistic equation and k = 3.75. A 2 per cent error in the input value (Fig. 14.1b) leads to major discrepancies after fewer than ten iterations. By about the 15th iteration, there's no apparent relation at all between the two trajectories. Therefore, predictions for the long term (and 15 iterations isn't long) are highly sensitive to the initial condition (values of k and x_0).

The key feature about extreme sensitivity to initial conditions, then, is that a seemingly tiny difference in input conditions becomes amplified or compounded over time and eventually can lead to a large difference in the predicted output. (Again, to invoke that type of behavior, the control parameter has to be large enough to put the system into the chaotic domain.)

Minor differences or errors in input values are absolutely unavoidable, both in the practical world of measurement and in the mathematician's world of computer calculations. In practice, nobody ever measures anything with infinite accuracy. Any measurement inevitably includes some uncontrollable and indeterminate error. Such error might be due to human mistakes, human frailties, or instrument

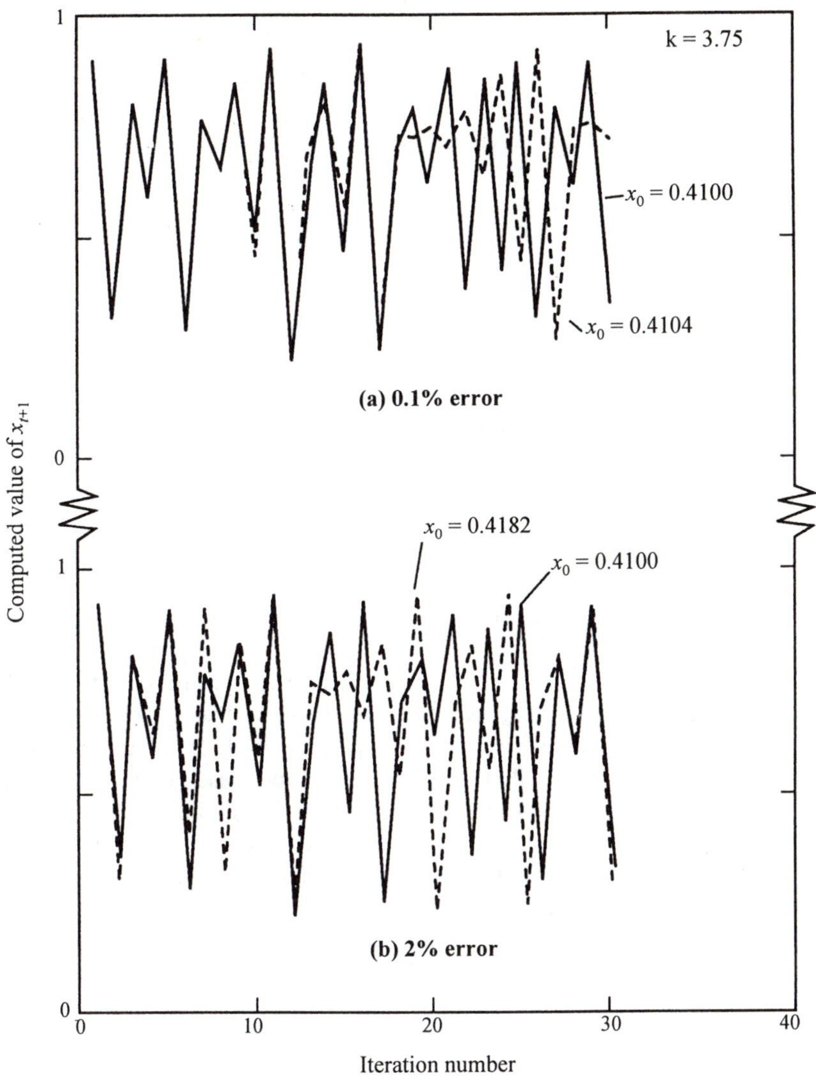

Figure 14.1 Variation in computed iterations of the logistic equation owing to minor differences (error) in initial input value x_0. (a) Error of 0.1 per cent. (b) Error of 2.0 per cent.

imprecision. Even if infinite accuracy were possible, we'd have to round off the value to a reasonable number of decimal places to make practical computations, so there'd be some roundoff error. In addition, there are background fluctuations or noise. Because of these considerations, a "datum point" on a graph is really a "datum region," owing to the finite uncertainty associated with each value.

Such inaccuracies may not affect short-term predictions. Long-term predictions, on the other hand, may be greatly affected and therefore meaningless. Given the inaccuracy of the input value, the value of a computed iterate at a far-off future time can be anywhere within a wide range. In fact, it can be anywhere on the attractor. The size of the attractor in phase space therefore represents an irreducible margin of error for any long-term prediction for a chaotic system. (At the same time, it's also a bound on the error.)

In conclusion, iterating an equation gives specific predictions for any given input values, but original input values are never completely accurate. It's an extension of the garbage-in, garbage-out theme: inaccuracies in, greater inaccuracies out.

Computer inaccuracy and imprecision

It's tempting to assume that describing a measurement more accurately, as computers enable us to do, might resolve the problem of sensitivity to initial conditions. However, greater accuracy just extends the time or "grace period" until the uncertainty expands to the same magnitude as the attractor. Furthermore, computers are a perfect example of inescapable sensitivity to initial conditions. The next few paragraphs explain why.

Ranked by increasing power of the machine, computers are of three types. (The classification is only approximate, as there can be some overlap between types.)

- *Microcomputers or personal computers ("PCs")* These are the friendly little things that sit on our desks or on our laps. Typical users of these computers include individuals, small and large businesses, and schools.
- *Minicomputers* Minicomputers usually have more than one user and are often found at colleges and universities. Some fit on a desktop, others are floor-mounted.
- *Mainframes, including "supercomputers"* A mainframe is a large multi-user system used by major establishments. It's usually located in a large air-conditioned room.

All of these computers operate on a so-called **binary system**. A computer's binary system reduces everything to some combination of zeroes and ones. The computer uses binary principles to represent numbers, store and manipulate data, and so on. There are many possible ways to use the binary system to achieve those purposes. All of them have one thing in common—they are imperfect. One reason is that, because of its binary system, a computer doesn't interpret an input number exactly. Another reason is that a computer only has a finite number of discrete states. For instance, a computer's memory capacity is always a finite limit on the "accuracy" of any input value. Also, a computer doesn't carry out mathematical operations exactly. Instead, it does so only approximately. (The approximation, of course, is plenty good enough for most purposes.) In other words, to some usually minor degree, computers are inherently inaccurate.

As a simple example of computer inaccuracy, let's take the logistic equation $x_{t+1} = kx_t(1-x_t)$, assign k = 3.8 and insert 0.4 for x_t. Using pencil and paper, you and I can calculate the first iterate (x_{t+1}) to be 3.8 (0.4) (1–0.4) = *exactly* 0.912. Suppose we want the answer expressed to 15 decimal places. The correct value then is 0.912 000 000 000 000. A computer, interestingly, won't give that result! Instead, it reports an answer such as 0.911 999 940 872 192 or 0.912 000 000 476 837. The error is, of course, tiny and probably negligible for most purposes. However, under iteration the error can snowball and become significant, as shown below.

The amount of inaccuracy varies with computer brand and model (IBM PC, Apple MacIntosh, IBM mainframe, Cray, etc.), computer operating system (MS-DOS, UNIX, Primos, VMS, etc.), type of computation (e.g. simple addition, iteration), compiler, software instructions, and probably other features. Thus, although we may think a computer calculates with accuracy to a certain number of decimal places, the real accuracy is something different from that, and it depends on all of the factors just mentioned.

Instructing the computer to calculate to a specified "precision" (e.g. single versus double "precision") or to a specified number of significant **digits** doesn't help. Results are still different because of the factors just mentioned. For instance, Rothman (1991: 58) describes computer simulations that Tomio Petrosky ran at the University of Texas. In the experiments, Petrosky routed a comet from outside the galaxy between the planet Jupiter and the Sun. The idea was to calculate how many times the comet would orbit the sun before sailing off again into outer space. With the computer set for six-digit accuracy, the model predicted 757 orbits; setting the computer to seven digits produced a prediction of 38 orbits; at eight digits, it gave 236; at nine, 44; at ten, 12; at 11, 157; and so on. The answer varied wildly, depending on the prescribed accuracy of the computations. Peitgen et al. (1992: 531–5) give an interesting discussion of the general problem.

The problem may be insignificant for some purposes but critical for others. It's insignificant, for instance, in the nonchaotic domain of the logistic equation. Within that domain, differences between successive iterates for conditions other than period-doubling become smaller (nearly zero) as iteration continues, regardless of the prescribed "precision." The same is true for any member of a period-doubling sequence. That's because iterates in the nonchaotic domain gravitate toward a single point or a repeated sequence of points.

The logistic equation's chaotic domain, on the other hand, is quite sensitive to initial conditions. Minor differences, such as those associated with computer "precision," are systematically amplified (Jackson 1991: 212). A good example is iterations done at single versus double precision on a Prime minicomputer with k = 3.8. Those iterations show a difference only in the eighth decimal place for the first iteration, but later differences build on that discrepancy and increase (Fig. 14.2). After only 35 or 40 iterations the absolute differences, attributable solely to the prescribed "precision," are as large as 0.8. That means, for instance, that the computations at one "precision" might generate a value of $x_{t+1} = 0.9$, whereas those at the other might report $x_{t+1} = 0.1$ for the same iteration number. Considering that the

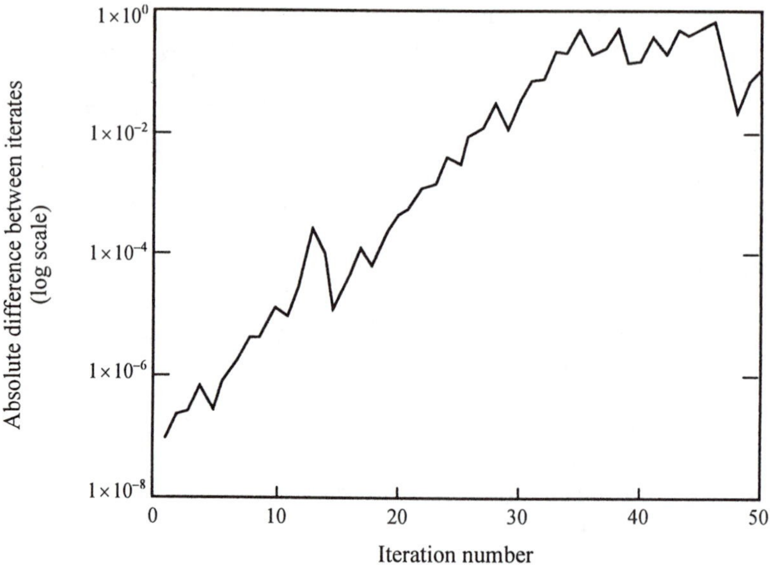

Figure 14.2 Temporal growth of difference between iterates obtained with single and double precision on a Prime minicomputer, using the logistic equation with k = 3.8.

values can range only from zero to 1, that maximum possible difference of 0.8 is immense. Differences remain large but fluctuate irregularly once the early stage of systematic increase is over. Any other type of computer gives similar results.

The above examples involved comparisons done on one computer. Different results arose because of different original conditions, in this case the number of digits to the right of the decimal in the computations. Such differences are a matter of accuracy (correctness or exactness). Besides that problem, iterates differ according to the particular computer. Differences attributable to brand or model of computer are a matter of precision or reproducibility rather than accuracy. There's no way to control imprecision. We're stuck with it regardless of the quality of our measurements.

As a quick example of imprecision, let's look at iterations of the logistic equation done on two different machines—a Data General Aviion Work Station (here called "DG") and a Prime minicomputer ("Prime"). I used Fortran language, "single precision," and $x_0 = 0.4$ on both computers. In both the nonchaotic and chaotic domains, results are similar to those described above. For instance, differences in the nonchaotic domain quickly become constant and negligible. In contrast, differences in results in the chaotic domain (k = 3.8) systematically increase (Fig. 14.3). After 35–40 iterations, x_{t+1} as given by one computer differed by as much as 0.8 from that of the other. For further iterates, differences in x_{t+1} between the two computers fluctuate irregularly.

Out of curiosity, I made similar chaotic-domain computations on my Hewlett–

Packard HP-97 desk calculator and compared the results to those of the other two machines. Each of the three machines gave a different series of iterates. Differences between iterates for any two of the three machines are similar and behave in the same way, soon becoming quite significant (Fig. 14.3).

The problems just described have led to an active research area: the interaction or relationships between finite computer arithmetic (such as the orbits a computer generates) on the one hand and the true dynamics of the theoretical system on the other.

Because of the inaccuracy and imprecision inherent in computers, we must treat iterations with care and suspicion. The inaccuracy and imprecision won't be significant for a whole range of parameter values (0 to about 3.56 for the logistic equation) and hence for a wide range of problems. The chaotic domain, on the other hand, shows sensitivity to initial conditions. Here differences between any two iteration sequences can be substantial, depending on many computer-related factors. However, ratios of iterates, quantitative measures based on iterates, and trends in iterates (or in their differences) probably are reliable. In other words, even though iterates may differ, they nonetheless fall on the same attractor. The same quantitative measures of the attractor's properties also emerge.

Where iterations simulate real data, carry the values only to a realistic number of digits. All digits that the computer produces beyond that significant number are garbage. Drop those numbers after each iteration; they can (in the chaotic domain) affect the results because of sensitive dependence on initial conditions.

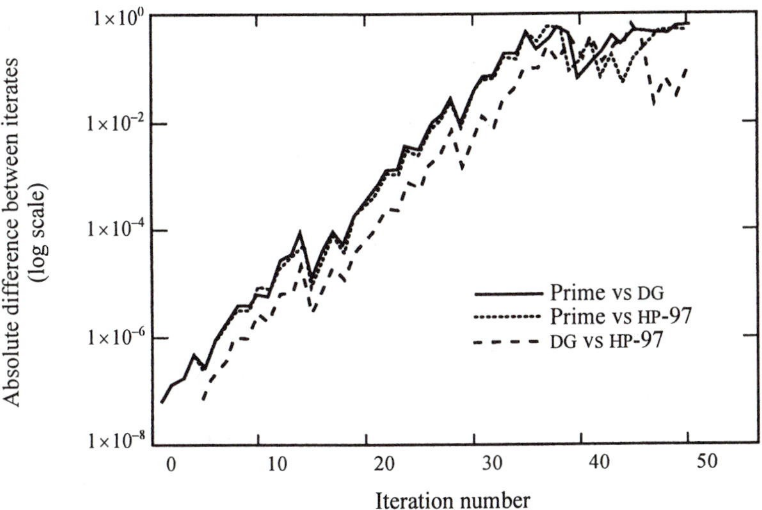

Figure 14.3 Temporal growth of differences between iterates done on different computers, using the logistic equation with k = 3.8. "Prime" = Prime minicomputer, "DG" = Data General Aviion Work Station, and "HP-97" = Hewlett–Packard-97 desk calculator.

Interpretations

Unpredictability is the most common interpretation of sensitive dependence on initial conditions. "Unpredictable" here refers to the system's future state, condition, or behavior.[1] Stated more formally, extreme sensitivity to initial conditions, combined with the inevitable measuring errors, roundoff errors, and computer precision, imposes limits on how accurately we can predict the long-term temporal behavior of any chaotic process. Beyond a certain time, as discussed in Chapters 15 and 25, long-term behavior looks random, is indeterminable, and cannot be reliably predicted.

This has important practical implications. Measurements of any variable over time may not help in accurately forecasting its future behavior beyond a brief time range. That is, the behavior may be chaotic, so that the long-term future evolution can't be predicted except to within certain broad limits (the size of the attractor). That could explain why weather forecasters often can't correctly predict the weather beyond a day or two (sometimes hours or minutes). Similarly, knowing the record of large floods over the past century or two may be useless in reliably predicting the frequency of future floods.

A second way of interpreting sensitive dependence on initial conditions is that a tiny historical accident can bring about a radically different eventual outcome. In other words, an infinitely small change in any external force can lead to radically different and unexpected long-term behavior. Or, an apparently steady pattern can be highly unstable to small disturbances. For example, continuing the flood analogy, seemingly minor changes in climate might lead to a major change in future flood frequencies. For species population, two iterations of the model, using the same parameters and almost the same first populations, might lead to vastly different predicted eventual populations. The timeworn and farfetched example from the chaos literature is the so-called "**butterfly effect**." It says that a butterfly flapping its wings in, say Brazil, might create different "initial conditions" and trigger a later tornado elsewhere (e.g. Texas), at least theoretically.

Sensitivity to initial conditions is a main feature of chaos. Some chaologists say it's the fundamental cause of chaos. Rediscovering and emphasizing that sensitivity is one of chaos's chief contributions to science. However, extreme sensitivity to first conditions doesn't necessarily, by itself, lead to chaos. For instance, even if we're just playing with numbers at our desk we still need an equation that can lead to chaos. Also, sensitivity to initial conditions can occur with random data (Wegman 1988). That's especially true if we interpret sensitivity loosely as any separation at all over time. How commonly it occurs with random data hasn't yet been resolved.

Sensitivity to initial conditions also points out the folly of placing too much

---

1. We can predict numbers, of course. It's just that the numbers are meaningless. For example, we just saw that the same person using two different computers can come up with widely disparate long-term predictions.

emphasis on the idea of determinism. On the one hand, natural phenomena are subject to rigid physical laws, and some equations (at least theoretical ones) may be exact. In practice, however, noise, perturbations, and measurement limitations are always present and can reduce determinism to a physically meaningless concept, regardless of any underlying mathematical validity. Many physical systems behave so erratically that they are indistinguishable from a random process, even though they are strictly determinate in a mathematical sense.

Summary

Sensitive dependence on initial conditions means that a seemingly insignificant difference in the starting value of a variable can, over time, lead to vast differences in output. (Some authors prefer a more limited definition, namely that the differences increase exponentially.) Measurement error, noise, or roundoff in the data values can cause such tiny differences in input. In that sense, sensitivity to initial conditions implies sensitivity to measurement error and to noise. Computers, in particular, are susceptible to sensitivity to initial conditions because of their binary system, operating system, software details, the way they are built, and so on. Sensitivity to initial conditions is a very important characteristic of chaos, but it's not a foolproof indicator; random data can also show it. One practical effect of sensitivity to initial conditions is that, even though a system may be deterministic and the governing laws known, long-term predictions are meaningless.

Chapter 15
The chaotic (strange) attractor

Nonchaotic attractors generally are points, cycles, or smooth surfaces (corresponding to static, periodic, and multifrequency systems, respectively). Their geometry is regular. Small initial errors or minor perturbations generally don't have significant long-term effects. (We saw this in following the routes of various trajectories as they went to a point- or limit-cycle attractor.) Also, neighboring trajectories stay close to one another. Predictions of a trajectory's motion on nonchaotic attractors therefore are fairly meaningful and useful, in spite of errors or differences in starting conditions. Now, just the opposite characteristics describe chaotic attractors, which I'll define as attractors within the chaotic regime (but see next paragraph). As of yet, there isn't any universal agreement on a definition of a chaotic attractor.

Because of the unexpected and quite different features just mentioned, you'll often see the term strange attractor in place of "chaotic attractor." The attractor is strange (in the sense of unfamiliar, poorly understood, or unknown) in terms of both dynamics and geometry. However, Grebogi et al. (1984) and some other authors distinguish between "chaotic" and "strange" attractors. To them, "chaotic" refers to the trajectory dynamics on the attractor. In particular, a chaotic attractor in their view is one on which two nearby trajectories diverge *exponentially* with time. Such chaotic attractors also have fractal properties (Ch. 17). In contrast, those authors use "strange" in connection with the attractor's geometrical structure. That is, they define a strange attractor as one having fractal (Cantor set) structure (Ch. 17) or certain other special geometrical properties. "Strange attractors" in their definition don't show sensitivity to initial conditions (exponential divergence). In other words, their chaotic attractors are strange (have fractal properties), but their strange attractors aren't chaotic (don't show exponential divergence of neighboring trajectories with time). In many cases it's either impossible or unnecessary to confirm one or both of sensitivity and strangeness. I'll use the terms "chaotic attractor" and "strange attractor" synonymously.

Similarities with other attractors

Before examining the characteristics of a chaotic attractor, let's see why it still qualifies as an attractor. Here are several features it has in common with all attractors.

- It's still the set of points (but in this case an infinite number of points) that the system settles down to in phase space.
- It occupies only certain zones (and is therefore still a shape) within the bounded phase space. All data points are confined to that shape. That is, all possible trajectories still arrive at and stay "on" the attractor. (As with non-chaotic attractors, a trajectory technically never gets completely onto a chaotic attractor but only approaches it asymptotically.) In that sense, a chaotic attractor is a unit made up of all chaotic trajectories. Figure 15.1 shows a two-dimensional view of the well known Rössler (1976) attractor, a three-dimensional strange attractor designed as a simplification of the Lorenz attractor.
- A chaotic attractor shows zones of recurrent behavior in the form of orderly periodicity, as explained below.
- It's quite reproducible.
- It has an invariant probability distribution, as explained in the following section.

Invariant probability distribution of attractor

First of all, don't be intimidated by the formidable-sounding name of this section. This item is no big deal—a sheep in wolf's clothing. All it does is indicate where, in phase space, a trajectory likes to spend its time while cruising around on an attractor. If the system has only one variable, a simple histogram of the time-series data gives a crude idea of those locations.

For a given value of the control parameter, a trajectory goes to its attractor and stays there forever. The attractor consists of phase space cells, each of which has its own unique address. The trajectory visits those cells as the system evolves. It might visit some cells more often than others. That preferential visiting is the feature we'll look at right now.

A periodic trajectory repeats the same circuitous phase space route. It therefore visits only a relatively few phase space cells or addresses. For example, the logistic equation with k = 3.4 iterates to an attractor of two fixed x^* values, specifically 0.452 and 0.842. Suppose we let that trajectory get to its attractor, then sample it for a long time and compile a frequency distribution of its wanderings. As usual, we'll assume relative frequency equals probability—the probability of finding the system in some particular phase space location at any given time. A graph of the relative frequency distribution or probability distribution for this example is simply a plot of probability versus x value for the possible range of values (here the

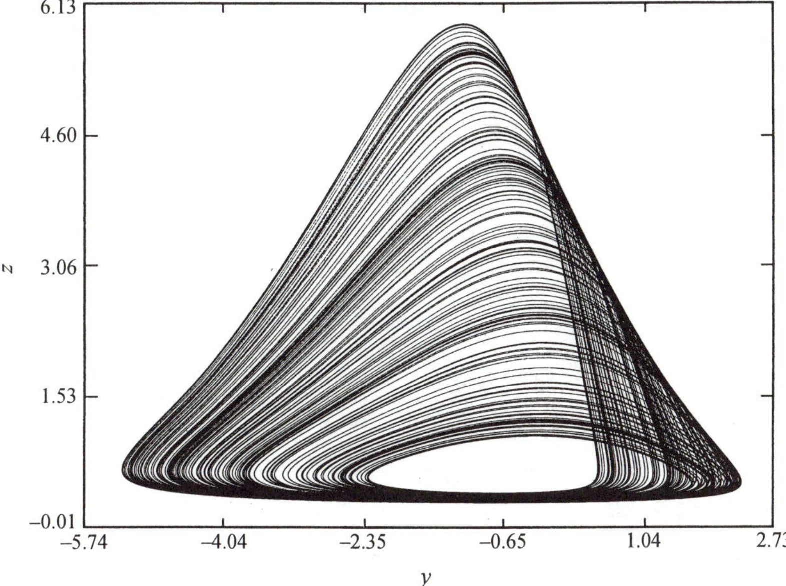

Figure 15.1 Two-dimensional projection of the Rössler strange attractor. Computer-generated graphics by Sebastian Kuzminsky.

interval $0 \leq x \leq 1$). The system only visits two phase space sites (here values of x), and those in regular alternation. The frequency distribution therefore consists simply of 50 per cent of the visits at $x = 0.452$ and the other 50 per cent at $x = 0.842$. A plot of that distribution (Fig. 15.2a) consists of two equal-height spikes (one at $x = 0.452$, the other at $x = 0.842$). Each spike extends upward to an ordinate (probability) value of 0.50.

That's all there is to it. We have just built a probability distribution for this particular attractor or variable, for the given parameter value. Furthermore, that distribution is time invariant (doesn't change with time). That's because, assuming a thorough sampling of the attractor (i.e. a long time series) and a fixed binning scheme, we get the same distribution, no matter where we start in the time sequence. Hence, "**invariant probability distribution** of attractor." You might also see it called an **invariant measure**, **natural probability measure**, or simply a probability distribution. All such terms mean the relative frequency with which a trajectory visits the different phase space regions of an attractor.

The sampling duration ideally ought to be long enough to characterize the system's long-term average behavior adequately. In truth, the distribution usually obtained for a chaotic attractor isn't exactly invariant. Rather, it becomes asymptotic toward some limiting distribution as time or number of observations goes to

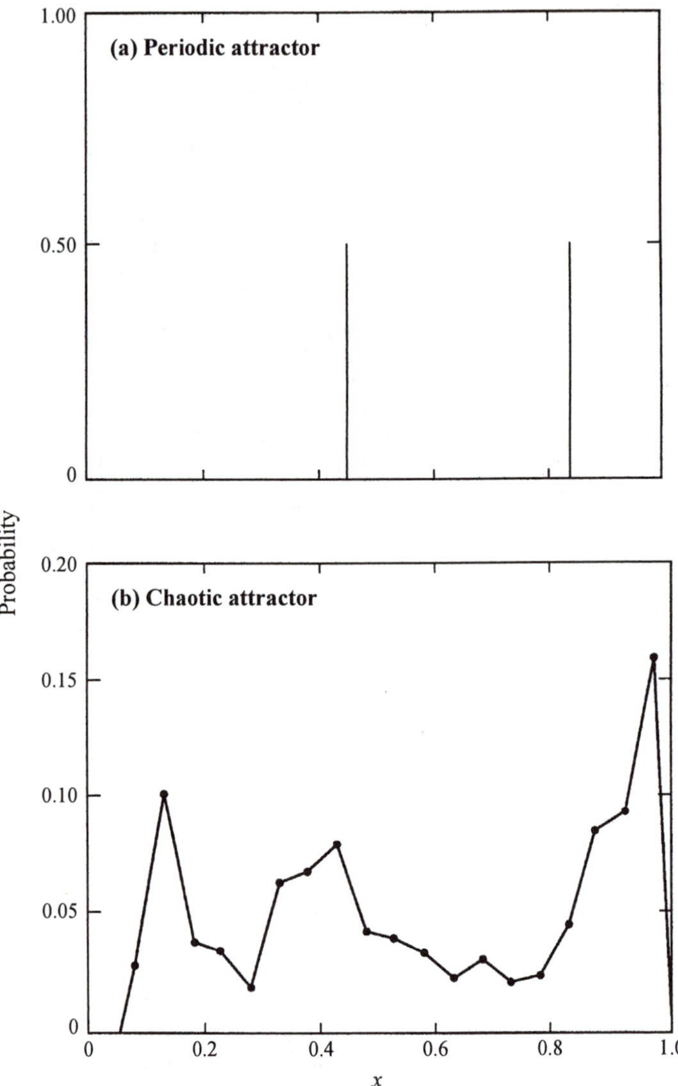

Figure 15.2 Invariant probability distributions based on the logistic equation, using k = 3.4 (upper) and k = 3.9 (lower).

infinity. As a result, an invariant probability distribution really is the distribution that the data approach as time goes to infinity.

The nonchaotic regime by definition includes only fixed points, limit cycles (regular periodicities), and quasiperiodic motion. The attractor's probability distribution for a given parameter value within the nonchaotic regime shows one or more pronounced spikes. For example, the "trajectory" of a fixed point is always at the same phase space address. The probability distribution for a fixed point therefore is a spike located at the point's value along the abscissa and extending vertically to a probability of 1.0. If the trajectory instead alternates between two phase space addresses, the plot shows two spikes (Fig. 15.2a). In general, the number of spikes corresponds to the periodicity (two spikes indicating period two, etc.). For quasiperiodic motion the spikes may not be as well defined.

The logistic equation's *chaotic* trajectory, in contrast, goes to many locations within phase space, that is, to relatively many sites on its attractor. True, it shows some favoritism for certain x values, for a given parameter value. True, also, it becomes periodic at some parameter values (Ch. 16). In general, however, the trajectory eventually gets to most x values within its possible range. The frequency distribution for a chaotic trajectory (Fig. 15.2b) doesn't have the pronounced spikes that typify the nonchaotic trajectory. However, we still compiled it in the same way, namely by partitioning the interval into classes or bins and counting the number of points that fall into each bin.

The frequency distribution for a given dataset varies with the number of classes (Fig. 6.3). We want to designate the number of classes (class width) such that the maximum number (narrowest width) of the bins gives a bin width greater than the noise level (accuracy) of the data. For example, if the data values range from 0 to 1 and are only accurate to within 0.05, it's senseless to set up boxes narrower than 0.05, such as with widths of 0.001. Furthermore, dividing a dataset into many classes tends to produce fewer observations within any one class. Some classes may not get a representative number of observations; others may get a number small enough to be statistically insignificant. (A relatively large dataset, say many thousands of observations, helps combat the problem of few observations in a class.) At the other extreme, a small number of classes (say fewer than about five) tends to obscure important details of the distribution. There isn't any simple formula to specify the appropriate number of classes to use for a given case. It's an arbitrary, subjective judgement. Figure 15.2 uses 20 classes.

A frequency distribution can be compiled for a system that involves virtually any number of variables. When more than one variable is involved, it's a *joint probability distribution* (Ch. 6). The idea is always the same, namely to define compartments by prescribing various class limits for each compartment, no matter how many variables the system has. Graphically, a system of two variables has two axes, such as x and y; the compartments then are rectangles that together make up a grid on the graph. For three variables (the most we can draw on a graph) the compartments are cubes. For four or more variables the compartments are called **hypercubes**. We have to use our imaginations to figuratively cover the attractor with hypercubes of a certain size; analytically, it's a straightforward matter to compile joint frequencies by having the computer count the frequencies in hypercube A, hypercube B, and so on.

Besides data consisting of one variable or of several variables, a third type of data for which chaologists estimate an invariant probability distribution (in this case a joint probability distribution) is lagged data of one variable. In that case the frequency distribution reflects x_t and subsets based on specified lags. For instance, in two dimensions (x_t and x_{t+m}) with x ranging from 0 to 1 we might set up bins having intervals of 0.1 for both "variables." The first bin might consist of x_t ranging from 0 to 0.1 and x_{t+m} ranging from 0 to 0.1, and so on. Then we count the number of observations falling into each bin, similar to the scheme of Figure 6.6. That kind of invariant probability distribution is popular in chaos analyses. The distributions for all three types of data show important global features of the frequency with which orbits visit various regions of the attractor.

In theory, then, a plot of the attractor's long-term probability distribution shows whether a system is in a nonchaotic or chaotic regime. In practice, that doesn't necessarily work. First, it may not be possible to collect enough data to estimate the probability distribution reliably. Secondly, a broad band of frequencies without pronounced spikes carries the same identification problem that a Fourier analysis does. That problem is that such a pattern *might* indicate chaos, but it might instead indicate a random process (Glass & Mackey 1988: 42–7). In other words, the broad band on a Fourier analysis is a necessary but not sufficient criterion of chaos.

Distinctive features of chaotic attractors

The distinctive features of chaotic attractors are:

- A trajectory within the chaotic regime (e.g. Fig. 15.1) is usually more complex than just a simple, regular loop. At some values of the control parameter, it supposedly never repeats itself (never stabilizes). (A trajectory that never repeats itself is an **aperiodic** or **nonperiodic** trajectory.) Many other parameter values seem to bring periodic orbits of long periodicity. (For example, iterating the logistic equation within the chaotic regime on my computer can produce periodicities that range from a few hundred to several thousand iterations, depending on k.) In practice it might be very difficult to distinguish cycles of long periodicity from aperiodic trajectories. Other parameter values (called windows) within the chaotic regime have special cascading (period-doubling) periodicities, as discussed in the next chapter.
- Trajectories on a chaotic attractor do not cross. If they did, then the system could behave in very different ways whenever the conditions at the crossing point recur.
- Two trajectories that at one time are quite close together diverge and eventually follow very different paths. That's because of the sensitivity to initial conditions that characterizes the chaotic regime. Topologists look upon such phase space divergence of neighboring trajectories as a stretching, as mentioned in Chapter 10.
- The phase space path of a chaotic trajectory also does a folding maneuver. That occurs when the trajectory reaches its phase space boundary (at limiting

values of one or more variables) and rebounds or deflects back in its plotted pattern. Stretching and folding are really just imaginary topological actions but are helpful notions nonetheless. In fact, Abraham & Shaw (1983: 107) write that "the basic dynamical feature of chaotic attractors is bounded expansion, or divergence and folding together of trajectories within a bounded space."

- A chaotic attractor has a complex, many-layered internal structure. The reason is that "folding" happens over and over again. That internal structure is usually (but not always) fractal. (A fractal [Ch. 17] is an elegant geometric pattern that looks the same regardless of any change in scale.)
- The external appearance is elaborate and variable compared to the loops or smooth-surfaced tori of the nonchaotic attractor. To date, many chaotic attractors have been found. Many more probably will be discovered.

  It's instructive to look at chaotic attractors from different angles in phase space. Modern computer-graphics packages enable us to look at attractors and other graphed objects from virtually any direction. That's a big advantage, especially for an attractor in three-dimensional phase space.
- Its dimension doesn't have to be an integer, such as 2 or 3. (We'll discuss dimensions in later chapters.) The noninteger and usually fractal nature of chaotic attractors led Mandelbrot (1983: 197) to recommend calling them fractal attractors rather than chaotic or strange attractors.

Definitions of a chaotic attractor

Based on the above features, chaologists proposed several definitions of a chaotic (strange) attractor. Two of the best are:

- A chaotic attractor is a complex phase space surface to which the trajectory is asymptotic in time and on which it wanders chaotically (Grebogi et al. 1982).
- A chaotic attractor is an attractor that shows extreme sensitivity to initial conditions (Eckmann & Ruelle 1985, Holden & Muhammad 1986).

According to Ralph Abraham (Fisher 1985: 31): "The chaotic attractor emerged in mathematical theory in 1932 or so, then came into view in science with Lorenz in 1971. And now it's cresting in a wave of fanatical popularity in all the sciences."

Summary

As with all attractors, chaotic (strange) attractors:

- are the set of conditions that a system can take on
- occupy certain zones (or have a certain geometric shape) in phase space

- have zones that are more popular, sometimes with periodic visits
- are reproducible
- have an invariant probability distribution.

The invariant probability distribution of an attractor is simply a histogram or relative frequency distribution representing the long-term relative frequency with which the system visits its various possible phase space locations, for a given value of the control parameter. In the nonchaotic regime, that distribution shows spikes. The number of spikes equals the periodicity. In the chaotic regime, the distribution tends to lack pronounced spikes. Random data also lack pronounced spikes, so an attractor's invariant probability distribution won't distinguish chaotic data from random data. The main characteristics of a chaotic or strange attractor are:

- an irregular, erratic, trajectory that can be periodic or nonperiodic
- a total absence of crossing trajectories
- stretching (divergence) of trajectories that originally were close together (sensitivity to initial conditions)
- folding, owing to confinement within the bounded phase space
- complex, many-layered, usually fractal internal structure
- elaborate or unusual outer geometry
- noninteger dimension.

Two definitions of a chaotic attractor are that it is an attractor that shows extreme sensitivity to initial conditions or a complex phase space surface to which the trajectory is asymptotic in time and on which it wanders chaotically.

Chapter 16
Order within chaos

Chaos *looks* erratic. However, it isn't just a vast sea of disorder. Against a general backdrop of apparent randomness are some curious features. For example, continued increases of a control parameter, such as k in the logistic equation, don't necessarily bring increased degrees of chaos. Each systematic increase in the control parameter, no matter how slight, does bring about a different trajectory. (And hence, even within chaos, the parameter is king!) However, within the chaotic domain such an increase often leads to some type of regularity, a kind of order. All of those regularities are quite reproducible, for a given value of the control parameter. Curiously, the order usually lasts only over brief ranges of the control parameter.

The orderly features as a group are striking enough that some people emphasize them more than the chaos. At the very least, chaos consists of various types of order camouflaged under random-like behavior. In that sense, order within chaos is the rule rather than the exception.

When chaos comes from iterating an equation, typical forms of order within chaos are windows (regions of periodicity), various routes to chaos (e.g. period-doubling), the chaotic attractor itself, zones of popularity that the trajectory prefers to visit more frequently, and fractal structure in a chaotic attractor.

Windows

One orderly peculiarity within chaos is **windows**—zones of k values for which iterations from any x_0 produce our friend, the periodic attractor, instead of a chaotic attractor. In other words, scattered throughout the chaotic realm, at known k values, are stable conditions where there actually isn't any chaos. Instead, there's a nonchaotic attractor consisting of a fixed periodicity.

Grossman & Thomae (1977) describe the chaotic regime as a "mixed state where periodic and chaotic time development are mingled with each other" or as a "superposition of a periodic motion and a chaotic motion in state space." Wolf (1983) calls it a complex sequence of periodic and chaotic trajectories. Chaotic trajectories can be difficult to identify; we may not be able to tell whether they are indeed chaotic or just periodic but with a very long period.

As with many other features of chaos, windows were discovered and studied only by iterating an equation, naturally using a computer. The apparent number of windows within the chaotic domain depends on computer precision. Iterations on powerful computers suggest that the chaotic regime of the logistic equation (Fig. 16.1), and probably of other iterated equations, includes an infinite number of windows of periodic behavior (Wolf 1986: 276; Grebogi et al. 1987; Peitgen et al. 1992: 635).

The most prominent window generated on the average personal computer using the logistic equation is at $k \approx 3.83$, where a period-three cycle turns up. On Figure 16.1 that window is a broad, vertical white band; the three black lines crossing the band are the only values the system takes on, within that interval of k. For the quadratic map (Eq. 10.2), the most prominent window within chaos is at $k \approx 1.76$ (Grebogi et al. 1987), again with a period-three cycle (Fig. 16.2).[1] Olsen & Degn (1985) mention that there are windows for all uneven periodicities (3, 5, 7, etc.) from 3 to infinity.

Any regular periodicity within the chaotic regime usually lasts for just a brief range of k values. Most windows, for example, are so narrow that they are invisible on a graph except under very high magnification. Also, the periodicity can be difficult to detect. For one thing, the periodicity usually is high. For another, the

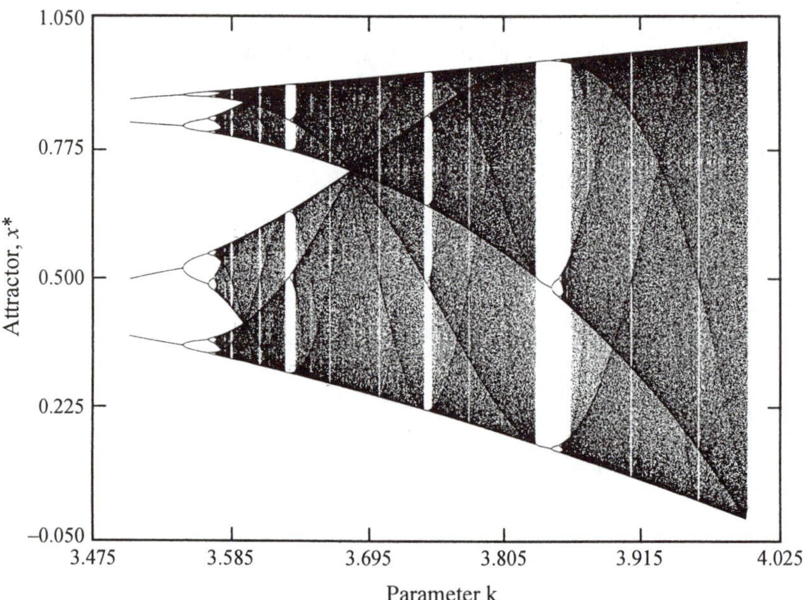

Figure 16.1 Chaotic domain of logistic equation. Computer-generated graphics by Sebastian Kuzminsky.

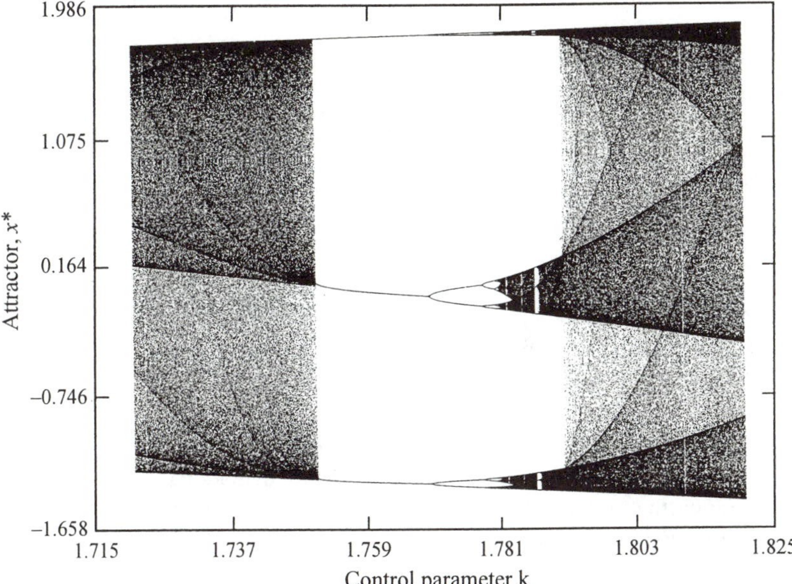

Figure 16.2 Window consisting of a period-three cycle, occurring at about k = 1.76 for the quadratic equation (after Grebogi et al. 1987). Computer-generated graphics by Sebastian Kuzminsky.

trajectory often looks chaotic (Hofstadter 1981, 1985). Even so, windows typically have a global scaling structure (Yorke et al. 1985; Peitgen et al. 1992: 636).

Real-world data typically defy any attempts to study such detailed phenomena. Windows aren't of much use in detecting chaos from such data because of noise and because they occur only over narrow and special conditions (Schaffer et al. 1986).

Routes between windows and chaos

A second kind of order within chaos is the routes to chaos, such as period-doubling and intermittency. The period-doubling goes on in the same cascading way that it does in the pre-chaotic domain. With the period-three attractor at k = 3.83 in the

1. Figures 16.1 and 16.2 were generated by starting with the lowest k value shown, then:
   1. choosing x_0
   2. iterating the equation thousands of times
   3. throwing out the first few hundred iterates as possibly atypical
   4. plotting the remaining points on the graph
   5. increasing k by a tiny amount and repeating steps 1–4 with that new k
   6. repeating step 5 until the desired range of k values had been covered.

logistic equation, for example, slight increases in k bring bifurcations to cycles of period 6, 12, 24, and so on (Jensen 1987).

Within chaotic-regime windows, period-doubling sequences themselves have orderly (in fact universal) features (besides the cyclical cascades). The logistic equation and quadratic map (Eqs 10.1 and 10.2) (Yorke et al. 1985) are examples. Here the width of any window bears an approximately constant relation to the distance from the start of the window to the first bifurcation. The general relation is

$$\lim_{n\to\infty} \frac{k_\infty - k_0}{k_2 - k_0} \to 2.25 \qquad (16.1)$$

where n here is the periodicity of the attractor, k_0 is the k value at which the window first appears, k_2 is the k value at which the starting periodicity undergoes its first doubling, and k_∞ is the k value at which chaos becomes re-established. As with the other universal numbers associated with period-doubling (Ch. 12), the ratio in Equation 16.1 comes closer and closer to the limiting number (2.25) as the periodicity gets higher, as the limit term suggests. The calculation is only approximate for windows that start at a low periodicity. For instance, for the period-three window of Figure 16.2, the window begins at about k = 1.75, bifurcates into period six at about k = 1.768, and ends its period-doubling in chaos at about k = 1.79. Plugging these values into Equation 16.1 gives (1.79–1.75) / (1.768–1.75) = 2.222.

Intermittency within chaos works in a way opposite from within the pre-chaotic domain, as the control parameter increases. That is, rather than moving from order to chaos, the scenario now goes from chaos to order. At a k value just below that associated with a window, periodicity alternates with chaos as iterations continue. As k increases, the periodic parts become longer (i.e. range over larger intervals of x) relative to the chaotic bursts. At a critical value of k, the duration (interval) of the periodic behavior becomes infinitely long. In other words, the periodic attractor then becomes fixed or firm (Grebogi et al. 1987).

A special route to chaos, known as a **crisis** (Grebogi et al. 1982), shows up only within the chaotic domain. A crisis is an abrupt change in a chaotic attractor, such that with increase in the control parameter the attractor suddenly enlarges greatly or gets destroyed. We don't yet know how common such events are. They occur in nonlinear circuits and in lasers (Grebogi et al. 1983).

According to Grebogi et al. (1983), transitions between order and chaos are reversible, at least within the chaotic domain. That is, within critical zones of the control parameter (edges of windows), chaos can be created or destroyed, depending on increases or decreases in the control parameter.

The chaotic attractor

A third structural aspect or underlying framework of chaos is the chaotic attractor. During a chaotic system's long-term evolution, the variables don't haphazardly occupy every possible location within the phase space. Instead, the same chaotic attractor eventually takes shape, regardless of where the iterations began (value of x_0). Although different values of x_0 do lead to different trajectories on the attractor, the trajectories never leave the attractor, and the attractor occupies only part of the phase space.

Zones of popularity on the attractor

The fourth orderly structure of chaos is zones of relatively greater popularity on each chaotic attractor. These are zones a chaotic system is more likely to visit during its evolution. On the graphs of Figures 16.1 and 16.2, areas the trajectory visits the most are proportionally darker. The darkest zones are most likely to be visited, for a given k. In contrast, the trajectory never goes to white zones at all. The zones of greater popularity could provide some guidance in devising a statistical theory for accurately predicting the likelihood of x_t taking on a particular value (Jensen 1987). A fifth orderly feature of chaos is the fractal structure, discussed in the next chapter.

All this regularity indicates that order is a basic ingredient in chaos. That's why some people look upon chaos as order disguised as disorder. Some authors believe that pure chaos sets in only when the control parameter is at a maximum (such as k = 4 in the logistic equation).

Self-organization, complexity, and emergent systems

Judging from the logistic equation, most orderly features of chaos depend strictly on the control parameter. For a given value of that parameter, the order develops spontaneously, that is, without external cause. Such spontaneous development seems to be a special class of an interesting process called **self-organization**. Self-organization is the act whereby a self-propagating system, without outside influence, takes itself from seeming irregularity into some sort of order. It seems to reflect a tendency for a dynamical system to organize itself into more complex structures. The structure can be spatial, temporal, or operational (functional). The time over which the structure lasts varies from one case to another.

Examples of self-organization are the organizing of birds into an orderly flock, of fish into a clearly arranged school, of sand particles into ripple marks, of weather elements (wind, moisture, etc.) into hurricanes, of water molecules into laminar flow, of stars into the spiral arms of a galaxy, and of the demand for goods, services, labor, salaries, and so on, into economic markets. Briggs & Peat (1989) cite, as other examples:

- the lattice of hexagonal cells that form after the onset of chaos by heating a pan of liquid from below

- successions of ordered or oscillatory regimes that follow chaos in various chemical reactions
- termite nests resulting from random-like termite activity
- rush-hour traffic patterns following less-busy, earlier random-like traffic
- organized amoeba (slime mold) migrations appearing after random-like aggregation.

Self-organization is a main feature of a kind of behavior called **complexity**. In this specialized sense, complexity (Lewin 1992, Waldrop 1992) is a type of dynamic behavior that never reaches equilibrium and in which many independent particle-like units or "agents" perpetually interact and seek mutual accommodation in any of many possible ways. The units or agents spontaneously organize and re-organize themselves in the process into ever larger and more involved structures over time. "Complex" dynamic behavior has at least six ingredients:

- A large number of somewhat similar but independent items, particles, members, components or agents.
- Dynamism—the particles' persistent movement and readjustment. Each agent continually acts on and responds to its fellow agents in perpetually novel ways.
- Adaptiveness: the system conforms or adjusts to new situations so as to insure survival or to bring about some advantageous realignment.
- Self-organization, whereby some order inevitably and spontaneously forms.
- Local rules that govern each cell or agent.
- Hierarchical progression in the evolution of rules and structures. As evolution goes on, the rules become more efficient and sophisticated, and the structure becomes more complex and larger. For instance, atoms form molecules, molecules form cells, cells form people, and thence to families, cities, nations, and so forth.

Because of those characteristics, complex adaptive systems are called **emergent** systems. Their chief characteristic is the *emergence* of new, more complex levels of order over time. Complexity, like chaos, implies that we can't necessarily understand a system by isolating its components and analyzing each component individually. Instead, looking at the system as a whole might provide greater—or at least equally helpful—insight.

Authors used to categorize dynamical behavior either as orderly (e.g. having a fixed-point or periodic attractor) or random (Crutchfield & Young 1989). Chaos and complexity are important additional types. So, as of today, dynamical behavior might be classified into order, complexity, chaos, and randomness. In fact, those four types might form a progression or hierarchy, in the order just listed. Also, the four types aren't mutually exclusive. Orderly, chaotic, and random-like behavior, for instance, all have elements of determinism. Similarly, complexity and chaos both contain order and randomness.

To the extent that there is order within the chaotic regime, the word chaos (since its normal usage implies utter confusion or total disorder) is a misnomer. But it's too late now.

Summary

In spite of its name and appearance, chaos has a great deal of order or regularity. Examples are windows (regions of periodicity within the chaotic domain), various routes to chaos from within a window, the unique geometric shape (chaotic attractor) within the phase space, zones of greater popularity on the attractor, and the attractor's fractal structure. Because of those orderly features, you might see chaos mentioned in the context of two newly emerging concepts. One is self-organization—the process whereby a self-propagating system, without any apparent external influence, takes itself from seeming irregularity or uniformity into a pattern or structure. The other is complexity—dynamic behavior characterized by continuous give-and-take among many independent agents, resulting in an adaptive, self-organizing system that forms larger and more complex structures over time.

Chapter 17
Fractal structure

A fifth geometric characteristic or type of order within many chaotic domains is *fractal structure*. I'll discuss this within a general treatment of fractals and their characteristics.

Definitions

A fractal is a pattern that repeats the same design and detail or definition over a broad range of scale. Any piece of a fractal appears the same as we repeatedly magnify it. For instance, a twig and its appendages from the edge of some species of trees form a pattern that repeats the design of the trunk and main branches of the tree. Such repetition of detail, or recurrence of statistically identical geometrical patterns as we look at smaller-upon-smaller parts of the original object, is the unifying theme of fractals. Fractal patterns don't have any characteristic size.

The definition just given is general. Experts don't agree on a more explicit or mathematical definition.[1]

Fractals are all around us. Examples are:

- branching configurations (e.g. in bronchial tubes, blood vessels, coral reefs, mineral deposits, river networks, trees, ferns)

---

1. The word "fractal" comes from a Latin word *fractus*, which loosely means "to break into irregular fragments." "Fractal" was coined in 1975 by the acknowledged "father of fractals," **Benoit B. Mandelbrot** (1924–) of International Business Machines, Yorktown Heights, New York. Mandelbrot was born in Warsaw, educated mostly in Paris, and has held many university positions in Europe and the USA. As an applied mathematician, he has contributed to such varied subjects as statistics, economics, engineering, physiology, geomorphology, languages, astronomy, and physics. He is a self-designated "nomad-by-choice" in regard to the topics he works on (Gleick 1987: 90). Many mathematicians and theoretical physicists at first disregarded his work on fractals, but fractals now seem to be here to stay. (As a separately defined concept, fractals are only a couple of decades old, although their mathematical antecedents go back many decades.) Mandelbrot is now a Professor of Mathematical Sciences at Yale University and a Watson Fellow at the IBM Thomas J. Watson research center in Yorktown Heights, New York.

- rough surfaces in general (landscapes, mountains, outcrops of rock)
- fracture networks and cracks in a surface
- objects that undergo fragmentation (coal and rock, soil, asteroids and meteorites, volcanic ejecta)
- objects that result from aggregation or disorderly growth (soot, dust)
- viscous fingering (e.g. injection of water or air into a subsurface oilfield)
- earthquake features (distribution of their magnitudes, spatial distribution, frequency distribution of aftershocks)
- flow patterns (turbulence, eddies, wakes, jets, clouds, smoke, other mists, streamlines of water particles)
- the irregular trace of a water/land interface (edge of a mud puddle, shoreline of a continent)
- galaxy distributions, size distribution of craters, rings of Saturn, fluctuations in interplanetary magnetic fields
- lightning
- changes in stock prices
- incomes of rich people.

Figure 17.1 shows a few photographs of fractals.

Fractals and chaos

Fractals deal with geometric patterns and quantitative ways of characterizing those patterns. Chaos, in contrast, deals with time evolution and its underlying or distinguishing characteristics. Fractals are a class of geometric forms; chaos is a class of dynamical behavior.

Fractals and chaos are closely intertwined and often occur together. For instance, most chaotic attractors have a fractal striated texture. Points on such attractors plot as a set of layers that look the same over a wide range of scales. The layered structure can be difficult to see because of noise and the small number of points in some datasets. The points then seem to plot on a single curve. Figure 17.2 is a chaotic attractor (Eq. 13.11) on which the same amount of detail appears as we repeatedly zoom in on smaller and smaller sections (and enlarge and refocus them for appropriate viewing). In general, "the chaotic attractors of flows or invertible maps typically are fractals; the chaotic attractors of noninvertible maps may or may not be fractals. The chaotic attractors of the logistic equation, for example, are not fractals" (Eubank & Farmer 1990). Other chaos-related geometric objects, such as the boundary between periodic and chaotic motions in phase space, also may have fractal properties (Moon 1992: 325). Because of those close relationships, fractals can help detect chaos.

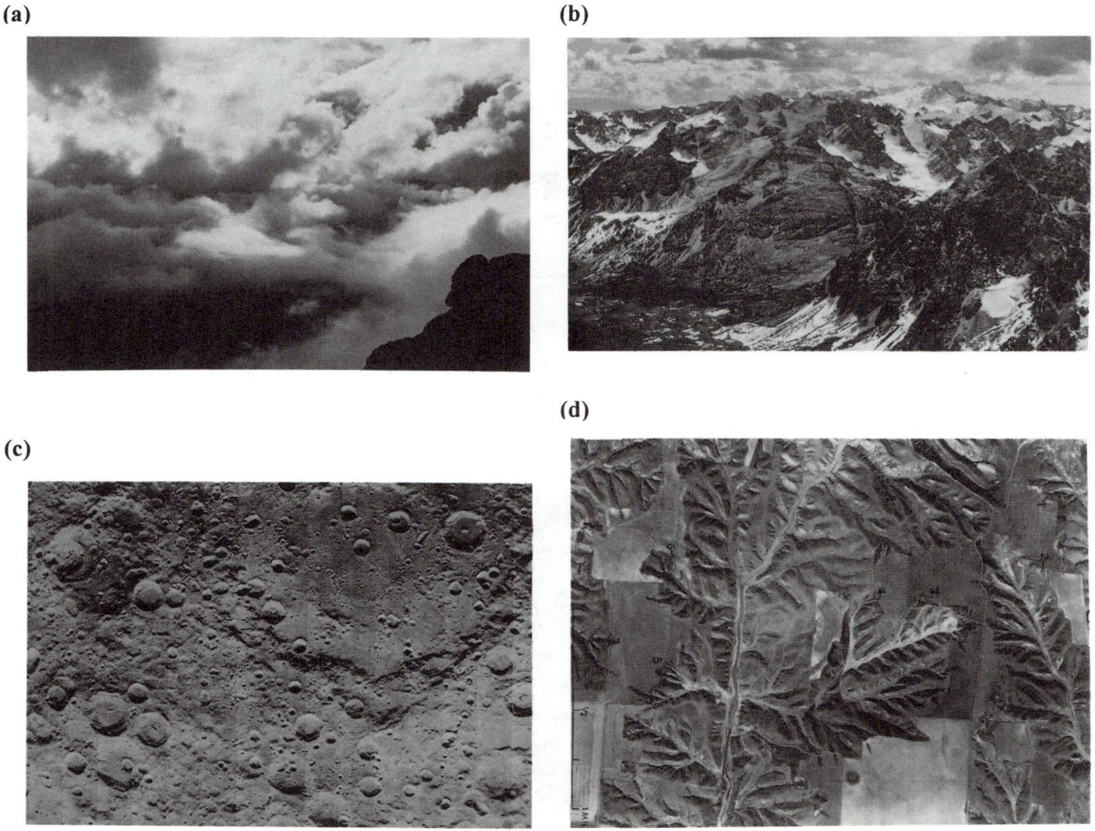

(a)

(b)

(c)

(d)

Figure 17.1 Photographs of natural features that have given rise to fractal patterns. (a) Clouds (photo: Nolan Doesken). (b) Earth surface topography (photo: Bob Broshears). (c) Asteroid impact craters, here on the Moon (photo: NASA/USGS, Lunar Orbiter I, 38M). (d) Gullies (photo: James Brice).

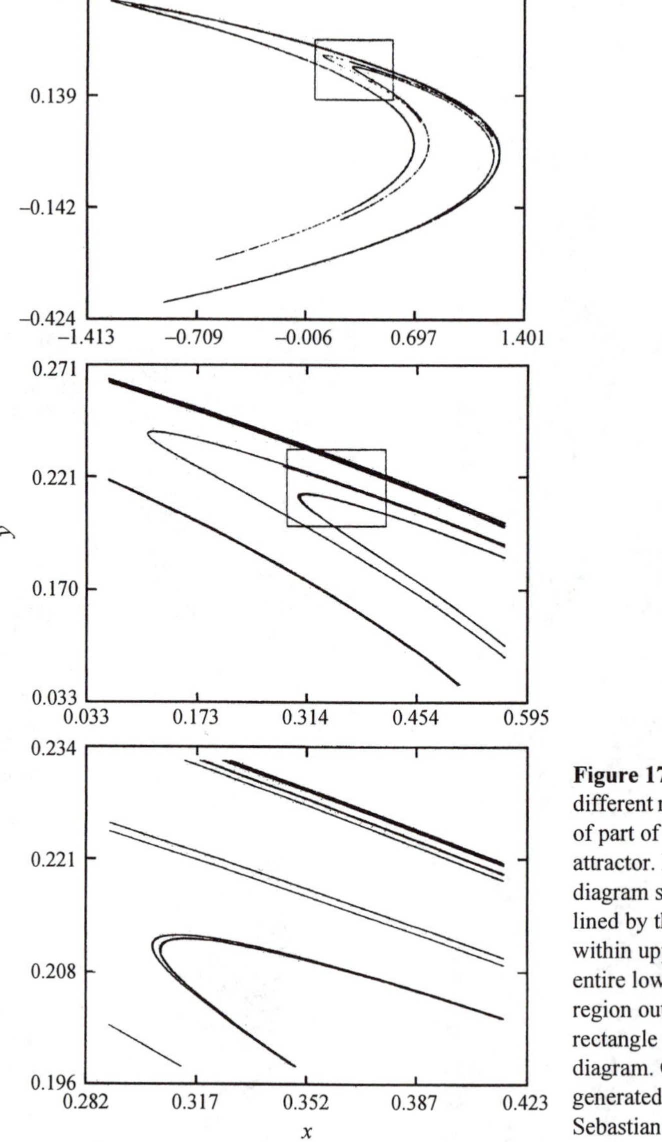

Figure 17.2 Three different magnifications of part of the Hénon attractor. Entire middle diagram shows area outlined by the rectangle within upper diagram; entire lower diagram is region outlined by the rectangle within middle diagram. Computer-generated graphics by Sebastian Kuzminsky.

Characteristics

All fractals have four common characteristics.

- They have a statistical geometric regularity. The amount of detail, or the geometric structure, look the same at one length scale as at another length scale. The recurrence of the same pattern over a range of scales is called **self-similarity** or **scale invariance**. People define these terms in slightly different ways, but the two terms mean the same thing.

  Self-similarity means that any part of the object, enlarged and refocused to whatever extent necessary for comparison, looks like the whole. The central theme is the preservation of detail upon magnification. Objects that aren't self-similar don't keep the same degree of detail upon magnification. Theoretically at least, a self-similar object has infinite detail.

  Scale invariance means that the object lacks a preferred or characteristic size (length, radius, etc.). Something scale-invariant is independent of size (looks pretty much the same at all sizes). Many geologic patterns, for example, are scale-invariant. (And that's why geologists place something of known scale in photographs of rock outcrops. Geology has few characteristic lengths.)

  Fractals don't have to be exactly self-similar. Some are only approximately or statistically self-similar, as explained below. Others—called **self-affine**—are only self-similar if some rotation, stretching, squashing, or other distortion (i.e. a direction-dependent rescaling) accompanies the magnification. (For instance, in two dimensions, one dimension might be stretched more than the other. In three dimensions, the two dimensions in the horizontal might retain the same relative scaling, but the vertical dimension would be scaled differently.)
- A characteristic number, in the form of a noninteger dimension (discussed in later chapters), quantifies the scaling of their pattern or complexity over a range of scales.
- Fractals usually are generated by many repetitions of a given operation. The operation might be natural, such as the disintegration of rock fragments, or it might be a mathematical exercise, as in solving the same equation over and over (iteration). If done mathematically, the many repeated calculations ordinarily require a computer.
- Fractals aren't smooth. They generally look rough, broken, disorganized, jagged, bumpy, or shaggy. Iteration, for example, doesn't produce a smooth object. Instead, it produces a line or surface that has detail over a wide range of scale. Consequently, there's no simple algebraic equation that can specify a particular spot on a fractal object.

An object can have one of the above features and still not be a fractal. For instance, a straight line, square, and cube are all self-similar but aren't fractals. Also, iteration doesn't necessarily yield a fractal.

Fractals are of two types (Saupe 1988: 72), as follows:

- **Deterministic** (ordered or exact) **fractals** These fractals look exactly the same (even in minute detail) at all levels of magnification. Theoretically at least, their range in terms of size is unlimited or infinite; they range over all scales, with no upper and lower limits. They are more common in mathematical and theoretical worlds. Popular examples in the literature are the von **Koch snowflake**, **Cantor set**, and Sierpinski gasket.
- **Natural** (approximate, statistical, or stochastic) **fractals** These are the fractals found in nature. We can compose them according to the same rules as deterministic fractals but with the additional element of randomness or noise included. The random element is critical for reproducing natural features, such as landscapes, branches, and coastlines. Such features, when scaled down or up, never look completely alike in detail; they only look generally or statistically alike. (All differences between the original and scaled versions are attributable to chance.) A second important distinction between natural and deterministic fractals is that natural fractals, since they simulate natural objects, are limited to a range of sizes. They exist only over a finite range of sizes. A rock particle, for example, can't be smaller than an atom and can only be so large, even though we don't know exactly how large. Because of that size limitation, there's an upper and lower limit beyond which self-similarity no longer holds for a natural fractal.

Fractals first became established in a spatial context (in analyzing static shapes, such as trees). Further work (West 1990, Plotnick & Prestegaard 1993) has begun to make a case for considering them in statistical and dynamical (temporal) contexts as well; certain diagnostic statistics recur at all levels of magnification. In a fractal dynamical sense, a time series looks the same over a wide range of timescales.

Value of fractals

Like chaos, fractals generate controversy within the mathematics community (Pool 1990a). Some mathematicians charge that fractals lack real mathematical content, theorems, and proofs and that fractals emphasize pretty pictures or computer-generated designs. In addition, fractals, like chaos, don't provide physical explanations. Defenders rejoin that fractals are a new and efficient way of mathematically describing natural objects (e.g. Burrough 1984). That is, fractals are a mathematical way of describing the variability of irregular features over a range of scales.

Fractals have geometric features that make them useful as idealized models for our natural world. Examples include the human circulatory system (Barcellos 1984), soil pollution, cave systems, the seeping of oil through porous rock (Peterson 1988), drainage networks, and coral reefs. The general principle in such modeling usually is to add the same structure at progressively smaller scales.

Scientists see fractals partly as a vehicle for opening up a whole new realm of questions about the physical world. Why is an object fractal? What physical processes produce fractals? Why is the same scaling valid over a wide range of scales? Mathematicians can use fractals to develop an intuition for certain mathematical problems. Such intuition in turn brings new conjectures and new approaches to solving important theorems.

Society in general (especially the television and motion picture industries) has taken a keen interest in fractals. Fractals, used with computer graphics, provide beautiful and realistic forgeries of many natural shapes, such as rocks, mountains, landscapes, stars, planets, and the like. The Star Trek television programs and motion pictures are a notable example of applying fractals in that way. Furthermore, fractals are easy and fun to generate, at least for someone who likes personal computers. Commercially available computer software programs show fractals in fascinating color and detail.

These and other important and practical applications came about only as unexpected by-products, after the concept of fractals had crystallized and been developed. That sort of thing happens quite often in the world of research. The applications show that basic research—research that may not promise any obvious benefit when first undertaken—can lead to practical and useful ends.

Summary

A fractal is a line, surface, or pattern that looks the same over a wide range of scales. In other words, a fractal is self-similar (any part of the object, suitably magnified and refocused, looks like the whole), and scale-invariant (lacks a characteristic size). Other typical features of fractals are that a single number (a noninteger dimension) describes their scaling, they can be formed by many repetitions of the same operation, and they are never smooth (being instead rough, jagged, etc.). A chaotic attractor usually is a fractal. The two main types of fractals are deterministic and natural fractals. Deterministic fractals are generated by iterating an equation or by following a specified numerical recipe. Therefore, they are exact. Natural fractals include noise and therefore are only approximately similar under change of scale. Natural fractals simulate natural objects, such as rocks, surface textures, and landscapes.

PART V

PHASE SPACE SIGNATURES

Pseudo phase space graphs are basic tools in studying chaos. Two reasons for their prominent role are that they require only one measured variable and they often indicate determinism hidden within an erratic outward appearance. Let's now look at several common and useful varieties of such graphs.

Chapter 18
Uncovering determinism

Poincaré sections

Some tools, such as hammers and saws, are favorites because of their simplicity, importance and usefulness. The Poincaré section is an example. A Poincaré section (named after French mathematician Henri Poincaré)[1] is merely a geometric picture of a trajectory's activity at a cross section of the attractor. The cross section is a slice or section through the attractor, transverse to the "flow" or bundle of trajectories (Fig. 18.1). "Transverse" here means "not parallel." The idea is to orient the imaginary cross section such that trajectories pierce it approximately normal to the surface of the section. (In fact, a Poincaré section sometimes is called a **surface of section**.) The main purposes of a Poincaré section are to provide a different view of the system's dynamics and possibly to help identify the type of attractor.

A Poincaré section is like a paper screen set up across the paths of the trajectory (Stewart 1989b). Every time the trajectory comes back around to pierce that section

---

1. **Poincaré, Jules-Henri (1854–1912)** Poincaré was a French mathematician, theoretical astronomer, and philosopher of science. He was probably the greatest mathematician of his time (around the turn of the twentieth century). As a child, he was ambidextrous but otherwise poorly coordinated, nearsighted, and at one time seriously ill with diphtheria. He excelled in written composition in elementary school and wrote a five-act tragedy in verse at age 13. Poincaré always had an exceptionally retentive memory, especially for numbers and formulas. Throughout his life he was able to do complex mathematical calculations in his head. Curiously, however, he supposedly was a poor chess player and virtually incapable of adding up a column of numbers correctly. He wrote his PhD thesis on differential equations, receiving his degree in 1879 from the Ecole Nationale Supérieure des Mines. He spent most of his professional career at the Université de Paris, where he became a full professor at the age of 27. There he combined his vast talents for mathematics, physics, and writing to produce nearly 500 papers and several books. "Changing his lectures every year, he would review optics, electricity, the equilibrium of fluid masses, the mathematics of electricity, astronomy, thermodynamics, light, and probability. Many of these lectures appeared in print shortly after they were delivered at the university . . . as he could quickly write a paper without extensive revisions" (*Encyclopedia Britanica*). Several of his major books dealt with mathematical methods in astronomy and the philosophy of science. He was also much interested in the role of the subconscious mind in mathematical creation and discovery. Poincaré married at age 27 and fathered three daughters and a son. At the relatively young age of 32 he was elected to the French Academy of Sciences. He died in 1912 at age 58, of an embolism following prostate surgery.

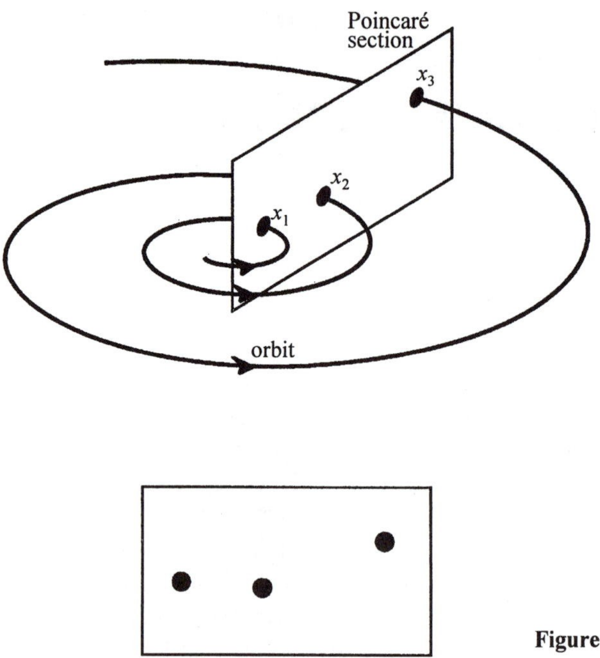

Figure 18.1 Poincaré section.

or hoop, it leaves a dot on the section (a hole through the paper). The section therefore shows a pattern of dots or points. Each point stands for a return visit by the trajectory. Usually, only trips that enter the section from one chosen side count. Penetrations made from the opposite direction make up either a separate and different Poincaré section, or, if combined with those entering from the opposite side, a two-sided Poincaré section.

By implication, the Poincaré section isn't just the slice through the attractor but also what's on the slice, namely the dots. Each dot has an address—the phase space coordinate values of the intersections of the trajectory with the slicing plane. It's possible to label the dots on a Poincaré section chronologically. However, the section itself is a purely geometric picture, described in terms of its phase space location and orientation.

Slices through four or more dimensions aren't useful, since the resulting sections can't be graphed on paper. Therefore, the Poincaré section concept applies only to attractors of three or fewer phase space dimensions. We can take the section either in standard or pseudo phase space. A computer is the standard tool for constructing a Poincaré section.

Although we can choose the section's location, it may not intersect any datum point exactly. In those cases, we get a trajectory's value at the section by interpolation (Ch. 4). An interpolation implies some sort of regular progression from one point to the next, as with continuous flows. That means that we can properly take a Poincaré section only for a *continuous* system, even if we sample that continuous system at discrete times. (With discrete systems such as the logistic equation, the successive observations bounce specifically from one observation to the next, without traversing a continuum of points in between. Therefore, interpolating between two sequential observations and constructing a Poincaré section for a *discrete* system is meaningless and misleading.)

The time that elapses between successive returns to the section is irrelevant in a Poincaré section. Those time intervals are constant in some cases, but in others they vary by a lot.

Don't confuse a Poincaré section with a **Poincaré map**. The Poincaré section is a surface (line or plane) through phase space, transverse to the trajectories. In contrast, the Poincaré map is a rule, function, graph or model that tells where the trajectory will next cross the Poincaré section, given the location of the previous crossing at the selected Poincaré section. The Poincaré map sometimes is called a return map, since it gives information about the trajectory's returns to that particular section.

A Poincaré section in general defines a certain range of conditions (values of phase space variables) that the system revisits from time to time. It also shows, by the density of dots on the plot, the relative frequency with which the system goes to various subareas within that range of conditions. Hence, the section is a way of sampling or probing the system to get details of the revisits.

Many authors refer to a Poincaré section as a **stroboscopic** map. (A stroboscope is an instrument for observing a moving body by making it visible intermittently, at some constant time interval.) Such references aren't always correct. Successive points on a one-dimensional stroboscopic map can be anywhere within the phase space. That is, sampling (or plotting selected data) at equal time intervals often doesn't correspond to sampling in the same plane through the attractor. There are exceptions, as with periodic motion (limit cycles), quasiperiodic motion, or so-called **forced oscillators** (devices to which some extra energy is periodically added). In such special cases, sampling at equal time intervals can correspond to sampling in a plane. Dots on the Poincaré section then represent a series of stroboscopic snapshots of the plane. We can then treat a stroboscopic map and Poincaré map as synonymous. In most cases (including chaotic systems), however, the two aren't the same; the route and duration of the trajectory between visits to a particular cross section can be very different from one trip to the next.

The Poincaré pattern of different attractor types

Patterns on Poincaré sections fall into categories according to the type of attractor. Ignoring the case of the point attractor, there are three main groups. These correspond to periodic, toroidal, and chaotic attractors.

- *Periodic attractors* The periodic attractor or limit cycle, as mentioned earlier, is a two-dimensional figure that always follows the exact same route (a closed loop) in standard phase space (Fig. 18.2). A cross section or slice (Poincaré section) through one side of such a two-dimensional attractor is just a line. It shows one or a few dots. The number of dots usually corresponds to the period of the attractor. For example, an attractor of period one shows one dot, because the trajectory perpetually repeats the same single loop (returns to the same unique spot on the slice) in phase space. An attractor of period two shows two dots (Fig. 18.2); and so on. In practice and for various reasons, the limit cycle might not be a perfect loop. Several points close together then might represent any one period.

- *Toroidal attractors* The torus, as you recall, is an imaginary surface shaped (in its simplest form) like a three-dimensional doughnut (Fig. 11.7). A winding or coiled-wire action of a trajectory in phase space forms the torus. All the

Attractor **Poincaré section**

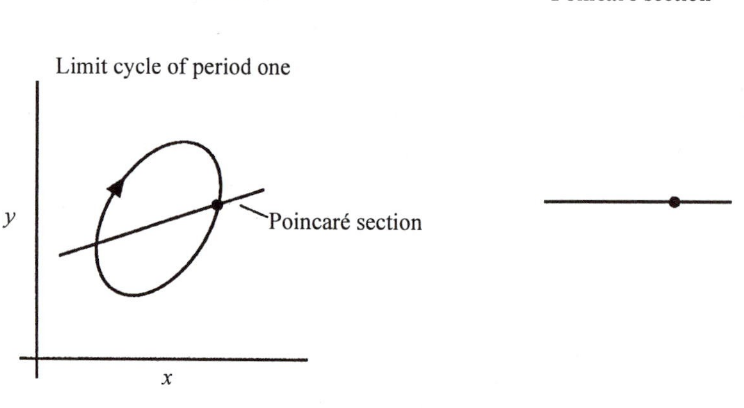

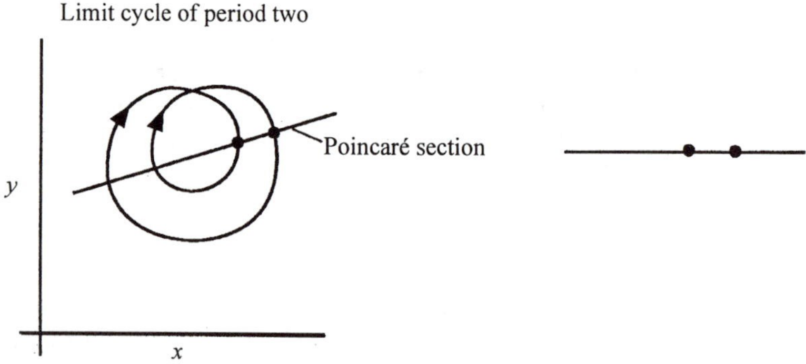

Figure 18.2 Limit cycles of period one (upper left) and period two (lower left) with associated Poincaré sections (right column).

action takes place on its surface, none in the interior. Recurring visits to a cross section taken through the doughnut produce a pattern like a dotted ring or circle (Fig. 18.3). In practice, the section rarely looks like a nice circle. Instead, it's usually a closed but deformed loop. As such, it has many distorted shapes.

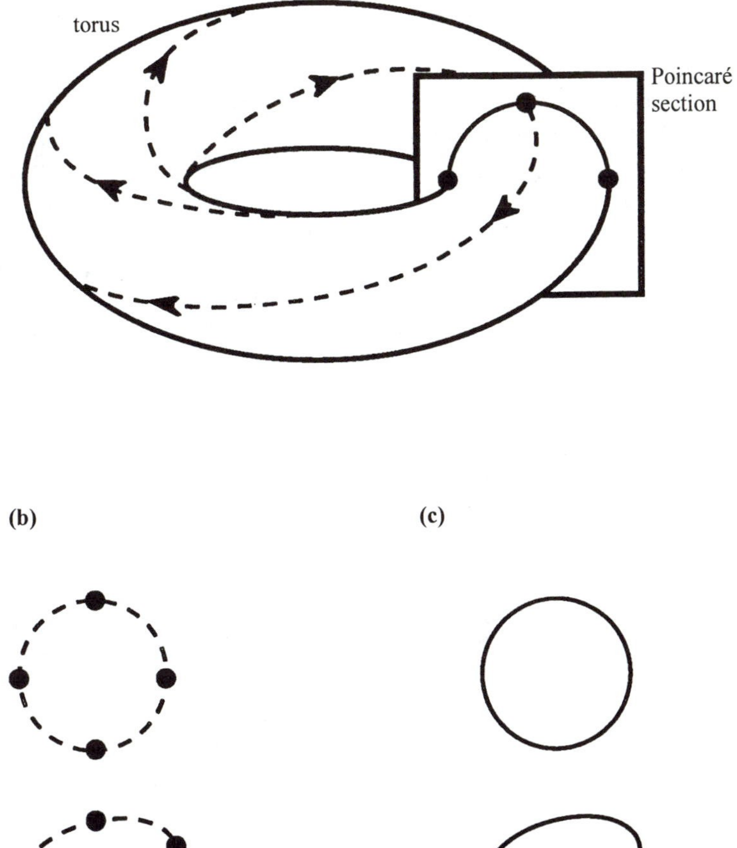

Figure 18.3 (a) Periodic toroidal attractor with four Poincaré mapping points (after Thompson & Stewart 1986). (b) Poincaré sections for periodic toroidal attractors (upper example corresponds to torus of (a)). (c) Poincaré sections for quasiperiodic toroidal attractors. In (b) and (c) the ring can have any of various shapes, depending on the angle at which the section is taken through the attractor.

The Poincaré section for a torus is either of two types (Bergé et al. 1984: 70). Both types are most understandable for a 2-torus (a torus made up of just two frequencies, as with Fig. 18.3). The two classes depend on whether the ratio of the two frequencies is rational (the ratio of two integers) or irrational (not expressible as the ratio of two integers).

A rational ratio means that the toroidal attractor is periodic, that is, the trajectory repeats itself at even time intervals. Therefore, it only visits certain points within the cross section. The ring on the Poincaré section then consists only of those points. Hence, we've got to fill in the rest of the ring with our imagination (Fig. 18.3b).

An irrational ratio of the two frequencies, on the other hand, means that the toroidal attractor is quasiperiodic. In that case, the trajectory never repeats itself exactly. Given enough time, it would pass through every point on the cross-sectional ring sooner or later. The Poincaré section then would show a ring that's completely filled or continuous (Fig. 18.3c). Finding such a ring isn't reliable proof of quasiperiodicity, however, as it sometimes represents a chaotic attractor (Bergé et al. 1984: 66). Also, in practice there may not be enough time for the ring to become continuous. Figure 18.4, for example, shows the Poincaré section of a quasiperiodic torus measured in a laboratory experiment on fluid flow.

- *Chaotic attractors* Poincaré sections through chaotic attractors usually show some type of organization but with relatively complex geometry (Fig. 18.5). In addition, an infinite number of points usually makes up a Poincaré section through a chaotic attractor (in contrast to the finite number of points on a section through most periodic attractors). Thirdly, there's often a distinctive fine structure—an orderly and fractal-like texture that typically has layers, like a flaky pastry or a slice through a head of lettuce. Scattered here and there are large zones with no points—conditions that the system avoids. The structure becomes better and better defined as more and more points appear (i.e. as the system evolves). Curiously, however, new points usually don't come forth in any regular spatial order. Instead, they seem to pop up at haphazard locations all over the attractor. Poincaré sections through chaotic attractors can also look like a heterogeneous cloud with no apparent order (Moon 1992: 63).

Where to take the slice

A Poincaré section theoretically can be along any plane in the phase space. However, taking the section along a particular axis of the phase space (so that one variable is constant) simplifies the analysis. For instance, the slice of Figure 18.5 is along the x axis and shows what happens in the x–z plane; the variable y is constant. Which variable to hold constant is up to us. The most information probably comes from taking the slice where it will show many penetrations. Hence, with the attrac-

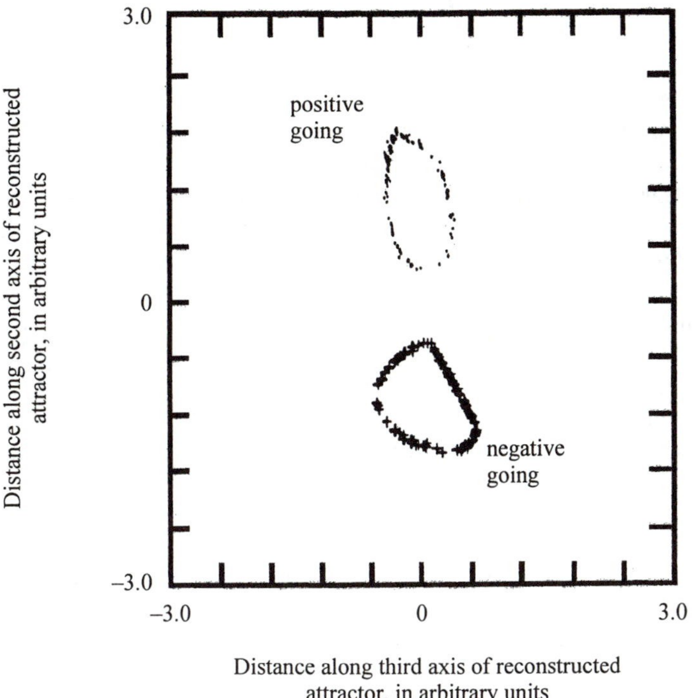

Figure 18.4 Poincaré section of a quasiperiodic torus for laboratory measurements of fluid flow (adapted from Read et al. 1992:610, with permission from Cambridge University Press).

tor of Figure 18.5, it's best to slice the attractor vertically along the x axis (as shown) or along the y axis; taking the section horizontally (holding z constant) probably wouldn't reveal much useful information.

For various reasons, people often don't take the section along a particular axis of the phase space. As a result, the vertical and horizontal figure boundaries on a separately drawn Poincaré section often don't correspond to the axes on the ancestral phase space plot.[2] As a result, the straight borders around many Poincaré sections don't represent phase space axes and don't have labels. The values along the borders (Poincaré section axes) in fact are in *arbitrary distance units*, as measured from some arbitrary origin. Such units often aren't indicated on the plot, however. In practice, it's a good idea to take many sections, because each section is unique and gives different information.

2. They do for a line if we don't rotate that line relative to the two phase space axes. They do for a plane only if we take the slice parallel to one of the phase space axes and don't rotate the slice relative to the other two phase space axes, in drawing the Poincaré section.

Steps in taking a Poincaré section

Drawing from the above comments, here are general steps in constructing a Poincaré section.

1. Plot the raw data in phase space or pseudo phase space.
2. Decide on the location and orientation of your Poincaré section. (The only two choices for the type of section are a line [if the phase space is two-dimensional] or a plane [if the phase space is three-dimensional].)
3. Get the parameters for the equation of your chosen section. Do this by using the type equation and any two points (on a line) or any three points (on a plane). Put the equation in the form $y-bx-c=0$ (for a line) or $ax+by+cz+d=0$ (for a plane).
4. Using the equation (including parameter values) for your section, insert the coordinates of each datum point in turn and solve the equation. The resulting number is positive if the datum point is on one side of the section and negative if it's on the other (Fig. 18.6).
5. Decide whether you want your Poincaré section to represent points entering from the positive side or negative side.
6. If you want the section to represent the positive-to-negative direction, find all

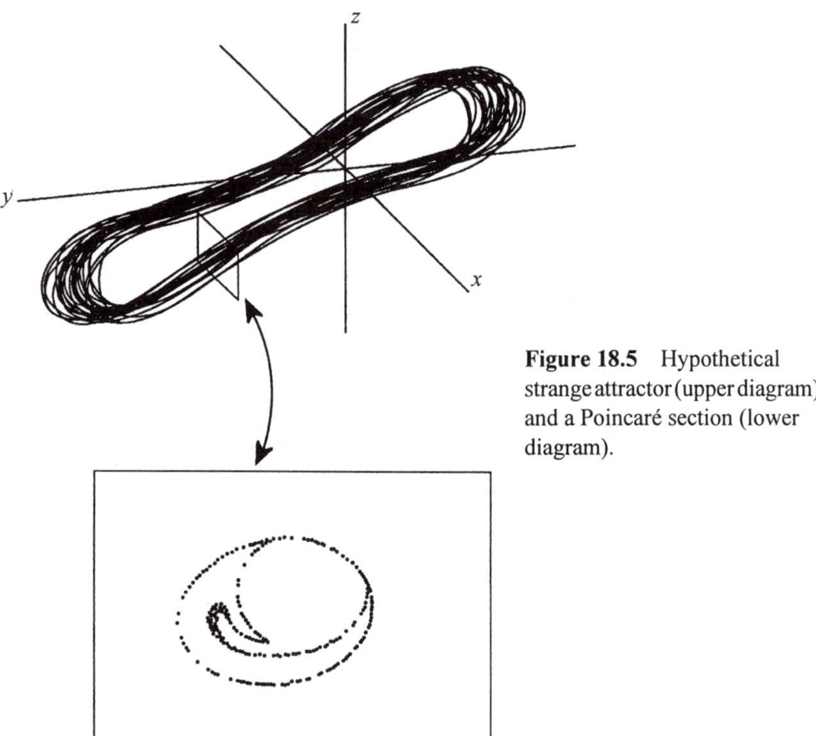

Figure 18.5 Hypothetical strange attractor (upper diagram) and a Poincaré section (lower diagram).

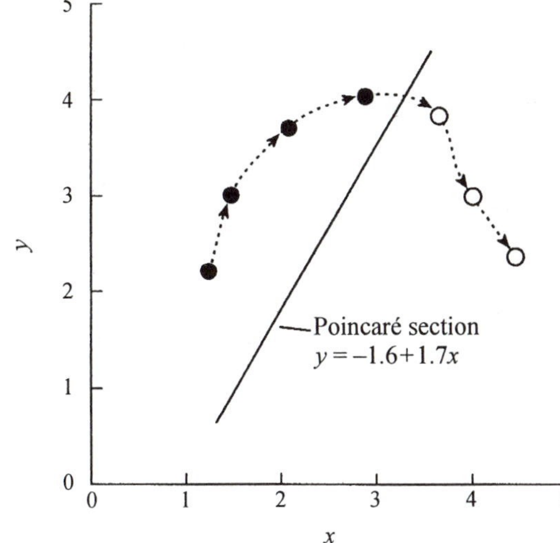

Figure 18.6 Hypothetical trajectory and Poincaré section. Inserting the coordinates for each point into the equation for the Poincaré section identifies a point's general location as being left of the section (positive solutions) or right of the section (negative solutions).

pairs of successive points in the step-four computations that have +– signs, respectively. If, on the other hand, you want a negative-to-positive representation, then locate all –+ pairs. These points represent observations that bracket the Poincaré section and that go in your desired direction.

7. For each qualifying pair, interpolate (Ch. 4) to get the coordinates where the trajectory intersects the Poincaré section.
8. Plot the Poincaré section as a separate figure.

Advantages of Poincaré sections

Under ideal circumstances (e.g. little or no noise), a Poincaré section can indicate the type of attractor, as described above.

A second advantage is that the section can reveal much useful information about the attractor's internal structure. The Poincaré section is a different way of looking at an attractor. Because it deals with just one cross section through the attractor, it tells us absolutely nothing about the rest of the attractor (the attractor's shape, general location in phase space, and so on). On the other hand, it gives information that usually isn't apparent on the phase space plot. In particular, it shows us the attractor's internal or cross-sectional make-up. Also, it brings that anatomy out in much greater detail and clarity compared to a standard phase space plot. For instance, for the three-dimensional object of Figure 18.5, going to a Poincaré section means losing the information or perspective provided by the three-dimensional view, especially the information about what the trajectory is doing in between returns to the section. On the other hand, the Poincaré section greatly enhances the information for that section through the attractor. It shows much more clearly the cross-sectional zones where most of the action is.

Thirdly, a Poincaré section simplifies the study of the trajectory and attractor. It does so because it has at least one fewer Euclidean dimensions than the attractor's phase space. For instance, say an attractor lies in a phase space of two dimensions, such as y versus x. A slice through that phase space or attractor is a line (Figs 18.2, 18.6), and a line as a geometric object has a Euclidean dimension of one. (That line's orientation on the plot doesn't matter—the Poincaré section is still a line, with dimension one.) Thus, the phase space dimension is two and the Euclidean dimension is one. The next degree of complexity is a phase space of three dimensions (x, y, z). A slice through that phase space or attractor is a plane (Figs 18.1, 18.3, 18.5), and a plane as a geometric object has two dimensions. (Again, that plane's orientation on the plot makes no difference. As a plane, it still has only two dimensions.) Here the phase space dimension is three, and the Euclidean dimension is two.

In some cases, the Poincaré section has two fewer dimensions than the phase space. One example is a limit cycle of period one. The phase space for a limit cycle has two dimensions. However, a limit cycle of period one shows up as a single dot (dimension zero) along the line that makes up the Poincaré section (Fig. 18.2). So, the Poincaré section has reduced the number of variables from two to zero. In another example, a three-dimensional attractor might be just a plane in certain regions of its phase space. A Poincaré section through that planar region is a line (dimension one). Again, therefore, the Poincaré section has two fewer dimensions than the phase space.

A fourth advantage to a Poincaré section is a vast reduction in the size of the dataset. That's because we can ignore almost all points along the trajectory and analyze only the points on or near the Poincaré section. Finally, the analysis doesn't need nearly as much computer time and capacity, because of the fewer points involved.

In summary, there are five advantages to taking a Poincaré section:

- The pattern might indicate the type of attractor.
- Information not otherwise available emerges about the attractor's internal structure. The attractor's cross-sectional dynamics emerge much more clearly.
- The number of coordinates (dimensions) decreases by at least one.
- The quantity of data to be managed is significantly less.
- The required computational time and computer capability are much less.

Return maps

A return map is a rule, function, or model that tells where the trajectory will next cross a particular Poincaré section, given the phase space location of a previous crossing at that section. Such a rule or map has at least two purposes or uses: (a) in graphical form, it reveals any temporal order to the trajectory's repeated visits to the phase space cross section; (b) if definable, it can help us predict successive trajectory excursions (successive states of the system) at the section.

Any local region of a phase space attractor might include several trajectory paths close together. The Poincaré section for such a neighborhood shows several dots close together. Those paths or dots appear only in a spatial or geographic relationship. That spatial arrangement doesn't tell us anything about the chronology at the section. The oldest dot, for instance, might be any dot in the group. Figure 18.7a shows the standard phase space plot for a hypothetical limit cycle of period eight. Figure 18.7b is a Poincaré section. The numbers next to the Poincaré section dots in Figure 18.7b indicate the chronological order. Thus, a Poincaré section by itself does a nice job of showing the spatial structure but provides no easily discernible information about the temporal order. Constructing a **first-return map** gives a better picture of the temporal history at the section.

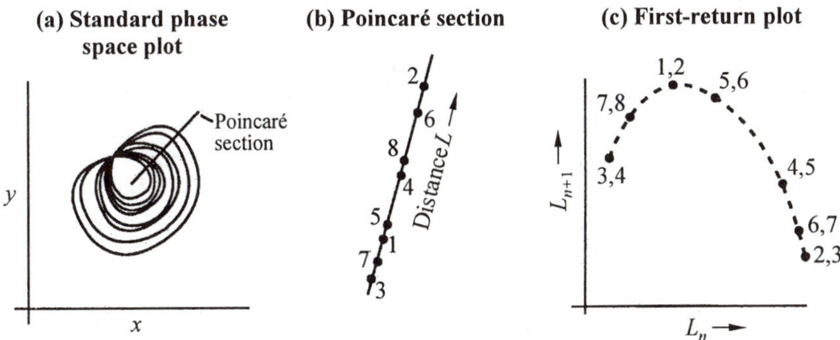

Figure 18.7 (a) Phase space plot, (b) Poincaré section, and (c) first-return map for a hypothetical limit cycle of period eight (after Cvitanovic 1984: 9–12). Numbers on (b) and (c) represent chronological order of trajectory's intersections with the Poincaré section.

A first-return map is a model (usually a graph) that rearranges the spatially ordered Poincaré section points into their temporal sequence. It does so by plotting each observation at the section versus the next observation at the section (Fig. 18.7c). (The system visits other locations in the phase space, in between returns to our Poincaré section. However, those intervening observations don't matter here.) The words "first return" mean that each plotted point is defined in terms of its value at a given penetration of the cross section and its value as of its first return or next visit to that section. In more general terms, a first-return map is the action, rule, function, equation, graph, or model that takes each point on a Poincaré section to its next (first-return) point on that section. Because it applies to a particular Poincaré section, some authors refer to it as a **Poincaré return map** or Poincaré map.

As with a Poincaré section, values along the borders in Figure 18.7c are in arbitrary distance units along the section, as measured from some arbitrary origin on the section. Hence, I'll symbolize any given penetration by L_n, where L is distance in any chosen units and n is the nth penetration at the section (Fig. 18.7b). The associated value or distance along the line as of the next return then is L_{n+1}.

A first-return map shows whether there's some kind of systematic relation between one observation and the next at the section. In other words, it tells whether there's a recognizable history in the data sequence. If there's no systematic relation, plotted points scatter randomly on the first-return graph. A systematic relation (e.g. a well defined curve, as in Fig. 18.7c), in contrast, means that the trajectories on the attractor follow some kind of hidden rule and have some kind of order. The system's long-term evolution, in other words, is somewhat orderly rather than random-like. To some extent, future predictions are reliable.

Plotting an observation L_n and the second return L_{n+2} gives a **second-return map** or a **second-order return map**. Similarly, L_{n+3} versus L_n is a third-return map or third-order return map; and so on. Technically, a return map needn't be a graph.

Although the first-return map involves the sequential ordering of the points on the Poincaré section, that doesn't by any means imply that the time intervals between successive observations are equal (constant). In fact, they usually aren't equal (unless the map is stroboscopic), since the trajectory's route usually varies with time. In general, a first-return map doesn't imply or require a constant time interval between successive observations. It only involves a chronological ordering.

Next-amplitude (successive-maxima) plots

Sometimes a feature oscillates, although not necessarily regularly. The peak value of an oscillation sometimes is called a *maximum* or an **amplitude**.[3] It's easy to identify relatively high values in the basic data. They are the peaks on the variable's plotted time series, or they plot in an outer zone of the attractor along the variable's axis in phase space. Select a variable that best exemplifies the occasional peaks or oscillations. Then make a list of the successive peak values (ignoring the length of time the system took between each trip). Plotting each value (on the abscissa) versus its successor produces a **next-amplitude plot** or a plot of **successive maxima**.

The peak values plotted on the graph almost certainly haven't all occurred at the same phase space location. Thus, in contrast to a Poincaré section or return map, a next-amplitude plot doesn't represent behavior at a particular phase space cross section.

In chaos the next-amplitude plot stems from Lorenz (1963). His attractor (Fig. 18.8) has three dimensions (x, y, and z) and consists of two lobes. The trajectory makes many circuits around one lobe, then switches to the other. Lorenz noticed (1963: 138) that the trajectory apparently leaves one lobe only after exceeding some critical distance from the center. Also, the amount by which it exceeds that distance apparently determines the point at which it enters the other lobe. That location in turn seems to determine the number of circuits it executes before changing lobes again. He therefore deduced that some single feature of a given circuit,

3. Amplitude, strictly speaking, is the extreme range of a fluctuating quantity, such as a pendulum, alternating current, tidal cycle, etc. Hence, the chaos usage is a bit loose.

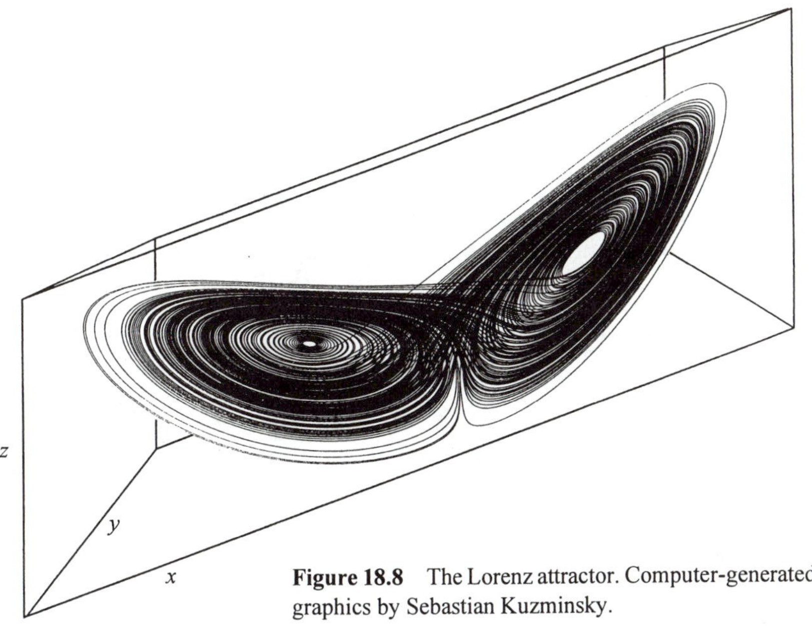

Figure 18.8 The Lorenz attractor. Computer-generated graphics by Sebastian Kuzminsky.

namely the maximum value of z, would predict the same feature (maximum value of z) of the following circuit. From the long string of iterates, Lorenz picked out only the occasional high values of z. He then drew a graph (map) with each high z on the abscissa versus the following high z (ordinate). Sure enough, a well defined line emerged, in this case shaped like a tent (Fig. 18.9). Lorenz concluded that "an investigator, unaware of the nature of the governing equations, could formulate an empirical prediction scheme from the 'data' of [the figure.] From the value of the most recent maximum of z, values at future maxima may be obtained . . ."

A next-amplitude graph needn't show a sharp peak. In fact, it can plot as anything from a well defined curve to an amorphous cloud of points (Olsen & Degn 1985). Figure 18.10, for example, shows such graphs for the time series of newborn babies' cries and for a laboratory experiment of carbon monoxide and oxygen. A well defined curve often means chaos. However, the pattern on a next-amplitude plot won't necessarily prove chaos, because of noise and other factors.

Authors commonly label the axes on next-amplitude maps and also on return maps as x_{n+1} and x_n (or comparable nomenclature), as on other pseudo phase space graphs. Consequently, a glance at the labels on the axes of a pseudo phase space graph won't reveal exactly what the plotted data represent.

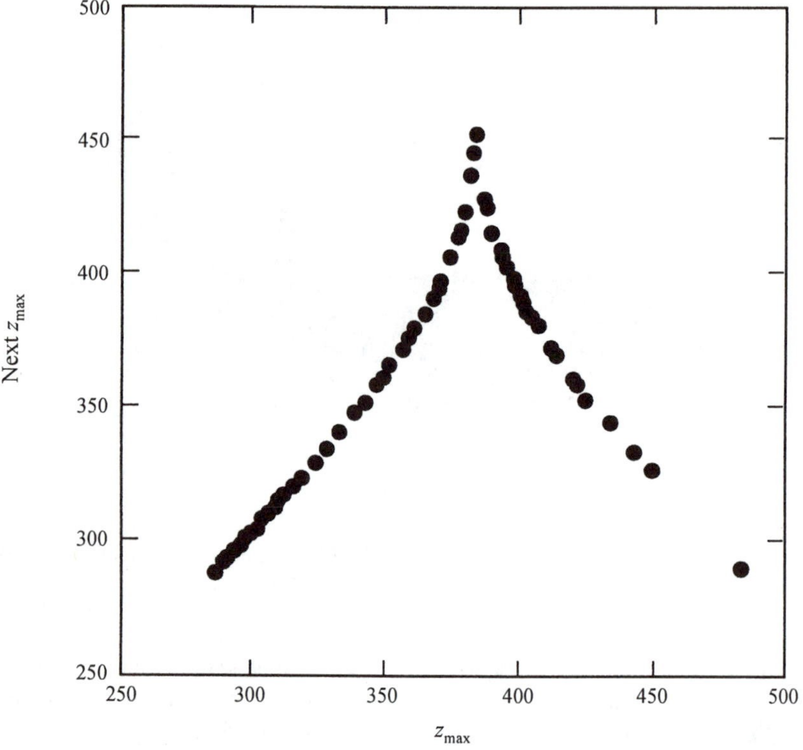

Figure 18.9 Next-amplitude map of Lorenz data for his variable z (from Lorenz 1963: table 2 and fig. 4).

Difference plots

A plot of differenced data (Ch. 9) is (naturally) a **difference plot**. The coordinates on the graph are lagged differences between successive observations. The abscissa is any computed difference Δt_n and the ordinate is the next difference, Δt_{n+1}, where n here is the next difference in the transformed series. (And so the general idea is very similar to the graphical version of a one-dimensional map.) Thus, if the lag is one, the graph has $x_{t+1}-x_t$ on the abscissa and $x_{t+2}-x_{t+1}$ on the ordinate. Similarly, graphing differenced data using a lag of two is just a matter of plotting $x_{t+2}-x_t$ (abscissa value) and $x_{t+3}-x_{t+1}$ for the first point; $x_{t+3}-x_{t+1}$ (abscissa value) and $x_{t+4}-x_{t+2}$ for the second point; and so on.

Early work suggested that difference plots might be useful in identifying chaotic time series. That is, they might help distinguish highly deterministic chaotic data

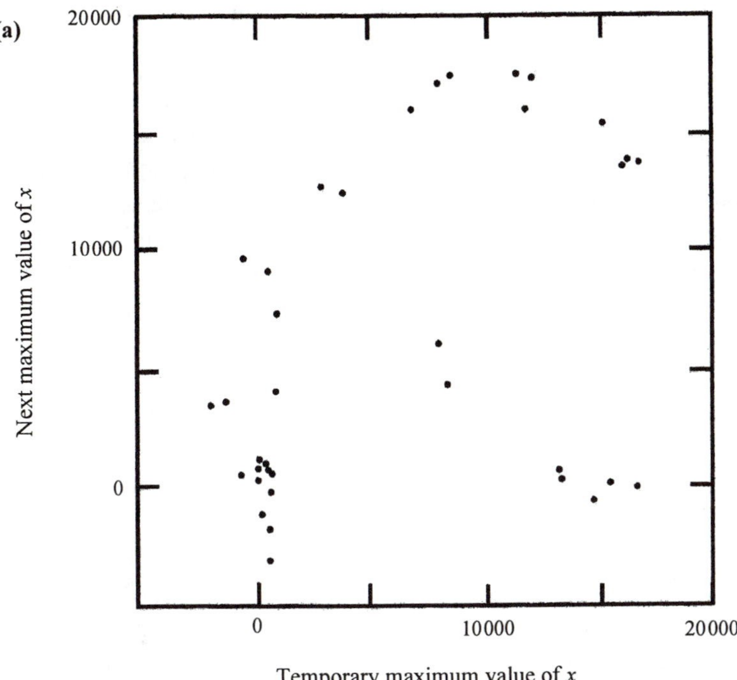

Figure 18.10 Next-amplitude maps for (a) cries of newborn babies (adapted from Mende et al. 1990, with permission from Elsevier Science – NL) and (b) laboratory reaction of carbon monoxide with oxygen (adapted from Johnson & Scott 1990, with permission from the Royal Society of Chemistry).

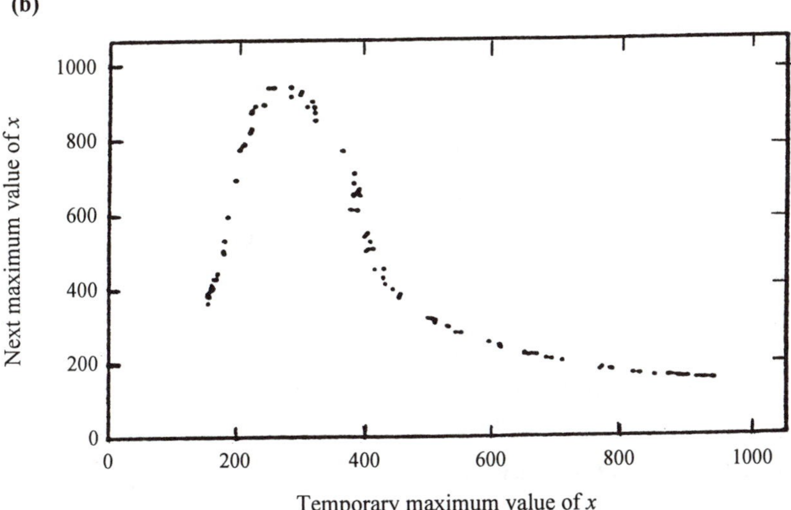

from negligibly deterministic ("random") data. Figure 18.11a is a time series of random numbers (from a random number table, and hence sufficiently random for our purposes). Figure 18.11c is a time series of chaotic logistic-equation (highly deterministic) iterates. There isn't any apparent pattern or structure to the plotted points on either graph. Now we first-difference both sets of data, using a lag of one, and make the corresponding **first-difference plots**. The first-difference plot for the random data (Fig. 18.11b) still shows considerable scatter (although perhaps a faint tendency for large negative differences to follow large positive differences; see also Fowler & Roach 1991: 73). In contrast, first differences of the chaotic data (Fig.

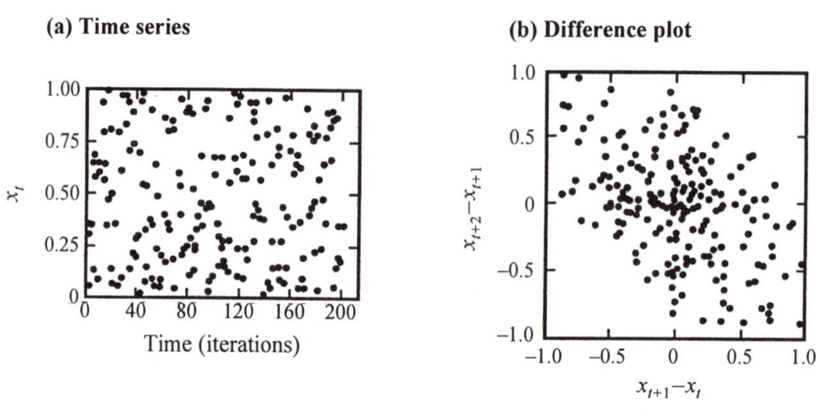

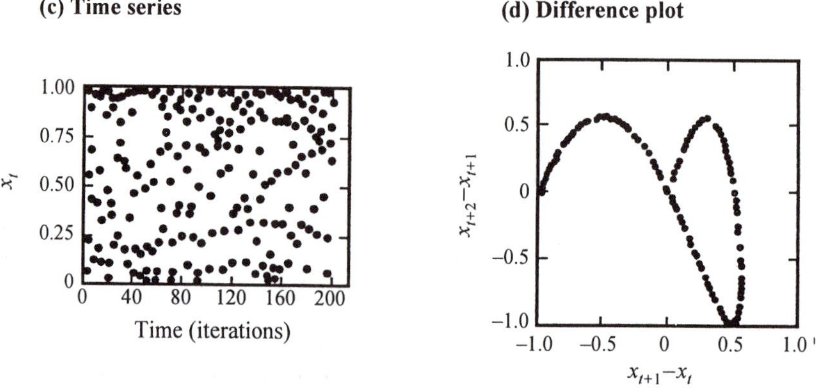

Figure 18.11 Use of differencing to distinguish between random and deterministic noise-free data.

18.11d) plot as a smooth (although slightly complex) curve. Any given difference between two successive iterates (any $x_{t+1}-x_t$) for the chaotic data leads to just one or two possible response differences (the associated $x_{t+2}-x_{t+1}$). The orderly, smooth curve is evidence of the high degree of underlying determinism.

Another example is Robert Shaw's (1984) classic dripping-faucet experiments, carried out during his graduate student days at the University of California at Santa Cruz. Briefly, Shaw turned on a faucet just enough for it to barely start dripping. The drips broke a light beam that recorded the time at which each drip occurred. At most faucet settings, the drips fell at uneven rates. Shaw translated the raw data into intervals or differences of time by adopting a lag of one and subtracting the time at which a given drop fell from the time at which its successor fell. The first interval (Δt_n) was the duration between drops one and two. That duration equals the time at which drop two fell minus the time at which drop one fell. The next interval—that between drops two and three—was time of occurrence of drop three minus time of occurrence of drop two = Δt_{n+1}. The third item of data was the amount of time between drops three and four. And so on. This procedure generated a new (transformed) time series consisting of successive differences or intervals. With increasing drip rates (faucet settings), those intervals showed periodicity, then period-doubling, and eventually chaos. The plot of first-differenced data for the chaotic regime, not reproduced here, revealed a humpbacked or parabolic curve, although with some scatter (Shaw 1984: 9). The curve looked much like a fuzzy version of the one-dimensional-map curve for iterates of the logistic equation (Fig. 10.1). Even though drop intervals varied with time, any one interval approximately indicated the next.

Differencing for purposes of graphically identifying chaos falls into the same category as many other potentially useful tools: it often doesn't work in practice, because of noise and other factors.

Difference plots, Poincaré sections, return maps, and next-amplitude plots are all based on some rule that we choose, specifying how x goes to a new x. As such, they are all variants of the one-dimensional map. The one-dimensional map (a pseudo phase space plot of x_{t+m} versus x_t) is a fifth way of uncovering determinism. We'll discuss that method in the next chapter. Poincaré sections and return maps pertain to a particular phase space cross section, whereas the one-dimensional map, next-amplitude plots, and difference plots don't.

Summary

A Poincaré section is a cross section of an attractor, taken transverse to the flow. It consists of a bunch of dots. Each dot represents the trajectory's intersection with the cross section. Ideally, unique Poincaré section patterns distinguish periodic, toroidal, and chaotic attractors. The pattern doesn't indicate the relative timing of the various points. However, the section does show the relative frequency or popularity of the various zones along the section. Other advantages (compared to analyzing an entire dataset) are that the number of variables to deal with usually decreases by at least one, and the amount of data to analyze, the computer time, and the required computer capability all are much less.

A first-return map is a pseudo phase space plot (L_{n+1} vs L_n) involving only the points at a Poincaré section, with those points rearranged into their temporal sequence. It reveals whether there's a systematic relation between one visit and the next, at the section.

A next-amplitude map is worthwhile for systems that occasionally reach peak values (whether or not the peaks are equally spaced in time). Such a map is just a plot of successive peak values of a selected variable, for instance a plot with x_{max} on the abscissa and the next x_{max} on the ordinate. A well defined pattern often indicates deterministic chaos.

A difference plot is a type of pseudo phase space graph on which the abscissa is the difference between an observation x_t and some later (lagged) observation. In symbols, that's $x_{t+m}-x_t$. The second axis is a similarly computed difference one lagged measurement removed ($x_{t+2m}-x_{t+m}$). A third axis (if desired) is a similar difference two lagged measurements removed. The purpose of a difference plot is to see whether successive differences show any systematic relation over time.

All of the above plots can be helpful with noise-free data. However, noise reduces their usefulness with real-world data.

Chapter 19
Attractor reconstruction

We've talked about attractors and pseudo phase space earlier. They are two of the most important and basic ideas in chaos. It's time now for a closer look.

We saw in the preceding chapter that the idea of pseudo phase space is important and useful. One advantage is that it only requires measurements of one variable, even though the dynamical system might involve several or many variables. (Measuring many variables at the same time is often difficult or impossible. A *standard* phase space graph in those cases is incomplete at best.) Another advantage is that pseudo phase space plots can help distinguish highly deterministic or chaotic data from negligibly deterministic (random) data, in systems that have low (three or fewer) dimensions. For example, the preceding chapter discussed four ways to use pseudo phase space to help find determinism in data. In this chapter, we'll see yet another way.

A third use of pseudo phase space is that, again for low-dimensional systems, it can provide a picture of the attractor. We'll see how that works in a few moments. Finally, as later chapters will show, a pseudo phase space analysis is the first step in calculating important measures in chaos theory (the correlation dimension, Lyapunov exponents, and others).

Making lag plots with just one of the measured features in pseudo phase space is known as "reconstructing". Technical literature describes it variously as **reconstructing the phase space**, *reconstructing the trajectory*, *reconstructing the phase portrait*, **state-space reconstruction**, and the like. As we'll see, such plots sometimes recover the essence of a multidimensional attractor. If they do, the procedure **reconstructs the attractor**. That's what we're going to concentrate on in this chapter.

At the outset here, let's be clear about when this idea of attractor reconstruction *doesn't* work. Any graph in pseudo phase space can only have up to three orthogonal axes. Therefore, as far as seeing an attractor on a graph is concerned, the methods we'll discuss in this chapter work only with dynamical systems whose attractors can be acceptably reproduced on a lag-space plot of three or fewer axes. Attractors can't be reproduced there in at least four circumstances. One is the obvious case of random data, in which there isn't any attractor anyway. Secondly, the system may have a high-dimensional attractor that doesn't show any definable structure in three or fewer dimensions. Thirdly, noise may obscure any attempted reconstruction. And fourthly, the data may be **nonstationary**. In all of these cases, there isn't any low-dimensional surface to model or reconstruct (Weigend 1994). And in any case, the whole ballgame depends on a subjective, visual inspection.

Conceptually, however, more than three axes (variables, in a sense) can be involved. Thus, a mathematical analysis (as opposed to a graph) can involve any number of dimensions (axes). For example, later chapters include situations for which a chaos analysis requires the computation of certain quantities in 2, 3, 4, 5, 6, or more dimensions. Researchers currently are trying to develop sophisticated ways to detect high-dimensional structure (attractors), even though analyses in two or three dimensions may not show any such structure.

The arbitrarily designated number of axes or dimensions is the *embedding dimension*, introduced in Chapter 3. In the present context, some authors also call it the "*reconstruction dimension*" (e.g. Fraser 1989a).

Chaos theory has developed three different ways to extract and show, graphically, the attractor of a nonlinear multivariable system from measurements of just one of the variables. Those methods are time-delay, **singular system analysis** (also known as **singular value decomposition** and by other names), and time derivatives. As mentioned, the three methods apply mainly to systems that are low-dimensional. Even then, there's no blanket guarantee that they work for all such systems and for all dynamical properties. As long as the measuring interval is constant, the procedures apply both to discrete evolution (including iterations) and to continuous evolution (whether measured or sampled at discrete intervals, on the one hand, or represented by **equations of motion**, on the other). The particular method we use can make a big difference in the quality of the reconstructed attractor.

The third method—time derivatives—requires the equations of motion of the system. Those equations usually aren't available with real-world data. Hence, that method isn't very practical and won't be discussed here.

Before exploring the other two methods, we need to spend a few moments on a couple of important concepts or tools. The first is how points in pseudo phase space can appear to get farther apart as we increase the embedding dimension. The other is the use of so-called surrogate data.

Growth of point distances in embedded space

Suppose you are sitting at the far end of the stadium (the end zone) at ground level during a football game and your depth perception deserts you. That means, looking down the field, you see all the players as if they were distributed along a single line that stretches across the width of the field. In your view, many of the players are very close together. If you measure the distances between them in the view that you have, those distances are small. To see the true spatial distribution of the players, you'd need a two-dimensional view of the field, as if from a helicopter. Most measured distances between players from that view are larger than from your end-on view. The phase space analogy of all this is as follows. Embedding a two-dimensional attractor in one dimension causes many points to appear closer together than they really are. Calculated distances between points in that case are short. Relaxing the embedding by embedding the data in two dimensions lets many points move to their true positions, thereby increasing their distance from one another. Points that seem to be close to a given point in a low dimension and are farther away in the correct embedding dimension are **false nearest neighbors**.

The same thing happens if we look at a three-dimensional attractor as if we see it only in two dimensions. Say the two dimensions are the length and width of this printed page you are reading and that the third dimension goes from the page to your eyes. The attractor really consists of a group of points between the page and your eyes. Forcing the images of all points of our three-dimensional attractor onto a two-dimensional plane (the page) increases their apparent closeness or density. Again, many points that seem to be close to another point can be false nearest neighbors; their true or desired distance can be much greater.

When the embedding dimension is low (say, two), there isn't enough space for a higher-dimensional attractor to express itself fully, so to speak. Our pseudo phase space plot crams the poor attractor into fewer dimensions than it wants to be in. In that case, computed distances between points tend to be small. Increasing the embedding dimension releases more and more of the attractor's points from the straight-jacket constraint we've imposed (i.e. the low number of embedding dimensions). In other words, they move closer to their real relative locations, even if still partly confined by the embedding dimension. So, distances between points increase with increase in embedding dimension. With further increases in embedding dimension, an attractor gradually assumes its true spatial configuration. Computed distances between points then stop increasing and become constant. That notion is basic to various proposed methods for estimating lag and true embedding dimension.

Surrogate data

Surrogate data are nondeterministic, artificially generated data that mimic certain features (although not *all* features) of a measured time series but are otherwise stochastic. (**Stochastic** means having no regularities in behavior, and thus characterizable only by statistical properties.) Surrogate data, for instance, might have the same mean, variance, Fourier power spectrum, autocorrelation, and so on as the measured time series. The idea is to use such artificial, nondeterministic data as a control and compare them to the measured time series. Theiler et al. (1992) and Prichard & Price (1993) suggest adaptations of this long-established idea to nonlinear dynamics; Theiler et al. (1993) review other applications in nonlinear dynamics and chaos theory.

The object is to see whether the observed time series, similar in many respects to the specially designed nondeterministic data, is likely to have essentially the same value of some chosen nonlinear measure (embedding dimension, for instance). If it has the same value (to within limits that you specify), then the real time series doesn't differ from the artificial (nondeterministic) data in regard to that measure. For example, if we were testing some value of embedding dimension, we'd conclude that there's no advantage to that embedding. That, in turn, would mean that the next lower (and hence easier to deal with) embedding might be just as good for analyzing the actual time series. Or, testing for determinism, we might conclude that the observed data probably aren't deterministic. On the other hand, if the observed time series has a significantly different value of the chosen measure, then there's something special about that time series (e.g. the embedding is an improvement, the time series might be deterministic, etc.).

The general steps in the surrogate data technique are:

1. Generate many surrogate datasets, each similar to the original time series. "Similar" means the artificial datasets are of the same length as the measured time series, and the artificial data are statistically indistinguishable from the observed time series in regard to characteristics that you specify (e.g. mean, standard deviation, etc.). The purpose of creating a *group* of surrogate datasets rather than one set is so that the test can be repeated on many qualifying surrogate datasets. That gives a *distribution* of results and lets you evaluate the comparison in terms of statistical significance. The number of surrogate datasets in practice seems to range from about 5 to 100.
2. Compute a discriminating statistic for the measured time series and for each of the surrogate datasets. The discriminating statistic is any number that quantifies some aspect of the time series. Typical examples in nonlinear dynamics are the correlation dimension, largest Lyapunov exponent, and forecasting error.
3. Hypothesize that there's no difference between the discriminating statistic of the measured time series, on the one hand, and the same statistic for the members of the ensemble of plausible alternatives (the surrogate datasets), on the other. (Such a hypothesis of no difference is called a **null hypothesis**.)
4. Ask whether the discriminating statistic computed for the original data differs significantly from those computed for each of the surrogate datasets (see, for instance, Kachigan 1986: 177). If it does, then reject the hypothesis of no difference. That is, conclude that there is a discernible difference.

Time-delay method

The time-delay technique is by far the most popular approach to reconstructing attractors. (It's also called the **delay,** delay-vector, or **delay-coordinate method**.) The method stems from Yule (1927), who introduced it to analyze sunspot periodicities. It consists basically of just an ordinary pseudo phase space plot.

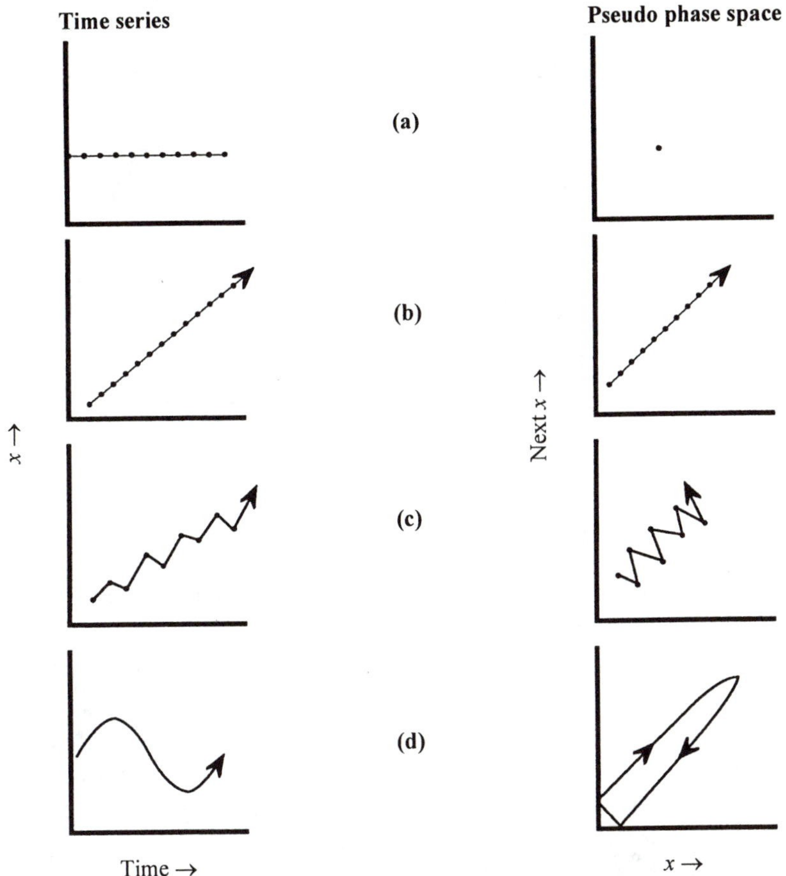

Figure 19.1 Deterministic time series and corresponding pseudo phase space plots.

The left column of Figure 19.1 shows four types of nonchaotic, deterministic time series involving just one variable. The corresponding graph in the right column shows the associated pseudo phase space plot. Figure 19.1a, the easiest case, is a quantity that remains constant with time. Every pair of successive values is the same. The plot of x_{t+1} versus x_t shows only a single dot, at $x_{t+1} = x_t$. The time series of Figures 19.1b,c,d are other deterministic relations. The associated lag space plot in each case is a recognizable and characteristic pattern—the attractor.

Figure 19.2a is a time series (same as Fig. 18.11c) for the logistic equation in the chaotic domain (k = 3.99). Figure 19.2b shows the lag-space plot of the same data. The lag space plot is the familiar parabola—the one-dimensional map—that we saw in Figures 10.2, 10.4, and 12.3. That parabola is the attractor, as reconstructed from the time series of Figure 19.2a. And that's all there is to the basic procedure of the **time-delay method**.

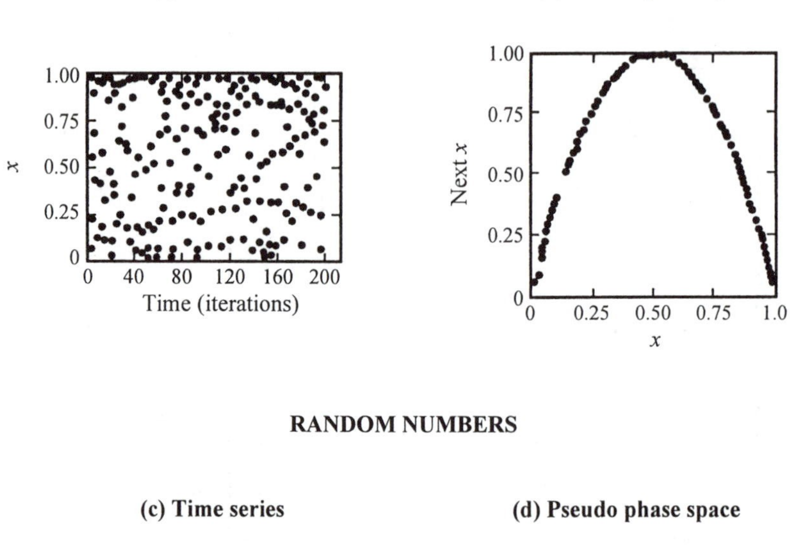

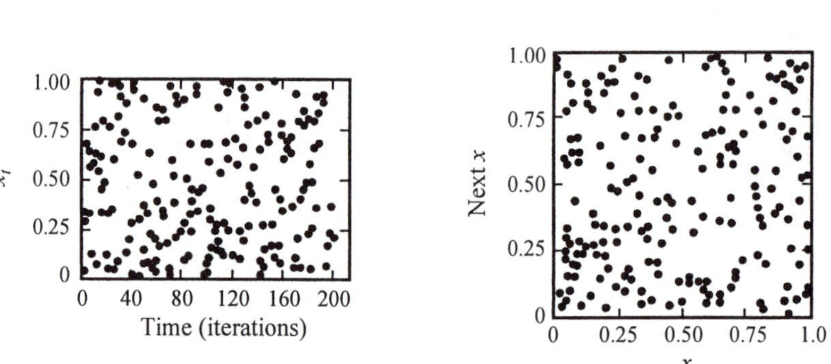

Figure 19.2 Use of pseudo phase space to distinguish between deterministic, noise-free data (a, b) and random data (c, d).

Something else is important about the method (indeed, about singular system analysis as well). That is that it can help distinguish uncorrelated from chaotic data, at least for noiseless and low-dimensional conditions. In other words, it can show whether there's any determinism—any recognizable relation—between one observation and the next, in a time series. To see that, let's look again at Figure 19.2. The time series of Figure 19.2a looks irregular and gives the impression that the system has no underlying law. However, the plot of x_{t+1} vs x_t (Fig. 19.2b) is the familiar well defined smooth parabola. Any one value (x_t) very clearly indicates the next

(x_{t+1}), at least in this relatively noise-free example. The plot therefore shows that the random-*looking* time series of Figure 19.2a has a high degree of determinism. That is, the data follow some kind of rule.

What kind of pattern do we get if, instead, the data have no mutual relation? A lag space plot in that case usually shows no discernible pattern. Figure 19.2c is a time series of random numbers from a random-number table (in fact, they are the same values as in Fig. 18.11a). Figure 19.2d is a lag space plot of x_{t+1} vs x_t for those same random numbers. For any value of x_t, the next random number (x_{t+1}) can have virtually any value within the possible range of 0 to 1. Both random-number graphs (Fig. 19.2c,d) show a space-filling scatter with no structure to the pattern.

The same idea can work in three dimensions or axes, too. Just as a plot of x_{t+m} versus x_t for a time series of random numbers fills the entire square (Fig. 19.2d), a three-dimensional plot of x_t, x_{t+1}, and x_{t+2} (or any other lagged values) for random numbers fills the entire cube. Conceptually continuing in four, or more dimensions, points for such uncorrelated data completely fill the entire **hyperspace**. In the limit, therefore, data that have no mutual relation are infinite-dimensional.

Summarizing the previous three paragraphs, deterministic data often yield a distinct geometric shape, at least if they are noiseless and the system isn't more than three-dimensional. Uncorrelated data, in contrast, usually don't show such a shape. (However, random numbers treated in certain mathematical ways can also plot as orderly structures. Hence, a distinct geometric shape in lag space isn't always a reliable indicator of low-dimensional determinism or of chaos.)

Researchers have pursued various other avenues to identify determinism in real-world time series. One promising direction examines the lagged observations in terms of continuity rather than structure (Kennel & Isabelle 1992, Wayland et al. 1993, Kaplan 1994). The reason for this approach is that, for a continuous and deterministic process, points close in lagged phase space will "map to" points that also are close. The technique first chooses two observations that are arbitrarily close in the embedding space. It then asks whether the next two respective observations also are close. That is, given two points x_i,x_j that are close to one another in the embedding space, are points x_{i+1},x_{j+1} also close to one another? If they are, the data are judged to follow a rule or to be deterministic. The degree of closeness is arbitrary. It is measured by a statistic that represents "prediction error," "translation error," or simply an average separation distance.

In using a pseudo phase space plot to reconstruct an attractor, we often have to throw out the first few observations. The justification for this is that they can be atypical, such as when we begin to iterate equations under certain conditions. Atypical initial iterates are called **transients**. Transients are early start-up values that aren't part of the relatively different pattern that eventually emerges. They are aberrations in that they don't describe the trajectory's long-term behavior. The number of early iterates to label as "transients" is strictly arbitrary; it may range from virtually none to several thousand.

Given a raw time series produced by the logistic equation (a string of iterates, with transients discarded), plotting x_{t+1} versus x_t immediately gives a nice picture of the underlying attractor. Of course, in this case there's only one variable (x) involved, and the equation dictates that all solutions within the chaotic regime fall on the parabola (the attractor). What about a more complex system, say one having two variables? Can a plot of x_{t+1} versus x_t (thus involving only one of the two variables) show us anything about that more complex attractor?

It can. Let's look at a couple of stripped-down examples. The first is another idealized shark–shrimp attractor (Fig. 19.3a, a limit cycle similar to Fig. 11.2). Suppose we don't have the shark data of Figure 19.3a but that we do have the shrimp data. Using only the time-series measurements of the one variable (shrimp populations, not listed here), we now plot those data in the form of x_{t+1} versus x_t. (In other words, the lag in this example is one.) Thus, the first plotted point has an abscissa value equal to the first measured shrimp population and an ordinate value equal to the second. Then we move up one value in the list and plot the second shrimp population against the third, and so on until all data are plotted. Figure 19.3b is the finished graph.

Figure 19.3b is a limit cycle or loop, as is the real attractor of Figure 19.3a. The "recovered" loop (Fig. 19.3b) is kind of squashed and rotated, so it doesn't look exactly like the real phase space attractor. That is, some features aren't the same; the new picture or description is incomplete. Yet the general nature, topology, and geometry of the real attractor (Fig. 19.3a) show up clearly. (Actually, it's worthwhile carrying such an analysis one step further, although we won't do it here. That step is to rotate axes and possibly also translate the attractor, to see it from different angles or points of view.)

The fact that the second version (Fig. 19.3b) is a distorted view of the real attractor (Fig. 19.3a) isn't important for our purposes. The important thing is that, at least for this example, the two versions are topologically equivalent. That means they both have the same dynamical properties. In particular, they have approximately

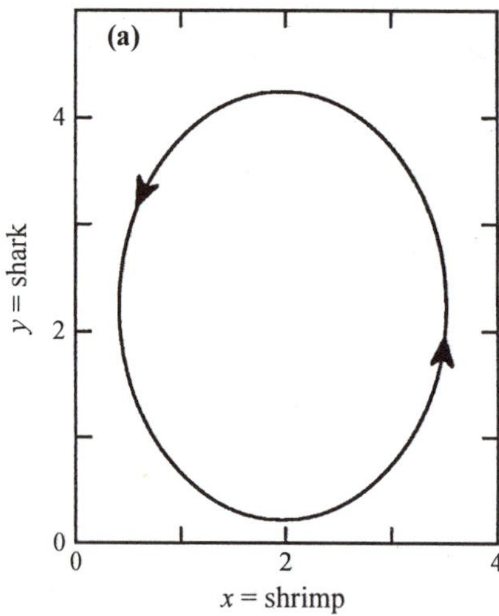

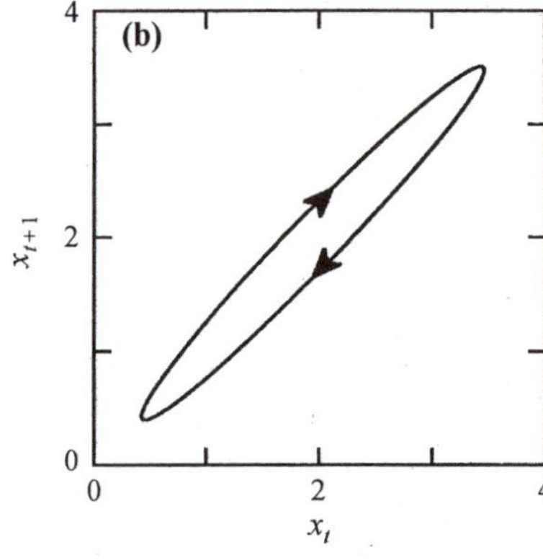

Figure 19.3 (a) Hypothetical shark–shrimp attractor and (b) the same attractor recovered from only the shrimp data.

the same values of key indicators of chaos, such as Lyapunov exponents, correlation dimension, Kolmogorov–Sinai entropy, and mutual information (all discussed in later chapters). Therefore, such properties are **invariant**,[1] or unchanged, by making the pseudo phase space plot. In other words, when two geometric figures are topologically equivalent, a particular dynamical measure (e.g. Lyapunov exponent) is essentially unchanged (invariant) whether we represent and analyze the system in real phase space or in pseudo phase space.

A second two-dimensional example of using a lag space plot to "see" or recover an attractor is the well known Hénon attractor. It's based on two iterated equations (Eq. 13.11) that together have the same basic properties as the Lorenz attractor. (However, it doesn't look like the Lorenz attractor in phase space. Also, it doesn't represent any particular physical system.) With the usual constants inserted, the equations are

$$x_{t+1} = y_t + 1 - 1.4x_t^2$$

and

$$y_{t+1} = 0.3x_t.$$

To iterate the equations, we first pick any arbitrary starting values of x_t and y_t. (The possible range for both is roughly from –1.3 to 1.3.) We then compute the new values x_{t+1} and y_{t+1}, use those as new starting values, and repeat for subsequent iterations. That gives a long list of paired values—an x_{t+1} and y_{t+1} for each iteration. With transients discarded, a plot of the two iterates (x_t, y_t) at each time shows the attractor (Fig. 19.4a). Now suppose we only have values of one of the variables, say x. Again using a lag of one, plotting x_{t+1} versus x_t produces Figure 19.4b. But for a rotation of 90°, the two attractors look very much alike.

Why does this work? Because "the evolution of [a variable] must be influenced by whatever other variables it's interacting with. Their values must somehow be contained in the history of that thing. Somehow their mark must be there." (J. D. Farmer, as quoted by Gleick 1987: 266). In other words, one variable can be so strongly related to another or others that, by itself, it carries dynamic information about those others as well (Ch. 27 deals with this notion). Our baby of Chapter 3 is an example. His height and weight are strongly and positively correlated; as one increases, so does the other. Therefore, simply plotting weights in the form of x_{t+1} versus x_t also shows a generally increasing relation. Another example is the Hénon equations above. There x_{t+1} depends directly on y_t as well as on x_t. That is, the value of y_t is represented within x_{t+1}. Consequently, the time series of just x values (Fig. 19.4b) contains information about both the x and y variables.

1. You may have noticed, by the way, that chaos theory uses "invariant" in several senses. A probability distribution as well as other characteristics of an attractor can be time invariant. Fractals, on the other hand, are scale invariant. Finally, many attractors have features that are invariant in that the features are unchanged by transformations, such as constructing a lag space graph from raw data.

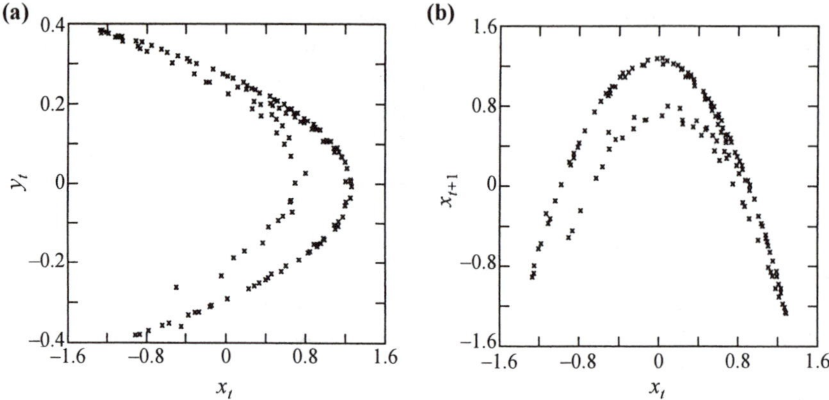

Figure 19.4 (a) Hénon attractor with variables y_t (ordinate) and x_t. (b) Reconstructed attractor using only the sequential values of one variable, x_t.

All conditions under which a pseudo phase space representation keeps the same dynamical properties as the original phase space aren't yet fully known. Problems can arise from the values assigned to parameters, such as the lag in the above examples. In addition, choosing a different variable for the pseudo phase space can produce very different results. For example, Schaffer et al. (1992: 3.27ff.) analyzed the Rössler system. It involves three variables, say x, y, and z. Results varied significantly depending on whether the pseudo phase space was built with x or with z. Thus, in some cases not just any old variable will do. Real-world data carry the additional problems of being few in number and noisy.

The relatively straightforward attractor reconstructions shown in Figures 19.1–19.4 illustrate the general time-delay idea. At the same time, those examples are a bit misleading. That's because they represent ideal conditions rather than real-world data. Reconstructing an attractor for real-world data can be very difficult, regardless of the method of reconstruction. (In many cases, of course, there may not be an attractor to reconstruct.) For one thing, a low- (two- or three-) dimensional plot may be too simplistic to reveal an attractor. Secondly, and this is especially important with the time-delay method, the techniques can have difficulties with noise and small datasets.

What this means is that merely looking at a two- or three-dimensional pseudo phase space plot often won't tell us with assurance that we've reconstructed an attractor. Besides, we really need reliable numerical criteria rather than having to depend on a subjective visual inspection. Unfortunately, there are presently no objective numerical indicators to identify a genuine reconstruction.

Figure 19.5, from Glass & Kaplan (1993), shows a lag space reconstruction using two axes. The reconstruction is based on actual measurements of the heartbeats of a sleeping person. Other variables interact with the one chosen for analysis (heart rate).

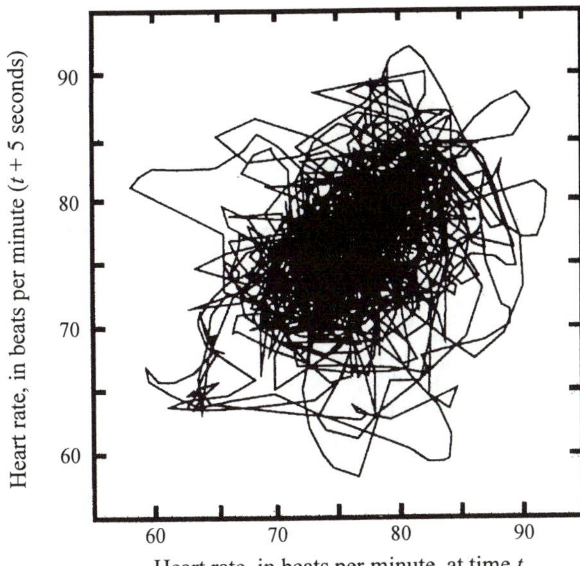

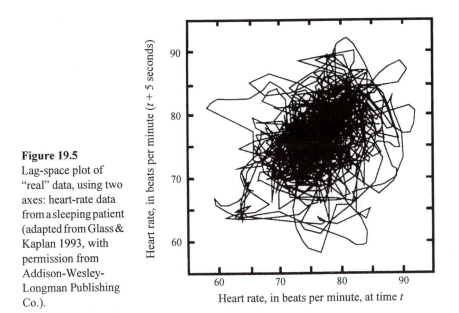

Figure 19.5 Lag-space plot of "real" data, using two axes: heart-rate data from a sleeping patient (adapted from Glass & Kaplan 1993, with permission from Addison-Wesley-Longman Publishing Co.).

Figure 19.6 shows two time-delay plots based on real-world data, this time using three axes. Figure 19.6a reflects occurrences of the childhood disease of chicken pox in St Louis during 1934–53 (from Schaffer et al. 1990). Figure 19.6b is based on the volume of water in the Great Salt Lake, Utah, during 1848–1992 (Abarbanel & Lall 1996).

The notion of using just one feature (x) of a time series to make a pseudo phase space plot and see an attractor seems almost too good to be true. Yet it has a solid foundation and mathematical justification. Some of the leading contributors to those theoretical aspects have been Whitney (1936), Packard et al. (1980), Mañé (1981), Sauer et al. (1991), and Takens (1981).

Reconstructing an attractor with the time-delay method rests mainly on choosing the lag and embedding dimension. Knowing those two parameters in advance would be a big help in getting a genuine, full-fledged reconstruction. There are many proposed methods for choosing their values.

Choice of lag

Most of our above examples used a lag of one. Longer lags are common. With real-world data, lags usually are in actual time units, such as seconds, months, and so on.

In theory, any lag is fine when you have unlimited amounts of noiseless data. In practice, of course, that Utopian situation doesn't occur. It turns out that, in practice, different-looking attractors (or no attractor) and different values of the "invariants" can result from different lags.

Let's see how the lag value influences a reconstruction attempt. We'll deal with just two phase space axes (x_t, x_{t+m}). One extreme is a lag of zero. A lag of zero says, in effect, that we're not going to allow enough time (in our analysis) for the system to get to its next state. With lag $m = 0$, we have $x_{t+m} = x_{t+0} = x_t$. Thus, the ordinate value (x_{t+m}) equals the abscissa value (x_t) for each plotted point; it's a plot of each value in the time series against itself. In other words, the two values that define each point are perfectly related. Therefore, each plotted point contains highly redundant information (one value predicts a maximum amount of information

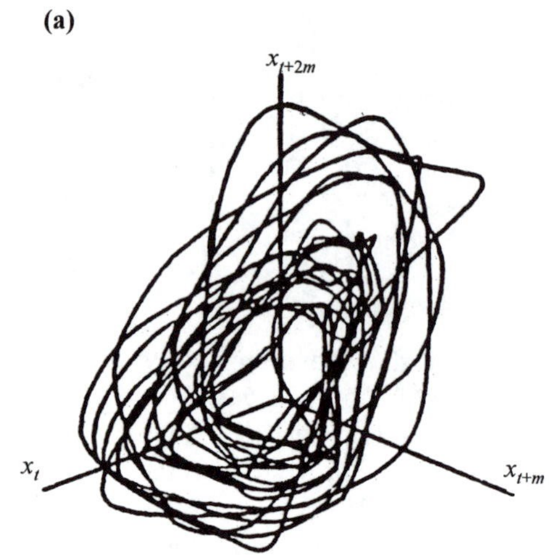

Figure 19.6 Time-delay plots of real-world data using three axes. (a) Occurrences of chicken pox in St. Louis during 1934–53 (adapted from Schaffer et al. 1990, with permission from the American Association for the Advancement of Science). (b) Volume of water in the Great Salt Lake, Utah, 1848–1992 (adapted from Abarbanel & Lall 1996, with permission from Springer-Verlag).

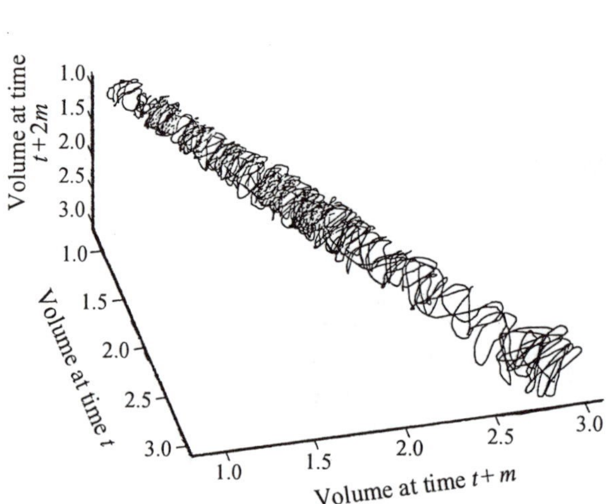

about the other, but it's useless information). No matter what shape the attractor really has, plotting many such perfectly related pairs of values results in nothing more than a meaningless 45° straight line (an identity line or "diagonal"). That miserable excuse for a "reconstructed attractor" should fill much more of the pseudo phase space than it does. A larger lag is called for.

What that example really shows is that, in making any graph, we have to keep an obvious principle in mind: the values we plot on the axes must be sufficiently "independent" from one another that we don't plot a value against itself. In standard phase space, this issue doesn't come up. We know we're plotting different entities (height versus weight, stock price versus time, etc.). In pseudo phase space, we have to be more careful. With an iterated equation, for instance, we just saw that a lag of zero would be ridiculous. The shortest reasonable lag for an iterated equation is 1. With real-world data, however, a lag can be very close to 0 in terms of actual time. That could easily happen for a continuous evolution sampled at discrete time steps, in which the time steps are very short relative to the rate at which the physical process unfolds. An example is measuring weather features (temperature, humidity, etc.) at, say, intervals of every 0.1 sec. For the relatively small lag of 0.1 sec (or one time interval), a variable undergoes virtually no change from one measurement to the next. That means that the coordinates of any point we want to plot are essentially equal. Any attempt to reconstruct an attractor in lag space shows nearly a one-to-one relation on the graph. For that example, such a relation would be meaningless. (It wouldn't be meaningless for those nonchaotic systems that either remain constant or evolve toward a constant value. However, a glance at the raw data is usually enough for us to identify those cases. The rest of this discussion doesn't pertain to those cases.) Another problem with real-world data and short lags is that the noise level may hide any local structure on our lag space plot.

In summary, with both an iterated equation and real-world data, there's a choice of lag for reconstruction purposes. Lags that are too small cause us to plot each value against itself. We need to avoid that. We do so by choosing a lag large enough that, except for the occasional coincidence, the values of x_{t+m} are each significantly different from the associated values of x_t. We'll talk about how to do that below.

Just as a very small lag (zero or nearly so) is counterproductive, so too is a lag that is "too large." Iterates of the Hénon equation are a good example. Figure 19.4b shows that a lag of one does a very good job of recovering the attractor. Is a lag of two any better? Figure 19.7a is the same basic data plotted with a lag of two. There's still plenty of structure. However, the structure is now somewhat more complex. The figure (even from a topologist's view) no longer looks like the real attractor. Further increases in lag (not shown here) for the same Hénon data show two tendencies. First, structure persists but becomes more and more complex. Secondly, structure loses quality or definition. That is, the figure becomes more diffuse; the dynamical relation between points becomes obscure and eventually disappears. For instance, a lag of ten shows random-like scatter (Fig. 19.7b). Any observation at that lag is so many generations removed from its successor that the graph no longer shows a relation.

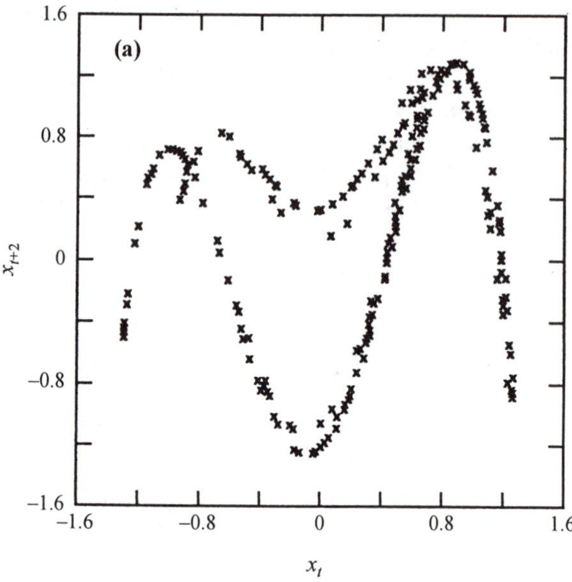

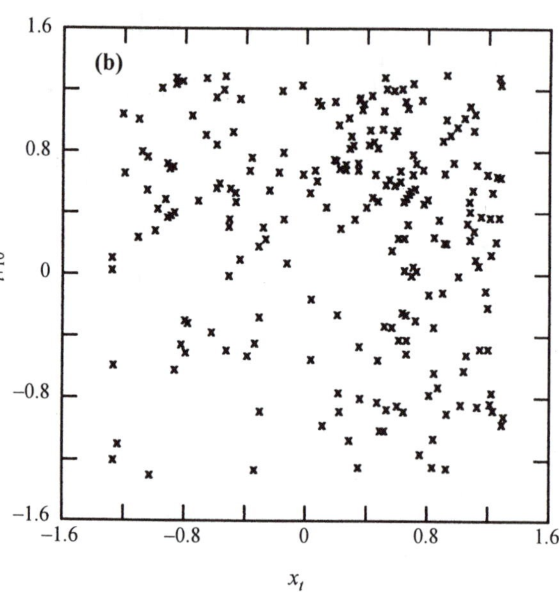

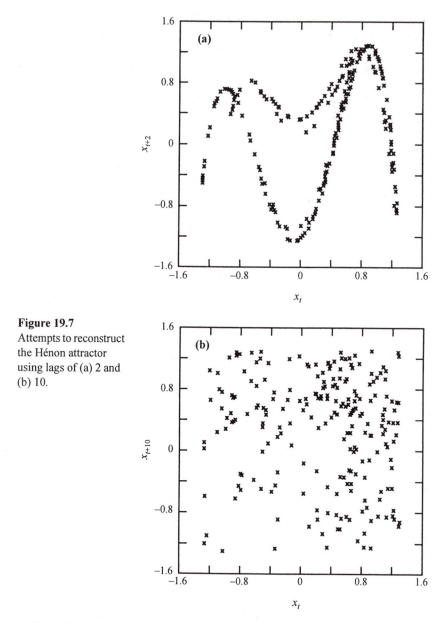

Figure 19.7 Attempts to reconstruct the Hénon attractor using lags of (a) 2 and (b) 10.

Iterated examples, then, show several important features. One is that a lag that is too small (zero for iterated equations) is useless. Another is that a lag can also be too big (Fig. 19.7b). Thirdly, for any given attractor, there's some optimum lag, neither too small nor too large.[2] And fourthly, once we know that the values of x_{t+m} are significantly different from their associated values of x_t, lag space plots using an appropriate lag can show a regular pattern (an attractor) if there is one (Figs

19.2a,b, 19.4). If there isn't any underlying law, as with random numbers, then lag space plots usually show random-like scatter (Figs 19.2c,d).

How, then, do we choose the optimum lag for real-world data? The answer at present is that there isn't any foolproof way. Several methods (discussed below) have been proposed. No single method seems to apply to all conceivable conditions. Rather, the choice of method depends on size of dataset, available computer power, required level of sophistication of the analysis, effort you are willing to invest, funding, and probably other factors (see, for example, Grassberger et al. 1991). For instance, the best lag for estimating a dimension might not be the best for prediction.

Actually, the concept of an "optimum lag" (in the sense of "resulting in a reconstruction closest to the true attractor") may turn out to be useful mainly in regard to low-dimensional attractors and essentially noise-free data. Real-world data rarely fulfil those requirements. Hence, with real-world data the best we might be able to do is look at each of many lags and compare the various results. One way to do that might be to look at a discriminating statistic as a function of lag.

I'll mention here five proposed methods for determining lag. In general, they are all based on the principles we saw above. Those principles are that

- the worst possible job of attractor reconstruction occurs when all points plot on a 45° straight line in lag space
- a lag that is too big is equally useless
- there is an optimum lag somewhere in between.

Five prominent proposed methods for determining lag are autocorrelation-function indicators, mutual information, redundancy, correlation integral, and space-filling methods.

- *Autocorrelation-function indicators* Autocorrelation (Ch. 7) in the case of lag space measures the degree to which earlier values are numerically similar to later values, within a lagged time series. In other words, it measures the degree of correlation of a variable at one time with itself at another time. At a lag of zero, the coordinates of each plotted point are equal ($x_{t+m} = x_t$), and autocorrelation is a maximum (1.0). It decreases with increase in lag. For most datasets it eventually fluctuates within some small range due only to noise rather than to lag. Hence, some chaologists just compute the autocorrelation at successive lags (thus obtaining the autocorrelation function) and take the optimal lag to be the lag at which that function goes to some small value. Examples that people use for such a "small" lag are:
  - The lag corresponding to an autocorrelation of zero.
  - The lag corresponding to the so-called **autocorrelation time** (the time required for the autocorrelation function to drop to 1/e (= 1/2.718 = 0.37).
  - The earliest lag at which the autocorrelation function stops decreasing (the

---

2. For an iterated equation [map] with no noise, this optimum lag seems to be 1. Abarbanel et al. (1993: 1345), for example, say that "without much grounds beyond intuition," we use a lag of 1 or 2 for map data.

"first minimum" of the function).

- A subjectively determined lag smaller than the autocorrelation time.
- The lag corresponding to 1/10 to 1/20 the time associated with the first minimum of the autocorrelation function.

Figure 19.8 shows autocorrelation as a function of lag for the volume of water in the Great Salt Lake, Utah during 1848–1992 (Abarbanel & Lall 1996). Autocorrelation goes to zero (criterion one above) at a lag of about 400. In contrast, the first minimum in the autocorrelation function (the earliest dip in the plotted curve) (criterion three above) occurs at a lag in the range of 12–17. The two methods therefore give vastly different estimates for optimum lag to use in phase-portrait reconstruction.

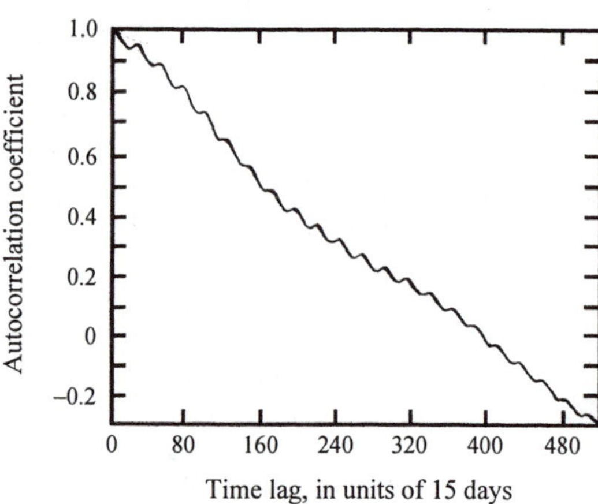

Figure 19.8 Autocorrelation as a function of lag for volumes of the Great Salt Lake, Utah, 1848–1992 (adapted from Abarbanel & Lall 1996, with permission from Springer-Verlag).

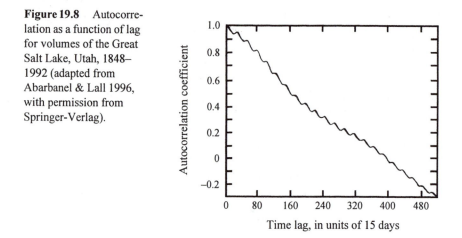

Autocorrelation evaluates the strength of a mutual relation on a linear basis. That basis is not necessarily appropriate for nonlinear systems (Abarbanel et al. 1993).

- *Mutual information* (Fraser & Swinney 1986, Fraser 1989a; this book, Ch. 27) **Mutual information**, like autocorrelation, tries to measure the extent to which values of x_{t+m} are related to values of x_t, at a given lag. (However, mutual information has the advantage of using probabilities, rather than a linear basis, to assess the correlation.) If values of x_{t+m} are strongly related to values of x_t for a given lag, mutual information is relatively high. If, instead, values of x_{t+m} are only weakly related to values of x_t at a particular lag, then mutual information at that lag is relatively low.

Here's another way to look at that idea. When coordinates of a point have the same value (as with a lag of zero), they represent the same information. With two coordinates, for example, one predicts the other exactly. In other words, the amount of information the one gives about the other (the mutual information) is a *maximum*. For attractor reconstruction, therefore, we don't want x_t to provide a lot of information about x_{t+m}. That means we want a *minimum* of mutual information

(Gershenfeld 1988: 325). At the same time, we want to use a reasonably small lag. Mutual information (computed with either Eq. 27.23 or Eq. 27.27) decreases as the coordinates of a lag space point depart from having the same value, that is, as they become more decorrelated. Thus, as with the autocorrelation function, mutual information decreases with increase in lag. Eventually, it stops decreasing. That lag—the first minimum in the mutual information—therefore is another possible criterion for optimal lag. (Mutual information for maps, such as the logistic equation and Hénon map, does not have a positive minimum. Rather, the relation goes to zero and stays there, as we increase lag. Cf. Abarbanel et al. 1993: fig. 10.)

Figure 19.9 shows mutual information as a function of lag for the same data as Figure 19.8. The first minimum in the plotted relation occurs at lags of about 12–17 (in this case, the same result obtained by using the first local minimum in the autocorrelation function in Fig. 19.8).

One drawback to mutual information as a criterion is that it requires very large datasets (many thousands of points) (Atmanspacher et al. 1988). That's because the method renders many data points unusable.

- *The redundancy approach of Fraser (1989a)* This method, discussed in Chapter 27, is a sophisticated extension of the mutual-information idea.
- *The correlation integral* (explained in Ch. 24 of this book) According to

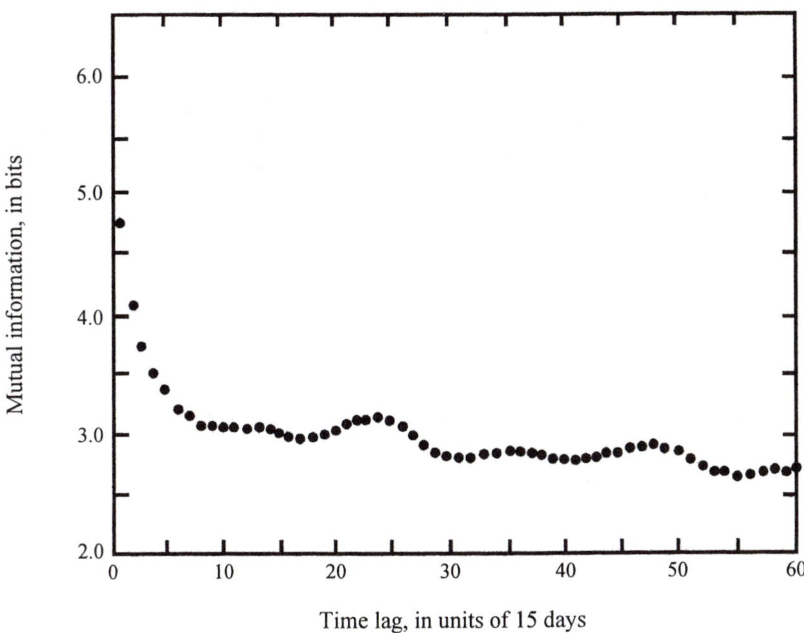

Figure 19.9 Mutual information as a function of lag for volumes of the Great Salt Lake, Utah, 1848–1992 (adapted from Abarbanel & Lall 1996, with permission from Springer-Verlag).

Liebert & Schuster (1989), the correlation integral is easier to compute, and the first minimum of its logarithm is a good criterion for the choice of lag.

- *The "space-filling" method* (Fraedrich 1986) This is simply a trial-and-error procedure of empirically plotting different lags in pseudo phase space and seeing what happens. The method is subjective (judgemental) and lacks a firm numerical criterion. We start with a small lag, namely one that produces something akin to a noisy identity line. (Again, that line means that the coordinates for any plotted point are almost or exactly the same value.) Then we try successively longer lags, each on a new graph. With each new lag, the trajectory's domain enlarges and fills more of the pseudo phase space. Such expansion of the zone where points plot signifies an erosion of that undesirable numerical similarity between the coordinates of each point. Eventually, a new lag doesn't bring any further increase in the amount of filled space. That stage indicates that the coordinates of each point have lost any correlation associated with short lag. That value of *m* is the correct choice of optimum lag. Bigger lags are counterproductive, as discussed above. And, as mentioned, the only feasible dimensions in which to try the method are two and three dimensions.

The techniques can be used jointly rather than alone. For instance, Zeng et al. (1993) suggest using the space-filling method together with autocorrelation.

Choice of embedding dimension

Once we've chosen a lag, the next item of business is usually the embedding dimension. For computational costs, simplicity of interpretation and other reasons, we'd like to reconstruct an attractor in a *small* embedding dimension. (After the attractor is reconstructed, larger dimensions also suffice, but we want the minimum possible.) There's no theory or even a rule-of-thumb available. None of the many proposed ways to estimate the minimum embedding dimension is yet widely accepted. Here is a brief list of the more prominent proposals, in no particular order of importance. (Some of them overlap in regard to overall concept.)

- correlation-dimension computation (Ch. 24) using the Grassberger & Procaccia (1983a,b) algorithm
- minimum mutual information (Fraser & Swinney 1986)
- minimum marginal (incremental) redundancy (Fraser 1989a)
- false nearest neighbors; variations on this theme turn up in Liebert et al. (1991) and Kennel et al. (1992)
- local approximation of a reconstructed attractor by its tangent (Froehling et al 1981, Broomhead et al. 1987)
- predictability (Casdagli 1989, Kennel & Isabelle 1992)
- true vector fields (Kaplan & Glass 1992)
- reconstruction of continuous flow vectors (Cremers & Hübler 1987)
- minimum information from reconstructed equations of motion (Crutchfield & McNamara 1987)

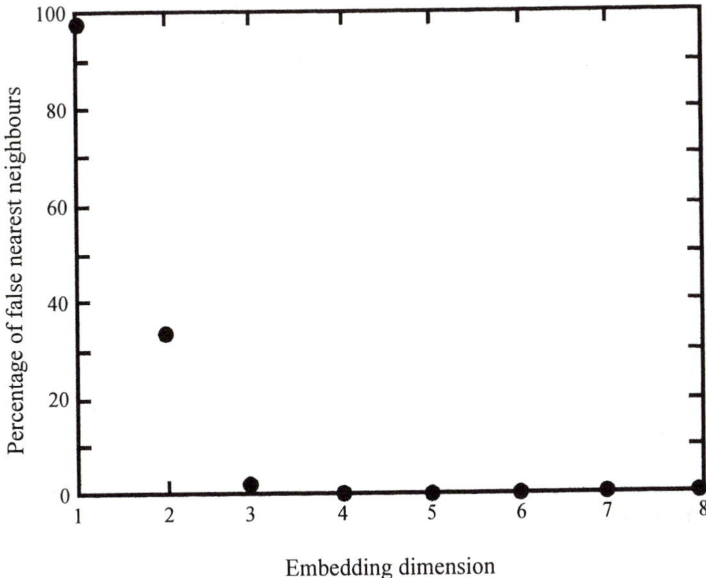

Figure 19.10 Percentage of false nearest neighbors volumes of the Great Salt Lake, Utah, 1848–1992 (adapted from Abarbanel & Lall 1996, with permission from Springer-Verlag).

- single valued continuous mapping (Aleksic 1991).

Chapter 24 discusses the correlation-dimension method, Chapter 27 the mutual information and incremental redundancy methods. Here I'll add a few words about the false nearest neighbor and predictability methods.

Nearest neighbors (points closest to any chosen datum point in pseudo phase space) are called true or false (Liebert et al. 1991, Kennel et al. 1992). True points ("true nearest neighbors") lie at their true phase space distance from the chosen central datum point. False nearest neighbors, in contrast, merely seem to be closer because the embedding space is too small, as explained in the football stadium analogy above. Each point is identified as true or false on the basis of whether its distance from the central point remains constant (true points) or increases (false points), as we systematically increase the number of embedding dimensions and compute distances to nearest neighbors from each datum point. Distances to false nearest neighbors continue to increase as long as the embedding dimension is too small. The correct embedding dimension in this approach is that for which the number (or percentage) of false nearest neighbors decreases to approximately zero. The Liebert et al. (1991) and Kennel & Isabelle (1992) methods actually estimate not just the optimum embedding dimension but also the optimum lag, at the same time.

Figure 19.10 plots the percentage of false nearest neighbors (computed over the entire attractor) against embedding dimension for the same Great Salt Lake data analyzed in Figures 19.8 and 19.9 (Abarbanel & Lall 1996). The percentage drops to about zero at an embedding dimension of about 3 or 4 and stays there for all higher dimensions.

Following Casdagli's (1989) work on prediction, Kennel & Isabelle (1992) proposed comparing the short-term nonlinear predictability (prediction error) of an observed time series to that for an ensemble of surrogate datasets. As with all surrogate data procedures, they compute a diagnostic statistic and test a hypothesis of "no difference." They compose a prediction formula and apply it to every datum point to predict the next value. Comparing the predicted to the true value for each point gives a set of prediction errors. The optimum embedding dimension (and also time delay) give maximum predictability compared to the surrogate datasets.

A good idea in practice is to use more than one method on the same data. Tsonis et al. (1994), for example, used both the correlation dimension and false nearest neighbor approaches. That gives an independent check on results.

Singular system analysis

The problems of noisy data, best lag and best embedding dimension led to a search for a better way to reconstruct an attractor. Two independent 1986 papers—the most detailed one by Broomhead and King, the other by Fraedrich—proposed essentially the same new method. It has many names, the two most common being singular system analysis and singular value decomposition.[3] I'll call it SSA for short.

SSA is more sophisticated than the time-delay method but has the potential of providing more information. Before being adopted for nonlinear dynamics, SSA was already firmly established in signal processing and in pattern analysis. It gives qualitative and quantitative information about the deterministic and stochastic parts of a time series, even when the series is short and noisy (Vautard & Ghil 1989). In fact, getting around the problem of noise was one of the chief goals in applying it to attractor reconstruction.

Like the time-delay method, SSA creates pseudo phase space axes on which we try to reconstruct an attractor. The axes are not lagged values of the chosen variable, however. SSA does build on some of the concepts of the time-delay method. For instance, SSA uses the notions of lag, lagged vectors, embedding space, and embedding dimensions. The main idea is to maximize the mean squared distance between points on the reconstructed attractor. Mathematical details (Broomhead & King 1986) are too lengthy, sophisticated and technical to review here, but we can at least go over the main points.

The fundamental ingredient in the method is a so-called trajectory matrix. The rows in the matrix are the N_d-dimensional vectors (lagged values of x), just as we

3. A few of its other aliases are singular spectrum analysis, principal component analysis, principal value decomposition, Karhunen–Loeve decomposition, and empirical orthogonal function analysis.

would use with the time-delay method. We have to estimate N_d, the total number of embedding dimensions, by means of one of the rules of thumb mentioned above. The length of such a row of the matrix is called the window length. Each row in the matrix corresponds to a sequential data window.

The next major ingredient is a new set of vectors generated from those of the trajectory matrix. The new vectors, called right singular vectors, are computed to be optimally aligned with the vectors of the trajectory matrix. The right singular vectors are linearly independent, orthogonal, and normalized (unit vectors). In other words, they are **orthonormal** (Ch. 5). As such, the right singular vectors make up the coordinate system (the axes of an embedding space) onto which we project the raw time series in hopes of reconstructing the attractor. The values plotted in that new embedding space are the corresponding singular values (the global singular spectrum). They are actually the root mean square projections of the trajectory onto the coordinate set of right singular vectors (Broomhead et al. 1987). The idea is to extract from the raw data the best possible projection of the attractor. From another point of view, each plotted point therefore is a weighted version of the sampled point in the time series (Mullin 1993). Figure 19.11 shows two examples. Part (a), from Read et al. 1992, is a quasiperiodic torus that reflects temperatures in a laboratory model of fluid flow. Part (b) is based on fluid velocities between two concentric cylinders in a laboratory (Mullin 1993: 37).

That is the essence of the method. However, it offers some potential side advantages. For instance, it includes a filter or technique for distinguishing values that are not significant from those that are. In other words, it can filter or reduce noise. Although the singular values include noise, the technique produces a signal-to-noise ratio for each singular vector. That ratio lets us identify those singular values that are mostly noise. We can then separate the singular values into two groups—those that are noise-dominated and those that, instead, are significant.

Another potential benefit is that the number of significant values is an upper limit on the embedding dimension of the attractor. That is, the method doesn't give the exact (or even approximate) embedding dimension but does indicate an upper limit. Such information can be useful in modelling and for other purposes.

Proponents of SSA have suggested many additional advantages. However, there is disagreement as to the conditions under which such advantages are valid. A few that seem to be generally accepted are:

- The procedure for getting the coordinates of the embedding space in SSA is more general than in the time-delay method in that all linear projections are considered (Fraser 1989b).
- The analysis is easy to implement. (Programs are available in many libraries, and the analysis is relative fast.)
- Very large datasets are not necessary.

Some possible disadvantages of SSA are:

- Its use of the autocorrelation function introduces *linear* independence as opposed to a more desirable *general* independence (Fraser 1989b).
- The noise level may appear at different numerical values depending on the

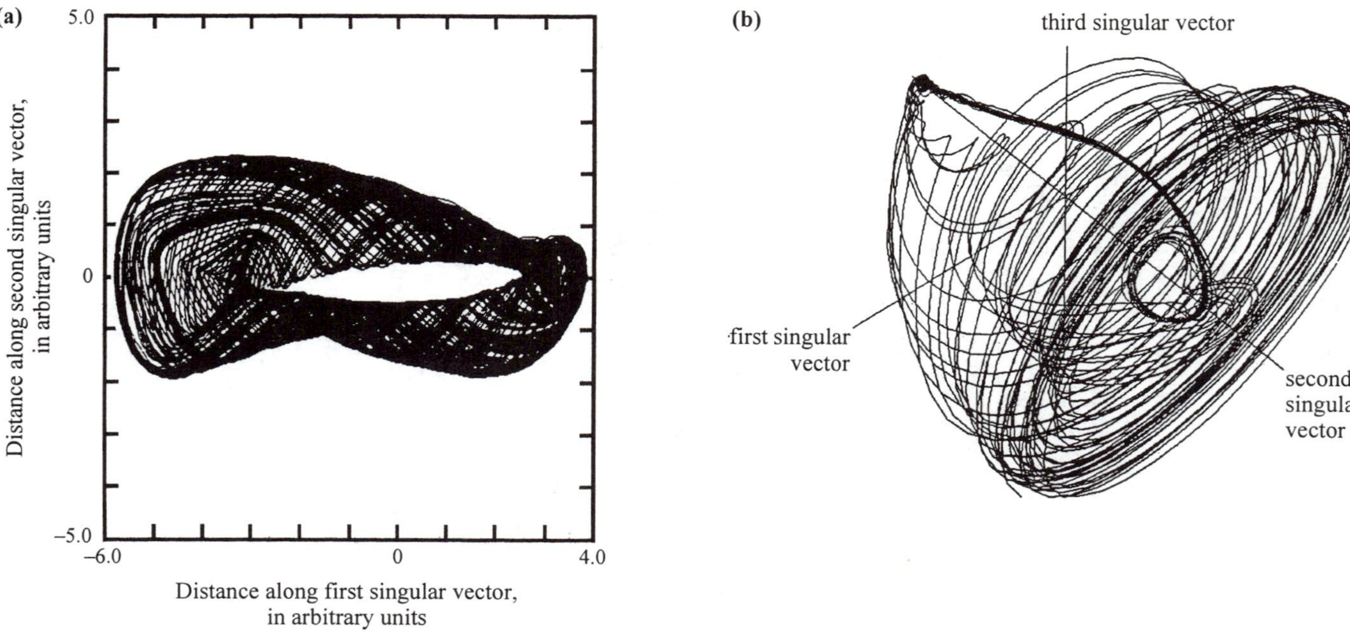

Figure 19.11 Reconstructed attractors using singular system analysis. (a) A two-dimensional example showing a quasiperiodic torus representing temperature data in laboratory fluid-flow experiments (adapted from Read et al. 1992, with permission from Cambridge University Press). (b) A three-dimensional example based on laboratory fluid-flow velocities between two concentric cylinders (adapted from Mullin 1993, with permission from Oxford University Press).

computer we use (Abarbanel et al. 1993).

At this time there isn't enough evidence to say whether SSA is superior to the time-delay method or vice versa. Sometimes one seems to work better, sometimes the other.

Summary

The idea of reconstructing an attractor implies that we want to see what it looks like, at least on a graph. That limits the analysis to three or fewer dimensions. Current methods, when we have only a measured time series, are the time-delay method and singular system analysis. Both methods use pseudo (lagged) phase space plots, with associated lags and embedding dimensions. In other words, the methods require data for only one variable, even though many others may be involved in the dynamics. Besides potentially showing an attractor, they can also distinguish data of negligible determinism (essentially random data) from highly deterministic data, at least for noiseless and low-dimensional conditions.

The time-delay method is relatively easy and straightforward. As a result, it is by far the more popular. However, a major problem is what value to use for the lag. Too small a value yields a meaningless identity line instead of an attractor; too large a value produces undecipherable scatter on the plot. Five proposed methods for choosing the optimal lag are based on autocorrelation, mutual information, incremental redundancy, the correlation integral, and space-filling, respectively. A second major problem with the time-delay method is determining the optimal embedding dimension (number of axes used in reconstructing an attractor); as with the value for lag, there's presently no reliable way to find it.

Singular system analysis is somewhat more technical and involved than the time-delay method. However, it has a broader justification in terms of procedure. Also, it includes a filter for reducing the noise that field data almost always have. Furthermore, it gives an upper limit on the attractor's embedding dimension.

Sometimes one method works better, sometimes the other, and sometimes neither. For real-world data (which are not only noisy but also tend to be few in number and of high dimensionality), there's presently no reliable way to reconstruct an attractor. That is, regardless of the reconstruction method, there aren't any reliable numerical criteria to enable us to recognize an adequate reconstruction.

PART VI

DIMENSIONS

The idea of "dimension" is a basic ingredient of chaos theory. A minor problem is that authors use the term to mean different things. Let's take a look at those various meanings. Then we'll take several chapters to examine some of the most important versions in more detail.

Chapter 20
Background information on dimensions

A dimension in an everyday sense is any measurable quantity (length, width, etc.). Under that umbrella definition, however, at least seven subtle variations or usages of "dimension" cavort on the chaos playing field. We've already touched upon some of these notions. Now it's time to put all of them into perspective.

Concepts of "dimension"

The seven general ways in which chaos theory uses the word "dimension" are:

- *The **Euclidean dimension*** (the number of measurable coordinates or variables needed to describe a Euclidean shape) For example, one measured coordinate or dimension (namely length) describes a straight line. Two dimensions (length and width) describe a basketball court (area), and three (length, width, and depth) describe a cabinet drawer. Also, as mentioned earlier, a point has a dimension of zero. Euclidean dimensions usually are envisioned as directions or axes at right angles to one another (Fig. 20.1a). Therefore, specifying the pertinent values in one, two, or three dimensions locates any point within a line, area, or volume, respectively.
- *The **topological dimension*** The topological dimension of a body is 1 + the Euclidean dimension of the simplest geometric object that can subdivide that body. For instance, think of a line as a series of connected points. Removing any point from the chain (except for the two endpoints) subdivides that line into two smaller lines. So, a point (dimension zero) is the simplest geometric object that subdivides a line. According to the formula, a line's topological dimension therefore is $1+0=1$ (Fig. 20.1b, left side). Now, what about a surface? To subdivide a surface into two smaller surfaces, removing a point from that surface won't do the job (it just creates a small hole in the surface). Instead, we need a line. A line has a Euclidean dimension of 1, so a surface's topological dimension is $1+1=2$ (Fig. 20.1b, right side). Similarly, a line won't subdivide

(a) Euclidean dimension

y *z*

x *x* *x* *y*

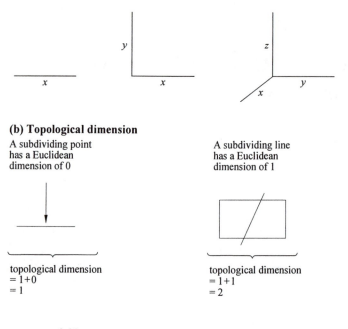

(b) Topological dimension

A subdividing point has a Euclidean dimension of 0

A subdividing line has a Euclidean dimension of 1

topological dimension = 1+0 = 1

topological dimension = 1+1 = 2

(c) Phase space variables

Seven-dimensional weather system

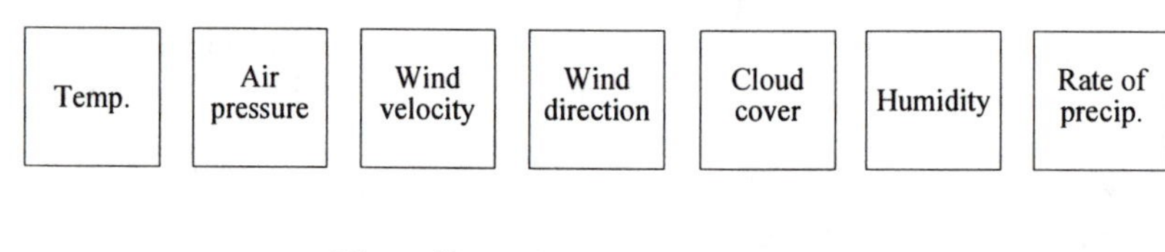

Three-dimensional system of weights

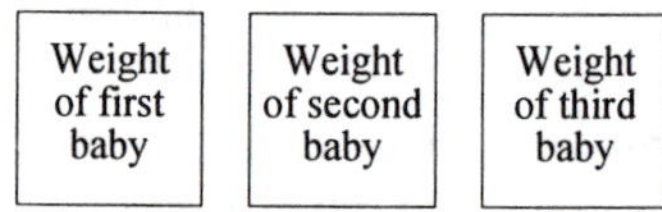

Eight-dimensional height–weight system

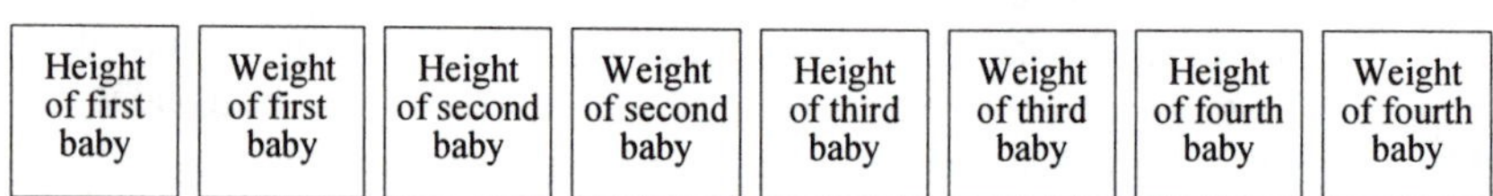

Figure 20.1 Usages of "dimension" in chaos theory, where the dimension is an integer.

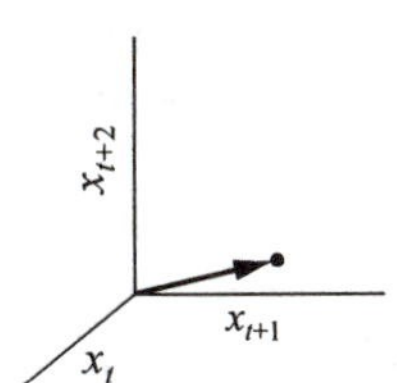

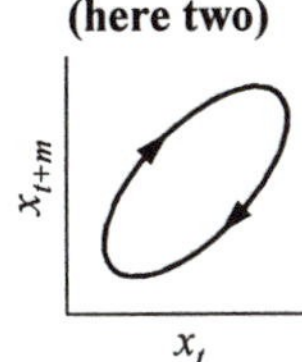

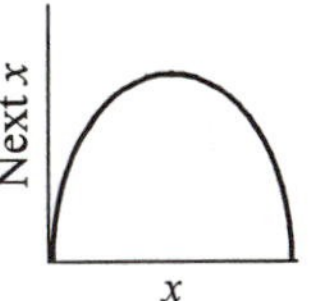

Figure 20.1 Continued.

a three-dimensional solid into two solids (it just bores a hole through the solid). Rather, only an entire plane (dimension 2) divides it. A solid's topological dimension therefore is $1+2=3$. The topological dimension is always an integer and usually has the same value as the Euclidean dimension.

- *The number of variables or axes in standard phase space* This popular usage has at least three categories, depending on how we define a "variable." In all cases, the variable is really just a measurable quantity. These three slightly different concepts of a variable or dimension are:
  - *An ingredient, component, or interacting element of a system* For example, someone might describe the weather in terms of seven variables or dimensions—temperature, air pressure, wind velocity, wind direction, cloud cover, humidity, and rate of precipitation. Measuring the values of each at a particular time provides a point in the conceptual "seven-dimensional" phase space (the state of the weather at that time) (Fig. 20.1c, first row).
  - *The number of objects or specimens on which to measure a particular feature* This usage derives from the previous one by expanding the idea of a "system." For instance, say you have two babies that you weigh once a week. Your "system" then consists of the weight evolution of both babies. You therefore set up a phase space with two dimensions or axes—one for the weight of the first baby, another for the weight of the second. At any given instant, their two weights plot as a single point in the two-axis (and hence "two-dimensional") phase space. That point stands for the state of the system (the two babies' weights) at that time. Weighing them on later occasions shows how the system develops over time. Carrying the analogy further, we might weigh three babies (Fig. 20.1c, second row), or all the babies in the world, and conceptually portray those weights as a single point in phase space. The number of dimensions in that space, N_d, equals the number of babies weighed. Subsequent periodic weighings of the same group of babies then describe a trajectory in the N_d-dimensional phase space.
  - *The number of objects or specimens times the number of mutual features being measured* An example is the babies' heights as well as their weights. Two babies require N_d = four phase space dimensions or axes—one axis each for height of baby no. 1, weight of baby no. 1, height of baby no. 2,

and weight of baby no. 2. (And so this interpretation extends the idea of a "system" still further.) For the heights and weights of four babies, the system requires eight dimensions or axes (four babies × the two features for each) (Fig. 20.1c, third row). The "dimension" again is the number of measured quantities, here the number of particles (babies) × number of common characteristics being measured. Such a concept can lead to a colossal number of dimensions. For instance, portraying the velocity and position of each of many gas molecules requires an astronomical number of dimensions. (Strictly, the preceding interpretations above are special cases of this one.)

For all three subgroups of this phase space interpretation of "dimension," chaologists like to speak of "low-dimensional systems" (i.e. systems having only a few phase space variables—usually three or less), "high-dimensional systems" (systems having many variables), and so forth.

- *The "N_d-dimensional vector"* This expression is just the previous concept as applied to pseudo phase space. (Actually, it could as well be used in standard phase space, too.) For example, as mentioned earlier, a two-dimensional vector is a pseudo phase space point defined on a graph of two axes or components, namely x_t and x_{t+m}; a three-dimensional vector is any point on a graph of x_t, x_{t+m}, and x_{t+2m} (Fig. 20.1d); and so on.
- *The embedding dimension (Chs 3, 19)* Closely related to the N_d-dimensional vector, the embedding dimension is the number of lagged values (x_t, x_{t+m}, etc.) used in a pseudo phase space plot, usually for the purpose of reconstructing an attractor (Fig. 20.1e).
- *The one-dimensional map*—a discrete equation or function that gives the value of a variable as a function of its value at the previous time (Fig. 20.1f) This isn't a dimension, of course, but it's another popular usage of the word.

The "dimension" of the six variants described thus far is always an integer. Our final variation differs in that respect.

- *A scaling exponent in a power law* This concept of dimension has given birth to a large group of variants, including the box-counting, similarity, capacity, Hausdorff–Besicovich (or simply Hausdorff), information, correlation, fractal, generalized, cluster, pointwise, Lyapunov, and nearest-neighbor dimensions (and others!). We'll discuss several of these in the following chapters; right now let's take a brief look at the basic philosophy.

  Suppose you are given some straight object, such as a stick or pole, and told to use it to estimate a distance from one place to another. You know the object's length. If it's relatively long, you'll only need to place it on the ground a few times to traverse the total distance. The shorter it is, the more placements or increments you'll need. For instance, you might measure the length of your back yard with a pole and decide that the yard is ten pole lengths long. Using a shorter stick, however, you might need 100 or more increments to cover the length of the yard. In addition, the results for any given pole or stick length can vary with the roughness of the terrain; the same stick needs fewer increments to traverse a smooth-surface yard than a rough, bouldery one. A general

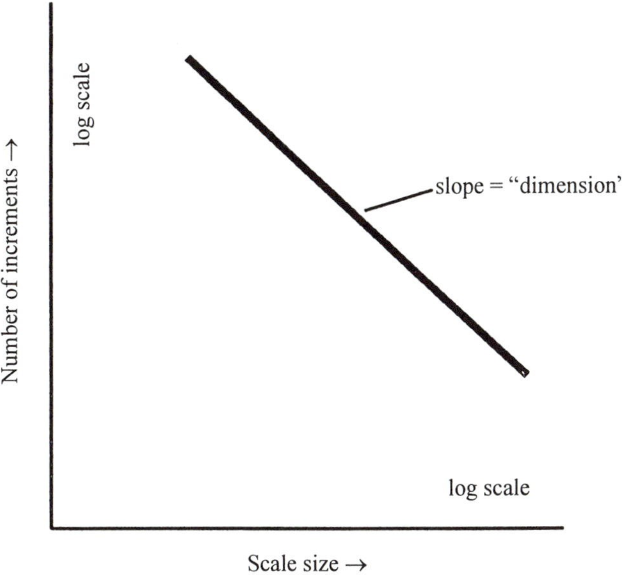

Figure 20.2 Power-law relation between number of increments and scale (measuring-tool) size.

empirical rule is that the number of increments often varies with scale size (e.g. the length of our pole or stick) raised to some power. That means we get a straight line on a logarithmic plot of number of increments (vertical axis) versus scale size (Fig. 20.2). The slope of that straight line is the power or exponent in the logarithmic proportionality. That power or exponent is called the dimension, for reasons to be explained. As we just saw, the number of increments increases as stick or scale size diminishes (Fig. 20.2). That inverse relation means that the exponent (the "dimension") in the power law is negative. Writing that idea in terms of a proportionality:

$$\textit{number of increments} \propto (\textit{scale size})^{-\textit{dimension}} \tag{20.1}$$

where ∝ means "is proportional to." Because dimension is an exponent in this relation, some authors call it an "exponent dimension." Dimension is an important characteristic in a mathematical sense here because it specifies the rate at which number of increments varies with scale size. The same general relation (Eq. 20.1) is valid in two or more phase space dimensions.

We usually need Equation 20.1 in a form that gives the exponent (the dimension) directly. To get that form, we take logs of both sides of the equation and rearrange terms. Putting Equation 20.1 into log form gives:

$$(\text{log of } \textit{number of increments}) \propto (-\textit{dimension}) \times (\text{log of } \textit{scale size}).$$

Rearranging to get dimension yields:

$$(-\textit{dimension}) \propto (\text{log of } \textit{number of increments}) \,/\, (\text{log of } \textit{scale size}).$$

To eliminate the minus sign attached to dimension, we arbitrarily make the log of scale size negative. Finally, rewriting –(log of *scale size*) as log of (1/*scale size*) (Appendix) gives

$$\textit{dimension} \propto \frac{\text{log of } (\textit{number of increments})}{\text{log of } (1/\textit{scale size})} \,. \tag{20.2}$$

"Dimension" as a scaling exponent (Eqs 20.1, 20.2) has a noninteger value, in most cases.

Types and importance of exponent dimensions

An exponent dimension often remains constant over a range of scale sizes. To that extent, it's an invariant quantitative measure. As we'll see in the following several chapters, it tells us how complex a system is. It's a quantitative numerical way of evaluating or comparing the geometric complexity of objects or systems of different size, shape, and structure. Therefore, it's new and potentially useful information.

Aside from being a useful concept, dimensions as scaling exponents are important in chaos theory for two reasons.

- Measuring the dimension (the exponent, or slope of a straight line on a log plot) can help distinguish between chaos and randomness. An object's dimension characterizes its geometry and is a fundamental physical trait. As explained in Chapter 19, a chaotic attractor has a shape in phase space. That shape has a finite and measurable dimension. In contrast, if a system operates randomly, it theoretically can visit the entire phase space sooner or later, and there probably won't be any definable configuration (attractor) within the phase space. The dimension then equals the embedding dimension and so becomes infinite as we continue to increase the embedding dimension (although there are exceptions, according to Osborne & Provenzale 1989). In practice, noise and scanty data often complicate this approach to identifying chaos.
- An attractor's dimension reveals the number of variables or measurable quantities we'll need to describe or model the system. For instance, a dimension of zero indicates a point in phase space (a constant value). To describe a line, we need one dimension; to model a rectangle, we need two variables—length and width. Modeling any system or object therefore requires at least as many variables as its dimension.

The Los Angeles temperature and wind-velocity attractors, discussed in Chapter 11, exemplify these qualities of dimensions. The point attractor (same temperature and wind velocity all the time) is the simplest possible type of attractor. It has a dimension of zero—we don't need any variables to describe it, because the weather is always the same. The next simplest type of attractor is the loop attractor (same daily cyclicity, day after day). It has a dimension of one, because a single variable, namely the time of day, can characterize such weather. Most weather evolution, of course, is fairly complex. Describing or modeling that evolution takes quite a few variables. In general, a chaotic attractor's dimension is a quantitative measure of its "strangeness" or of how chaotic it is.

There are many types or variations of dimension as a scaling exponent. The most useful type in chaos theory hasn't yet been determined. Some that are mentioned quite commonly, in no particular order of importance, are:

- similarity dimension
- capacity
- Hausdorff–Besicovich (or simply Hausdorff) dimension
- information dimension
- correlation dimension
- fractal dimension.

The following chapters discuss the first five on that list; we'll talk about the fractal dimension later in this chapter. Some candidates that, for brevity, I have left off the list are the generalized, cluster, pointwise, Lyapunov, and nearest-neighbor dimensions. References that give further information include Farmer et al. (1983), Holzfuss & Mayer-Kress (1986), Mayer-Kress (1987), Rasband (1990), Theiler (1990a), and Moon (1992).

The various types of exponent dimensions are all interrelated. Some even have the same numerical value for certain conditions. As a result, terminology tends to be loose, especially if the technical paper discussing "dimension" deals more with fractals than with chaos. The only way to be sure of the particular exponent dimension authors use is to look carefully at their mathematical definition and computational procedure. That's especially true if they mention the Hausdorff or fractal dimensions.

Most exponent dimensions fall into one of two categories. The first category measures only the attractor's geometry. That is, it takes no account of how often the trajectory visits various neighborhoods in the phase space. Examples are the similarity dimension, capacity, and Hausdorff–Besicovich dimension. The other category considers not only geometry but also probabilistic or informational aspects of the attractor. It takes into account that a trajectory may visit some phase space neighborhoods more often than others. Examples in this group are the information and correlation dimensions.

The correlation dimension presently seems to be the most popular in chaos theory. Actually, more than one of the above types might be helpful for a particular problem. Abraham et al. (1989) point out that some people use several measures of dimension to describe a chaotic time series.

As with many aspects of chaos, determining an attractor's dimension is an inexact procedure (Peitgen et al. 1992: 744) and is the subject of much research and rapid development. Glass & Mackey (1988: 52) point out that some topics particularly in need of more attention are the effects of noise, the requirements for the size of the dataset, and the effects of attractor geometry.

Fractal dimension

Fractal dimensions appear quite often in the technical literature. This popular member of the dimension family isn't really a separate dimension, however. Instead, "fractal dimension" is a general or basket term that might refer to any of the exponent dimensions mentioned above or even to other dimensions as well.

There are at least two possible reasons for attaching "fractal" to the various dimensions. First, if the estimated lengths of such objects follow a power law when plotted against scale size, the objects are often fractal. In other words, the same geometric rule or relation holds at lower and lower measuring scales. Secondly, the dimension value for the analytical types of dimension usually is noninteger, and people tend to associate a noninteger dimension with a fractal. Because of that noninteger quality, some people unintentionally introduce further confusion by using not only the term "fractal dimension" but also **fractional dimension**. In general, "fractional dimension" and "fractal dimension" are the same thing.

Early (*c*. 1980) usage of the term "fractal dimension" usually meant one or more main types prevalent at that time—similarity dimension, Hausdorff dimension, and/or capacity. Grouping those three dimensions together tacitly assumes or implies that a straight-line relation between number of increments and scale size (Eq. 20.2) can legitimately be extrapolated to a scale size of zero. If that assumption is correct (and there's no way to prove or disprove it), then at a scale size of zero the similarity dimension, capacity, and Hausdorff dimension all have the same value, since they all derive from the same basic formula. That's why some people think of them as one family.

When other measures of dimension were developed or refined, people began applying the term "fractal dimension" to the other types as well. Hence, "fractal dimension" and "fractional dimension" might mean any or all of similarity dimension, capacity, Hausdorff, pointwise, correlation, information or nearest-neighbor dimension, and perhaps others. Bergé et al. (1984), Holzfuss & Mayer-Kress (1986), and Rasband (1990), among others, point out the ambiguity. Whenever you see the terms "fractal dimension" or "fractional dimension," you've got to choose between two possible interpretations:

- The author is using the term in a general way to refer to most or all exponent dimensions as a group (and with luck this will be evident from the context). Example: "The fractal dimension usually is noninteger."
- The author really means a particular exponent dimension. In that case you have to examine the fine print and try to figure out which one.

Summary

There are at least seven slightly different meanings or usages of "dimension" in chaos theory:

- The Euclidean dimension—the number of coordinates or variables needed to describe a Euclidean object.
- The topological dimension, equal to 1 + the dimension of whatever standard divides the object.
- The number of measurable quantities or phase space variables in a system.
- The N_d-dimensional vector (a point in pseudo phase space, based on N_d successive iterates or measurements).
- The embedding dimension (Chs 3, 19). Closely related to the N_d-dimensional vector, the embedding dimension is the number of lagged values (x_t, x_{t+m}, etc.) used in a pseudo phase space plot, usually for the purpose of reconstructing an attractor.
- The one-dimensional map.
- The noninteger ("exponent") dimension relating the number of increments needed to measure an object to the size of the scale used in the measurement.

This last concept of dimension can help distinguish chaos from randomness and gives a rough idea of the number of variables needed to model the object. The exponent dimension includes a large group of special measures, such as the "fractal" dimension, correlation dimension, and many others.

The so-called "fractal dimension" is a basket term that can refer to any of a wide variety of exponent dimensions. Its meaning is clear when it refers to the various dimensions as a group; otherwise, it can be ambiguous. The reason for the name is that the object being measured is often a fractal. Also, people tend to associate a noninteger dimension with a fractal. The noninteger aspect leads some people to use a similarly broad and vague term, "fractional dimension," which means the same thing as "fractal dimension."

Chapter 21
Similarity dimension

The similarity dimension's main importance is that it involves principles that are basic to all types of exponent dimensions. It's a way of designating an object's geometry by a single number. It applies to objects that consist of, or that can be approximated by, straight lines. Examples are other lines (straight or wiggly), squares, cubes, and triangles.

Scaling

The underlying theme of the similarity dimension is the repeated subdivision of an object into smaller replicas of the original. Such subdivision is a form of scaling. Scaling here means to enlarge or reduce the representation of an object or pattern according to a fixed proportion or ratio, using facsimiles of the original. Chaos analyses almost always reduce or "scale down" the object, rather than enlarging it. The idea is to find out how certain results change as the length of the measuring device becomes smaller and smaller.

Suppose we take a straight line and chop it up equally into four identical smaller pieces. Each new piece is a miniature facsimile of the original line. In particular, each miniature replica is one fourth the size of the original. As represented by any one of these replicas, the original line then has been *scaled down* by a *ratio* of one fourth. The **scaling ratio** is the relative or fractional *length* of each new little piece.

Simple geometric forms and their dimension

If we repeatedly scale something down, the little replicas retain the same shape and appearance for each successive scaling. That is, the original line and four little lines are *self-similar*—each little piece, suitably magnified, looks just like the mother. We'll look at just a few examples here. The examples all involve geometric objects, such as a line or a triangle. I'll divide the process into three levels of increasing complexity, which I'll call the basic level, deletion level, and the delete-and-substitute level. The three levels all involve two common steps:

1. Subdivide an object into equal subparts, according to some arbitrary scaling ratio. An example is the straight line that we just chopped up into four smaller straight lines.
2. Repeat step one on each new little facsimile.

Basic level

At the simplest level, each new facsimile at every subdivision becomes a contributing member of the new pattern. Each new piece remains oriented in the same way as the original. Our subdivided straight line again is an example (Fig. 21.1, top row).

Now we're going to look for a relation between the scaling ratio and the number of new pieces, at any one subdivision. Let's arbitrarily assign our original straight line a "unit" length (a length of "one unit"). The four new little lines for our first subdivision in the above example then have a length of ¼. Conversely, the

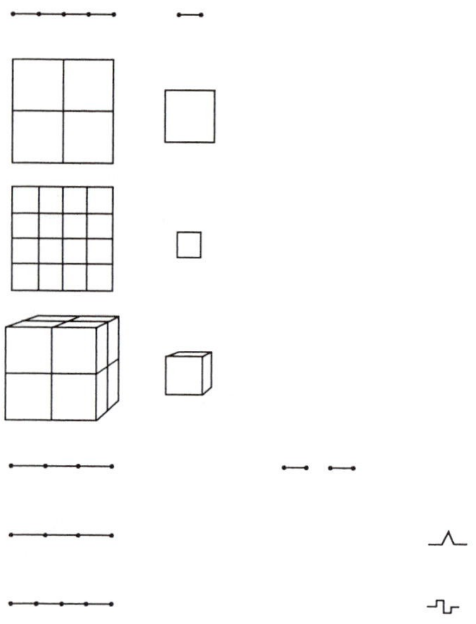

Original object	Segments via processing: Basic level	Deletion level	Delete and substitute level	Scaling ratio, r	Number of new facsimiles, N	Dimension, $D = \frac{\log N}{\log (1/r)}$
				1/4	4	1
				1/2	4	2
				1/4	16	2
				1/2	8	3
				1/3	2	0.63
				1/3	4	1.26
				1/4	8	1.5

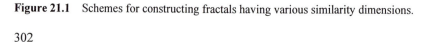

Figure 21.1 Schemes for constructing fractals having various similarity dimensions.

mother's unit length of 1.0 is simply the relative length of each subsection (the scaling ratio r, here ¼) times the number of subsections (N, here 4). Putting that statement in equation form:

unit length of original = scaling ratio × number of subsections

or, in symbols,

$$1 = \mathrm{r}N. \tag{21.1}$$

The straight line has dimension $D = 1$. Having established how the unit length is related to r and N (Eq. 21.1), let's now ask how the unit length is related to r, N, and D. With the above example of $N = 4$, $\mathrm{r} = ¼$, and $D = 1$, we can write $4[(¼)^1] = 1$, or

$$N\mathrm{r}^D = 1 \tag{21.2}$$

where, as before, 1 stands for the unit length.

Let's see if that relation holds in two dimensions. We'll subdivide a square of unit area into smaller squares. Here's something tricky: the scaling ratio doesn't apply to the area of the square. Instead, it indicates the relative *length* of each *side* of the new little squares. Thus, a scaling ratio of 1/2 means we break one side of the unit square into halves, so that the length of each new little square is 1/2 the length of the mother. Reducing a unit square by a scaling ratio r of 1/2 produces N = four little squares (Fig. 21.1, second row, first column). A square has dimension $D = 2$. Hence, the quantity on the left in Equation 21.2, $N\mathrm{r}^D$, is $4[(½)^2]$. This equals 1 (the original object's size or unit area), so Equation 21.2 is valid for that example also.

What if we use a scaling ratio of ¼ instead of ½ on the unit square? That means we divide each side of the square into fourths. In other words, each new little square has a relative or fractional *length* r of ¼. That divides the original big square into $N = 16$ little squares (Fig. 21.1, third row). $N\mathrm{r}^D$ therefore is $16[(¼)^2]$ or 1, so again Equation 21.2 holds.

A three-dimensional example is to scale a unit cube by half along each of its sides. That means $\mathrm{r} = ½$. That creates $N = 8$ little cubes (Fig. 21.1, fourth row). $N\mathrm{r}^D$ is $8[(½)^3]$ or again equal to one, as Equation 21.2 indicates.

In the examples thus far, r has been the same in *all* directions. For instance, when scaling the unit square down by a ratio r of ½, we divided its *length* by one half and also its *width* by one half, producing $N = 4$ new little squares. Similarly, scaling the unit square down by a ratio of ¼ entailed dividing the length into quarters and the width into quarters, yielding $N = 16$ new squares. We're not going to go into the more complicated case where the scaling ratio is different for different directions (e.g. one ratio along the x direction, another along the y direction). Dimension D still has a constant value for those situations, but Equation 21.2

becomes more involved. Voss (1988: 60, 62) and Feder (1988: ch. 2) give more details.

As mentioned in deriving Equation 20.2, the usual goal is a value for D. To get an expression for D, we rearrange Equation 21.2, as follows. Taking logs of both sides gives $\log N+D(\log r) = \log 1$. Since the log of 1 is zero, that relation reduces to $\log N+D(\log r) = 0$. Dividing all terms by logr and rearranging to get D gives $D = -(\log N/\log r)$. Arbitrarily applying the minus sign to the logr part gives $D = (\log N)/(-\log r)$. A rule from the Appendix lets us redefine the denominator, $-\log r$, as $\log(1/r)$. Hence, Equation 21.2 when rearranged to define D is:

$$D = \frac{\log N}{\log(1/r)} \tag{21.3}$$

$$= \frac{\text{log of number of pieces}}{\text{log of (1/scaling ratio)}}.$$

D as defined in that manner (Eq. 21.3) is the **similarity dimension** (Mandelbrot 1967, Voss 1988). (Peitgen et al. (1992: 205) call it the **self-similarity dimension**.) It conforms to our model Equation 20.2. It gets its name from being based on self-similar objects. Equation 21.3 is in a form applicable only to objects made up of straight lines.

Deletion level

Now for a slightly more complex level of scaling down. Here we impose a simple rule on the new facsimiles, as we subdivide a straight line. The rule is that, each time we subdivide, we throw out (delete) one of the new little pieces. For our example, we'll choose a scaling ratio of ⅓ and divide our straight line into thirds. Then we'll throw out the middle piece. For instance, Figure 21.1, fifth row, shows a line divided into three equal parts, and the diagram to the right shows the two end segments with the middle third deleted. Figure 21.2, first two rows, shows the same thing.

In keeping with the general scheme or algorithm, we next repeat that process on the surviving pieces. For instance, Figure 21.2 (third row) shows the results of doing it to the two survivors of the first step; the fourth row in turn does it again; and so on. Repeating this process over and over produces a series of infinitely small, separate pieces. That series, reached in the limit where the little lines become points, is called a Cantor set.

What's the dimension of a Cantor set? Since the original line (dimension one) ends up being severely decimated, the dimension must be less than one. Yet, some tiny remnants or points of the line still remain, at least in theory, no matter how many times we subdivide. That means the dimension must be greater than zero. Let's see what Equation 21.3 gives. *For any given line or facsimile*, the number of

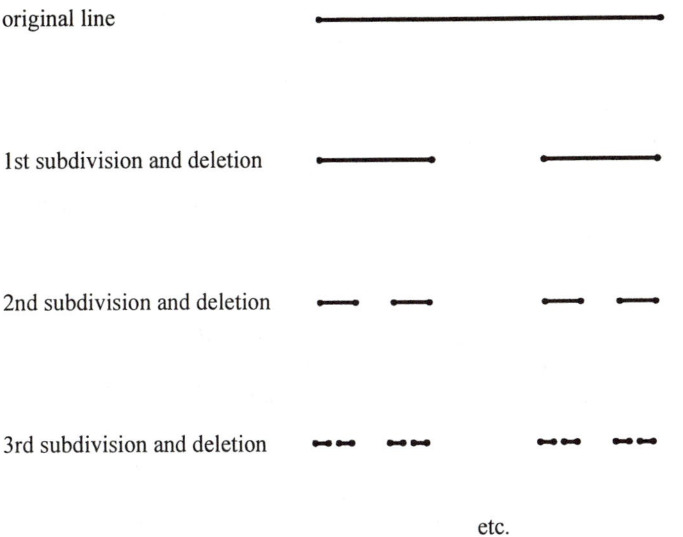

Figure 21.2 Procedure for deriving the Cantor set.

offspring (after deleting the middle piece) is two. Hence, $N = 2$. The scaling ratio r is ⅓ (we're dividing any line into three equal little lines, each being ⅓ the original). The denominator in Equation 21.3 is log(1/r). Since r = ⅓, 1/r is one divided by ⅓ or 3, and log(1/r) = log3. Thus, Equation 21.3 says that $D = \log 2/\log 3 = 0.631$. And so the similarity dimension needn't be an integer! The same is true of other exponent dimensions.

Delete-and-substitute level

For the third level of complexity in scaling down, we subdivide and delete, as before, but now we also substitute two or more little lines into the gap left by the deletion. The new pieces aren't necessarily oriented in the same way as the pieces of the previous step. A good example is the von **Koch snowflake**, named after its inventor, Swedish mathematician Helge von Koch (1870–1924). Forming a true von Koch snowflake involves doing the procedure I'll describe on each side of an equilateral triangle. (The side of the triangle is what is self-similar, not a piece of a side or the entire triangle as a whole.) Here's the procedure:

1. Choose one side of the equilateral triangle and divide that straight line into thirds. Thus, r = ⅓.
2. Throw away the middle third. (So far, this is exactly what we did with the Cantor set. But now comes the difference.)
3. Replace the gap just created with two segments of the same length as the other

little segments, in the form of a peak (Fig. 21.1, sixth row, fourth column).

The new arrangement then has a total of $N = 4$ miniature segments of the same size. Each segment by itself looks like—but may not be oriented like—the original straight line.

4. Repeat steps 1–3 on each of the N new lines, forever.

Having created a piece of the snowflake, let's now compute its similarity dimension (Eq. 21.3). The numerator is log$N$. For any one line, the number of new little pieces (after imposing the first three rules just mentioned) is four. Thus, the numerator in Equation 21.3 is log4. The denominator, the log of the reciprocal of the scaling ratio, is log(1/[⅓]) or log3, as with the Cantor set. Hence, Equation 21.3 becomes D = log4/log3 = 1.26. Again, therefore, the similarity dimension isn't an integer. In this case, it's greater than 1 but less than 2.

Choosing other scaling factors and other total numbers of little facsimile lines leads to other examples of the von Koch snowflake. For instance, Figure 21.1 (bottom row) shows a straight line divided into four equal segments rather than three segments. Thus, r now is ¼ rather than ⅓. Here we define the middle part—the part to throw out—as the second and third segments. Furthermore, we'll replace each of those two segments with three segments of equal size, in the form of a box with one side missing. The new pattern has a total of $N = 8$ little segments. Here the computed similarity dimension D = log8/log4 = 1.5.

One way to interpret the noninteger dimension of the von Koch line is that the final line takes up more space than a straight line. It occupies part (but not all) of a two-dimensional area. In an extreme case, a line conceivably can fill an area, in which case the line's dimension becomes two. In the same way, a line can occupy part or virtually all of a cube. The line's dimension then can range from slightly greater than 1 to nearly 3. By the same reasoning, a surface can have a dimension ranging from 2.0 (a perfectly smooth, flat plane) to 3.0 (a surface that's infinitely crumpled, such that it entirely fills a volume).

An object's dimension is never greater than the topological dimension of the space it occupies (the embedding space). For instance, a wiggly line drawn on a page occupies some part of a two-dimensional space. As the line becomes more and more serpentine, it's dimension increases. However, as long as the line stays on the page, it can never attain a dimension greater than 2. In the same way, a crumpled surface can't have a dimension greater than 3.

The similarity dimension (or other type of exponent dimension) gives a rough idea of how complex a system is, at least when that dimension exceeds 1.0. For instance, a line becomes more and more complex as it goes from straight (dimension 1.0) to slightly curved to extremely curved (dimension approaching 2.0). Thus, the calculated dimension is a capsulized, quantitative, descriptive measure of the line's complexity. The same idea applies to a surface and conceptually also to higher-dimensional forms.

This alternate (noninteger) way of considering dimension opens up the possibility of a continuum of values for dimension, rather than just integer values. How large is the possible range of that continuum? Complex processes can have many variables or dimensions. Mandelbrot (1967: 636) says that "the scientist ought to consider dimension as a continuous quantity ranging from zero to infinity."

The dimension of a wiggly line

The simple cases examined thus far (Fig. 21.1) have at least four features that usually aren't fulfilled in nature:

- Regular or Euclidean objects (e.g. squares and cubes) with nice straight lines. The natural world, in contrast, consists mostly of irregular objects.
- Exact as opposed to approximate similarity. In nature the little segments usually look only approximately (rather than exactly) like each other or like their parent. The concepts of the similarity dimension still apply.
- The arbitrary designation of length as 1. In practice, length is in length units (miles, millimeters, etc.) and rarely equals 1. (The same idea goes for areas and volumes.)
- Expressing the size of the little replicas as a fraction of the original (i.e. as scaling ratio r). In practice, we measure the little replicas in length units.

The next step up in complexity therefore is a line that's more common in nature, namely a wiggly line (one that has no apparent straight sections). Let's see how to get its similarity dimension. We'll no longer arbitrarily assign a value of 1 (unit length) to the length of the wiggly line. Instead, its value is in length units, is unknown, and has to be determined or estimated as a necessary step in calculating the dimension.

Suppose we impose a new caveat: we're not able to measure the wiggly line's length exactly and instead have to *estimate* it by using small, constant-length straight-line increments (with a permissible partial increment at the end). The tool for getting such an estimate is a "yardstick" or ruler of arbitrary length (or a compass or set of dividers of known gap). We place one end of the ruler at the beginning of the line and the other end wherever it intersects the line. We move the zero end of the ruler to that point of intersection and measure another ruler length of the line, and so on. Then the estimated length of the line is the ruler length times the number of increments needed. To use a straight-line example, suppose my wife asks me to determine the length of the dining-room table. If the length ε (Greek epsilon) of my measuring stick is one meter (ε = 1 meter) and I find that the table is exactly N = 2 stick-lengths long, then the table length obviously is 1 meter × 2 stick-repetitions = 2 meters. The same idea applies to a curved line, except that we only get an estimate of the length of such a line.

There are a couple of important points here. One, we now get our scaling standard not from an analytic scaling ratio r but rather from an empirical method using the ruler length ε. Two, as we'll see, the estimated length of a curved line depends on the particular ruler length we use. (To show that dependence, we'll symbolize that estimated length as L_ε.) Thus, the estimated length of a curved line is

$$L_\varepsilon = \varepsilon N \qquad (21.4)$$

where N, as before, is the number of little segments which, added together, make up the whole. Equation 21.4 is the wiggly line counterpart of Equation 21.1.

With our earlier, more simplified approach, $N = 1/r^D$ (by rearranging Eq. 21.2). An object made of curved lines requires three changes in that expression for N. The first is to insert a constant of proportionality for going from unit length to a measured length. Hence, $N = 1/r^D$ becomes $N = a\,(1/r^D)$ where a is that constant. The second change uses the new scaling standard ε (ruler length) in place of r, giving $N = a\,(1/\varepsilon^D)$. Thirdly, we rewrite the part within parentheses by again applying the rule that $1/\varepsilon^D = \varepsilon^{-D}$. Those changes give

$$N = a\varepsilon^{-D}. \qquad (21.5)$$

Equation 21.5 has the form of model Equation 20.1.

Our goal is to relate estimated length L_ε to ruler length ε and dimension D. Toward that end, we insert the right-hand side of Equation 21.5 in place of N in the relation $L_\varepsilon = \varepsilon N$ (Eq. 21.4).

That gives

$$L_\varepsilon = \varepsilon N = \varepsilon(a\varepsilon^{-D}) = \varepsilon a(\varepsilon^{-D}) = a\varepsilon^1(\varepsilon^{-D})$$

or

$$L_\varepsilon = a\varepsilon^{1-D}. \qquad (21.6)$$

Equation 21.6, a modified version of Equation 21.3, is a form used for natural objects (objects not made up of straight lines). Putting Equation 21.6 into log form and rearranging to solve for D yields $D \propto \log L_\varepsilon / \log(1/\varepsilon)$, analogous to Equation 21.3 and model Equation 20.2.

Equation 21.6 says that the estimated length of an irregular line isn't a fixed constant quantity. Instead, it depends on the measuring standard or ruler length (ε). That makes sense. A ruler long enough to span the straight-line distance exactly from one end of the wiggly line to the other yields the smallest possible estimate of the line's length (straight line being shortest distance between two points). However, such a long ruler would skip over many meanders and bends in the wiggly line's route, so the true length is somewhat longer. Choosing a shorter yardstick includes some or all bends, in which case our estimated length becomes longer. So, there's a quantifiable and inverse relation between the measured (estimated) length L_ε and ε: as ε decreases, L_ε increases. Furthermore, that inverse relation (Eq. 21.6) doesn't depend on the particular ruler lengths used.

Equation 21.6 also says that the different estimated lengths L_ε as related to ruler length ε plot as a power law. Data that follow a power law plot as a straight line on log paper, with the dependent variable (here L_ε) on the ordinate and with ε on the abscissa. (Incidentally, log paper by definition is graph paper on which both axes are logarithmic. Terms presently in vogue, namely "double log paper" and "log–log paper," are redundant.)

Finally, Equation 21.6 provides a simple way of determining the similarity dimension D for a wiggly line. We first make various estimates of the length of that line by several trials, each involving a different ruler length. Then we plot the resulting data on log paper (L_ε versus ε). The points must fall approximately on a straight line (indicating a power law), or the technique doesn't work. A power law's exponent (the quantity $1-D$ in the case of Eq. 21.6) is the slope of the straight line on the graph. Since the slope of the line is $1-D$, the similarity dimension $D = 1-$slope. Thus, a straight-line relation between L_ε and ε enables us to determine the object's similarity dimension by measuring the slope of the line on the log plot.

Unless your data (L_ε and ε) are particularly cooperative, they'll show some scatter. For the similarity dimension as well as for other dimensions, there are several problems in fitting a straight line to points on a graph. I'll discuss those problems in the chapter on the correlation dimension.

Field examples

The "ruler length" method is the traditional way of finding the similarity dimension of irregular lines, such as coastlines. Examples are the coastline of Britain (Richardson 1961, Mandelbrot 1967) and the coastline of Norway (Feder 1988). Let's look at a couple of coastlines right now (Fig. 21.3). (Chaos, of course, usually involves estimating the dimension of an attractor.) The overall procedure involves four easy steps:

1. Make successive estimates of the distance along the coast from point A to point B, using different but constant straight-line increments (ruler lengths or divider widths) for each estimate.
2. Plot the data (estimated distance L_ε versus ruler length ε) on log paper.
3. Fit a straight line to the points, if they plot in an approximately straight pattern.
4. Determine D as 1 minus the slope of the line.

(Incidentally, you'll sometimes see that plot of L_ε versus ε on log paper called a **Richardson plot**, since Richardson (1961) used it in his work on coastlines.) For this chapter, I'll use a less than fully rigorous version of those four steps (more comments on this later).

For step one, begin by arbitrarily choosing a ruler length ε. Place one end of the ruler at the starting point and the other end wherever it first intersects the coastline. The point where it intersects the coastline then becomes the starting point for the next increment, and so on. The last increment usually is a partial one, to enable us to arrive exactly at the desired finishing point. Then repeat the exercise several times, using a different constant ruler length in each case.

Figure 21.3a shows part of the coastline of Gatun Lake in the Panama Canal

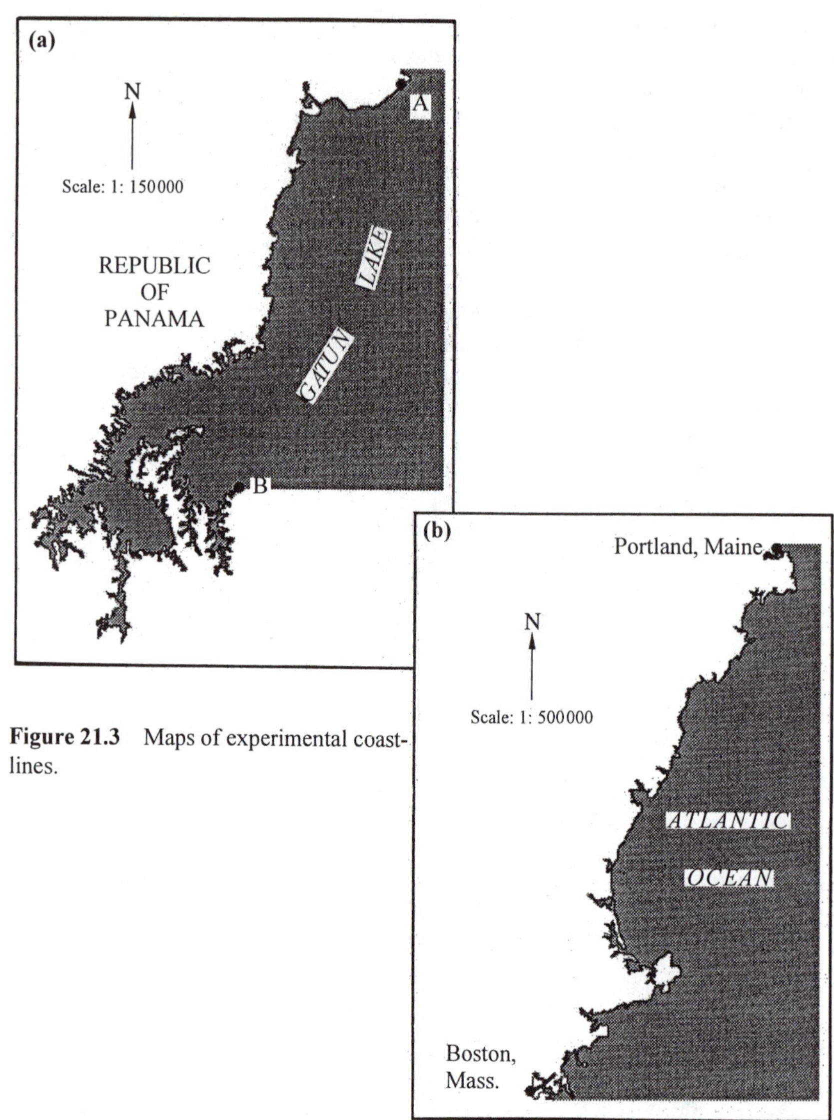

Figure 21.3 Maps of experimental coastlines.

Zone. On my enlarged map, I used a ruler length (divider width) ε of 16 cm to make the first estimate (Table 21.1). That step length corresponds to an increment of 16 km on the ground. It took 2.61 increments, using that yardstick, to go from point A to point B (Fig. 21.3a) along the lake's perimeter. The estimated shoreline distance therefore is the number of increments times the length of an increment, or 2.61 increments × 16 km = 41.8 km (Table 21.1).

The first ruler length was relatively long—16 cm. It leaped over many eccentricities of the coastline. To follow the coastline more closely, I next chose a shorter

Table 21.1 Experimental data for determining the length of selected shorelines.

Compass width (cm)	Ruler length (km)	Number of ruler lengths, N	Estimated shoreline length L (km)
	Part of Gatun Lake, Panama Canal Zone		
16	16	2.61	41.8
8	8	6.78	54.2
4	4	19.00	76.0
2	2	49.90	99.8
1	1	136.4	136.4
0.5	0.5	396.0	198.0
	Atlantic coast from Portland, Maine to Boston, Massachusetts		
31.5	157.5	1	157.5
16	80.0	2.01	160.8
8	40	4.09	163.6
4	20	8.54	170.8
2	10	20.05	200.5
1	5	60.8	304.0
0.5	2.5	135.4	338.5

ruler length of 8cm (or 8km of ground distance). With an increment of that length, I needed 6.78 ruler lengths (54.2km) to go along the lake's perimeter from A to B (Table 21.1). By halving the ruler length, then, my estimate of the coastline distance went from 41.8km to 54.2km. In succeeding trials, I used the shorter and shorter ruler lengths of 4, 2, 1, and 0.5cm. Each of those ruler lengths enabled me to adhere more and more accurately to the coastline. Each length produced a successively longer estimate of the shoreline distance (Table 21.1).

If Equation 21.6 holds, the estimated shoreline length L_ε versus ruler length ε plots as a power law. It does plot that way (Fig. 21.4a). The slope of a line fitted to the points is the quantity $1-D$, which is the exponent in Equation 21.6. In this example the data plot in a downward trend on the graph, so the slope is negative. The measured slope of that "best-fit" line (called a structural line, per Ch. 24) is –0.45. Similarity dimension D is 1–slope, so $D = 1-(-0.45)$ or $D = 1.45$. Such an answer ties in nicely with the notion that the similarity dimension D of an irregular line (e.g. part of a von Koch snowflake or of a coastline) has a value between 1 and 2 (cf. the 1.26 and 1.5 in Fig. 21.1).

Figure 21.3b shows a second example, namely the Atlantic coast from Portland, Maine to Boston, Massachusetts. Table 21.1 lists the ruler lengths I used on my enlarged map for this example, along with the associated number of increments and estimated coastline distances. Figure 21.4b is a plot of the data. Here, for whatever reason, the points follow a power law only approximately. The fitted line has a slope of –0.20, so $D = 1.20$.

Figure 21.4b shows that not all coastline data (L_ε versus ε) fall on a nice straight line on log paper, in spite of the impression that other studies might give. Where

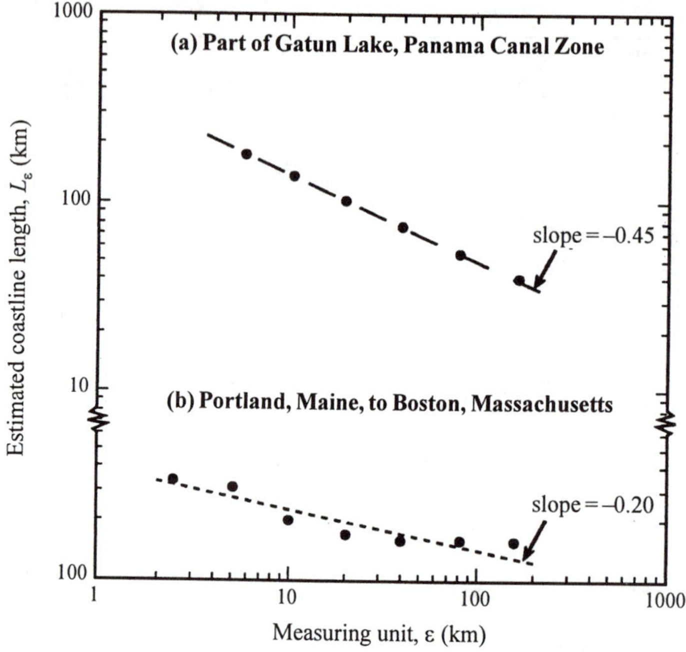

Figure 21.4 Log plots of coastline lengths as functions of length of measuring unit.

the data don't adhere closely to a straight line, we have to make a subjective judgement as to whether we can justify putting a straight line through all the points. Some alternatives are to fit a straight line to a subrange of the points or to fit a curve to all the points. (For tutorial purposes here, I'll use a single straight line for all the points on Fig. 21.4b.)

Some data show several zones or subranges of ε, each characterized by a differently sloping straight-line relation on log paper. Goodchild & Mark (1987) review several examples from the literature. Mandelbrot (1983: 148) discusses a related problem under the heading "slowly drifting log*N*/log(1/r);" see also Orford & Whalley (1987). Hence, in some cases the similarity dimension won't be constant over a wide range of ruler lengths; in others, the scatter of points is so great that the object won't even have a measurable similarity dimension. Goodchild & Mark (1987: 267) conclude that "most real coastlines and other spatial entities aren't fractals in the pure sense of having a constant *D* but in a looser sense of exhibiting the behavior associated with noninteger dimensions."

Certainly, as a general principle, the shorter the standard increment or ruler length, the more closely it can follow the coastline's configuration. In fact, the only condition that reproduces the coastline precisely is when the length of the ruler decreases to zero. That limiting condition isn't possible; a ruler of length zero doesn't exist. Even if it did, the measured coastline then would be infinitely long.

Typically, therefore, the straight-line relation between L_ε and ε no longer holds near the longest and shortest possible ruler lengths ε. (If it did, the shoreline of an island would be infinitely long, even though it encloses a finite area!) Instead, the relation becomes a curve in those regions. Chapter 24 discusses problems that such "tail regions" bring to determining the slope of the straight segment.

The minimum range of ruler lengths for us to confidently define a power law, as mentioned in Chapter 24, probably is 1–3 log cycles. The number of required data points within that range depends on the purpose of the exercise. For instance, if the purpose is simply to get a general idea of the existence of a power law, only a few (say five or six) points might suffice. At the other extreme, if you're going to place a great deal of importance on the numerical value of the estimated dimension, then you might need at least 50 or 100 points. (And for all these reasons, the examples in this chapter only demonstrate the general principles, not the rigorous procedure.)

For the range over which the power law holds, the similarity dimension is a constant mathematical property or inherent structural feature that doesn't change with change of scale. Such a property can be useful, for example in modeling.

Equation 21.6 and its associated log plot are likely to provide an exact measure of similarity dimension only for mathematical fractals. If used on natural fractals, the technique provides only an approximate dimension. There are three reasons:

- Natural fractals are more prone to measurement error.
- The starting and finishing points along the coastline or target object usually are arbitrary, and different endpoints often yield slightly different results. The standard procedure for reducing that potential error is to do several repetitions with a given ruler length, starting each repetition at a slightly different point, and then to average the results.
- Fitting the straight line on the graph (including identifying the appropriate terminal points of that line) can pose problems. Many repeated determinations of dimension reveal some variability. However, they cluster around an average or central-tendency value.

Rulers and boxes

Our measuring or scaling tool in all the above examples was one-dimensional—a ruler or yardstick. Any other object of known size also can be a suitable standard. Common examples are squares, circles, boxes, discs, or spheres. Each has a characteristic diameter or radius. Thus, we can measure the length of a coastline by using squares or boxes instead of a ruler to cover the coastline. A related popular method is to form a grid of squares covering the entire coastline and to count the number of squares that contain any part of the coastline (e.g. Feder 1988). Longley & Batty (1989) describe techniques that are more sophisticated.

Of the measuring devices just mentioned, early chaologists liked the box (which they imaginarily extended to more than three dimensions if necessary). You'll often see the similarity dimension as determined with a grid or lattice of boxes called the **box dimension** or **box-counting dimension**.

The similarity or box-counting dimension is somewhat impractical for most strange attractors. The reason is that we need an unrealistically large dataset to make sure that some phase space regions are really empty rather than just visited rarely.

Summary

Scaling means to reduce or enlarge an object by some factor, known as the scaling ratio. For Euclidean objects, the similarity dimension D relates number of required measuring increments N to scaling ratio r by the formula $D = \log N/\log(1/r)$. For natural objects, we first make a log plot of estimated length of object (L_ε) versus scaling ratio ε (ruler length). Then we compute D, empirically or graphically, as 1 minus the slope of the best-fit straight line on that plot. In other words, plotted points for natural objects often follow a power law of the form $L_\varepsilon = a\varepsilon^{1-D}$. A single straight-line relationship may not apply to all the points on such a plot. The value of D usually isn't an integer. People use squares, boxes, circles, discs, and spheres, as well as rulers, to estimate an object's length.

Chapter 22
Capacity and Hausdorff dimension

Capacity

Our discussion from here onward deals with exponent dimensions that estimate the dimension of irregular objects, such as fractals. For such irregular or natural objects (as opposed to mathematically determined objects), it's probably impossible to get an exact measure of dimension. We can, however, get closer and closer to an exact measure by using smaller and smaller measuring tools. To indicate that possibility, the relation $D = \log N/\log(1/r)$ (Eq. 21.3) needs a way to show that the equation gives a closer and closer answer as the ruler length gets smaller and smaller and approaches zero. Chaologists do that with a limit term. With such a limit term added and with ruler length ε written in place of scaling ratio r, Equation 21.3 becomes

$$D_c = \lim_{\varepsilon \to 0} \frac{\log N}{\log(1/\varepsilon)} . \qquad (22.1)$$

I'll call the dimension D_c as defined in that fashion the **capacity**, following Farmer et al. (1983). (Some people call it the capacity dimension or the limit capacity.) Technically, capacity is only definable in the range of extremely small ruler lengths ε, specifically in the limit as ε goes to zero.

The latter qualification is important. It means that in practice we can't measure capacity the way we did a similarity dimension. How do people determine capacity? Since measurements using a ruler length (or comparable device for multiple dimensions) near zero are impossible, capacity is something we can only estimate. The general approach is to extrapolate a plotted relationship of estimated length, area, and so on versus ε (e.g. Fig. 21.4), all the way down to $\varepsilon \approx 0$. If the plotted relationship being extrapolated is a power law, then the similarity dimension and capacity are the same. In other words, the "capacity" that many authors claim to measure is, for practical purposes, the same as the similarity dimension as defined here. (Besides the limit term, the similarity dimension and capacity also differ in certain minor technical and philosophical aspects.) Alternatively, it might be possible to use any other known relation of L_ε as a function of ε (in other words, not just a power law) and to evaluate the function mathematically at the limit where ε = 0.

Hausdorff dimension

We come now to a dimension that's frequently mentioned in the chaos literature. It's also one of the least useful in analyzing real-world data. As you've probably noticed, the idea of dimensions is inextricably bound up with the idea of measuring lengths, areas, volumes, and so on. Any value determined for such quantities is a **measure**. We really only estimate these measures (using rulers, squares, boxes, spheres, etc.) instead of determining them exactly. An example is the estimate of the length of the coastline with a ruler of length ε. Our measure of the object's total length was number of rulers or segments times the length of the ruler, or $L_\varepsilon = N\varepsilon$ (Eq. 21.4).

Equation 21.4 is really a simplified version of the generalized relation obtained from Equation 21.2, namely $N\mathrm{r}^D = 1$. With size of scaling tool ε in place of scaling ratio r, Equation 21.2 becomes $N\varepsilon^D = 1$. Also, estimated length L_ε can replace unit length 1, so $1 = N\mathrm{r}^D$ becomes

$$L_\varepsilon = N\varepsilon^D. \tag{22.2}$$

(If $D = 1$, $\varepsilon^D = \varepsilon^1 = \varepsilon$, so $L_\varepsilon = N\varepsilon$, which is Eq. 21.4.) To generalize Equation 22.2, we'll use the more general estimated measure M_ε instead of the one-dimensional length, L_ε:

$$M_\varepsilon = N\varepsilon^D. \tag{22.3}$$

As with the coastline example, a large ε produces a low number of segments needed to measure an object, whereas a small ε has a correspondingly larger number of segments. An assumption that now comes into play is that the interrelations between N and ε remain valid even as ruler or scale size ε becomes infinitely small and approaches zero (ε→0). In that region, the estimate M_ε gets very close to the true value.

The **Hausdorff**[1] **(or Hausdorff–Besicovich) dimension** applies only at the theoretical limit of ε→0 (like capacity). Also, it's a critical value, marking the boundary between two regimes. The best way to see that is to look at some sample computations of M_ε. For that purpose, we'll divide the dimension D into two component dimensions, called D_H and d, so that $D = D_H + d$. In Equation 22.3, D is an exponent. That exponent now becomes $D_H + d$. A law of exponents (Appendix) says that $x^{(c+d)} = x^c x^d$. Here exponents c and d are counterparts to our D_H and d.

Hence, we write ε^D in Equation 22.3 as ε^{D_H+d}, and an alternate way of expressing that is $\varepsilon^{D_H}(\varepsilon^d)$. Substituting that identity into Equation 22.3:

$$M_\varepsilon = N\varepsilon^{D_H}(\varepsilon^d). \tag{22.4}$$

Since we've defined D as D_H+d, rearrangement gives $d=D-D_H$. Making that substitution into Equation 22.4:

$$M_\varepsilon = N\varepsilon^{D_H}(\varepsilon^{D-D_H}). \tag{22.5}$$

To simplify our sample calculations, we'll arbitrarily set component $D_H = 1$. Equation 22.5 with $D_H = 1$ becomes

$$M_\varepsilon = N\varepsilon^1(\varepsilon^{D-1}). \tag{22.6}$$

Since $\varepsilon^1 = \varepsilon$, the ingredient $N\varepsilon^1$ in that equation is just $N\varepsilon$. As our ruler or **scale length** ε gets smaller and smaller, our estimate of a measure gets closer and closer to the true value. $N\varepsilon$ then becomes the true measure M_{tr}. The true measure is simply the number of increments times the scale size, with no error. Thus, for that idealized limiting condition, we replace $N\varepsilon^1$ of Equation 22.6 with M_{tr}. Equation 22.6 becomes

$$M_\varepsilon = M_{tr}(\varepsilon^{D-1}). \tag{22.7}$$

As mentioned, the Hausdorff dimension represents a critical boundary between two regimes. Let's call them the "zero regime" and the "infinity regime." We'll use Equation 22.7 to identify those two regimes and the transition between them.

---

1. **Hausdorff, Felix (1868–1942)** Felix Hausdorff was a German mathematician and philosopher who made important contributions not only to the theory of dimensions but also to topology in general and to other areas of mathematics. After studying mathematics and astronomy at Leipzig, Freiburg and Berlin, he graduated from Leipzig in 1891. For at least a dozen or more years after his graduation, he devoted himself to a range of interests far broader than just mathematics. For instance, although he published several papers on mathematics during that period, he also published four papers on astronomy and optics. For the most part, however, he occupied himself with philosophy and literature during this period, and his friends were mainly artists and writers. By 1904 he had published two books of poems and aphorisms, a philosophical book, several philosophical essays, articles on literature, and a play. His stature in the academic world and in mathematics grew steadily, nonetheless. His later publications were mostly in mathematics, especially in topology and set theory. He became docent at Leipzig in 1896, associate professor at Leipzig in 1902 and at Bonn in 1910, and full professor at Greifswald in 1913. In 1921 he returned to the university at Bonn, but in 1935 he was forcibly retired from that university. As soon as the Second World War began, the Nazis scheduled him (a Jew) for an "internment camp." When that internment became imminent, Hausdorff committed suicide with his wife and her sister on 26 January 1942.

Zero regime

The main feature of one regime is that the computed value of M_ε, using Equation 22.7, is zero. That value results when $D>D_H$. To see that, let's compute M_ε for $\varepsilon \to 0$, using different values of D. In all cases, we'll keep D greater than D_H. Since we have arbitrarily set $D_H = 1$, that means $D>1$.

1. For the first trial, let's arbitrarily take $D=2$. The exponent $D-1$ in Equation 22.7 then is $2-1=1$. Equation 22.7 becomes $M_\varepsilon = M_{tr}(\varepsilon^1)$. The situation we're interested in is ε becoming zero. If ε is zero, $\varepsilon^1 = 0^1$. Zero raised to any positive exponent is zero, so $0^1 = 0$. Equation 22.7 therefore says that if $D=2$, $M_\varepsilon = 0$. The value of 0 for M_ε is important.
2. For a second trial, suppose D is smaller but still greater than 1, say 1.001. Then the exponent $D-1$ in Equation 22.7 is $1.001-1 = 0.001$. Equation 22.7 becomes $M_\varepsilon = M_{tr}\varepsilon^{0.001}$. As $\varepsilon \to 0$, $\varepsilon^{0.001} = 0^{0.001}$. As before, 0 raised to any positive exponent (here 0.001) is 0. Hence, Equation 22.7 again gives $M_\varepsilon = M_{tr}(0) = 0$.
3. The two trials we've just done are enough to show the general principle. We've had $D>D_H$ or, with $D_H = 1$, $D>1$. That means the power to which we raise ε in Equation 22.7 is some positive constant. We're interested in the limit as $\varepsilon \to 0$. When ε becomes 0, the term ε^{D-1} in Equation 22.7 becomes 0^{D-1}. That, in turn, is 0 as long as $D-1$ is some positive constant. Hence, the entire right-hand side of Equation 22.7 (or M_ε) is 0 whenever $D>D_H$.

Transition

A transition occurs when $D = D_H$. Since we've set $D_H = 1$, $D = D_H$ means that D also is 1. The exponent $D-1$ in Equation 22.7 then becomes $1-1$ or zero. The quantity ε^{D-1} in Equation 22.7 therefore is ε^0. At the theoretical limit where ε is zero, the ε term in Equation 22.7 becomes 0^0. Mathematicians call that quantity "indeterminate." What's important here is that the computed values of M_ε suddenly stopped being zero. A discontinuity occurs, right where $D = D_H$.

Infinity regime

Thus far, we've seen that one regime occurs when $D>D_H$ and that a sharp transition occurs at $D = D_H$. The other regime, as we might guess, occurs when $D<D_H$. For instance, suppose $D = 0.98$. (And so with $D_H = 1$, we now have $D<D_H$.) Then, for use in Equation 22.7, $D-1=0.98-1=-0.02$. Equation 22.7 becomes $M_\varepsilon = M_{tr}\varepsilon^{-0.02}$. The important feature here is that the exponent, -0.02, now is negative. A law of exponents for that case is $x^{-d} = 1/x^d$. Hence, the term $\varepsilon^{-0.02}$ is $1/\varepsilon^{0.02}$. In other words, we can write Equation 22.7 in an equivalent form as $M_\varepsilon = M_{tr}(1/\varepsilon^{0.02})$. As ε gets smaller, the entire denominator $\varepsilon^{0.02}$ also gets smaller. The reciprocal, or

$1/\varepsilon^{0.02}$, therefore gets larger. In the limit where ε becomes infinitely small and goes to zero, $1/\varepsilon^{0.02}$ becomes infinitely large. The entire right-hand side of Equation 22.7, namely $M_{\text{tr}}(\varepsilon^{D-1})$, therefore also becomes infinitely large when $\varepsilon \rightarrow 0$ at $D = 0.98$.

Other values of D that are less than 1 (less than D_H) produce the same result. That is, the exponent in Equation 22.7 becomes negative, and when $\varepsilon \rightarrow 0$ the computed M_ε is infinitely large.

Recapitulation

In summary, our computations showed two general regimes for the measure M_ε. The two regimes exist only in the limit where $\varepsilon \rightarrow 0$. In the first regime, D is greater than D_H ($D>1$ for our example), and $M_\varepsilon = 0$. In the other regime, D is less than D_H ($D<1$ for our example), and $M_\varepsilon = \infty$. Therefore, the dimension D_H (1 in the example) is critical. Such a critical dimension is the Hausdorff (or Hausdorff–Besicovich) dimension. *The Hausdorff dimension is that critical dimension at which computed values of the measure jump from zero to infinity* (Fig. 22.1). Its two characteristics are that it's a critical value, representing a sharp boundary or discontinuity, and it's local, applying only in the limit of $\varepsilon \rightarrow 0$.

Defining the Hausdorff dimension (D_H) in terms of an equation leads to Equation 22.1, the capacity. However, the Hausdorff dimension differs from capacity in ways other than the theoretical development. For one thing, the Hausdorff dimension allows for measuring tools (boxes, etc.) of different sizes (ε) within any one traverse of the object. (The representative ε then is the largest of the various sizes used in the estimate.) Capacity (and similarity dimension) don't allow such flexibility. In practice, using different-size measuring tools within a traverse can lead to different estimates of the Hausdorff dimension and capacity for the same object (Essex & Nerenberg 1990). Secondly, changes in the coordinates (data transformations) don't affect the Hausdorff dimension but do affect the estimate of capacity.

No matter how loosely authors might use the terms, both the Hausdorff dimension and capacity really only have any meaning in the limit where $\varepsilon \rightarrow 0$. Consequently, they are difficult to estimate for real-world data because there's always a risky extrapolation involved. Mathematically oriented writers like to quote or use the Hausdorff dimension because, to them, it has interesting theoretical features.

Box-counting methods for determining the similarity dimension, capacity, and Hausdorff dimension have several drawbacks. In particular, the techniques are arduous, time-consuming, and impractical for systems in which the dimension is greater than about two or three (see, for example, Greenside et al. 1982). For instance, a system might only rarely visit some regions of an attractor. Consequently, in generating the attractor and getting an accurate count of boxes in a high-dimensional system, much computer time can be eaten up getting to the improbable boxes. However, further developments (Liebovitch & Toth 1989) offer some promise of resolving these problems.

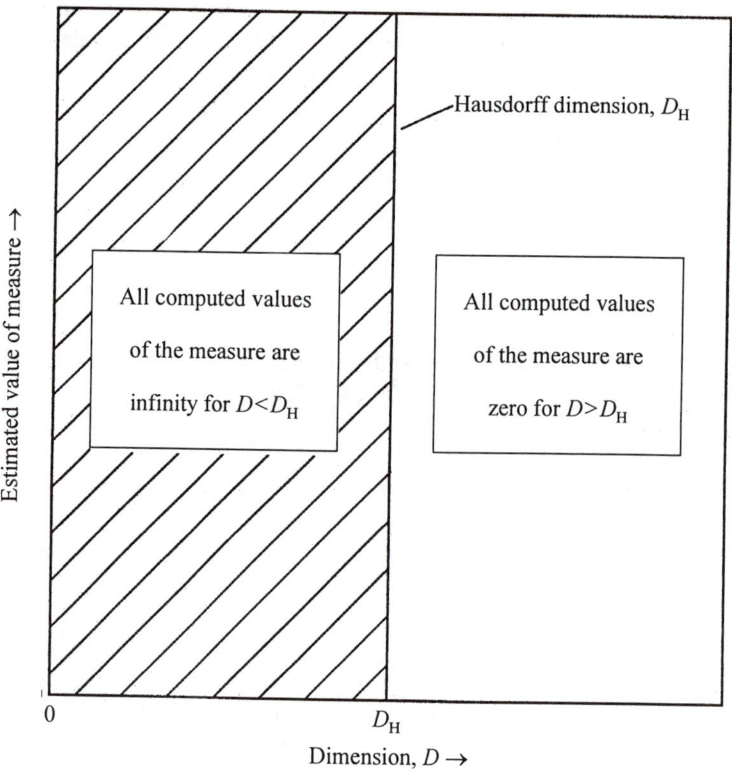

Figure 22.1 Definition of Hausdorff dimension.

Summary

Capacity is definable only in the limiting condition of scaling ratio (ruler length) approaching zero. In practice, there's obviously a limit as to how small the ruler length can be. Therefore, capacity can only be estimated, rather than measured exactly. Typically, people make that estimate by extrapolating an empirical best-fit relation between measured length L_ε and ruler length ε down to where $\varepsilon \approx 0$. If we can assume that the extrapolated relation really stays a power law down to $\varepsilon \approx 0$, then capacity is the same as similarity dimension.

The Hausdorff dimension is that critical dimension (a sharp boundary or discontinuity) that distinguishes an estimated measure of zero from an estimated measure of infinity, in the limit of $\varepsilon \rightarrow 0$. It's useful in a theoretical, mathematical context but not in practice.

Chapter 23
Information dimension

A trajectory on a chaotic attractor can go in any of many directions. It can also visit some zones on the attractor often and others rarely. The similarity dimension, capacity, and Hausdorff dimension consider only the geometry of the attractor (its apparent size and shape). They make no attempt to account for the frequency at which a trajectory visits different regions on the attractor. The next two measures—the **information dimension** and the **correlation dimension**—try to reflect not only the attractor's geometry but also the frequency with which the system gets to various phase space sectors.

In several ways, the procedures for the information and correlation dimensions are like those of the dimensions discussed earlier. For example, all methods start by choosing a measuring standard (circle, sphere, cube, etc.) and assigning it a size. We cover the attractor with the chosen measuring standard, such as a grid having a certain box size. Then we repeat with successively larger (or smaller) sizes of that standard. With the counting procedure, however, the information and correlation dimensions as a group take on a different philosophy. Specifically, they don't use a count of the number of increments needed to cover the object (a purely geometrical assessment). Instead, they involve a count of the number of data points that fall within each cell of the chosen measuring device. In three dimensions, for example, we might count the number of points within a covering of spheres (or cubes). (In four or more dimensions, the counterpart measuring object—used mathematically rather than graphically—is a **hypersphere** or hypercube [Ch. 15].) Such measurements reflect the frequency with which a trajectory gets close to its previous path, or the frequency with which it visits a given subspace on an attractor.

We get the best possible measurement when the trajectory covers the attractor to the fullest extent possible. That is, our data ought to include phase space zones that the system visits only rarely, as well as the more popular regions. In practice, that usually means the dataset should be large and should cover a long time span, so as to sample as broad a range of conditions as possible. As usual, I'll assume for this discussion that the original measurements were taken at equal time intervals.

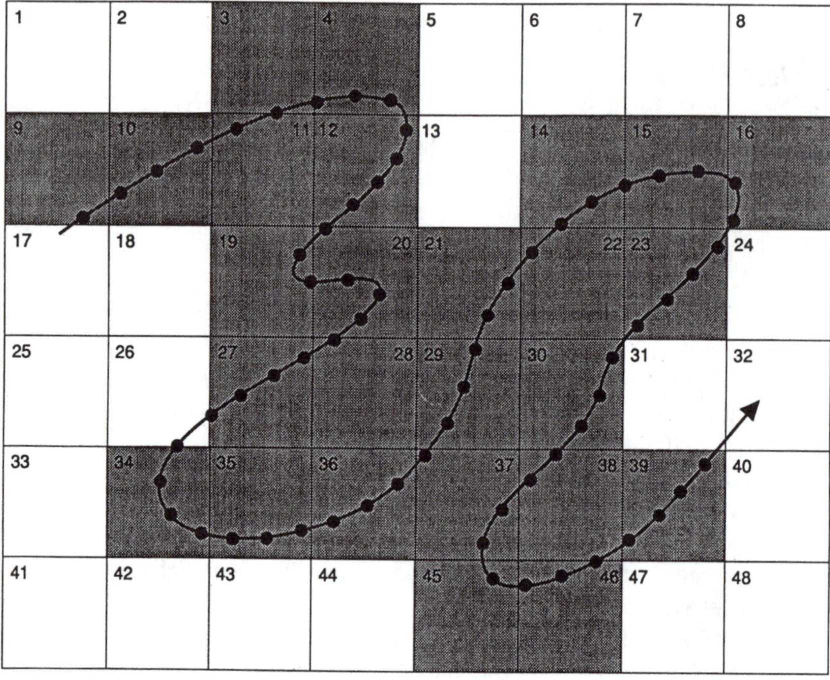

Figure 23.1 A phase space divided into 48 possible states and a trajectory sampled at successive, discrete times. Some states are visited frequently, others rarely, and still others not at all.

Development

The attractor to analyze for information dimension can be in standard or lag space. We'll use a generic two-dimensional attractor in this discussion.

The first step is to cover the attractor with a grid of characteristic size ε (Fig. 23.1).[1] Next, we count the *number of data points* in each bin. The counts for the series of bins show a frequency or relative popularity of the different cells or phase space regions. It's common to normalize such counts by dividing each one by the total number of points on the trajectory (a constant). The resulting values then are a relative frequency or an ordinary probability. In other words, we now think in terms of frequency distributions and probabilities.

The data for this two-dimensional example (Fig. 23.1) apply only to the particular cell size ε used. Doing the counting operation with a different bin size leads to a different histogram or frequency distribution. If the new cell size is smaller, we

1. A grid approach is not absolutely required but is common.

get a more accurate or refined picture of how the points are distributed within the phase space. We gain, as they say, "higher **resolution**."[2] In a sense, we *gain information*. And so, in general, we gain information as cell size ε becomes smaller (at least up to a certain vague limit, as discussed below). Chapter 6 explains the $\Sigma P\log(1/P)$ approach for quantifying such information into a number. We can do the same for the *rate* at which we gain information with decrease in bin size. The definitions and relations involved are as follows.

Let's use the symbol I_ε for the amount of information needed to describe the trajectory to within an accuracy ε (size of measuring tool). The reason for including ε as a subscript in I_ε is to emphasize that the computed information depends strongly on the scale of resolution ε. Expressing information I_ε in terms of a probability distribution (see Eq. 6.17b):

$$I_\varepsilon = \sum_{i=1}^{N_s} P_i \log_2(1/P_i) \qquad (23.1)$$

where N_s is the total number of states of size ε that have a nonzero probability, i refers to the successive states, and P_i is the probability determined for the ith state. Equation 23.1 says that, once we have estimated the probability of each state, there are two mathematical steps in determining I_ε. First, for each state, calculate probability P_i times the log (base 2) of $1/P_i$. Secondly, add up those products for all states.

As mentioned, information increases as cell size decreases. In other words, I_ε increases with 1/ε. Empirically, researchers found that, within an intermediate range of cell size ε, information I_ε is approximately proportional to the log of 1/ε. The data in that range of ε therefore plot as a straight line on semilog paper (graph paper on which one axis is arithmetic, the other logarithmic). (Fig. 23.2 shows an example which will be explained below.) The straight line appears with I_ε as the dependent variable (the vertical axis—here the arithmetic scale) and with 1/ε as the indicator or independent variable (the horizontal axis—here the log scale). With the arithmetic and log axes arranged in that particular way, the relation is a **logarithmic equation**. Because it's a straight line, it has the form $y = c+bx$ or, in our context,

$$I_\varepsilon = a+D_I\log_2(1/\varepsilon) \qquad (23.2)$$

where parameter a is the intercept and parameter D_I is the slope of the straight line. The slope of the line, D_I, is the information dimension.

On the plot of I_ε versus 1/ε, let's think about what happens as box size ε gets larger. Eventually, ε becomes big enough that just one box contains all the points. When that happens, the probability P_i for that all-inclusive box is 1.0. (We know

2. Resolution is the act of breaking something up into its constituent parts or of determining the individual elements of something.

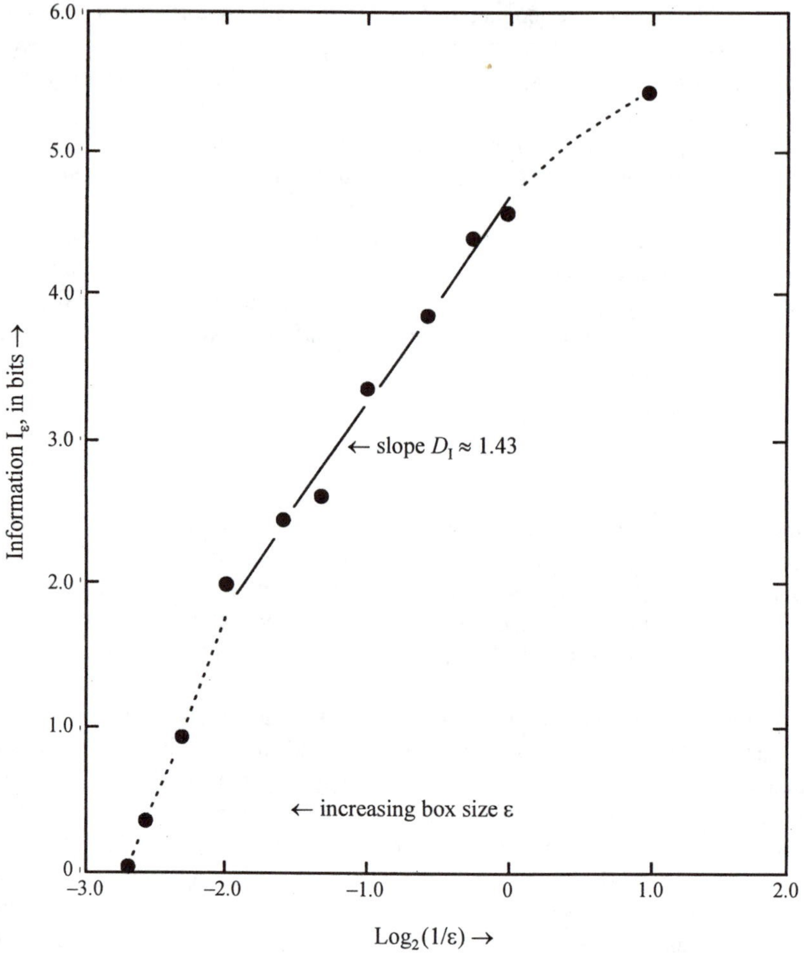

Figure 23.2 Plot of information I_ε as a function of $\log_2(1/\varepsilon)$, for various box coverings of the trajectory of Figure 23.1. Estimated information dimension D_I (slope of straight-line relation) is 1.43.

with certainty which box the system will be in at any time.) The information I_ε for that case is zero. (We get no information when we're told of an absolute certainty.) Mathematically, we get zero for I_ε because $1/P_i$ in Equation 23.1 is 1/1 or 1, and the log of 1 is 0, so the product $P_i \log_2 (1/P_i)$ is zero. On Figure 23.2, that point plots at the lower left end of the relation.

At the other extreme, box size can become small enough that any one box contains no more than one observation. Any boxes smaller than that limiting small ε then also include no more than one observation. For all boxes within this range of small ε, the number of boxes having one observation is constant and equal to the number of points on the trajectory. The probability P_i for each such box also is constant. Therefore, so are the quantities $P_i \log_2 (1/P_i)$. That means the sum of those quantities, or information I_ε, also remains constant for all ε within the range. So, at some limiting small value of ε, further increases in resolution (further decreases in box size) do not bring any further increase in information. On the graphical relation of I_ε versus 1/ε (e.g. Fig. 23.2), that range of ε occurs in the upper right part of the graph and is a straight horizontal line (with a slope of zero). (The cell sizes used in Fig. 23.2 haven't quite reached the limiting small size.) Noise also can obscure the true relation at small ε. These considerations mean that the smallest ε to include in measuring the straight-line slope D_I may well be somewhat larger than zero (Ruelle 1990, Lorenz 1991).

The two end zones or tail regions don't represent the true scaling properties of the attractor. We assume that only the straight middle segment reflects the true scaling properties. That middle segment therefore is called the **scaling region**.

Now let's go back to Equation 23.2. Rearranging that equation to define the information dimension D_I gives

$$D_I = \frac{I_\varepsilon}{\log_2 (1/\varepsilon)} - \frac{a}{\log_2 (1/\varepsilon)}.$$

A more economical expression for D_I would be nice. Toward that end, chaologists have tried to show that the relation does not apply to very large ε. In what may be overkill, they do so by inserting a limit term:

$$D_I = \lim_{\varepsilon \to 0} \left[\frac{I_\varepsilon}{\log_2 (1/\varepsilon)} - \frac{a}{\log_2 (1/\varepsilon)} \right].$$

Let's look at the term on the far right, a/$\log_2$ (1/ε). As ε becomes smaller and smaller, 1/ε and hence also its log becomes larger and larger. Since the denominator of the term gets larger and larger at constant a, the entire term, namely a/$\log_2$ (1/ε), becomes smaller and smaller. In the limit of ε approaching zero, the term a/$\log_2$ (1/ε) also approaches zero. In other words, it becomes negligibly small and therefore can be dropped, leaving

$$D_I = \lim_{\varepsilon \to 0} \frac{I_\varepsilon}{\log_2 (1/\varepsilon)} \qquad (23.3)$$

in which I_ε comes from Equation 23.1. The problem in that development, of course, is that the straight line doesn't apply at very small values of ε either, as just discussed. Nonetheless, Equation 23.3 is a common definition of D_I.

Because the information dimension D_I measures the rate of change of I_ε with respect to $\log_2$ (1/ε), it's the rate at which the information scales with increase in the measurements' rigor or precision (Farmer 1982). In other words, it's the rate at which we gain information from measurements of a dynamical system with increasing resolution (decrease in cell size).

A simplified example

Figure 23.1 shows a simplified example of the general procedure of analyzing a trajectory to determine its information dimension. The trajectory here has 67 points. Figure 23.1 (not to scale) shows the covering I did with one particular box size ε; on different runs I used larger and smaller box sizes. For each cell size, a count of the number of points in each successive (*i*th) cell divided by 67 (the total number of points) gives the probability P_i for that box or state. With the covering of Figure 23.1, only states 3, 4, 9, 10, etc. (the shaded cells) have observations. (For simplicity, I've labelled each state with a single number. Ordinarily, the distinguishing label of each state is the coordinates of the center point.) States 1 and 2 have no observations, so they each have a probability of 0/67 or 0. State 3 has one observation, so the probability for that state is 1/67 or 0.015. The quantity $P\log_2(1/P)$ then is $0.015 \log_2(1/0.015)$ or 0.091. Next is state 4. It has three observations. Its probability therefore is 3/67 or 0.045. Computing $P\log_2(1/P)$ gives $0.045 \log_2(1/0.045)$ or 0.201. We continue with each of the other states in turn, using the estimated P for each state to compute the associated $P\log_2(1/P)$. Equation 23.1 then says that the sum of all those values of $P\log_2(1/P)$ is the information I_ε for that box size. For the covering of Figure 23.1, that sum is 4.57.

The next step is to make a graph (Fig. 23.2). The vertical axis (arithmetic scale) has I_ε. The horizontal axis has $\log_2$ of the reciprocal of cell size ε. The slope of the straight line for intermediate box sizes is the information dimension, D_I. For this example, $D_I \approx 1.43$.

The above explanation of the information dimension (Figs 23.1, 23.2) used two dimensions, for simplicity. The computed D_I therefore applies only to two dimensions. In practice, we don't know the attractor's dimension in advance. It could, for example, be much greater than 2.0. To solve that problem, we have to analyze the data in successively higher dimensions, in separate tests. That is, we cover the attractor with cubes, spheres, hypercubes or hyperspheres of size ε, then use the distance formula to determine the number of points in each successive hypercube, and so on. For randomlike (nondeterministic) data, the computed dimension D_I should keep increasing in direct proportion to the number of dimensions we prescribe for our analysis (the embedding dimension in the case of lag space). For deterministic data, on the other hand, the prescribed number of dimensions in the analysis eventually should become large enough that the computed information dimension becomes asymptotic to some limiting, finite value. Figure 23.3 (from Mundt et al. 1991) is a graph for such an analysis for about 241 continuous years of sunspot observations (1749–1990). Deciding whether the computed values of D_I have reached an approximate plateau or limit can be rather subjective.

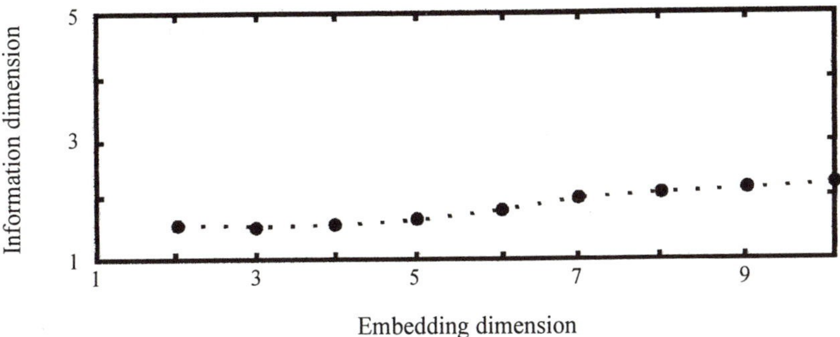

Figure 23.3 Information dimension computed for various embedding dimensions for 241 years of sunspot activity (adapted from Mundt et al. 1991, with permission from the American Geophysical Union).

The number of compartments can be large. For example, say we have only two phase space variables (x, y). Dividing x into ten possible classes and y into ten possible classes gives 100 phase space compartments (possible states). In fact, if we partition all variables into the same number of classes, then the total number of possible states N_s equals the partitioning raised to the power of the number of variables (a variation of our earlier Eq. 6.6). For instance, partitioning one variable into 10 classes gives 10^1 or 10 possible states; partitioning each of two variables into 10 classes gives 10^2 or 100 possible states; three variables using 10 classes for each gives 10^3 or 1000 possible states, and so on. Fortunately, the trajectory might never visit many of those states (Fig. 23.1). Even so, you can see why you'll often need large datasets.

Summary

The information dimension D_I takes into account the frequency with which the dynamical system visits different phase space regions over time. D_I equals

$$\lim_{\varepsilon \to 0} [I_\varepsilon / \log_2 (1/\varepsilon)].$$

In that expression, information I_ε is

$$\sum_{i=1}^{N_s} P_i \log_2 (1/P_i),$$

in which P_i is the probability that the system will be in the ith state. In practice, D_I is the slope of the best-fit straight line on a graph of I_ε (on the ordinate) versus

$\log_2(1/\varepsilon)$, on arithmetic scales. In other words, the information dimension reflects the rate of change of information with increase in measurement resolution, over intermediate length scales. Any such value pertains only to the number of dimensions in which we chose to analyze the data. We therefore have to compute D_I for each of many embedding dimensions (2, 3, 4, etc.) to see if an asymptotic value of D_I appears.

Chapter 24
Correlation dimension

The correlation dimension presently is the most popular measure of dimension. It's much like the information dimension but is slightly more complex. As mentioned, the information dimension usually is based on spreading a grid of uniformly sized compartments over the trajectory like a quilt. That's like moving the measuring device over the object by equal, incremental lengths. Analysis for the correlation dimension could also be done with that approach. Instead, however, the usual technique is to center a compartment on each successive datum point in turn, regardless of how many points a region has and how far apart the points may be.

Many types of exponent dimension are essentially impossible to compute in practice, either because they apply to some unattainable limit (such as $\varepsilon \rightarrow 0$) or they are computationally very inefficient. The correlation dimension avoids those problems. Also, for a given dataset, it probes the attractor to a much finer scale than, say, the box-counting dimension.

Two data points that plot close together in phase space are highly correlated spatially. (One value is a close estimate of the other.) However, depending on the trajectory's route between them, those same two points can be totally unrelated with regard to time. (The time associated with one point may be vastly and unpredictably different from the time of the other.) The correlation dimension only tests points for their spatial interrelations; it ignores time. (That's also true of the information dimension, but for other reasons it acquired a different name.)

Measuring procedure

The general principles of the correlation dimension (Grassberger & Procaccia 1983a,b) initially were explained for standard phase space, using the logistic, Hénon, and Lorenz attractors. However, in the same papers Grassberger & Procaccia also introduced, as "a variant" or "important modification," the lag space approach. They showed that, at least for those attractors, the calculated dimension is the same in either type of phase space. Today most algorithms use the easier and more practical approach of working with just one physical feature. Thus, let's think of all "points" or "data points" in the following discussion as lag vectors—pseudo phase space points defined by lagged values of one physical feature. Finding the appropriate lag is a rather uncertain matter, as discussed in Chapter 19.

The procedure for getting the correlation dimension involves not only lag but also the embedding dimension—the number of pseudo phase space axes. For any given practical problem, there isn't any way to determine in advance the correct embedding dimension. It depends on the attractor's true dimension in regular phase space, and that value is what we're trying to find. The correct embedding dimension emerges only after the analysis.

Once the lag is specified, the procedure usually begins with an embedding dimension of two (two-dimensional pseudo phase space). First, situate the measuring cell such that its center is a datum point in the pseudo phase space. Next, count the number of data points in the cell. After that, center the cell on the reconstructed trajectory's next point (in the ideal approach) and make a new count. Keep repeating that same procedure, systematically moving the cell's center to each successive point on the trajectory. (Some people choose center points at random to get a representative sample of the attractor, instead of going to every point on the trajectory.) Let's look at an example with just five data points.

Figure 24.1a shows five data points on an undrawn trajectory. Choosing point 1 to begin our procedure, we inscribe around that point a cell (in this two-dimensional example, a circle) with radius ε of arbitrary length. Our immediate task now is to determine how many points lie within the circle. Points 2 and 3 lie within the circle, whereas points 4 and 5 are outside the circle. (That's readily apparent on our sketch. In practice, however, we do the work on a computer instead of on a sketch. The computer identifies the qualifying points by calculating the distances from point 1 to each of the other points, using the distance formula [Eq. 4.3]. The computer then notes whether each distance is less than or greater than ε.) We therefore

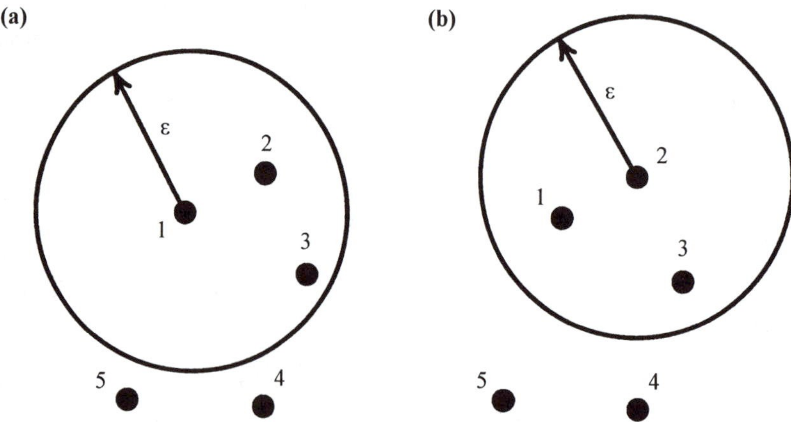

Figure 24.1 Two-dimensional sketch of five data points and use of arbitrary distance ε to identify qualifying nearby neighbors from (a) point 1 and from (b) point 2.

exclude points 4 and 5 from this stage of the analysis. Incidentally, our result of two qualifying points (numbers 2 and 3) is numerically equal to two qualifying *pairs* of points (1,2 and 1,3), where each pair involves the center point (here point number 1). Some authors use number of qualifying points, whereas others use number of qualifying pairs.

Next we center our circle on point 2, keeping the same radius as before (Fig. 24.1b). Within the circle at its new location, points 1 and 3 now qualify. (As before, the reference point doesn't count.) Keeping that same radius and systematically centering the circle on each point, in turn, we count the qualifying points within each circle. Once through the entire dataset with the *same* radius, we add up the total number of qualifying points for that radius. For example, Figure 24.1a has a total of eight points for the radius indicated. (We get two points when the circle is centered at point 1, two more when it is on point 2, two again at point 3, and one point each for centerings on points 4 and 5.) Having obtained the total for the radius chosen, we now work only with that total rather than with the numbers pertaining to any particular point. I'll refer to that total (eight in this example) in a general way as the "total number of points within radius ε" or the "total number of qualifying points."

The total number of points defining the trajectory (i.e. the size of the basic-dataset) obviously influences the total count for a given radius. For instance, the count of qualifying points for a given radius is much smaller for a trajectory made up of ten points compared to a trajectory of 10000 points. For comparison purposes, therefore, we normalize each count of qualifying points, to account for the total available on the trajectory. That means dividing each total of qualifying points by some maximum reference constant. That constant here is the maximum number of total points obtainable by applying the circling-and-counting procedure to each point throughout the dataset, for a given radius. The normalized result is the **correlation integral** or **correlation sum**, C_ε, for the particular radius:

$$C_\varepsilon = \frac{\textit{total number of points within radius } \varepsilon}{\textit{largest number of mathematically possible points}}$$

$$= \frac{\textit{total number of points within radius } \varepsilon}{N(N-1)} \tag{24.1}$$

in which N is the total number of points in the dataset (i.e. on the trajectory).

According to Equation 24.1, the largest number of mathematically possible points (the denominator) is $N(N-1)$. Figure 24.1 (a or b) shows why. Here N (the number of points in the basic data) is 5. The larger the radius, the more qualifying points around any center point. The extreme case is a radius large enough to encompass all points on the trajectory, regardless of where the center point is. Such a large radius on Figure 24.1 would yield four qualifying points for each of the five sequentially selected center points. For instance, those qualifying around point 4 are points 1, 2, 3, and 5. Adding four such points five times gives a grand total for the entire figure of 20 qualifying points, and we get that answer alternatively by taking $N(N-1) = 5(4) = 20$.

A special version of the ratio that defines the correlation sum (Eq. 24.1) comes from considering the limit as N becomes large. When N is very large, the 1 in $N-1$ becomes negligible. For all practical purposes, $N-1$ then becomes simply N. The quantity $N(N-1)$ (the denominator in the definition of correlation sum) therefore becomes $N(N)$ or N^2. Thus, in the limit of infinitely large N we can define the correlation sum (Eq. 24.1) as

$$C_\varepsilon = \lim_{N \to \infty} \frac{\textit{total number of points within radius } \varepsilon}{N^2}. \tag{24.2}$$

For practical purposes, that equation is good enough even when N is only as large as a few hundred, that is, far short of infinity.

In the technical literature, Equation 24.2 often appears in an imposing symbol form, as follows:

$$C_\varepsilon = \lim_{N\to\infty} \frac{1}{N^2} \sum_{i=1}^{N} \sum_{j=1}^{N} G(\varepsilon - |x_i - x_j|). \tag{24.3}$$

You're already on to some of that notation, so you know that this symbol version (just like many other equations you've seen in earlier chapters) isn't as formidable as it looks. Let's dissect it. As explained earlier, the best way to dissect many equations is to start at the far right and work leftward. The x_i in Equation 24.3 stands for a point on which we center our measuring device (e.g. our circle). x_j is each other point on the trajectory (each point to which we'll measure the distance from the circle's center point x_i). For each center point, the absolute distance between x_i and x_j is $|x_i - x_j|$. The distance formula gives that absolute distance.

The next thing Equation 24.3 says to do is to subtract that distance from radius ε. In symbols, that means compute $\varepsilon - |x_i - x_j|$. If the answer is negative, then the measured distance $|x_i - x_j|$ is greater than ε. That means point x_j is beyond the circle of radius ε and therefore doesn't qualify for our count. On the other hand, if $\varepsilon - |x_i - x_j|$ is positive, then $|x_i - x_j|$ is smaller than ε, and the point x_j is within the circle.

We now have to devise a way to earmark the qualifying points (the points within the circle). (In the highly unlikely event that the distance to a point *equals* the radius, we can count it or not, just so we're consistent throughout the analysis.) Equation 24.3's next ingredient (to the left of the distance symbols) is G. G is an efficient way to label each qualifying point, that is, each point for which $\varepsilon - |x_i - x_j|$ is positive (>0). In another sense, G acts as a sort of gatekeeper or admissions director. (The technical literature gives it the imposing name of the **Heaviside function**.) It lets all *qualifying* points into the ballgame for further action and nullifies all others. If $\varepsilon - |x_i - x_j|$ is positive, the point x_j has to be counted, as just explained. For all those cases the computer program assigns a value of 1 to the entire expression $G(\varepsilon - |x_i - x_j|)$. If, instead, $\varepsilon - |x_i - x_j|$ is negative, the point x_j is beyond the radius of the measuring device. For those cases, the computer program assigns a value of 0 to $G(\varepsilon-|x_i-x_j|)$.

The actual counting is the next step. That's just a matter of adding up all the one's for each center point x_i. Equation 24.3 indicates that operation by the two summation signs

$$\sum_{i=1}^{N}\sum_{j=1}^{N}$$

(the next two items proceeding leftward in the equation). Those two signs together simply mean that we go to the first center point x_i and sum the results of $G(\varepsilon-|x_i-x_j|)$ for all points (all x_j's), then do the same for the next center point x_i, and so on all the way through the total of N points. That gives the total count of qualifying points, for the radius being used (the numerator in Eq. 24.2).

Normalization is the equation's final job. As explained above, that's done by dividing the total number of qualifying points by the total number of available points. Strictly, the total number of available points is $N(N-1)$. Hence, we'd multiply the counted total by $1/[N(N-1)]$. Equation 24.3 uses the approximation (per Eq. 24.2) and so multiplies by $1/N^2$, with N^2 being the total number of available points or pairs on the trajectory in the abstract limit where N becomes infinitely large. And that's all there is to Equation 24.3.

Having determined the correlation sum for our first radius, we next increase the radius and go through the entire dataset with the new radius. The larger radius catches more points than the smaller radius did. That is, the new radius yields a larger total number of qualifying points (numerator in Eq. 24.1). The normalization constant N^2 depends only on the size of the basic dataset and so is constant regardless of the radius ε. Hence, the larger ε yields a larger correlation sum.

The idea is to keep repeating the entire procedure, using larger and larger radii. Each new radius produces a larger and larger total of qualifying points and a larger correlation sum. We end up with a dataset of successively larger radii and their associated correlation sums. Those radii and correlation sums apply *only* to the *two*-dimensional pseudo phase space in which we've been working. We now have to go to a three-dimensional pseudo phase space (an embedding dimension of three) and compute a similar dataset (or, rather, tell our computer to do it). All computed distances with the distance formula now involve three coordinates instead of two. Once the radii and associated correlation sums for three-dimensional pseudo phase space are assembled, we move on to four embedding dimensions, then five, and so on. A typical analysis involves computing a dataset for embedding dimensions of up to about ten. You might get by with fewer, or you might need more, depending on what a plot of the data shows. That plot is the next step.

Defining the correlation dimension

The computations provide, then, a batch of radii (ε) and their associated correlation sums ($C_ε$) for each of several embedding dimensions. Next step is to plot those data on *log* paper. Data for any one relation apply only to one particular embedding dimension. Correlation sum for a given radius ($C_ε$) goes on the ordinate, radius ε on the abscissa (Fig. 24.2). For any one embedding dimension, the lowest point worth plotting (and hence the beginning of any relation) corresponds to the shortest interpoint distance. (Smaller radii don't catch any points at all, so there's no relation.)

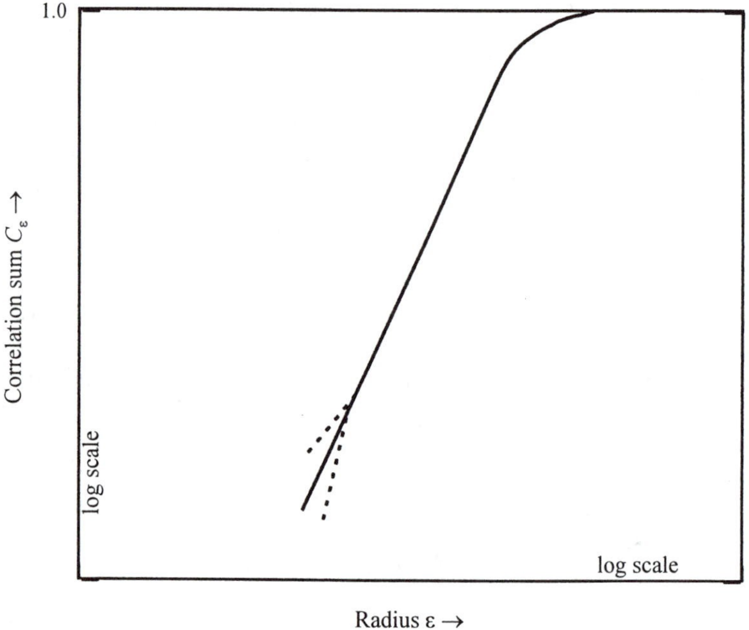

Figure 24.2 Idealized logarithmic plot of correlation sum $C_ε$ versus measuring radius ε, for a pseudo phase space reconstruction of a specified dimension.

Except for tail regions at the ends of the distribution, data for a given embedding dimension tend to plot as a straight line (a power law) on log paper (Fig. 24.2). Figure 24.3, involving the special case of *uniformly distributed* phase space points, shows the reason for the power law. One center point is enough to demonstrate the power-law relation. For a given embedding dimension and center point, we choose a radius and count the number of qualifying points ($N_ε$). Using the same center point, we then repeat for larger and larger radii. A plot of such data (not included here) shows that $N_ε \propto ε^{\text{dimension}}$, which is a power law. For instance, data for the one-

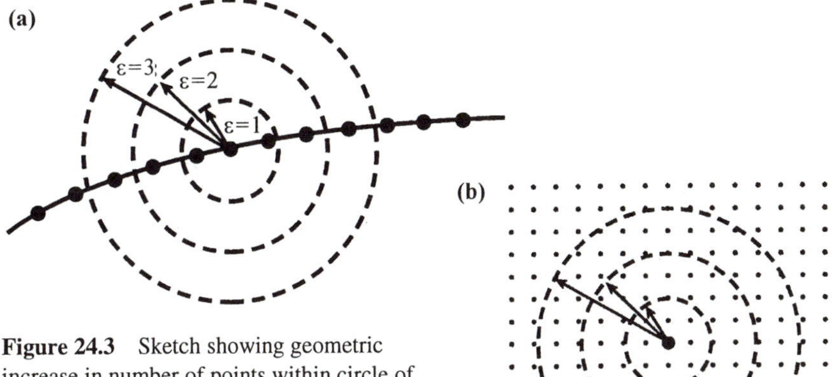

Figure 24.3 Sketch showing geometric increase in number of points within circle of radius ε for uniformly spaced points (after Bergé et al. 1984: fig. VI.36). (a) One-dimensional attractor (line). (b) Two-dimensional attractor (plane).

dimensional case (Fig. 24.3a) follow the relation $N_\varepsilon \propto \varepsilon^1$. In the two-dimensional case (Fig. 24.3b), the data adhere to the rule $N_\varepsilon \propto \varepsilon^2$, and so on. Also, that proportionality doesn't change if we deal with a correlation sum rather than just N_ε. The numerator in the correlation sum just increases by a constant multiplier that equals the number of center points; the denominator, based on the size of the dataset, is also a constant.

Why does Figure 24.2 have tail regions where the power law no longer holds? The general idea is the same as we discussed for the information dimension. As we increase our measuring radius ε, it eventually becomes so large that it starts to catch nearly all the available points. That is, it catches fewer and fewer new points. The numerator in Equation 24.1 (total number of points within radius ε) then increases at a lesser rate than at smaller radii. The plotted data then depart from a power law, and the relation becomes flatter (upper end of curve in Fig. 24.2). The radius finally becomes so large that it catches all possible points, no matter where it's centered. Thereafter, the total number of points remains constant at the maximum available in the data. The numerator and denominator in Equation 24.1 then are equal, and the correlation sum becomes 1 (its maximum possible value).

A tail region also occurs at small radii. The reason is that, in practice, data include noise and aren't uniformly distributed. Some small radius marks the beginning of a zone where measuring errors (noise) are of the same magnitude as true values. We can then no longer distinguish between the two. Furthermore, qualifying points become very scarce at small radii. In fact, even with noiseless data, our radius is eventually so small that it doesn't catch any points. All of these features lead to unreliable statistics. The result is that the plotted relation at the smallest radii might curve away from the straight line, in either direction (Fig. 24.2).

We have, then, a scaling region (middle segment of plotted line), just as we found on a related plot when deriving the information dimension.

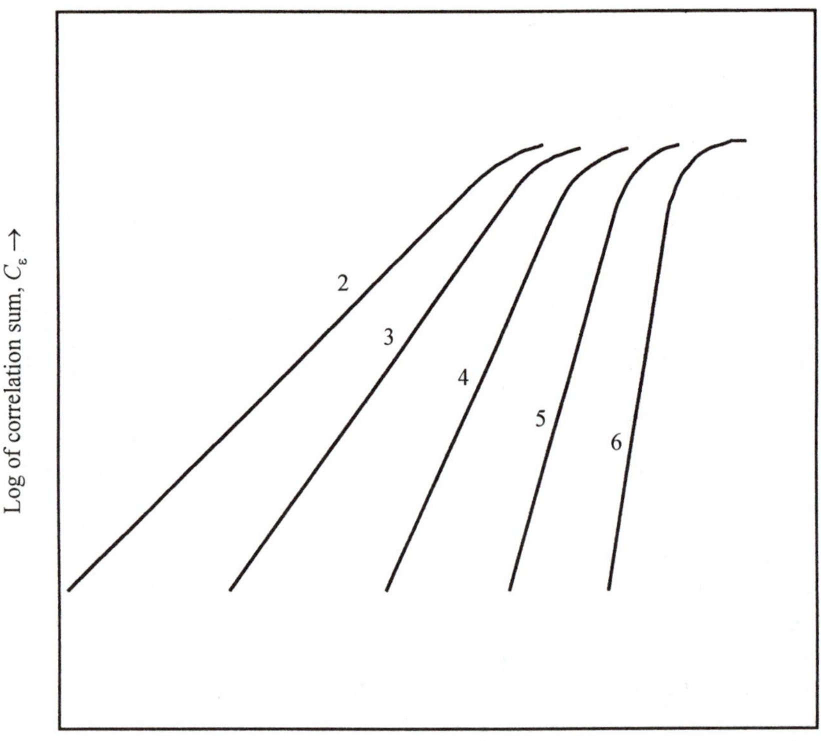

Figure 24.4 Idealized plot of correlation sum as a function of measuring radius. Number alongside each relation is embedding dimension.

In practice, one graph can include plots of the correlation sum versus radius relations for all embedding dimensions studied. Figure 24.4, with the lower tail regions omitted, shows what such a master graph ideally might look like. Figure 24.5, from Mundt et al. (1991), shows the same plot for field data (the sunspot data analyzed in Fig. 23.3).

The power-law relation for the scaling region says that

$$C_\varepsilon \propto \varepsilon^\nu \tag{24.4}$$

where ν (Greek nu) is an exponent (the slope of the straight line). As with Figure 24.3, the value of that exponent in a sense is the value of the object's (attractor's) dimension, at least for uniformly distributed data. The exponent ν therefore is called the correlation dimension or **correlation exponent**. In most cases, ν isn't an integer.

Figures 24.4 and 24.5 show that, for a given radius (abscissa value), correlation

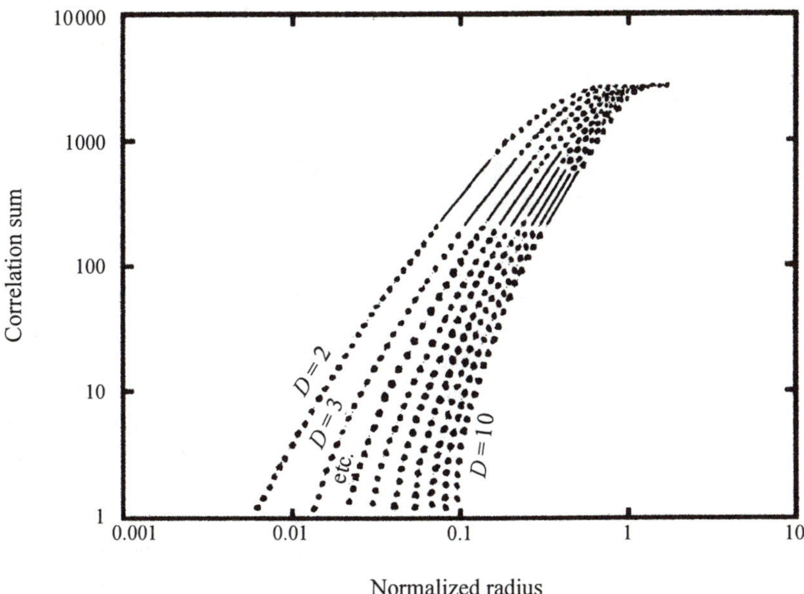

Figure 24.5 Correlation sum versus measuring radius for 241 years of sunspot activity (adapted from Mundt et al. 1991, with permission from the American Geophysical Union).

sum decreases as embedding dimension increases. Conversely, that sum increases as embedding dimension gets smaller. The reason is that low embedding dimensions produce many false nearest neighbors—points that are close to one another only because of the excessively low embedding. For example, in two dimensions a circle of a given radius (our measuring device) includes some points that are beyond that radius in three dimensions. Thus, for a given radius the correlation sum is larger for an embedding dimension of two than for an embedding dimension of three. By the same token, it's larger for an embedding dimension of three than for four, and so on.

The false-nearest-neighbors phenomenon and Figures 24.4 and 24.5 also show that, for a given correlation sum, the associated radius increases as the embedding dimension increases. For example, to get the same number of points (the same correlation sum) in three dimensions as in two requires a larger radius.

Deterministic and random correlation dimensions

As a group, graphical relations such as Figures 24.4 and 24.5 *in theory* ought to show whether the data are random (mutually unrelated), on the one hand, or deterministic, on the other. Here's how. The correlation dimension or exponent ν for a given embedding dimension is represented by the slope of the straight line on Figures 24.4 and 24.5. The steeper the slope, the greater the correlation dimension. The straight-line sections on those figures have a steeper slope as embedding dimension increases. Therefore, correlation dimension increases with increase in embedding dimension. If the data represent an attractor, the assigned embedding space eventually allows the attractor full self-expression. Thereafter, adding more embedding dimensions has no effect on the computed correlation dimension. For instance, a one-dimensional line is still one-dimensional whether embedded in a plane, in the atmosphere, or in any number of dimensions. Similarly, a plane is still two-dimensional whether embedded in three or more dimensions. What that means is that, as long as the embedding dimension is greater than the attractor dimension, the reconstructed attractor almost always has the same correlation dimension as the true attractor. (However, for practical considerations, we don't want an embedding dimension any larger than necessary.)

For highly deterministic or chaotic data, therefore, the correlation dimension initially increases with increase in embedding dimension, but eventually it becomes constant. Figure 24.6 (lower curve) shows an idealized relation for such data. The line for chaotic data flattens and becomes approximately horizontal at (or at least asymptotic toward) some final correlation dimension. The eventual, true value of the correlation dimension has an important practical implication: the minimum number of variables needed to describe or model the system is the next highest integer value above the correlation dimension. For instance, if the correlation dimension is 2.68, three variables might be enough to model the system.

Random data, in contrast, continually fill their allotted space as we increase the embedding dimension, at least for an infinite number of observations. Consequently, the slope (correlation dimension) continues to increase, without any indication of becoming asymptotic (Fig. 24.6).

As I said, that's how it's supposed to work in theory. In practice, data are limited. They are also noisy. The ideal patterns described above then don't necessarily happen. For instance, random numbers treated in certain ways can show an apparent stabilization of slopes on the plot of correlation sum versus radius (e.g. Osborne & Provenzale 1989). In such case, we'd be misled into declaring a low-dimensional determinism when there wasn't any. Furthermore, a plot such as Figure 24.6 for real-world data often shows a *tendency* to reach a plateau but doesn't become flat enough to give a very precise value of that potential asymptote. Figure 24.7, for instance, shows correlation dimension versus embedding dimension for (a) laboratory populations of the sheep blowfly (Godfray & Blythe 1990) and (b) a 144-year biweekly time series of the volume of the Great Salt Lake, Utah (Sangoyomi et al. 1996). A reliable value of the correlation dimension isn't very apparent. Other

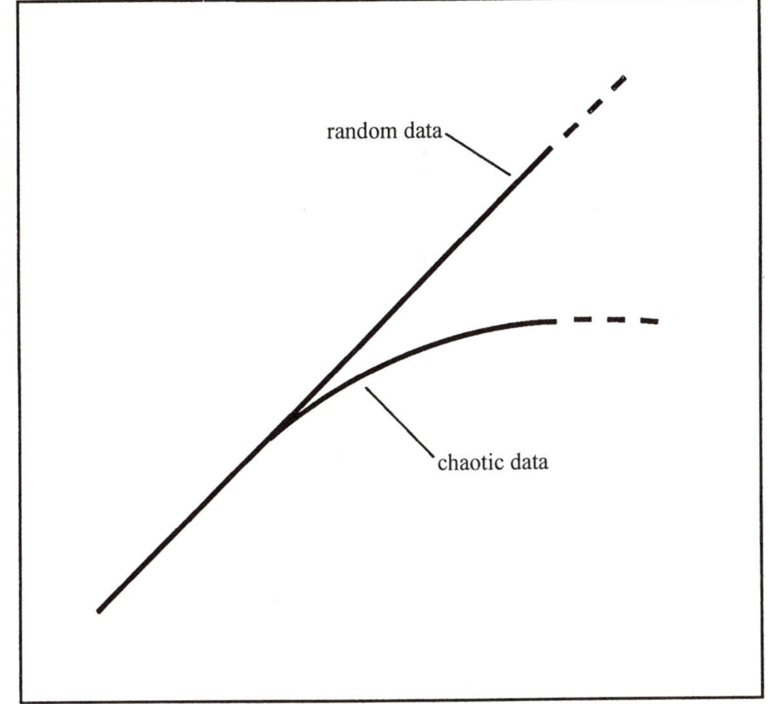

Figure 24.6 Hypothetical behavior (on arithmetic scales) of correlation dimension with increase in embedding dimension, for chaotic as compared to random data.

comments follow in the section "Practical considerations and problems", below.

Summary of computation procedure

1. Choose the time delay.
2. For each of successively larger embedding dimensions (2, 3, 4, etc.), compute the correlation sum for various radii ε.
3. Plot correlation sum versus radius on log paper, defining a separate relation for each embedding dimension.
4. Measure the correlation dimension as the slope of the (hopefully) straight middle zone (the scaling region) of each relation.
5. Plot correlation dimension versus embedding dimension on arithmetic paper.

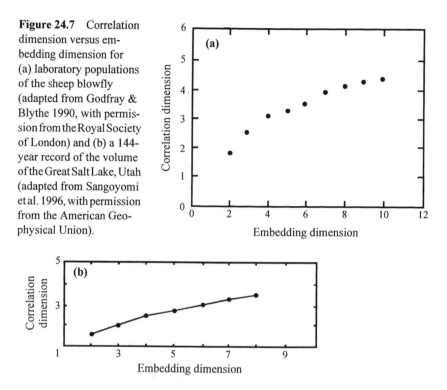

Figure 24.7 Correlation dimension versus embedding dimension for (a) laboratory populations of the sheep blowfly (adapted from Godfray & Blythe 1990, with permission from the Royal Society of London) and (b) a 144-year record of the volume of the Great Salt Lake, Utah (adapted from Sangoyomi et al. 1996, with permission from the American Geophysical Union).

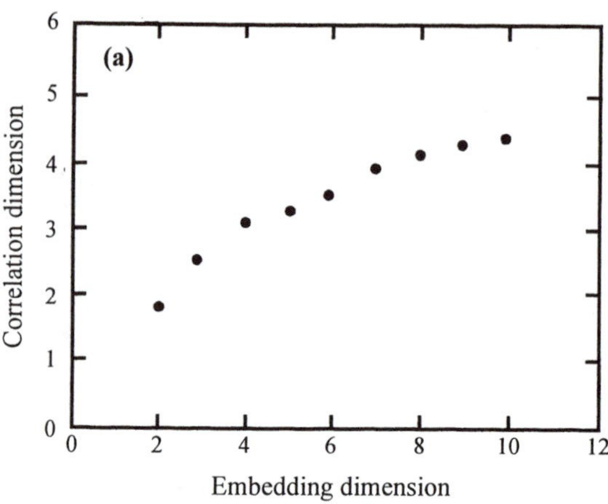

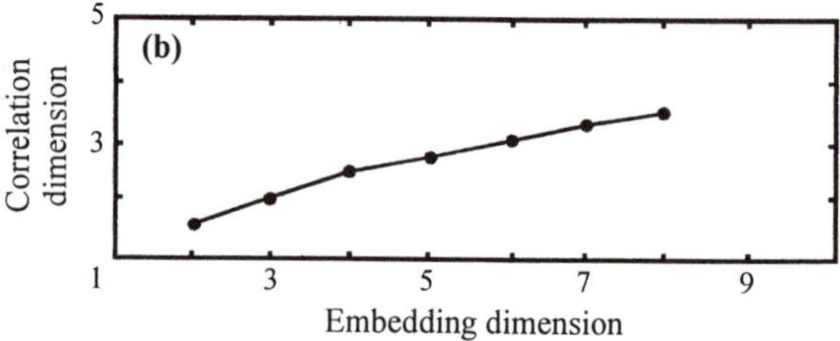

6. Decide if the latter relation seems to become asymptotic (hence possibly indicating an attractor) or keeps increasing in proportion to the embedding dimension (hence probably random data).

Practical considerations and problems

At present, the correlation dimension probably is the first thing most chaologists try to determine for their data. It has at least two favorable features. One is that it probes the attractor to a much finer scale than do other measures of dimension, such as the similarity dimension (Theiler 1990a). The second is that it's much more practical (easier and faster to compute) than other types of dimensions. On the other hand, it has certain elements of bias (Dvorák & Klaschka 1990). More importantly, there are big problems in computing it for a real-world time series, as discussed in the rest of this chapter.

Theiler et al. (1992: 164) say that "With a finite amount of data, and especially if the data are noisy, the dimension estimated by the algorithm will at best be approximate (and at worst, outright wrong)." Sometimes it's simply not possible to determine a correlation dimension. Albano et al. (1987: 218) summarize the problem by saying that "Calculating the correlation dimension is a deceptive numerical procedure. Getting a number is easy. Getting a dynamically meaningful number can be difficult, except in those instances where it is impossible . . ."

Two perennial problems are the appropriate embedding dimension and lag (Ch. 19). Both can affect the apparent correlation dimension. The usual way of estimating the embedding dimension is to plot measured slope ν versus embedding dimension and take the correct embedding dimension as the lowest approximate asymptotic value of a plateau (Figs 24.6, 24.7). Lag was discussed in Chapter 19. Further, choosing a different physical feature (variable) on which to do the embedding analysis in lag space can yield different estimates of the correlation dimension (Lorenz 1991), as we noted in Chapter 19 regarding attractor reconstruction. Now let's look at some additional problems.

Size of dataset

I, for one, come from a field in which a sample size of 50–200 is common. I was therefore a bit shocked and demoralized to find that, in chaos theory, a sample of that size is nearly worthless. Most tools presently available require many hundreds, more often many thousands, of measurements. The correlation dimension provides a good example of why this is so.

Determining the correlation dimension involves embedding the data in successively higher dimensions. Suppose we'd like to have at least six data points within each measuring radius or bin. We start with an embedding dimension of one—a straight line—and divide it into just two bins. Distributing our sample uniformly into two equal bins means we need a total of 12 data points to get six points per bin. Moving next to two embedding dimensions, we now have a plane instead of a line. Keeping the same bin length, there are now four compartments instead of two. To maintain a representation of six points per bin, our sample size has to be twice as large (24 points). In three embedding dimensions (a cube), there are eight bins. To maintain six points per bin, the sample size now must be four times as large (48 points). Hence, to keep the same representation (six points per bin), the trend or law is this: as embedding dimension increases arithmetically (from 1 to 2 to 3), required sample size increases geometrically (e.g. from 12 to 24 to 48). That's an exponential relation. In other words, even for the ideal case of a uniform distribution of points in space, the required sample size increases approximately exponentially with embedding dimension, other considerations being constant.

If, instead, we keep to a sample size of 12 points while increasing the embedding dimension, the average number of points per bin soon gets pretty sparse. For instance, an embedding dimension of two (i.e. four bins in pseudo phase space) leaves an average of three points per bin. For an embedding dimension of three, there'd be only one or two points (an average of 1.5) per bin. Worse yet, uniform distributions rarely occur in practice. As we embed the data in higher and higher dimensions, many bins that should contain some observations may not have any at all, because our small dataset is spread too thinly. In other words, if we don't take a large number of measurements, the data won't adequately represent the attractor when embedded in several dimensions.

Therefore, in some way that varies from case to case, sample size limits the number of embedding dimensions we can reasonably explore. The problem is worst at small radii. There the correlation sum is less than it should be because the small radius catches fewer points than it should. It might not even catch any points at all, whereas it would if the attractor were fully reconstructed. Erroneously low correlation sums lead to erroneously steeper slopes of the fitted line. The apparent correlation dimension then is higher than the correct dimension.

There's no general rule specifying the desirable size of a dataset, for purposes of estimating the correlation dimension. A few proposed rules-of-thumb suggest size of dataset once you already know embedding dimension D. Examples are:

- 10^D
- 42^D (Smith 1988)
- $2^D(D+1)^D$ (Nerenberg & Essex 1990: eq. 2.8).

In practice, those guidelines aren't of much help because we don't know D in advance.

Other authors give rules-of-thumb based on N. For instance, according to Ruelle (1990), "one should not believe dimension estimates that are not well below $2\log_{10}N$." Eckmann & Ruelle (1992) say that a correlation dimension is only reliable when it's less than the quantity $2\log_{10}N/\log_{10}(h_a/\varepsilon)$, where h_a is the diameter of the attractor. Typically, the largest usable radius ε is about 0.1 the diameter of the attractor, in which case h_a/ε is 10 and the denominator in their relation is 1. For those conditions, the reported correlation dimension must be less than $2\log_{10}N$, in accordance with Ruelle's 1990 rule. For example, if N is 1000, the reported correlation dimension must be less than 6.

Eckmann & Ruelle (1992) say that dimension estimates as high as 6 or 7 probably reflect the small number of data points rather than the dimension of an attractor. Rapp et al. (1989: 102) remark that if the dimension of an attractor is greater than about 5, the data requirements typically are unobtainable. Tredicce & Abraham (1990: 193) feel that "calculation of dimensions bigger than 3 or 4 is a tricky business, and reports of such results should be viewed with extreme caution." On the positive side, Havstad & Ehlers (1989) suggest several procedures that might minimize problems with small datasets.

It's tempting to try to enlarge small datasets by interpolating between measured values. Be careful doing this, because too many artificial values lead to autocorrelation. Also, according to Ruelle (1990), artificially increasing the number of data points by interpolation "produces a spurious slope of 1 at small radii."

What are some typical sample sizes in practice? Several papers report correlation dimensions calculated from 500–1000 points. However, the consensus seems to be that such low numbers are the exception. Fowler & Roach (1991: 78) recommend datasets of at least 5000 values. Atmanspacher et al. (1988) say that the accuracy of the computed correlation dimension is "generally not very high if N is substantially smaller than 10000."

Actually, as Abarbanel et al. (1993: 1355) discuss, it's not the raw number of data points that matters. Rather, what's important is "how many different times any particular locale of state space is revisited by an evolving trajectory." Ruelle (1994) calls this "recurrence." He says that many methods of measuring dimension don't apply when the attractor's dimension is large; the "recurrence time is then too large and the system is not observed twice near the same configuration" (1994: 28). In other words, while we do need a large amount of data, we should take pains to sample the entire attractor as fully and repetitively as possible. (Taking one million measurements of temperature over a couple of days can't be expected to tell us much about century-long climatic patterns.) Most chaos tools, including the correlation dimension, cannot easily tell the difference between phase space regions that are unoccupied because they aren't part of an attractor and regions that are part of an attractor but weren't sampled.

So far, we've discussed problems with datasets that are too small. Computer time can be a problem in the unlikely event that you have *too many* data. That is, the required computer time influences the maximum amount of data you can handle. For the usual algorithm, Smith (1988: 287) says to compute dimensions no larger than about 5 or 6. On the other hand, Abarbanel et al. (1993: 1355) say that several researchers have explored dimensions as high as 20 with the Grassberger–Procaccia algorithm. Tredicce & Abraham (1990: 193) comment that "On small laboratory computers it is generally prohibitive to try to calculate dimensions using more than about 10000 points."

Naturally, we'd like to reduce computer time. One time-consuming feature is the systematic shifting of the measuring device to each point in turn. Not only does that take an unnecessarily large amount of computer time, it's justifiable theoretically only if the points are statistically independent (Dvorák & Klaschka 1990). Parker & Chua (1989: 187–8) summarize several techniques that can reduce the number of distances to be calculated. For example, one approach that various authors (beginning with Theiler 1987) advocate is to compute only those distances that are less than some arbitrary radius. The arbitrary radius ideally corresponds to the upper break-point in the correlation-sum–radius relation, where the straight-line relation no longer holds. Another method is to sample the data randomly and compute distances only for those sampled points. However, Theiler (1990b) says that we should have nearly as many of such "reference points" as the number of points in the original dataset.

Noise

Noise is undesirable but may not always be serious. Theiler (1990a: 1068) says that in dimension estimation, the effect of low-amplitude noise often isn't as significant as other effects. The reason is that the global structure of a strange attractor can be robust enough that a trajectory can still reflect that global scaling even though plagued by noise.

Correlation dimension

Noisy data tend to produce a break in slope on the plot of correlation sum versus radius. The break occurs at that value of radius that equals the magnitude of the noise. Below that value, noise effects predominate and the data behave randomly. That means they keep filling the pseudo phase space for each new embedding dimension. The straight line in that case has a slope about equal to the embedding dimension, at least when the embedding dimension is less than or equal to about 5. Above the break, noise effects aren't too significant, and the slope of the straight line gives the correct correlation dimension.

One feature sometimes considered separately from noise is the finite precision (number of significant digits) to which the data were measured. Authors (e.g. Theiler 1990a, Tredicce & Abraham 1990) propose various ways to correct for that error. For instance, Tredicce & Abraham (p. 193) divide the precision of the measurements by 2, add the result to the measuring radius, and plot that sum on the abscissa (instead of plotting just radius) on the graph of correlation sum versus radius.

Theiler (1990a: 1068) recommends against using a low-pass filter to reduce the effects of noise. The reason he gives is that, based on other research, linear filtering can artificially raise the measured dimension of the time series.

Slope measurement

Chapter 4 explained several ways to measure the slope of a straight line, *once that line is drawn on a graph*. Putting the straight line on the graph in the first place is another matter. Typical situations in chaos theory that require such line-fitting include the similarity dimension (Ch. 21), information dimension (Ch. 23), and especially the correlation dimension.

There are really five problems in fitting a straight line to dimension-related data. First, the data may not always plot as a straight line. For example, Ramsey & Yuan (1989) say that the relationship between log of correlation sum and log of radius is only approximately linear over a relatively narrow range of radii. They add that the relation may be highly nonlinear and that we can't automatically assume it's linear. If it's not linear, the best we can do is to simulate the scaling by getting a rough approximation of the slope (Essex & Nerenberg 1991). In some cases the slope (correlation dimension) may be undefinable.

Secondly, the scaling region may have two or more subsections, each with its own straight line but of different slope. A break in slope in some cases can reflect the magnitude of the noise, as just explained. In other cases we may have to report a separate and distinct slope for each subsection.

Thirdly, to justify fitting a line at all, our straight line ought to cover some minimum range of abscissa values (radii or ruler lengths). The required range depends partly on the scatter and is a subjective judgement. In science and engineering in general, about one to three log cycles seem to be the most common recommendation. However, that may not be feasible for our purposes here. Many scaling regions cover less than one log cycle of radius values; some are so short as to be virtually undefinable. The shrinking of the scaling region as embedding dimension increases makes the problem worse.

Fourthly, where a tail region occurs at the upper and lower ends of the straight line, we have to decide which data points to include in the line-fitting. Most of the obviously aberrant points can be subjectively excluded at the outset. That leaves just a few questionable points at each end of the scaling region to worry about. One partly objective technique for those few points is to delete them one by one, fitting a straight line after each deletion to determine which deletions no longer significantly affect the computed slope. An alternative approach (Caswell & Yorke 1986) is to measure several slopes, each for a different range of radii.

Fifthly, there are many ways to fit a straight line to data (Troutman & Williams 1987). Some of the better-known methods are eye, least squares, group averages, structural lines, least normal squares, and the reduced major axis. (In other words, the eyeball method and least squares aren't the only techniques.) Most methods can be generalized or refined by transformations, weighting, and so-called robust procedures. In addition, other techniques have been devised for special purposes, including the correlation dimension.

More or less in order of simplicity (not necessarily validity), here are some of the methods used to measure the correlation dimension. Except for Takens's method, all of them involve fitting a straight line to the points in the scaling region on the log plot of correlation sum versus measuring radius.

- Decide (probably subjectively) on the range of the scaling region, connect the two endpoints of the scaling region with a chord, and take the slope of that chord as the correlation dimension. If the data plot as a curve, this crude method only approximates or simulates the correlation dimension (Essex & Nerenberg 1991).
- Compute, for each successive embedding dimension, a separate local slope for each successive pair of points on the log plot of correlation sum versus measuring radius (e.g. Atmanspacher et al. 1988). Then plot those local slopes against the log of radius. On such a graph the approximate true slope (true correlation dimension) should appear as a plateau or constant slope for a range of radius.
- Use least squares. Most of us learned (erroneously) that, to be "scientific" and objective, we should always fit a straight line by the least-squares technique. Actually, a least-squares line is designed for *prediction*. It gives the best estimate or prediction of the dependent variable, for a given value of the independent variable. In contrast, the structural line (a different method) provides the best estimate of such parameters as slope and intercept. It so happens that, for the special case where the independent variable has no error, the structural line mathematically is the same as the ordinary least-squares line. That might be the case in the similarity, information, and correlation dimensions, where the abscissa is the length or radius of the measuring device. However, least-squares techniques also have several other drawbacks, at least

for the correlation dimension (Theiler 1990a: 1066).

- Use Takens's (1985) equation. Takens invoked probability and maximum likelihood concepts to develop a way of estimating the correlation dimension. (Maximum likelihood is a standard statistical method for estimating parameters, such as slopes and intercepts.) His equation for the correlation dimension is:

$$\hat{\nu} = \frac{-1}{[\Sigma \log\left(|x_i - x_j|/\varepsilon_0\right)]/N} \tag{24.5}$$

where $\hat{\nu}$ is the estimated slope, ε_0 is the largest radius for which the straight-line relation holds and N is the total number of distances (qualifying points) in the entire straight-line analysis. Thus, in logs, the denominator of the equation normalizes each absolute distance $|x_i - x_j|$ by dividing by the largest applicable radius ε_0. It then averages the logs of all those normalized distances by adding them up and dividing that sum by the total number of distances involved (N). The standard error (the standard deviation of the errors) with Takens's estimation of ν is the minimum possible for the sample of points. Theiler (1990a,b) discusses the computation of that error.

- Use Ramsey & Yuan's (1989) "random coefficient regression model." That proposed model is an empirical best-fit line computed as

$$\ln C_\varepsilon = \mathrm{c} + \nu \ln \varepsilon + \mathrm{u} \tag{24.6}$$

in which ln stands for natural log (base e), u is an error term, and the constants c (the intercept) and ν (the slope or correlation dimension) both depend on sample size, embedding dimension, and experimental error. The authors give additional equations that explicitly take these last three factors into account.

Where the raw data (initial time series) represent computer-generated iterations and show very little scatter on the plot of correlation sum versus radius, two decimal places might be justifiable for the correlation dimension. With most real-world data, the correlation dimension (when measurable) probably can't be reliably estimated to more than one decimal place.

Convergence of slopes

Most analyses produce increasing correlation dimensions (slope of straight line within scaling region on the plot of correlation sum versus radius) with increase in embedding dimension (Fig. 24.4). Ideally, the slopes for chaotic data eventually taper off and become constant (Fig. 24.5). There's no good way to get a precise value for the true (asymptotic) value of the correlation dimension. Once the progressive increase in slope seems to stop, people commonly take an arithmetic average of the slopes of the last few embeddings as the true value. Various other unpublicized methods undoubtedly are in use.

Some researchers also place limits on how closely the slope values should agree over a specified number of embeddings. For example, Albano et al. (1987) require agreement to within 10 per cent over at least four embeddings.

False correlation dimensions

Several features can lead us into the trap of counterfeit correlation dimensions. For instance, even with data that are essentially random, the correlation dimension can become asymptotic (or seem to become asymptotic) to a constant noninteger value (Osborne & Provenzale 1989, Rapp 1992). How often this can happen with real-world data isn't yet known, but it certainly is unsettling. Autocorrelation, small datasets, and excess use of interpolated (unmeasured) "data" also can produce spurious or incorrect results. Lorenz (1991) suggests that analyzing different variables from multivariate systems leads to different values of the correlation dimension. Theiler (1990a) reviews various other possible sources of error in computing the correlation dimension. These unfortunate facts show two important principles:

- A constant value (especially a noninteger) for the correlation dimension doesn't prove chaos.
- Interpret any reported correlation dimension with great circumspection (Rapp et al. 1993). Slightly less extreme, Provenzale et al. (1992: 36) say to suspect correlation integrals (and hence also correlation dimensions) if a standard phase space plot and Poincaré sections show "messy" distributions with no discernible structure or if they show isolated, nonrecurrent patches of points.

There are at least three ways to reduce the chances of getting false correlation dimensions. First, use procedures or tests aimed at specific problems. For instance, the possible problem with certain random data might be reduced or eliminated by using Theiler's (1986) modified correlation integral. Provenzale et al. (1992: 31) suggest a simple test for detecting autocorrelation effects. Secondly, impose "reliability criteria" on the overall analysis. Albano et al. (1987: 210) require all of the following:

- The correlation sum must be linear in the scaling region (with local slopes not deviating by more than, say, 10 per cent).
- The scaling region must extend over a minimum range of two log cycles of radius.
- The correlation dimension eventually must remain constant as embedding dimension increases. They usually allow no more than 10 per cent variation in correlation dimension over four consecutive embeddings.

If their analysis doesn't fulfil all three criteria, the authors declare "dimension unresolved." Thirdly, apply surrogate-data tests. Such tests might show whether the correlation dimension could as well come from nondeterministic data.

Transients

Attractor reconstruction won't be nearly as easy as desired if the plotted points aren't on (haven't yet reached) the attractor. That is, there are enough problems already, without unknowingly including transients. Transients are relatively easy to identify with noiseless, computer-generated data. That may not be true with laboratory experiments. Field experiments often can't control such matters. In many cases we may have to assume that transients died away. However, their potential for causing difficulties is always present.

Grassberger et al. (1991) offer several other useful observations and suggestions about computing the correlation dimension in practice.

Summary

The correlation dimension is the most popular noninteger dimension currently used. It probes the attractor to a much finer scale than does the information dimension and is also easier and faster to compute. Like the information dimension, it takes into account the frequency with which the system visits different phase space zones. Most other dimensions involve moving a measuring device by equal, incremental lengths over the attractor (tantamount to placing a uniform grid over it). In contrast, the correlation dimension involves systematically locating the device at each datum point, in turn. The procedure usually begins by embedding the data in a two-dimensional pseudo phase space. For a given radius ε, count the number of points within distance ε from the reference point. After doing that for each point on the trajectory, sum the counts and normalize the sum. That yields a so-called correlation sum. Then repeat that procedure to get correlation sums for larger and larger values of ε. A log plot of correlation sum versus ε (for that particular embedding dimension) typically shows a straight or nearly straight central region. The slope of that straight segment is the correlation dimension. The next step is to repeat the entire procedure for larger and larger embedding dimensions. For chaotic data, the correlation dimension initially increases with embedding dimension, but eventually (at least in the ideal case) it asymptotically approaches a final (true) value.

In spite of its popularity, an accurate computation of correlation dimension is difficult and sometimes impossible in practice. Some of the reasons are:

- It requires large datasets (at least thousands of observations).
- We can't determine the correct final embedding dimension in advance; we get it only by trying successively larger embeddings and hoping that the measured correlation dimension soon becomes approximately constant.
- Noise complicates our interpretation of results.
- Measuring the correlation dimension itself is difficult. That difficulty stems from the absence (in some cases) of a straight line to measure, the occasional

presence of two or more straight lines rather than one, a possibly short range of radii over which the straight-line relation holds, a subjective judgement as to the applicable range of radii, and what line-fitting method to use.

- Deciding on the final (asymptotic) value of correlation dimension, as embedding dimension increases, usually is subjective.
- The value that eventually emerges for the correlation dimension may be wrong. Use several auxiliary techniques, including reliability criteria and surrogate-data tests, to increase the chances of obtaining a reliable value for the correlation dimension.
- Transients, possibly difficult to identify, can affect the results.

PART VII

QUANTITATIVE MEASURES OF CHAOS

Several statistics, when reliably computed, indicate chaos (at least in theory). Moreover, by their magnitude, sign, or trend, they suggest (again, at least theoretically) how chaotic a system is. The most important of such statistics at present are Lyapunov exponents (Ch. 25), Kolmogorov–Sinai entropy (Ch. 26), and mutual information or redundancy (Ch. 27).

Chapter 25
Lyapunov exponents

A Lyapunov[1] ("lee-ah-poo-noff") exponent is a number that describes the dynamics of trajectory evolution. It capsulizes the average rate of convergence or divergence of two neighboring trajectories in phase space. Its value can be negative, zero, or positive. Negative values mean that the two trajectories draw closer to one another. Positive exponents, on the other hand, result from trajectory divergence and appear only within the chaotic domain. In other words, a positive Lyapunov exponent is one of the most important indicators of chaos. A positive Lyapunov exponent measures or quantifies sensitive dependence on initial conditions by showing the average rate (as evaluated over the entire attractor) at which two close points separate with time. Some authors refer to a Lyapunov exponent as a **Lyapunov characteristic exponent** (LCE) or simply a **characteristic exponent**. As we'll see, a dataset can have several Lyapunov exponents.

The origin of this important component of the chaos game seems to go back to a nineteenth century Russian mathematician, Sofya Kovalevskaya (1850–91). According to Percival (1989), Kovalevskaya in 1889 mathematically defined dynamical instability as an average of a measure of the rate of growth of small deviations.[2] Russian mathematician Aleksandr Lyapunov made the definition of

---

1. **Lyapunov, Aleksandr Mikhailovich (1857–1918)** Lyapunov was one of the greatest Russian mathematicians. Born in Yaroslavl, he attended St Petersburg University in the late 1870s, where the famed P. L. Chebyshev had created a well known mathematics department. As a university senior in 1880, Lyapunov won a gold medal in a department-sponsored competition for his composition on hydrostatics. That composition was the basis for his first two publications, which appeared in 1881. He was awarded his Master's degree at St Petersburg in 1885, married his first cousin in 1886, and received his PhD in 1892. In 1895 he was appointed to the Chair of Mechanics at Kharkov University and began what was the happiest period of his life. In 1902 he returned to St Petersburg, having been elected in 1901 to the Chair of Applied Mathematics of the Russian Academy of Sciences. He now began a relatively secluded life, in which he placed much emphasis on his scientific work but also found time for his favorite hobby, plants (his apartment was decorated with rubber plants and palms which he had grown himself). During his productive career his general subject was mathematical physics—stability of motion, equilibrium of mechanical systems, potential theory, theory of probability, and the like. He is known as the founder of the modern theory of stability of motion. In 1918 his wife died of tuberculosis. Three days later Lyapunov himself died, leaving a note in which he asked to be buried in the same grave as his wife . . ..

dynamical instability more general in a lengthy treatise in 1892 (translated from Russian to French in 1907; both versions subsequently reprinted, e.g. Liapounoff 1949). Oseledec (1968) provided further theoretical basis.

Local convergence or divergence of neighboring trajectories

The simplest way to define a Lyapunov exponent and show its properties is with the one-dimensional map. (As we'll see later, a system has as many Lyapunov exponents as there are dimensions in its phase space. A one-dimensional case provides only one exponent. We'll eventually extend our one-dimensional case to many dimensions [many Lyapunov exponents].) Let's use our faithful friend, the logistic equation (Eq. 10.1): $x_{t+1} = kx_t(1-x_t)$. In particular, we'll examine the progressive displacement or gap, over time, between two trajectories that start close together (Fig. 25.1).

Trajectory convergence

Chapter 10 showed that, with the logistic equation and k<3 (hence in the nonchaotic domain), all trajectories for a given k value converge to the same fixed point. Let's inspect that convergence more closely. First we'll generate a so-called reference trajectory by choosing an appropriate k value and a value for x_0, then iterating the equation many times. (Some chaologists call a reference trajectory a **fiducial trajectory**; **fiducial** means "used as a standard of reference for measurement or calculation.") Armed with a reference trajectory, we'll then generate a nearby trajectory by starting the iterations over again, using the same k value but a slightly different x_0. Then we'll compare those two trajectories. In particular, for each iteration (each "new time"), we'll compute the difference δ (Greek delta) between the two predicted values of x_{t+1} (the difference between the two converging trajectories). Our goal is to see how the difference or displacement varies with time.

Let's begin with k = 0.95. All logistic-equation trajectories for k<1 go to a point attractor at the origin (Ch. 10). Say we start the reference trajectory at $x_0 = 0.40$ and the other trajectory at $x_0 = 0.41$. The absolute difference (hereafter simply called difference) between those two values, δ_0, is 0.41–0.40 = 0.01. For our comparison of displacement versus time, therefore, the first datum point is $\delta_0 = 0.01$ at $t = 0$.

For the first iteration of the logistic equation with k = 0.95, the reference trajectory starts at $x_0 = 0.40$ and goes to an x_{t+1} of 0.228. For the nearby trajectory, using that same k value but with $x_0 = 0.41$, $x_{t+1} = 0.229805$. The difference between

2. At the University of Stockholm in 1884, Kovalevskaya became Professor of Mathematics—probably the first woman ever to do so at a European university. For a fascinating account, see Kovalevskaya (1978).

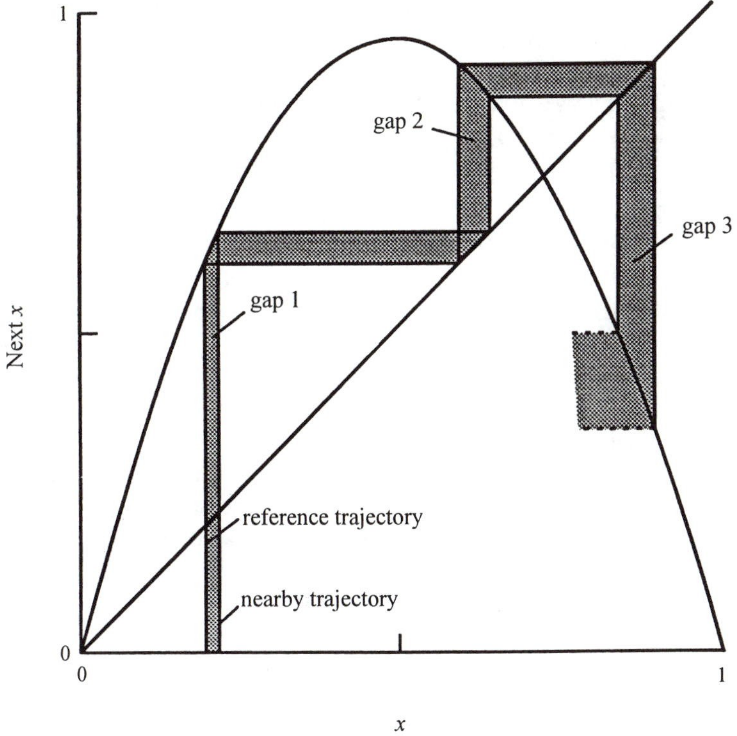

Figure 25.1 Logistic equation (parabola) and hypothetical evolution of two trajectories that start close together. In this example the trajectories diverge with time.

those two calculated values is 0.229 805–0.228 = 0.001 805. So, the second datum point is a trajectory separation or gap of $\delta_1 = 0.001\,805$ at $t = 1$. Next we continue the procedure for subsequent iterations. Still holding k constant at 0.95, the difference in calculated values for the second iteration ($t = 2$) is $\delta_2 = 0.000\,929\,73$. For the third iteration, the displacement is $\delta_3 = 0.000\,587\,04$, and so on.

Plotting separation δ on the ordinate against time shows that, except for transients, the calculated differences between two neighboring trajectories for our iterations plot as a straight line on semilog paper (Fig. 25.2). The particular type of semilog paper has the arithmetic scale on the abscissa and the log scale on the ordinate. Any straight line on that type of graph paper has the general form $y = y_0 e^{bx}$, where y_0 is the extrapolated value of the dependent variable y at $x = 0$, e is the base of natural logarithms (a constant, 2.718 . . .), b is a constant representing the slope of the straight line on the graph, and x is the independent or given variable. If the straight line slopes upward, b is positive; if it slopes downward, b is negative. This type of relation is an exponential relation, as defined in the chapter on chaotic equations.

The given variable x here is time (iteration number, n). The dependent variable y is the separation distance δ between the two trajectories. Extrapolating the straight line back to $n = 0$, we'll symbolize the intercept as δ_a and the finite value of δ at a particular time as δ_n ("delta n"). The general form of the exponential equation for the straight-line part of the plot therefore is:

$$\delta_n = \delta_a e^{bn}. \tag{25.1}$$

As our calculated values show, the gaps between trajectories for k = 0.95 become smaller (converge) with time. Consequently, the straight line in Figure 25.2 slopes downward (slope b is negative). Because the plotted points show virtually no scatter at all about the line, we can draw the straight line by eye. Its slope as measured graphically is any arbitrarily chosen ordinate distance divided by the associated abscissa distance. For the example of Figure 25.2, the slope measured in that fashion is about –0.05.

Let's now go to a larger k value (say, 2.8) and again iterate the logistic equation for two neighboring trajectories. We'll start the trajectories at the same two x values as before (0.40 and 0.41). Within the region $1<k<3$, all trajectories are still in the nonchaotic domain and go to a fixed-point attractor. However, the fixed point for this range of k isn't at zero. Figure 25.3 plots difference in calculated values (δ_n,

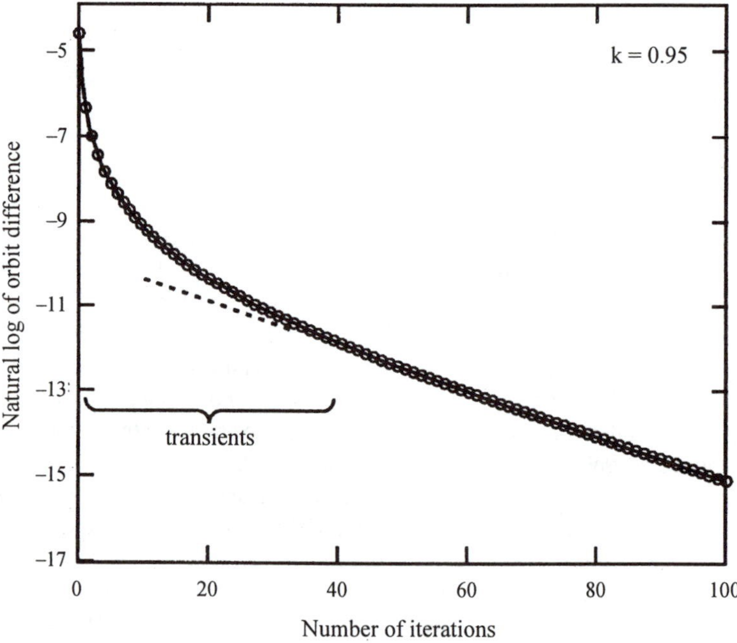

Figure 25.2 Graph of amount of separation of two neighboring trajectories over time for the logistic equation, with k = 0.95 (hence the nonchaotic domain with a fixed-point attractor).

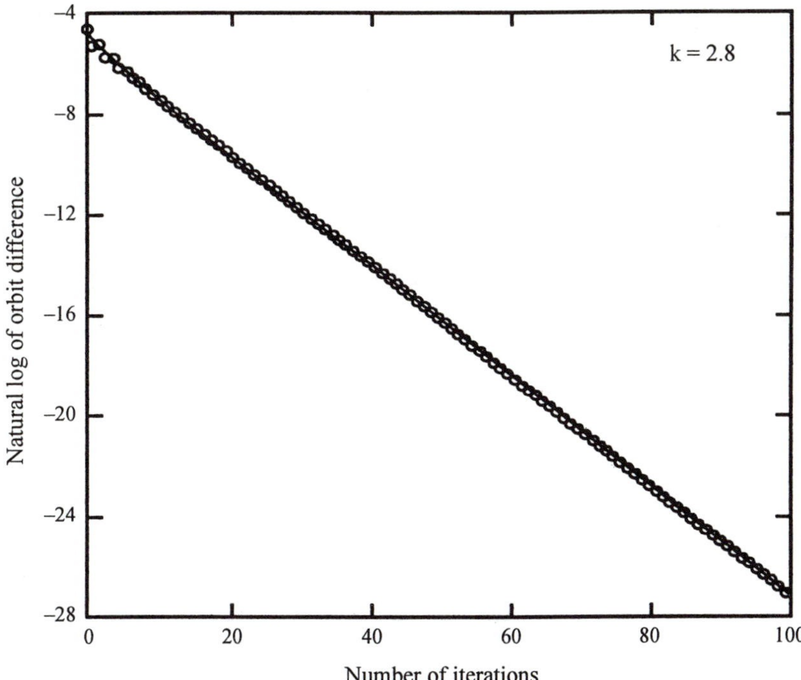

Figure 25.3 Graph of amount of separation of two neighboring trajectories over time for the logistic equation, with k = 2.8 (hence the nonchaotic domain with a fixed-point attractor).

on a log scale) versus time. Again the gaps get smaller with time. Comparing Figure 25.3 with Figure 25.2 shows that with k = 2.8 the slope of the straight line (the local rate of convergence of the two trajectories) again is negative. In this case the slope is now about −0.22. (On a minor note, with k = 2.8 the points began falling on a straight line after only about five or six iterations.)

Now we'll increase k to 3.4 and start the two trajectories at x_0 values of 0.060 and 0.062. Chapter 10 showed that the range 3 < k < 3.57 produces period-doubling (hence no chaos) and that k = 3.4 corresponds to an attractor of period two. The two alternating values of the attractor are approximately 0.451 96 . . . and 0.842 15 . . .. That is, each trajectory first approximates one attractor value, say 0.45. Then, on the next iteration, they approximate the other (0.84). They also get closer and closer to the true value with time (number of iterations). So, we have two meaningful comparisons at a periodicity of two. One comparison evaluates the gap between the two neighboring trajectories near the lower of the two attractor values (0.45 . . .). The other looks at the difference between the two trajectories near the other attractor value (0.84 . . .). It turns out that the gap associated with the lower attractor value is greater than that associated with the higher attractor value, for any approximate time (any pair of successive iterations). The plot of trajectory separation versus time (Fig. 25.4) therefore yields two separate straight lines—a higher one

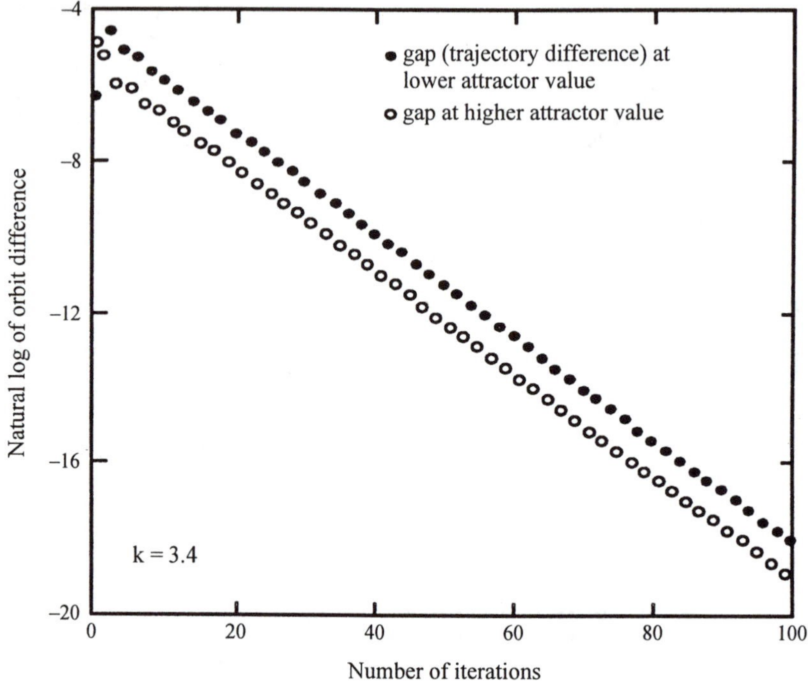

Figure 25.4 Graph of amount of separation of two neighboring trajectories over time for the logistic equation, with k = 3.4 (nonchaotic domain, period two).

for the gap near the attractor value of 0.451 96 . . ., and a lower one for the gap near the attractor value of 0.842 15 . . .. Both lines slope downward at about –0.14.

Three important features about all of the above nonchaotic examples are:

- The gap between neighboring trajectories has followed an *exponential* law in regard to time.
- The neighboring trajectories have *converged* toward their mutual attractor.
- The convergence means that gaps decrease with time, so *the slope of the fitted line is negative.*

Trajectory divergence

Sticking with the logistic equation as our model, let's now look at trajectory divergence. Divergence means that the gaps get larger with time. Neighboring trajectories with the logistic equation (and in fact with all dynamical systems) diverge only in the chaotic domain. Gaps as the two trajectories *approach* their mutual attractor are now less important; instead, we monitor those differences once the two trajectories are "on" their mutual attractor. Let's compute some differences and see how they change with time, as before.

We'll choose k and x_0 and begin iterating Equation 10.1, as usual. To make sure of getting "onto the attractor," we first iterate long enough to get past the transients. Let's say it takes about 100 iterations to get onto the attractor. We'll designate the 100th point as the new x_0 (a point on the attractor). We also arbitrarily choose a different but nearby x_0, as the starting point of a neighboring trajectory. (Strictly, we only know with certainty that just the first of those trajectories is approximately on the attractor, assuming our 100 iterations were enough. So, we'll assume that the other trajectory, beginning such a minute distance away, also is approximately on the attractor.) Finally, we determine the differences in the computed x_{t+1} (the amount of separation of the two trajectories) for each successive iteration and plot those differences against time, as before.

Suppose k = 3.8 and we begin iterating the reference trajectory at an x_0 of 0.06. On my computer the 100th iteration of the logistic equation (Eq. 10.1) gives x_{t+1} = 0.72035971. We assume that the trajectory is now on the attractor and take that value as a new x_0. If we choose a small and arbitrary difference δ_0 of 0.002, then the x_0 for the nearby trajectory begins at 0.72035971+0.002 or 0.72235971. (The number of decimal places here also is arbitrary.) We then calculate the trajectories that start at each of these x_0 values (0.72035971 and 0.72235971). Figure 25.5 shows computed differences in x_{t+1} plotted against time for 100 iterations. Figure 25.5 looks radically different from its trajectory-convergent counterparts (Figs 25.2–4). The three major differences are:

- The computed gap between trajectories changes in a systematic way only for a short time (about ten or so iterations in this case); thereafter, it becomes irregular. So, the divergence rate only applies to an early and restricted range along a trajectory. The irregular behavior beyond that range corresponds to the stage in which sensitivity to initial conditions has had such a cumulative and major effect that the two trajectories go off in totally independent, unrelated directions. (They may occasionally pass close to one another again, but that's purely by chance.)
- The general trend of systematic change (here over the first ten or so iterations) now slopes *upward* (has a positive value) instead of downward (negative value). The upward slope indicates that the gap between the two trajectories increases with time rather than decreases. The reason is sensitivity to initial conditions, the hallmark of the chaotic attractor. (And again, "initial" here means any place on the attractor where we might choose to begin monitoring the divergence of two neighboring trajectories.)
- There's now some scatter about the best-fit line. Hence, the straight line only represents the "average" rate of change.

The last deviation δ_n that adheres to the straight line represents the end of systematic divergence and is important in a practical way: it marks the approximate limit of predictability. For the time over which the systematic (exponential) divergence holds, the measured slope enables us to predict displacements with

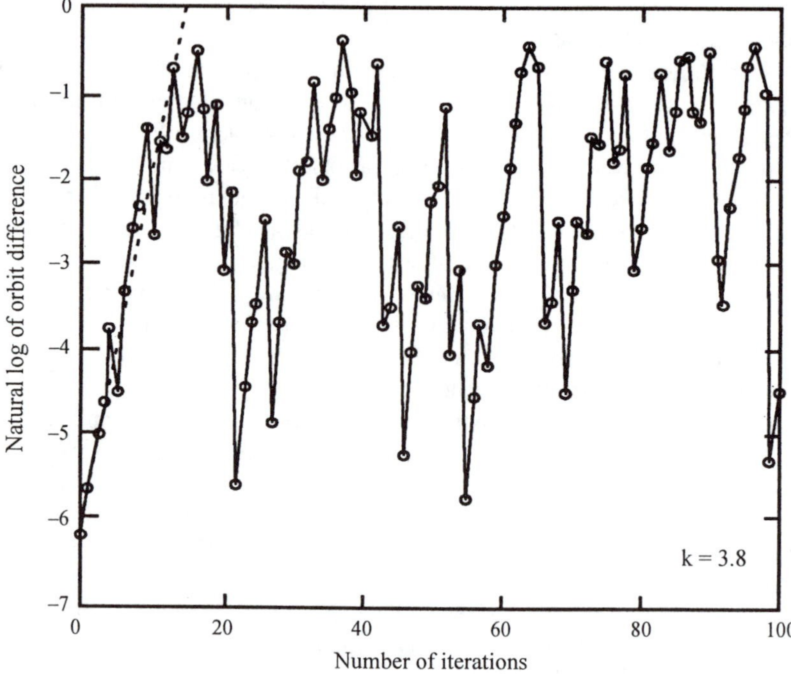

Figure 25.5 Graph of amount of separation of two neighboring trajectories over time for the logistic equation, with k = 3.8 (chaotic domain, transients excluded, starting gap 0.002).

reasonable accuracy. However, once the regular separation ends we might as well select values indiscriminately from within the range of possible values. In other words, short-term predictions are relatively reliable, but long-term predictions (defined as those for times later than the end of the systematic divergence on the graph) are meaningless. Furthermore, that's true even when we know the evolution equations exactly.

The predictable range for real-world data can vary widely, from a fraction of a second to millions of years. Weather predictability, for example, is on the order of a few days, at best. Predictability of the orbits of planets in the solar system (excluding Pluto) seems to be about 4–5 million years (Laskar 1989, Sussman & Wisdom 1992).

The notion of predictability we're discussing here, as well as most other nonlinear prediction methods, rests on a key assumption. That is that, for whatever set of conditions (values of the variables) a system might be in, it always responds in the same way (always goes to the same new set of values). In other words, the assumption is that phase space trajectories for such a system always follow approximately the same local route.

What if we had chosen a smaller starting gap, δ_0? For the same k (3.8) and with

x_0 for the reference trajectory again at 0.72035971, suppose $\delta_0 = 0.00000001$, instead of 0.002. (Such a radically smaller starting displacement might correspond, for example, to a much better measuring capability.) Using the smaller separation distance makes x_0 for the neighboring trajectory equal to 0.72035972. Subsequent computations now produce the graph shown in Figure 25.6. The smaller δ_0 is the cause of all differences from Figure 25.5. Slopes of the straight lines are not too dissimilar—0.42 on Figure 25.5, 0.48 on Figure 25.6. However, the straight line (range of systematic divergence and hence of somewhat reliable predictability) lasts much longer on Figure 25.6, namely to nearly 40 iterations. In other words, better accuracy on our measurements enables us to predict further into the future.

Other examples where tiny differences increase exponentially (although I didn't describe them as such at the time) appeared in the discussion of "Computer Inaccuracy and Imprecision" (e.g. Figs 14.2, 14.3). There I ignored the concept of transients (i.e. I used the first iterates), and yet the results weren't affected. Discarding transients therefore may not always be necessary.

In summary, the constant b in the equation $\delta_n = \delta_a e^{bn}$ reflects the local exponential rate at which trajectories converge (negative exponent b) or diverge (positive

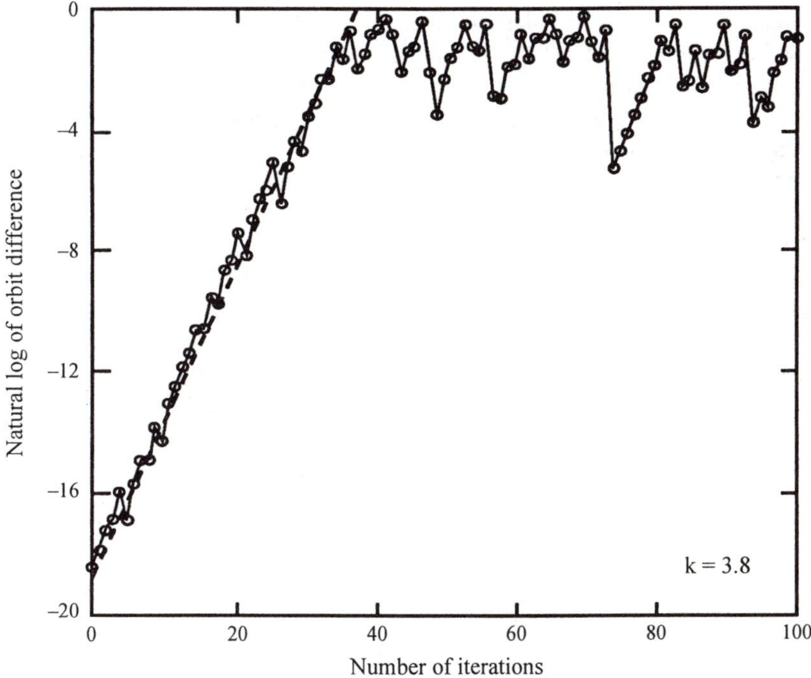

Figure 25.6 Graph of amount of separation of two neighboring trajectories over time for the logistic equation, with k = 3.8 (chaotic domain, transients excluded, starting gap 0.00000001).

exponent b). The gap between neighboring trajectories can also remain constant with time (as on a torus), in which case the constant b is zero.

Variability of local convergence and divergence rates over the attractor

Does the slope b vary with the two trajectories' location on the attractor? (Because a trajectory visits successive locations as time progresses, examining different locations on the attractor is tantamount to studying effects over time.) Continuing with the logistic equation, we'll fix k at 2.8 and conduct five separate experiments, corresponding to reference-trajectory starting values x_0 of 0.05, 0.20, 0.50, 0.80, and 0.95, respectively. In each case, we'll set δ_0 as 0.01. Starting values of the nearby trajectories therefore become 0.06, 0.21, 0.51, 0.81, and 0.96, respectively.

For all five experiments at k = 2.8, the rate of convergence (slope b) turns out to be the same, namely –0.22 (the plots aren't shown here). Therefore, the local rate of convergence seems to be the same, at least for k = 2.8 (a nonchaotic case).

Neighboring trajectories on chaotic attractors, on the other hand, don't necessarily behave so consistently. Slopes for the Lorenz attractor (Fig. 18.8), for example, range from –15 to +16, depending on the particular location on the attractor and other factors (Nese 1989). Two adjacent trajectories along the tops of the attractor's lobes can even converge, producing negative values of the slope b for that small zone. In contrast, adjacent trajectories monitored at the bottoms of the two lobes separate rapidly, yielding large and positive values of b. Therefore, we'll have to start referring to a "local slope" (a value of b that pertains to a particular local region on an attractor) from now on.

The divergence or convergence of neighboring trajectories at different rates depending on phase space location complicates chaos analyses. Nicolis et al. (1983) proposed to account for that variability with a **non-uniformity factor** (NUF). The NUF treats local slope as a statistical variable because it varies from place to place on a chaotic attractor (even though constant at any small region, per Eq. 25.1). The many different values can be compiled into a frequency distribution. The Nicolis et al. NUF is simply the standard deviation of that distribution of local-slope values. Thus, the NUF is a global indicator reflecting the spread of the local slopes about the mean local slope.

Local Lyapunov exponents

The exponent b—the exponential rate of trajectory convergence or divergence as derived for any one local region on an attractor—is a so-called local **Lyapunov exponent**. It characterizes a system's dynamics for that particular region, that is, over the associated brief timeframe. Because those local dynamics and hence exponents can vary with location on the attractor, it's important always to attach the adjective "local" when referring to such exponents. They aren't the same as the standard Lyapunov exponent that we're going to derive in the next section. A few representative papers dealing with local Lyapunov exponents include Fujisaka (1983), Nicolis et al. (1983), Grassberger & Procaccia (1984), Nese (1989), and Abarbanel (1992); see also Abarbanel et al. (1991a) and Abarbanel et al. (1992).

Local Lyapunov exponents can suggest attractor regions of potentially greater short-term predictability. For one thing, the duration over which the exponential relation holds tells us something about the time limits on predictability. Also, the value of the exponent reveals something about the reliability of predictions. For instance, a large and positive local Lyapunov exponent (fast divergence of neighboring orbits) suggests great sensitivity to initial conditions and poor predictability. Negative local exponents, in contrast, suggest convergence of neighboring trajectories and hence relatively good predictability. Measuring—and separately evaluating—local exponents over the entire attractor can tell us how they compare to the global average, their variability (which can be large) over the attractor, the special conditions for which predictions are reliable or unreliable, and many other things.

Definition of Lyapunov exponent

It's convenient to have a single value that represents all the local slopes over the attractor, that is, over time. Understandably, people adopted an average for that representative number. That average goes by the name of the first or largest Lyapunov exponent.[3] Now we're going to develop the basic exponential equation into an expression that represents an average local slope for the entire attractor. This will take several minutes, but it'll be worth it.

We start with Equation 25.1 ($\delta_n = \delta_a e^{bn}$). For simplicity, let's assume that there are no transients, so that $\delta_a = \delta_0$. Inserting δ_0 for δ_a and dividing both sides by δ_0:

$$\frac{\delta_n}{\delta_0} = e^{bn}. \tag{25.2}$$

Equation 25.2 highlights the ratio of the nth gap δ_n to the starting gap δ_0. That important ratio (the left side of the equation) is the factor by which the initial gap δ_0 gets stretched or compressed over the n iterations. Figure 25.7 shows a starting gap (indicated by δ_0 parallel to the abscissa) and a stretching over one iteration, in pseudo phase space. The reference trajectory in Figure 25.7 starts at x_0. The neigh-

3. "Exponent" applies because each local slope b is an exponent in the basic exponential equation, $\delta_n = \delta_a e^{bn}$. Also, as mentioned early in the chapter, there are usually several Lyapunov exponents, namely as many as there are dimensions in the phase space. We'll get to that topic in a later section.

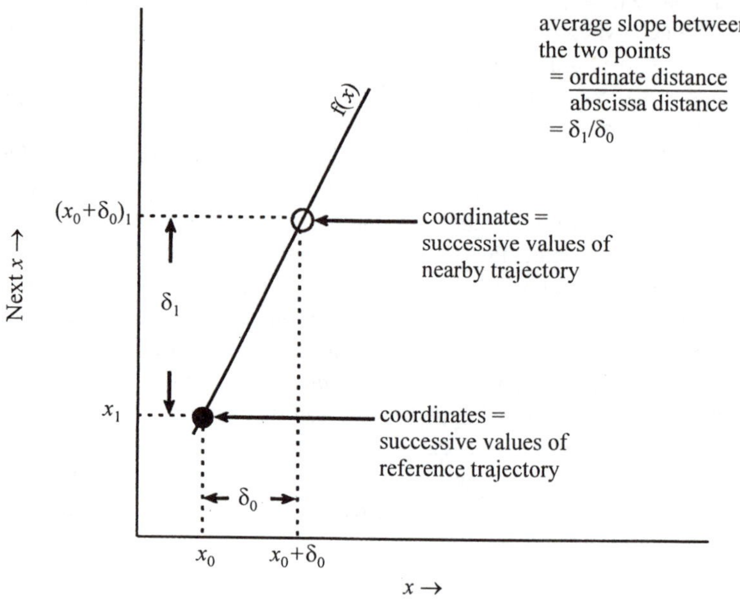

Figure 25.7 One-dimensional map (pseudo phase space plot) showing hypothetical progression of a gap between two adjacent trajectories over one time step. Gap at time 0 is δ_0, and gap at time 1 is δ_1. The gap changes (here grows) at a rate given by the function f(x).

boring trajectory also starts at time zero but at a distance δ_0 away. In symbols, the neighboring trajectory therefore starts at a value of $x_0+\delta_0$, as with our earlier examples. Values as of the next iteration (time one) are plotted on the ordinate. Those values are indicated by subscript 1. At time one the reference trajectory has a value x_1, the gap has become δ_1, and the neighboring trajectory has a value $(x_0+\delta_0)_1$.

The important quantity is the rate at which the separation distance changes. We'll designate the nature of the change by an undefined general rule that we'll call a "function of x." We write such a function in symbols as f(x) (called "f of x"). In Figure 25.7, the rate at which f(x) changes is approximated by the ratio δ_1/δ_0. Graphically, the ratio δ_1/δ_0 is an ordinate distance δ_1 divided by an abscissa distance δ_0 (Fig. 25.7). In other words, it's the average slope of the curve or straight line that connects the two sequentially plotted points. (The slope δ_1/δ_0 isn't the same as our local slope b. However, the two slopes are mathematically related in some way, as Eq. 25.1 implies.)

Mathematicians symbolize the nth iterate of any function f of x as $f^n(x)$. (Putting n in the position of an apparent exponent is an unfortunate convention, as it looks like a power. Remember, it's not a power when located right after the f; instead, it's just an identifying label.) To show that iterating the function produces x, they just put "x =" in front of the expression $f^n(x)$, yielding $x = f^n(x)$. Also, the applicable iteration number (1, 2, etc.) replaces n.

After n iterations the ratio of gap δ_n divided by original gap δ_0 gives an average rate of change over those n iterations. Now let's make the initial gap (δ_0) smaller and smaller (approaching zero). (In symbols, $\delta_0 \to 0$.) That means the second plotted point on Figure 25.7 approaches the first point. The computed rate of change for that condition is called a **first derivative** (or simply a derivative).[4] For that special limiting condition ($\delta_0 \to 0$), mathematicians replace δ with a lower-case d. The d still refers to the gap, just like the δ. To make sure everyone knows that the gap is in terms of x, they write the gap as dx.

Sometimes the gap is expressed in terms of some function such as f or $f^n(x)$, in which case it's written df or $df^n(x)$, respectively. Hence, df or $df^n(x)$ indicates the discrepancy (e.g. the "ordinate gap" in Fig. 25.7) between two successive iterations as δ_0 becomes infinitely small. The relative change for the special condition of δ_0 becoming infinitely small then is no longer simply δ_1/δ_0; instead, that change (ordinate distance/abscissa distance) becomes df/dx. Since our function f is $f^n(x)$, the df/dx is $df^n(x)/dx$. Mathematicians write $df^n(x)/dx$ as

$$\frac{d}{dx} f^n(x).$$

They call it the "derivative of the nth iterate of the function f with respect to x." (Again, a first derivative is merely a slope or relative change, measured at the limit where the starting gap δ_0 becomes infinitely small. It's therefore analogous to ratios such as δ_3/δ_0, δ_n/δ_0, and so on.) Summarizing: the rate of change as $\delta_0 \to 0$ is

$$\textit{rate of change} = \frac{d}{dx} f^n(x) = \lim_{\delta_0 \to 0} \frac{\textit{ordinate distance}}{\textit{abscissa distance}} = \lim_{\delta_0 \to 0} \frac{\delta_n}{\delta_0}. \quad (25.3)$$

As δ_0 gets smaller, the calculated rate for the first iteration (δ_1/δ_0) applies closer and closer to the point that started at x_0 (Fig. 25.7). Hence, we assign δ_1/δ_0 to the value x_0. Also, we'll adopt another standard symbol and indicate the derivative of the function f (i.e. the value df/dx) by $f'(x)$ (spoken as "f prime of x"). The relative change for the *first* iteration therefore is

$$\textit{rate of change at point } x_0 = \frac{\delta_1}{\delta_0} = f'(x_0). \quad (25.4)$$

For the next iteration, value x_1 takes the role of x_0, gap δ_1 takes the place of old gap δ_0, and new gap δ_2 takes the place formerly occupied by δ_1. Following the same pattern as in Equation 25.4, we symbolize the relative change associated with point x_1 as $f'(x_1)$ and calculate it as δ_2/δ_1:

4. A **derivative** is a value that derives from, comes from, or is traceable to a particular source, in this case the point on the curve.

$$\textit{rate of change at point } x_1 = \frac{\delta_2}{\delta_1} = f'(x_1). \tag{25.5}$$

(So, with derivatives written as $f'(x_0)$, $f'(x_1)$, and so on, the value within parentheses (x_0, x_1, etc.) indicates where, along the curve, to measure the rate of change.)

We now invoke a handy rule known as the **chain rule**. It defines the derivative of the function f, as taken at the value x_0. That is, it defines

$$\frac{d}{dx}f^n(x_0).$$

In particular, it defines that quantity as the product of two terms. One is the derivative of the function taken at the value of the preceding iteration x_{n-1}, which in symbols is $f'(x_{n-1})$. The other is the derivative of that preceding iterate of the function taken at the value x_0, or

$$\frac{d}{dx}f^{n-1}(x_0).$$

Writing the definition in symbols,

$$\frac{d}{dx}f^n(x_0) = f'(x_{n-1})\cdot\frac{d}{dx}f^{n-1}(x_0). \tag{25.6}$$

We're going to apply the chain rule to a series of iterates or successive observations. As an example, let's apply it to four sequential observations. That means writing Equation 25.6 four times, namely for $n=4$, $n=3$, $n=2$, and $n=1$. However, a term for $n=1$ will be part of the expression for $n=2$ and so doesn't have to be written separately. Hence, we only have to write Equation 25.6 three times, as follows:

Equation 25.6 with $n=4$ is

$$\frac{d}{dx}f^4(x_0) = f'(x_3)\cdot\frac{d}{dx}f^3(x_0). \tag{25.7}$$

Equation 25.6 with $n=3$ is

$$\frac{d}{dx}f^3(x_0) = f'(x_2)\cdot\frac{d}{dx}f^2(x_0). \tag{25.8}$$

Equation 25.6 with $n=2$ is

$$\frac{d}{dx}f^2(x_0) = f'(x_1)\cdot\frac{d}{dx}f(x_0). \tag{25.9}$$

The term on the far right in Equation 25.9, namely

$$\frac{d}{dx}f(x_0),$$

is an alternate way of writing $f'(x_0)$, so Equation 25.9 really is

$$\frac{d}{dx}f^2(x_0) = f'(x_1) \cdot f'(x_0). \qquad (25.10)$$

Equations 25.7, 25.8, and 25.10 have some common terms. Hence, we can make substitutions. For instance, the last term in Equation 25.7 is

$$\frac{d}{dx}f^3(x_0),$$

and that term is also defined by Equation 25.8. So, we can insert the right side of Equation 25.8 into the last term of Equation 25.7. Equation 25.7 then becomes

$$\frac{d}{dx}f^4(x_0) = f'x_3 \cdot f'x_2 \cdot \frac{d}{dx}f^2(x_0). \qquad (25.11)$$

Again the term on the far right,

$$\frac{d}{dx}f^2(x_0),$$

has been defined separately (Eq. 25.10). Making that substitution, Equation 25.11 becomes

$$\frac{d}{dx}f^4(x_0) = f'(x_3) \cdot f'(x_2) \cdot f'(x_1) \cdot f'(x_0)$$

$$= \text{the product of the } f'(x_i)\text{'s} \qquad (25.12)$$

where x_i is a general symbol for the successive values of the iterations and i here ranges from zero to $n-1$. (And so, with our little group of equations, the last term in each equation is linked, as in a chain, to the definition given by the next equation.) Equation 25.12 says that the quantity

$$\frac{d}{dx}f^n(x)$$

is just the product of all the first derivatives, that is, the product of the rates of change $f'(x_i)$, for i ranging from zero to $n-1$.

We can write the derivatives on the right-hand side of Equation 25.12 as ratios of gaps, as with Equations 25.4 and 25.5. For instance, $f'(x_0) = \delta_1/\delta_0$, per Equation 25.4. Continuing with our example of $n = 4$, we therefore substitute into Equation 25.12 the ratios δ_1/δ_0 for $f'(x_0)$, δ_2/δ_1 for $f'(x_1)$, δ_3/δ_2 for $f'(x_2)$, and δ_4/δ_3 for $f'(x_3)$:

$$\frac{d}{dx}f^4(x_0) = \frac{\delta_4}{\delta_3}\cdot\frac{\delta_3}{\delta_2}\cdot\frac{\delta_2}{\delta_1}\cdot\frac{\delta_1}{\delta_0}.$$

Cancelling like terms from the top and bottom on the right-hand side leaves

$$\frac{d}{dx}f^4(x_0) = \frac{\delta_4}{\delta_0}$$

or, in general,

$$\frac{d}{dx}f^n(x) = \frac{\delta_n}{\delta_0} \qquad (25.13)$$

where the value of n reflects the number of derivatives or ratios δ_i/δ_{i-1} involved. For instance, with $n = 4$ there are four derivatives, namely the ratios $\delta_4/\delta_3, \ldots \delta_1/\delta_0$.

Equations 25.12 and 25.13 together say that

$$\frac{\delta_n}{\delta_0} = \text{the product of the } f'(x_i)\text{'s}. \qquad (25.14)$$

Equation 25.14 brings our story full circle back to Equation 25.2, $\delta_n/\delta_0 = e^{bn}$ (with $\delta_a = \delta_0$). The ratio on the left (δ_n/δ_0), according to Equation 25.14, represents the product of a bunch of derivatives. Next we'll rearrange our transientless version of Equation 25.2 to isolate the exponent b, the separation rate of two neighboring trajectories. To get b, we first put Equation 25.2 into log form (base e):

$$\log_e(\delta_n/\delta_0) = bn\log_e e.$$

The ingredient on the far right, $\log_e e$, equals 1, so

$$\log_e(\delta_n/\delta_0) = bn.$$

Finally, dividing both sides by n gives an expression for a local slope, b:

$$b = \frac{1}{n}\log_e(\delta_n/\delta_0). \qquad (25.15)$$

Equation 25.15, by the way, reveals the units of b. The ratio δ_n/δ_0 is dimensionless, since both δ's are in the same units (units of x). The log of that ratio also is dimensionless (log units). Iterations usually represent time, so we let n have units of time. Thus, b has dimensions of 1/time (e.g. 1/seconds, 1/years, etc.). In practice, people often give the numerator a name that depends on the base of the logs. The most popular base of the logs is either e or 2. If taken to base e, local slope b is in **nats** per unit time; if to base 2, it's in bits per unit time (Ch. 6).

We're almost done. This next step is important. First, a word about the philosophy behind it. Local slopes on an attractor, as mentioned, can vary greatly. If we measure only one local slope, we've only examined one locality on the attractor—a locality that might be unrepresentative. In many cases we want to cover the entire attractor. The limitation of the exponential scaling zone (the straight-line relation, as in Figs 25.5 and 25.6) means that we can't use the same two trajectories, from one place to another along the attractor. The usual procedure is to track the same reference trajectory over the entire attractor but choose different "nearby" trajectories at different locations on the attractor. That gives us a bunch of local slopes that represent trajectory divergence (or convergence, or whatever) at different locations on the attractor.

Equation 25.15 describes the separation rate of one pair of neighboring trajectories at some specified location on the attractor. We'll now rewrite the equation in a form whereby it applies to a *group* of local slopes. In other words, it applies to the separation rates of many pairs of neighboring trajectories. We assume that Equation 25.1 and its log-transformed and rearranged version, Equation 25.15, still apply. However, the interpretation is slightly different when a *group* of local slopes is involved. For one thing, δ_n/δ_0 according to Equation 25.14 equals the product of the $f'(x_i)$ values. Hence, we can substitute the product of the $f'(x_i)$ values for δ_n/δ_0 in Equation 25.15. Equation 25.15 uses the log of (δ_n/δ_0), so we need the log of the product of the $f'(x_i)$ values. The log of a product, such as the log of $[f'(x_3)\cdot f'(x_2)\cdot f'(x_1)\cdot f'(x_0)]$ (our example from above with $n=4$), is the *sum* of the logs of each component (Appendix). In our example, that sum is log of $f'(x_3)$+log of $f'(x_2)$+log of $f'(x_1)$+log of $f'(x_0)$. The standard and general way of writing such a sum in concise symbol form (here we'll take logs to base e) is

$$\sum_{i=0}^{n-1} \log_e f'(x_i).$$

For instance, the summation for our values ranges from $i=0$ for $f'(x_0)$ to $i=3$ for $f'(x_3)$, and 3 is $n-1$ or $4-1$. Finally, although gaps or distances can be positive or negative depending on which value is subtracted, we're interested only in the absolute values of the gaps; it's not possible to take the log of a negative number. Hence, we want logs of *absolute values* of derivatives. In symbol form, that's $\log_e|f'(x_i)|$. With those revisions, $\log_e(\delta_n/\delta_0)$ of Equation 25.15 (applicable to just one local slope) becomes

$$\sum_{i=0}^{n-1} \log_e|f'(x_i)|$$

(applicable to a group of local slopes). Making that substitution into Equation 25.15:

$$b = \frac{1}{n}\sum_{i=0}^{n-1} \log_e|f'(x_i)| . \tag{25.16}$$

Equation 25.16 is just a fancy and shorthand way of telling us to add up the logs of the absolute values of the local slopes and then to get the average log by multiplying that sum by $1/n$. To show that the answer isn't just one local slope but is instead the average of many local slopes, we'll adopt the symbol λ_1 (Greek lambda, here with subscript 1) in place of b. Finally, to indicate that the slopes should come from all over the attractor (i.e. that the time span should be very long), we qualify the right-hand side of Equation 25.16 with the term $\lim_{n\to\infty}$. Those changes bring us to our overall goal—a symbol expression for the first, or largest Lyapunov exponent λ_1 (e.g. Wolf 1986: 275):

$$\lambda_1 = \lim_{n\to\infty}\frac{1}{n}\sum_{i=0}^{n-1} \log_e|f'(x_i)| . \tag{25.17}$$

Equation 25.17 shows that the first Lyapunov exponent λ_1 is a logarithm—the average log of the absolute values of the n derivatives or local slopes, as measured over the entire attractor. In mathematical terminology, a Lyapunov exponent is also the logarithm of a so-called **Lyapunov number**. In other words, a Lyapunov number is the antilog of the Lyapunov exponent, or the number whose logarithm, to a given base, equals the Lyapunov exponent.

Because it averages local divergences and/or convergences from many places over the entire attractor, *a Lyapunov exponent is a global quantity*, not a local quantity. We can interpret it in three ways.

- In n dimensions, λ_1 quantifies, in a single number, the average rate at which the fastest growing phase space dimension grows.
- It quantifies average predictability over the attractor.
- Because neighboring trajectories represent changes in initial conditions of (or perturbations to) a system, λ_1 is an average or global measure of how sensitive the system is to slight changes or perturbations. A system isn't sensitive at all in the nonchaotic regime, since any two nearby trajectories converge. In contrast, a system is highly sensitive in the chaotic regime, in that two neighboring trajectories separate, sometimes rapidly.

The value of λ_1 doesn't depend on where we start monitoring, because we're going to cover the entire attractor regardless. Therefore, λ_1 is a basic characteristic for the attractor and is invariant for any attractor. (Invariant here means invariant in regard to starting time and also "unaffected by transformations of the data.")

Lyapunov spectrum

A common approach in visualizing phase space motion is to imagine how a small length, area, volume or higher-dimensional element might evolve in time. Examples of such elements are a square or circle (in two-dimensional phase space) and a cube or sphere (in three dimensions). The center of such an element is an observed datum point. There's an elaborate terminology for describing such an element. Originating at every measured datum point, and making up the associated element's skeletal framework, are one or more equal-length orthogonal axes, called **principal axes**. Each axis ends at another phase space point, although not necessarily an observed (measured) point. Because those axes start at a common center point (a datum point) and each axis ends at some other point in phase space, chaologists call each axis a *vector* (Ch. 5), **axis vector**, or **perturbation vector**. The number of principal axes corresponds to the dimension of the phase space. (However, as we'll see, their directions continually change and have no relation to the directions of the phase space axes.) People usually normalize the orthogonal axes (make them orthonormal) so that the axes have unit (1.0) length.

In any dissipative (energy-losing) system, whether chaotic or not, a length, area, volume or higher-dimensional element shrinks over time, in phase space (Fig. 25.8a). The frictional pendulum and its point attractor of Figure 11.1b are a good example. On a phase space plot of velocity versus position, the dynamics describe a spiral. The size of the spiral gets smaller and smaller with time. Eventually the pendulum stops swinging altogether. The spiral then reaches its smallest possible size (a dot), representing a constant velocity and position. Conservative systems, in contrast, don't lose energy with time. In the graphical illustration of Figure 25.8b, the overall total area of the box remains constant over time, even though there may be some change in shape (because of unequal changes in axes lengths).

The phase space orientation of a set of orthonormal axes is arbitrary. In other words, the axes aren't necessarily parallel to the x, y, and z coordinates of the phase space plot. Each axis vector must simply be orthogonal to all other axis vectors.

Now we let the system evolve forward in time, that is, we go to the next measurement. In the process, the imaginary axes or vectors move. Because of the basic nonlinear nature of the system, their lengths change. The amount of change varies from one axis to another. In other words, our idealized element gets distorted. Its orientation in the phase space also may change. Principal axes have numerical labels according to how much change or distortion they undergo. The principal axis that's stretched the most (or reduced the least) is the **first principal axis**. That axis, in other words, corresponds to the direction in which the flow is most unstable. The axis that undergoes the next greatest amount of growth (or least shrinkage) is the second principal axis, and so on.

Now, what does all of this have to do with Lyapunov exponents? The two endpoints of each principal axis are considered to be neighboring points in phase space. We monitor the change in their gap, over time. That is, we measure the growth or shrinkage of each principal axis over the entire attractor, according to

(a) Dissipative systems lose energy

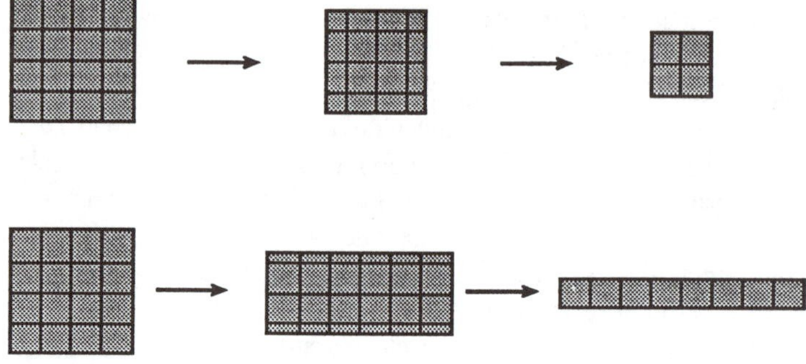

(b) Conservative systems do not lose energy

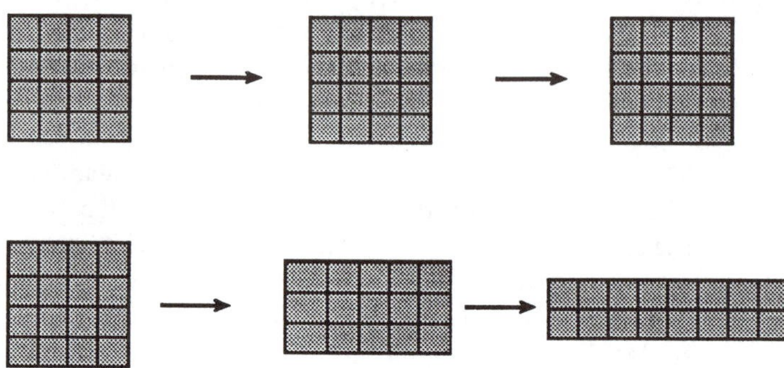

Figure 25.8 Phase space area and some possible changes over time (after Bergé et al. 1984: 23). Size (area) of main box (the one on the left, in each row) represents an amount of energy. In dissipative systems (the two examples in part (a)), area of box decreases over time. In conservative systems (the two examples in part (b)), area of box remains constant over time. With both systems, shape of area-element can change over time, and trajectories can diverge whether area stays constant or not.

whether its endpoints get closer together or farther apart. That means *we get a Lyapunov exponent for each principal axis*. The largest Lyapunov exponent measures the rate of expansion of the first principal axis—the one that shows the largest amount of growth (or the slowest rate of shrinkage) over the attractor. The second Lyapunov exponent measures the rate of change of the second principal axis, and so on down to the smallest Lyapunov exponent.

For a system to be chaotic, at least one principal axis must show an overall *divergence* of neighboring trajectories (Figs 25.5, 25.6). Therefore, at least one Lyapunov exponent has to be *positive* on a chaotic attractor. The Lyapunov exponents of an attractor, considered as a group and ordered by magnitude, are known as the attractor's *spectrum of Lyapunov exponents*.

A different Lyapunov exponent corresponds to each principal axis, and the number of principal axes equals the number of phase space dimensions. Consequently, the number of Lyapunov exponents equals the number of phase space dimensions. However, all principal axes continually change direction or orientation in phase space, depending on attractor configuration and on flow direction. Hence, a particular Lyapunov exponent doesn't correspond to a particular phase space direction or to a particular axis of the phase space graph.

Chaos can happen in dynamical systems of any number of dimensions. Thus, there can be any number of positive Lyapunov exponents (almost as many as the dimensions of the phase space itself).

Besides positive and negative exponents, a "zero" exponent also is possible. An exponent of zero corresponds to an axis along which the gap between trajectories either stays constant or increases at a rate less than exponential. (Some authors call the zero-exponent case **marginally stable**. "Marginally stable" means that any perturbation stays at about the same level, as the system evolves.) Mathematicians have proven that, except for the fixed-point attractor, at least one Lyapunov exponent is always zero (cf. Parker & Chua 1989: 321). Hence, there are three main types of trajectory behavior in dynamical systems:

- asymptotically unstable (exponential divergence of neighboring trajectories, positive Lyapunov exponent)
- marginally stable (zero exponent)
- asymptotically stable (exponential convergence of neighboring trajectories, negative exponent).

Authors describe a system's dynamics qualitatively in shorthand form by simply listing the signs of the Lyapunov exponents. For instance, (+,0,–) means a three-dimensional phase space in which neighboring trajectories as averaged over the entire attractor diverge exponentially along the first principal axis, are marginally stable along the second, and converge along the third. The Lorenz attractor (Fig. 18.8) is an example.

In three-dimensional phase space, the various attractor types have the following spectrum of Lyapunov exponents:

- A fixed-point attractor is (–,–,–). That label tells us that convergence goes on along all three principal axes of any phase space volume.
- A limit cycle is (0,–,–), because a volume element stays constant along the direction of the trajectory but contracts (the gaps converge) in the directions of the other two principal axes.
- Orbits on a torus, once they reach the attractor, keep the same distance thereafter. A volume element therefore stays constant for two principal axes but contracts along the third (0,0,–).
- Finally, on a chaotic attractor the two trajectories diverge in the direction of the first principal axis, yielding a positive exponent. The exponent for the second principal axis has to be zero because at least one axis must have a zero exponent. The exponent for the third principal axis has to be negative (and larger in absolute value than that of the positive exponent), because the over-

all sum of the exponents has to be negative. So, the classification code for a chaotic attractor is (+,0,–).

You'll often see the statement that the length of any phase space element (i.e. the first principal axis of our N_d axes) grows as $c^{\lambda_1 t}$, where c is a constant (usually 2 or 2.718 . . .). Similarly, the cross-sectional area of any element grows as $c^{(\lambda_1+\lambda_2)t}$; its volume grows as $c^{(\lambda_1+\lambda_2+\lambda_3)t}$; and so on. Here's a brief explanation of where those relations come from. They arise from the assumed (or real) exponential growth rate, as follows. As with the earlier derivation, we'll deal with local slope b, with the understanding that the same principle applies globally. With time t as the independent variable and length y as the dependent variable, the exponential relation for the length of the principal axis at time one (y_1) is

$$y_1 = y_0 c^{b_1 t} \tag{25.18}$$

in which y_0 is the length at time zero and b_1 is the slope associated with the first principal axis. The ratio y_1/y_0 then is the relative growth of the axis over one time unit. Equation 25.18 shows that y_1/y_0 grows as $c^{b_1 t}$.

Now for the two-dimensional case. The expression for growth of a cross-sectional area stems from the relation that the area of a disk or circle is π times the radius squared. Let's designate two orthogonal radii, y and z, as the first and second principal axes. The area of the circle then is πyz. That area at time zero is $\pi y_0 z_0$. By the next time increment (time one), the circle deforms into an ellipse. An expression similar to πyz applies to the cross-sectional area of the ellipse for time one, so the area at time one is $\pi y_1 z_1$. The relative growth of the area during the time interval is area at time one divided by area at time zero $= \pi y_1 z_1 / \pi y_0 z_0 = y_1 z_1 / y_0 z_0$. Next we insert expressions for y_1 and z_1 into this last ratio. Assuming that the axis y_1 grows exponentially, its value at time one is $y_1 = y_0 c^{b_1 t}$. Similarly, assuming that axis z_1 grows exponentially, its value at time one is $z_1 = z_0 c^{b_2 t}$, where b_2 is the slope associated with the second principal axis. Inserting these expressions into the relative-growth ratio $y_1 z_1 / y_0 z_0$ gives:

$$\textit{relative growth} = y_1 z_1 / y_0 z_0 = y_0 c^{b_1 t} \cdot z_0 c^{b_2 t} / y_0 z_0 = c^{b_1 t} \cdot c^{b_2 t}$$

or

$$\textit{relative growth} = c^{(b_1+b_2)t}. \tag{25.19}$$

In other words, cross-sectional area grows as $c^{(b_1+b_2)t}$. In the same way, volume grows as $c^{(b_1+b_2+b_3)t}$, and so on.

Practical calculation of Lyapunov exponents

Some of the proposed methods for calculating Lyapunov exponents from measured data are those of Wright (1984), Eckmann & Ruelle (1985) (explained in more detail in Eckmann et al. 1986), Sano & Sawada (1985), Wolf et al. (1985), Briggs (1990), Bryant et al. (1990), Stoop & Parisi (1991), Zeng et al. (1991), Nychka et al. (1992), and probably others. Wright's (1984) method applies only to relatively limited conditions, so I won't elaborate on it here. The basic idea behind all other methods is to follow sets of trajectories over short time-spans and compute their rates of separation, then average those rates over the attractor. These other methods have at least five features in common.

- They begin by reconstructing the attractor, often with the standard time-delay embedding procedure. If the attractor can't be reconstructed, then the game is over already; we can't measure a Lyapunov exponent. (And, as mentioned, there are few reliable criteria to tell us when we've successfully reconstructed the attractor.)
- Working in the reconstructed phase space, all methods focus on a fiducial trajectory over the attractor.
- They directly or implicitly treat each fiducial point as the center of a little area, sphere or hypersphere that has principal axes. As mentioned earlier, evolution of the system disrupts the orthogonality of such principal axes. For instance, a circle in phase space gets distorted into an oval; a sphere evolves into an ellipsoid. (In chaotic systems, the axes in fact tend to align themselves in the direction of most rapid local growth.) The loss of orthogonality is undesirable. For one thing, if the axes collapse toward a common direction they become indistinguishable. In that case we can only measure one trajectory gap. For another, probing the phase space is more thorough if the vectors are orthogonal. Thirdly, mutually perpendicular axes are much easier and more familiar to work with than axes that aren't mutually perpendicular. Because of these considerations, most methods periodically reset the axes to where they are again mutually perpendicular and of unit length (orthonormalized).

There are several ways to reset principal axes to an orthogonal orientation. Different techniques for determining Lyapunov exponents use different resetting methods. Probably the two most common methods are Gram–Schmidt orthogonalization and QR decomposition (Zeng et al. 1992). Adding the normalization step completes the impressive-sounding procedure called **reorthonormalization** (*re*: again; *ortho*: right angle; *normal*: of unit length). The set of principal-axis vectors travels around the attractor, anchored to the reference trajectory, and every so often we reorthonormalize those vectors.

- They're much better at measuring positive exponents than negative ones. At least with chaotic attractors, nearby trajectories diverge over much of the attractor. Most of such chaotic data lend themselves to positive exponents. Of course, one positive exponent is enough to show chaos, so if our only goal is to find out whether or not the system is chaotic, the problem becomes easier.

The Wolf et al. technique, in fact, provides only the single largest positive exponent.[5] The other methods, not necessarily successfully, try to find the entire spectrum of exponents.

As mentioned, however, negative exponents also are difficult to measure. They reflect trajectory convergence. Such convergence on chaotic attractors can take place normal to the flow, in a direction where the attractor is relatively thin and information relatively scarce. In addition, measurement of negative exponents depends on their magnitudes and on the relative amount of noise in the data (Sano & Sawada 1985). If trajectories converge to the extent that attractor thickness is less than the resolution of the data (the noise magnitude), then there's no way to get any further information about contraction rates. Measuring a Lyapunov exponent for those conditions produces an incorrect zero exponent, because noise on average neither expands nor contracts. Finally, in other situations, negative exponents for chaotic or nonchaotic attractors measure the rate at which neighboring trajectories that are off the attractor get onto the attractor (Abarbanel et al. 1990), as explained early in this chapter. We may need to determine such rates by analyzing transients, and transients usually last over a very short time or may not have been measured.

- Most methods produce extra (false or spurious) exponents that you have to identify and throw out. (The methods of Wolf et al. [1985] and Stoop & Parisi [1991] may be exceptions.) The extra exponents appear because the methods normally yield the same number of exponents as the number of dimensions of the phase space, and the phase space here is an artificial one whose embedding dimension necessarily is greater than that of the dynamical system it represents. Authors use various approaches, sometimes apparently successfully, to recognize false exponents. Bryant (1992) gives a good discussion with recommendations for how to identify them.

So much for what the methods have in common. Now let's take a closer look at three of them—those of Wolf et al., Eckmann & Ruelle, and Sano & Sawada.

The popular Wolf et al. (1985) method begins by taking the first point of the reconstructed (lagged) phase space to represent the fiducial trajectory (Fig. 25.9). The authors choose a nearby point as the first observation along the neighboring trajectory. That nearby point must be from a different temporal section of the dataset. They monitor the gap between the two trajectories over time. The length of time over which they do so can be until the occasion of the very next datum value, or until the gap exceeds an arbitrary distance ε, or until another trajectory becomes closer. When such a predesignated condition occurs, they consider that particular neighboring trajectory to have migrated a maximum allowable distance from the fiducial trajectory. At that stage, they compute a local slope as [final separation distance minus starting separation distance] divided by the associated time interval. Then they abandon that first neighboring trajectory.

The last point just analyzed on the fiducial trajectory stays on for additional duty

5. Those authors do, however, also propose a more elaborate and cumbersome technique for the two largest exponents.

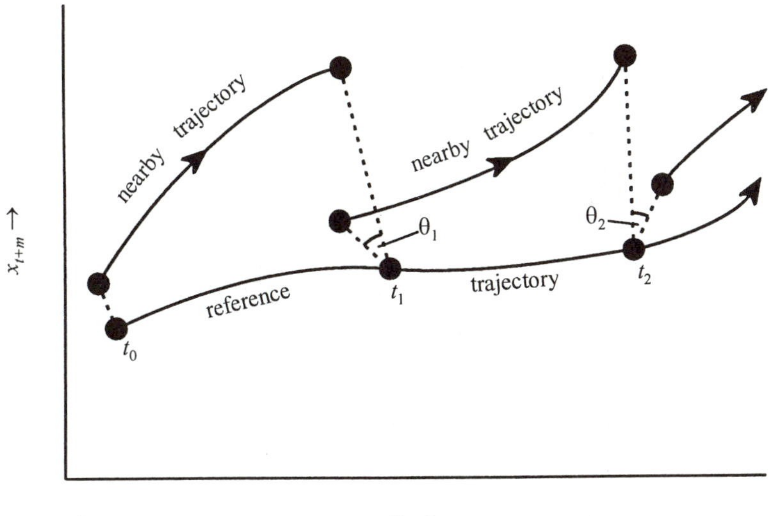

Figure 25.9 Wolf et al. (1985) approach to estimating largest Lyapunov exponent from measured data. At each replacement time, angle θ (between old length element and new element) as well as new length element are minimized.

as a new reference point. A new nearby point is chosen to represent a new nearby trajectory (Fig. 25.9). Then they measure the rate at which the new nearby trajectory diverges from the fiducial trajectory, as before. They repeat the process until the fiducial trajectory gets to the end of the data. At this stage, we assume that the data cover the attractor. Averaging the logs of the absolute values of the various local divergence rates gives the largest Lyapunov exponent (Eq. 25.17).

When a nearby trajectory becomes obsolete and is replaced with a new one in the Wolf et al. method, two criteria determine the starting point for the new nearby trajectory. First, the starting point should be as close as possible (in the reconstructed phase space) to the new fiducial point. Secondly, the spatial orientation of the last (now abandoned) separation vector should be preserved as closely as possible. That means making the angle θ on Figure 25.9 as small as possible. In practice, fulfilling those two criteria can be difficult, and the computer often must look at more than one nearby datum point to find the best compromise replacement point. Fortunately, such orientation problems usually don't have a major effect on the Lyapunov exponent. Also, divergence rates of replacement trajectories must be monitored closely. If they grow suspiciously quickly, the fiducial and nearby points probably aren't on the same part of the attractor and are going in different directions because of a fold in the attractor. In that case (very rapid divergence), go back and choose a different second point. Abarbanel et al. (1990) give further discussion.

The amount of data needed for calculating Lyapunov exponents with the Wolf et al. system depends on three factors (Wolf et al. 1985: 305). First is the number of replacement points that will be needed; second is the number of orbits around the attractor needed to assess the stretching adequately; and third is how well the attractor can be reconstructed with delay coordinates. In practice, the amount of data increases approximately exponentially with the dimension D of the attractor; the authors give 10^D–30^D as a general range of that desirable amount. About 10000–100000 points are typical requirements. Eckmann & Ruelle's (1992) rule of thumb yields one of the largest estimates: the required number of points is about the *square* of that needed to reliably estimate the correlation dimension. (Ch. 24 mentions their rule for estimating the required size of the dataset for determining the correlation dimension.) For example, if by their rule we need 1000 points to get the correlation dimension, we need 1000000 points to estimate a Lyapunov exponent!

Eckmann & Ruelle (1985) (also Eckmann et al. 1986) and Sano & Sawada (1985), working independently, proposed methods that are philosophically very much alike. Hence, in most of the following general description I'll treat them as one method. This technique follows *entire neighborhoods* (groups of points) near the fiducial trajectory, over a short time. In a stripped-down, simplified version, we begin by taking the first datum point as the first fiducial point. Then we locate all other points (x_{j1}, x_{j2}, etc.) within an arbitrarily prescribed radius ε (Fig. 25.10).

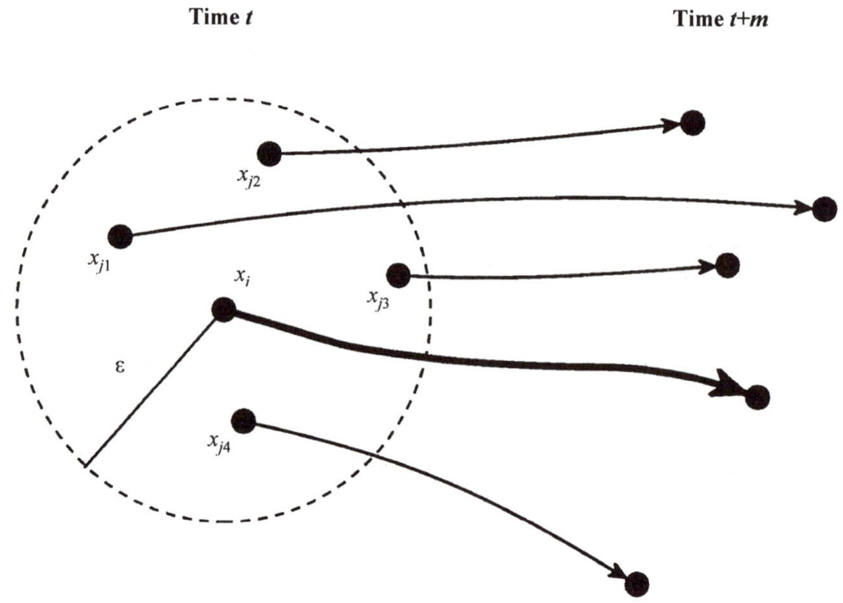

Figure 25.10 Sketch of Eckmann et al. concept for estimating a Lyapunov exponent. A circle of radius ε is centered at fiducial point x_i; four qualifying neighboring points are shown.

Those qualifying points are called "neighbors." Each neighbor has its own separation distance from the fiducial point. Moving along the fiducial trajectory to the second datum point, we next find how many of the original neighbors still qualify as neighbors. Those surviving neighbors have a new gap relative to the fiducial trajectory. We assume that the time between steps one and two is short enough that the change in gaps is linear rather than nonlinear. That is, we assume a linear relation between the gaps at time one and those at time two. In other words, the group relation is that gaps at time 1 equal a constant times gaps at time zero. Fitting a straight line to those gaps (Eckmann & Ruelle used least squares) gives the value of that constant. The same general procedure is repeated at each successive point on the fiducial trajectory. From the many values of the constant determined in the regressions, we compute the Lyapunov exponents from certain mathematical equations.

In practice, we might follow the neighbors over several data values before doing the linear fit. Also, the radius ε doesn't necessarily have to be constant from one fiducial point to the next. The selected value of ε is a compromise between the two conflicting requirements that ε be small enough that the flow can be considered linear (a limitation resulting from nonlinearity) and large enough to include enough neighbors to satisfy certain statistical requirements (a limitation caused by noise). The optimum choice often appears only by trial and error and linear fitting. Another problem is that finding neighbors can be difficult. The number of neighbors within a given radius depends on the size of the dataset, noise, and the fractal nature of the attractor.

Probably the main difference between the Eckmann & Ruelle (1985) approach and that of Sano & Sawada (1985) is the way of doing the orthonormalization. Abarbanel et al. (1991b) review other differences. Other methods mentioned above are largely variations or improvements of the Eckmann–Ruelle/Sano–Sawada approach.

Calculating Lyapunov exponents for an attractor reconstructed from a time series of discretely measured, noisy real-world data is a problem that has only partly been resolved. For instance, Theiler et al. (1992) found that it is possible to get positive Lyapunov exponents for data known to be nonchaotic. (In fact, even data that are essentially random conceivably can have a positive Lyapunov exponent.) Along the same line, Vassilicos et al. (1993) analyzed dollar to Deutsche mark exchange rates. They found a "marginally positive" largest Lyapunov exponent (and hence possible chaos) for a subsample of 15000 data points but a marginally negative exponent using the full sample of 20000 data points. In general, all the various calculation methods also have problems with quality of data (noise), how best to fit the function that describes local dynamics, and the size of the local neighborhoods (Glass & Kaplan 1993). In addition, the methods can't always reliably recognize false exponents, as mentioned. Finally, they have difficulty in high-dimensional systems, possibly because trajectories on those attractors rarely return to a given locality (a problem mentioned earlier in regard to dimension estimates). But then again, most chaos tools don't work well in high dimensions.

Summary

A Lyapunov exponent is a number that reflects the rate of divergence or convergence, averaged over the entire attractor, of two neighboring phase space trajectories. Trajectory divergence or convergence has to follow an exponential law, for the exponent to be definable. In other words, the separations must plot as a straight line with trajectory gap on the ordinate (log scale) and time on the abscissa (arithmetic scale). The general procedure is to determine the logarithms of local slopes of such straight-line relations for all regions of the attractor and then average those logs to derive the Lyapunov exponent. A negative Lyapunov exponent indicates an average convergence of trajectories, a positive exponent an average divergence. Convergence and hence negative Lyapunov exponents typify nonchaotic attractors. Divergence (and hence at least one positive Lyapunov exponent) usually happens only on chaotic attractors. However, convergence can also take place on chaotic attractors. Hence, local rates can vary considerably at different regions on the attractor.

A "local Lyapunov exponent" evaluates trajectory separation over short finite time periods (i.e. over local regions of the phase space attractor). Local Lyapunov exponents can be useful in the practical world in prediction over short timescales and in assessing how accurate a forecast might be.

The number of Lyapunov exponents for a set of data is numerically the same as the number of phase space axes. However, each exponent corresponds to a principal axis of a length, area, volume or higher-dimensional element, and the orientation of such principal axes varies with location on the attractor. Hence, Lyapunov exponents don't correspond to the phase space axes nor to any particular phase space orientation. The exponents of an attractor collectively are known as the spectrum of Lyapunov exponents and are represented in a shorthand form as a string of plusses, minuses, and zeros. There isn't yet a completely reliable way to compute Lyapunov exponents for attractors constructed from real-world data; the two most common techniques are the Wolf et al. (1985) method or a refinement of the Eckmann & Ruelle (1985) / Sano & Sawada (1985) method.

Chapter 26
Kolmogorov–Sinai entropy

We come now to a very elegant quantitative measure of chaos. To date, it's been useful mostly with computer-generated data, because it requires unusually large amounts of relatively noise-free values. It also needs enormous computer capabilities. It's popular in the chaos literature and involves many important principles of chaos theory.

The notion of **entropy** crops up in physics, mathematics, statistics, economics, computer science, literature, earth sciences, biology, and other fields. To many people, however, it's always been an abstract and confusing idea. In fact, mathematician John von Neumann supposedly recommended using the term because "no one knows what entropy really is, so in a debate you'll always have the advantage" (Tribus & McIrvine 1971: 180). Part of the confusion is because people apply "entropy," like "dimension," to many different concepts or measures.[1] Be that as it may, one particular variant of the group is an important indicator of chaos.

Selected milestones in the history of entropy

Clausius's thermodynamic entropy

Based in part on French physicist Sadi Carnot's groundwork of the 1820s, the German physicist Rudolf Clausius[2] introduced entropy in 1865 in his work on heat-producing engines. The general idea is that it's impossible to direct all of a system's energy into useful work, because some of that energy isn't available (escapes, for example). Entropy in this original thermodynamic (heat movement) sense is a measure or calculation of the inaccessible energy. High entropy means that much of a system's energy can't be used for work. In other words, only a small part is available for work. Low entropy means that only a small proportion of the system's energy is unobtainable, that is, most of it's available.

1. Within science, literature and other fields, this plethora of uses of "entropy" causes much unhappiness and hand-wringing. Denbigh & Denbigh (1985), Schneider (1988), and Hayles (1990) give more extensive discussions.

Kolmogorov–Sinai entropy

Boltzmann's statistical entropy

Austrian physicist Ludwig Boltzmann[3] contributed a key development in the "career" of entropy when he devised a statistical or probability measure of entropy *H*:

$$H = -\mathrm{K}\sum_{i=1}^{N_s} P_i \log P_i \tag{26.1}$$

where K is Boltzmann's constant (and depends only on the units) and P_i here is the ordinary probability of an element being in any one of the N_s phase space states. Equation 26.1 is very similar to our Equation 6.17c, reproduced below. (In fact, the rest of this chapter relies heavily on the probability material we discussed in Ch. 6.)

Shannon's information entropy

At least as early as the 1870s, some researchers (possibly including even Boltzmann himself) began looking upon entropy as a measure of information. In the mid-twentieth century, Nyquist, Hartley, Shannon and others in electrical communications, and Wiener in cybernetics, made major progress toward a numerical measure of information. Claude Shannon,[4] an engineer with the Bell Telephone

---

2. **Clausius, Rudolf (1822–88)** Rudolf Clausius, one of the greatest of theoretical physicists, made outstanding contributions to thermodynamics and kinetic theory during the 1850s and 1860s. After graduating from the University of Halle in 1848 he became a professor at the Imperial School of Artillery and Engineering in Berlin. Later he was professor at Zürich, Würzburg, and finally Bonn. His scientific accomplishments include formulating or restating fundamental laws of thermodynamics, putting the theory of heat on a sound basis, and developing the kinetic theory of gases. Much of this work began when he was in his late twenties and climaxed in his early forties. Although regarded as the founder of thermodynamics, he gave much credit to such predecessors as Sadi Carnot, Henri Clapeyron, Robert Mayer, and others. Max Planck, another great German physicist, perhaps best described Clausius's impact: "One day, I happened to come across the treatises of Rudolf Clausius, whose lucid style and enlightening clarity of reasoning made an enormous impression on me, and I became deeply absorbed in his articles, with an ever increasing enthusiasm. I appreciated especially his exact formulation of the two laws of thermodynamics, and the sharp distinction which he was the first to establish between them . . ."
3. **Boltzmann, Ludwig E. (1844–1906)** Austrian physicist Ludwig Boltzmann received his doctorate at the University of Vienna in 1866 and later held professorships in mathematics and physics at Vienna, Grad, Munich, and Leipzig. In 1894 he became Professor of Theoretical Physics at Vienna, where he spent most of the rest of his life. He was one of the principal founders of statistical mechanics. In the 1870s he published several papers in which he showed that the second law of thermodynamics was explainable by applying probability theory and laws of mechanics to the motions of atoms. His work was vigorously attacked and long-misunderstood, but he was eventually proved right. He himself was physically strong, sensitive, and had a fine sense of humor. However, he tended to have periods of silence and depression. During one of his depressed periods, he took his own life at age 62.

Laboratories, built on earlier work on mathematical approaches to information. In the late 1940s (summarized in Shannon & Weaver 1949) he largely created the then-new discipline of information theory. In particular, by considering probabilities, Shannon arrived at Equation 6.17c:

$$I_w = -\sum_{i=1}^{N_s} P_i \log P_i. \qquad \text{(6.17c, repeated)}$$

Any total (N, N_r, etc.) can replace N_s in that equation. (Norbert Wiener developed essentially the same expression for his own field, at about the same time.) The formula is, of course, nearly the same as Boltzmann's long-established equation for entropy (Eq. 26.1) as used in statistical mechanics (with Boltzmann's K = 1). The greatest strength of Shannon's formula is its broad applicability; whereas Boltzmann's formula was limited to statistical mechanics, Shannon's applies to any field and to many types of probability.

In producing his equation, Shannon simultaneously provided both a gigantic advance and massive consternation. The consternation arose because of the name he gave the equation. On the one hand, he said it actually measures information, choice, or uncertainty. However, to the dismay and confusion of many, he opted to call it "entropy" because his equation was essentially identical to Boltzmann's. Physicists today are still arguing over whether Shannon's entropy (dealing with the generation of averages of various statistical options) really measures the same thing as Boltzmann's entropy (dealing with the probabilities of accessible molecular states). In any event, many people began to treat entropy and information (in the Shannon sense) as the same thing. Chaos theory treats them as the same thing.

Important examples of equations that quantitatively measure both information and entropy are $I = \log_2 (1/P)$ (our Eq. 6.16a) and

---

4. **Shannon, Claude E. (1916–)** Claude Shannon is generally acknowledged as the father of information theory. After receiving his PhD in the mathematics of genetic transmission from Massachusetts Institute of Technology (MIT) in 1940, he worked at Bell Laboratories from 1941–56. During the Second World War he was part of a team that developed digital codes, including a code that Roosevelt and Churchill used for transatlantic messages. That work led to his theory of communication and to his introduction of entropy into information theory. His now-classic 1948 paper on that subject wasn't widely understood or accepted at the time and even received a negative review from a prominent mathematician. Today, however, "Shannon's insights help shape virtually all systems that store, process, or transmit information in digital form—from compact discs to supercomputers, from facsimile machines to deep-space probes such as Voyager" (Horgan 1992). In 1956 he returned to MIT as professor of communications science. His career includes many national and international honors and awards but is also noteworthy for his many side interests. Juggling is but one example; he reportedly rode through the halls of Bell Laboratories on a unicycle while juggling four balls. (He claims that his hands are too small to juggle more than four balls at once!) Shannon's home near Boston (called, naturally, "Entropy House") contains entire rooms devoted to his many inventions and hobbies. Those include a rocket-powered frisbee, a mechanical mouse that finds its way out of a maze, a juggling robot of the comedian W. C. Fields, and a computer called Throbac (Thrifty Roman Numeral Backward Computer) that calculates in roman numerals.

$$I_w = \sum_{i=1}^{N_s} P_i \log (1/P_i). \qquad \text{(6.17b, repeated)}$$

Since we're now talking about entropy, we'll substitute the symbol H for entropy, in place of I, and H_w (for I_w) for the weighted sum. Equation 6.16a (the equal-probability case) therefore becomes:

$$H = \log_2(1/P) \qquad (26.2)$$

and Equation 6.17b (the unequal-probability case) becomes:

$$H_w = \sum_{i=1}^{N_s} P_i \log (1/P_i). \qquad (26.3)$$

As explained in Chapter 6, the weighted sum H_w is an average value. Some authors call the entropy of Equations 26.2 or 26.3 the **information entropy**.

All relations developed in Chapter 6 between information and probability also apply to entropy and probability. Whether the context is one of information or entropy, *probabilities* are the foundation or essence of it all; they are the required basic data. As Equations 26.2 and 26.3 indicate, the entropy or information varies depending on the distribution of those probabilities. In other words, entropy doesn't depend at all on the actual values of the variable; rather, it's just a statistic that characterizes an ensemble of probabilities. For that reason, the entropy H_w properly is called the "entropy of the probability distribution P_i." An important feature of Equation 26.3 and entropy H_w is that the probabilities P_i can be any type of probability (conditional, etc.), not just the ordinary ones. We'll come back to that feature later.

Two variants or extensions of information entropy that are important in chaos theory are **Kolmogorov–Sinai entropy**[5] and mutual information. Both variants (discussed later) use, or are based on, Equations 26.2 and 26.3.

Interpretations of entropy

Information is but one of many interpretations of entropy (Table 26.1). Chronologically, it was a relatively late interpretation. There's some debate and confusion in the literature as to whether all interpretations are correct. Opinions also differ about the sign that the value computed from Equations 26.2 or 26.3 should have, for a particular interpretation.

The first entry in Table 26.1 is Clausius's thermodynamic interpretation. Other scientists developed the second interpretation—disorder—soon after Clausius's initial work. The rationale was as follows. Energy is dissipated and entropy therefore increases not only with heat loss but also with other actions. Examples are the

Table 26.1 Extreme or opposite states of entropy (modified and expanded from Gatlin 1972: 29).

High entropy	Low entropy
1. Large proportion of energy unavailable for doing work	Small proportion of energy unavailable for doing work
2. Disorder, disorganization, thorough mix	Order, high degree of organization, meticulous sorting or separation
3. Equally probable events, low probability of a selected event	Preordained outcomes, high probability of a selected event
4. Uniform distribution	Highly uneven distribution
5. Great uncertainty	Near certainty, high reliability
6. Randomness or unpredictability	Nonrandomness, accurate forecasts
7. Freedom (wide variety) of choice, many possible outcomes	Narrowly constrained choice, few possible outcomes
8. Large diversity	Small diversity
9. Great surprise	Little or no surprise
10. Much information	Little information
11. Large amount of information used to specify state of system	Small amount of information used to specify state of system
12. High accuracy of data	Low accuracy of data

change from solid to liquid to gas, chemical reactions, and mixing, as when hot and cold liquids mix or when a gas expands into a vacuum. Those processes involve not only an increase in entropy but also an accompanying decrease in the orderly

5. **Kolmogorov, Andrei Nikolaevich (1903–87)** It's difficult to name anyone who has matched Russian mathematician Andrei N. Kolmogorov's scientific, pedagogical, and organizational contributions. His scientific career alone was outstanding. In his approximately 500 publications, he made important advances in probability theory, Fourier analysis and time series, automata theory, intuitionistic arithmetic, topology, functional analysis, Markov processes, turbulence, theory of fire, statistics of mass production, information theory, random processes, and dynamical systems. A significant part of his scientific work dealt with applying mathematics to natural sciences, life sciences, and Earth sciences. In addition, he wrote papers on philosophy, history, methodology of science, mathematical education, and other topics. After being awarded his PhD by Moscow University in 1929, he was (besides his teaching and research duties) head of various departments and institutes at Moscow State University from 1933 until his death in 1987. Over the years he established several institutes and laboratories in Moscow. On top of all that, he founded and/or edited several technical mathematical journals. He received many national and international awards and honors.

   Equally important were his contributions in education. He was keenly interested in teaching, especially young people, throughout his life. In fact, school mathematics education was his main (but not only) concern during the last third of his career. During his life he organized national mathematical competitions for students, gave popular lectures to schoolchildren, led national committees that analyzed teaching methods, helped launch a boarding school (elementary school) for gifted physics and mathematics students, helped initiate and publish about 200 special (mathematical) issues of a student-oriented magazine, helped train teachers on new course content (by lecturing and writing articles), served on the editorial boards of pedagogical publications, chaired many committees on education improvement, and so on. Finally, he had a particularly broad range of other interests in life—art (especially Old Russian painting and architecture), nature (hiking and mountain skiing), travel (including trips to at least 20 foreign countries), human company, Russian poetry, and gardening.

arrangement of constituent atoms (an increase in disorder). For example, atoms and molecules supposedly are ordered (have low entropy) in a solid, less ordered in a liquid, and quite disordered (have high entropy) in a gas. Consequently, scientists began to look upon a measurement of entropy as a measurement of the system's degree of disorder or disorganization.

The third common interpretation in Table 26.1 is in terms of probability (Eq. 26.1). Boltzmann's probabilities dealt with particle motion, deviations from equilibrium, and similar notions; we'll take a simpler analogy. Say we dump the pieces of a picture puzzle on the table and completely stir them up so that they are in great disorder (high entropy). Selecting a piece at random, there's an equal chance of picking any particular piece. Similarly, rolling a six-sided die has an equal chance of producing any of the six numbers; the spin of a roulette wheel can produce one particular number as well as another. Just as entropy is a maximum for disordered conditions, it's also a maximum for equally probable events. That is, when all possible outcomes are equally likely, the probability of any one outcome is low, and entropy is high. In contrast, when some scoundrel rigs the roulette wheel or loads the die so that a particular number will come up, the outcome is far from equiprobable, and entropy is lowest. So, there's an inverse relation between entropy and probability, as Equation 26.2 shows.

A stripped-down example of two possible states shows how entropy varies with probability distribution. Equation 26.3 for two states is $H_w = P_1 \log_2 (1/P_1) + P_2 \log_2 (1/P_2)$. Choosing P_1 automatically determines P_2, since the probabilities always sum to 1. For instance, if $P_1 = 0.1$, then $P_2 = 0.9$, and $H_w = 0.1 \log_2 (1/0.1) + 0.9 \log_2 (1/0.9) = 0.332 + 0.137 = 0.469$ bits. Choosing any P_1 (and hence P_2) and computing the entropy, for different combinations of P_1 and P_2, produces the curve shown in Figure 26.1. Probabilities are most *unequal* when one of them is zero and the other is 1.0. Computed entropy then is zero (the lowest possible value). As the two probabilities become closer in value (i.e. as P_1 increases from 0 to 0.5), entropy increases. When the two probabilities become equal ($P_1 = P_2 = 0.5$), entropy reaches its maximum value. (And so, as mentioned above, any value for entropy applies only to a particular distribution of the probabilities P_i.)

The fourth interpretation, based on the probability notion, is that of uniformity of distribution of data. If the data are uniformly distributed among a certain number of compartments, the probability of getting each compartment in a sample or trial is equal. In that sense, a uniform distribution is a high-entropy condition (Fig. 26.1 at peak of curve, where $P_1 = P_2 = 0.5$). Conversely, a very nonuniform distribution (e.g. all observations in one compartment) means low entropy, because one bin has a probability of 1 and the other bins a probability of zero (Fig. 26.1 at either end of the parabola).

A fifth notion of entropy is uncertainty. The uncertainty can pertain to the outcome of an experiment about to be run, or it can pertain to the state of a dynamical system. For instance, when rolling a die, we really have no idea which number is going to result. In other words, our uncertainty is greatest. Therefore, we might as well assume complete disarray, equal probabilities, and a uniform distribution.

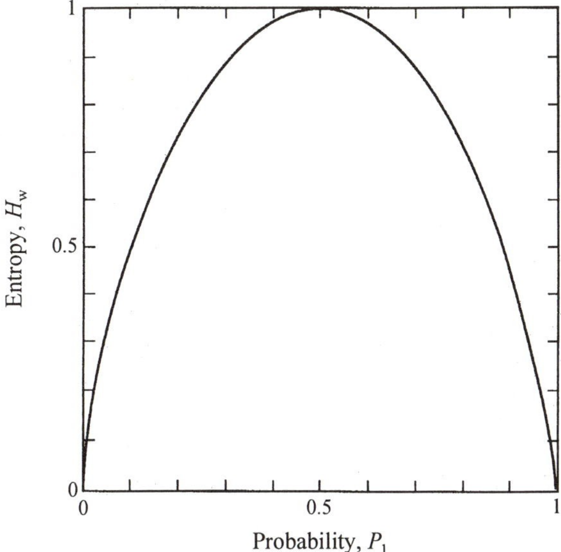

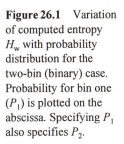

Figure 26.1 Variation of computed entropy H_w with probability distribution for the two-bin (binary) case. Probability for bin one (P_1) is plotted on the abscissa. Specifying P_1 also specifies P_2.

Thus, just as complete disarray and equal probabilities represent greatest entropy, so too does maximum uncertainty. The other extreme is when we're absolutely sure what an outcome will be. Then there's zero uncertainty and zero entropy. (For example, absolute certainty means probability $P = 1$, for which case the entropy in Equations 26.2 and 26.3 computes to zero.) Hence, entropy reflects degree of uncertainty—the lower the entropy, the less the uncertainty, and vice versa.

In concert with the above five variations is a sixth idea: that of randomly distributed observations versus reliable predictability (and so in this sense "random" means unpredictable). Where there's disorder (thorough mixing) and great uncertainty (high entropy), predictions can't be based on any known structure or pattern and can only be done probabilistically. In contrast, something well organized or nearly certain (low entropy) is usually very predictable.

Freedom of choice, or large number of possible outcomes or states, is a seventh way of looking at entropy. When constrained to one choice or outcome, there's no disorder, uncertainty, or unpredictability whatsoever. Probability is highest (1.0) and entropy is lowest (zero). As more choices become available, entropy increases. Maximum entropy comes about when the outcome can happen in the largest number of ways. The easiest way to see this is to use the identity derived in Chapter 6, namely the equal-probability case where $1/P$ = number of possible outcomes, N_s. Inserting N_s for $1/P$ in Equation 26.2 above shows that entropy $H = \log N_s$. So, entropy increases with N_s.

The idea of many possible outcomes suggests diversity, the eighth variation in Table 26.1. Conceptually, the analogy seems valid. However, please be careful when adapting it to practical problems. Some ecologists, for example, use Equation 26.1 to measure the diversity of species in a sample. Unfortunately, as Table

26.1 shows, high entropy can reflect not only large diversity but also uniform distribution. A single value of entropy doesn't distinguish between the two. For instance, a high entropy can indicate both a large number of species and a uniform distribution of those species within the sample. Looking at it another way, a sample that has only a few species but uniform distribution can yield the same value of entropy as a sample with many species that are very unevenly distributed (Pielou 1969: 222). Entropy therefore isn't a suitable measure of diversity because it can be ambiguous.

A notion that follows probability, uncertainty, and variety of choice is that of surprise, a ninth interpretation. In this sense, entropy (the number obtained by calculating Eqs 26.2 or 26.3) reflects the amount of surprise you might feel upon learning the outcome of an event or measurement. When the likelihood of a particular result is high and uncertainty therefore is lowest, you aren't surprised when the result comes out as you expected. Entropy for such a near-certain little-surprise situation is low (as with the no-snow-in-Miami-on-4-July analogy of Ch. 6). Conversely, where any event is as likely as another out of many possible events, uncertainty before the event is highest (the high-entropy condition). You are then more surprised at the outcome, whatever it is.

I've already mentioned the idea of interpreting entropy as information (entry 10 in Table 26.1). In this sense, the number given by Equations 26.2 or 26.3 is an information value that characterizes a particular group of probabilities. We say that the given ensemble or probability distribution represents an information value of so many bits. Furthermore, a relatively large number means a relatively large amount of information, and vice versa. For example, Chapter 6 showed that the two-choice equal-probability case (e.g. whether the baby will be a boy or girl) carries an information of one bit; that's not very much information, because we knew beforehand that there were only two possibilities. In contrast, Equation 26.2 tells us that rolling a six-sided die is characterized by the relatively larger amount of 2.59 bits of information (Ch. 6). To the extent that a new measurement changes the probability in Equation 26.3, entropy is the average amount of new information we've gained by making the measurement.

The 11th concept in Table 26.1 interprets entropy as the amount of information needed to specify the state of the system to a particular accuracy. The grids and computed entropies of Table 26.2[6] help explain this notion. The table pertains to cases where all possibilities are equally likely. Each grid represents a subdivision of phase space, using boxes of size ε. Suppose a system is in a particular state (a particular box within a phase space grid), and we want to guess that state or box. The procedure we'll follow in zeroing in on the target box is to ask yes-or-no questions (Schuster 1988: 236ff.). In particular, the game will be to locate the target box by asking the fewest possible questions about where it is. The most economical approach is to ask questions that subdivide the eligible boxes into two equal groups. In other words, we methodically narrow down the possible locations of the

6. Table 26.2 is merely Table 6.1, reproduced here for convenience.

target box by halving the remaining group of eligible boxes. A typical yes-or-no question might be, is the box in the lower half of the grid?

Let's see what happens to the value of information entropy as the required number of yes-or-no questions increases. For the top row of Table 26.2, we don't need to ask any questions at all because there's only one box to locate. In other words, to specify the state of the system to that very crude accuracy, the required number of questions is zero. Here's the important thing: the computed entropy (col. 4 of the table) is also zero. Now let's move to the next step in complexity: two boxes (second row of table). Suppose the variables that define the state of our system plot in the left-most of those two boxes. Just one question is enough to locate that box (i.e. to specify that condition). It makes no difference whether the answer is yes or no. For instance, we can ask "Is it the right-hand box?" The answer is no, so we'd know we want the left-hand box. One question suffices. Also, the computed value of H (assuming equal probabilities for the two boxes) is $\log_2 (1/P) = \log_2 (1/0.5) = \log_2 2 = 1$ bit. (Or, alternatively, $H = \log_2 N_s = \log_2 2 = 1$ bit.) In other words, again the required number of questions equals the computed entropy (here 1). The next level of complexity is the middle row of the table. The grid now has four boxes. Locating any particular target box takes two questions:

1. Is our mystery box in the upper half of the grid? (The answer tells us whether our target is one of the upper two boxes or one of the lower two.)
2. Is it the left-hand box?

Table 26.2 Probabilities and information for equally likely outcomes (after Rasband 1990: 196).*

1 Grid	2 Number of compartments or states N_S	3 Probability P $=1/N_S$	4 Entropy $H = \log_2(1/P)$ $= \log_2 N_S$ (in bits)
[1 box]	1	1	0
[2 boxes]	2	0.5	1
[4 boxes]	4	0.25	2
[8 boxes]	8	0.125	3
[16 boxes]	16	0.062	4

* This table contains the same information as Table 6.1.

Also, the computed entropy H is 2 bits. Again, therefore, the number of yes-or-no questions needed to locate the box is the same as the entropy. The same principle holds for the other grids in the table. In other words, the computed entropy H (in bits) tells how many yes-or-no questions it takes to guess the state of the system.

Guessing the state of the system is tantamount to specifying it to a particular accuracy. Entropy increases proportionally. To specify the system to the lowest possible accuracy, namely to within just the entire grid, requires zero bits; to the still somewhat crude accuracy of half the grid, 1 bit; to the greater accuracy of one-quarter of the grid, 2 bits; and so on. The higher the entropy, the greater the accuracy to which we can determine the state of the system.

The converse of this last statement leads to the final (12th) interpretation of entropy: entropy indicates the accuracy of the data. For instance, in Table 26.2 each grid pattern might represent the degree of accuracy to which our tools let us measure the system. A certain computed entropy (col. 4) pertains to each of those grids or accuracies. If we can only measure to within the entire largest box (row 1—the low-accuracy case), the associated entropy is low (zero). On the other hand, if we can measure to within a relatively small box (row 4—a relatively high accuracy), entropy is considerably higher. So, high entropy corresponds to high accuracy in the measurements and vice versa.

In practice the probabilities usually aren't equal and hence are weighted in arriving at the total entropy. However, the various interpretations of entropy as just described are still valid.

Kolmogorov–Sinai entropy defined

Shannon's entropy (information entropy), regardless of what types of probabilities we use in it, can't by itself identify chaos. Its value is always positive and finite (except when $P = 1$ and entropy is zero) and can vary by a lot, depending on the control parameter, number (size) of compartments, and other factors. However, the equations for information entropy (Eqs 26.2, 26.3) form the basis for another type of entropy—Kolmogorov–Sinai entropy—that at least theoretically can identify chaos.[7] For brevity, I'll usually just refer to it as K–S entropy. It has several aliases in the literature, including "source entropy," "entropy of the source," **measure-theoretic entropy**, "metric-invariant entropy," and **metric entropy**.

K–S entropy has the following three important features:

- First, K–S entropy requires *sequence probabilities* (the probabilities that the system will follow various routes over time; Ch. 6). The entropy equations (Eqs 26.2, 26.3) are very general and don't specify any particular type of probability. K–S entropy deals with the information or uncertainty associated with a *time sequence* of measurements or observations.

7. A. N. Kolmogorov (1959) proposed applying Shannon's entropy to dynamical systems. Later in 1959 Y. Sinai gave a refined definition and a proof.

A brief example shows how to use sequence probabilities in computing information entropy. We'll compute information entropy for longer and longer overall periods—first for a period of one time interval (e.g. week), then for a period that lasts over two consecutive intervals, and so on. Such computations are easy and straightforward. Equation 26.3, namely

$$H_w = \sum_{i=1}^{N_s} P_i \log(1/P_i),$$

is the equation. At each successive time we just plug in the first of the various probabilities (an example is given below), compute the associated product $P\log(1/P)$, repeat for each other probability in turn, and add up the products, as Equation 26.3 says. The probabilities to use (sequence probabilities) are the likelihood that the system will follow each of the various possible routes that finish at the chosen time. The summation then is over all routes represented in the data, or N_r. Changing the symbols slightly helps keep those distinctions in mind. $H_{\Delta t}$ replaces H_w, to show that we're computing entropy over a particular window of time (duration). (Here that duration usually will be in terms of number of observations rather than actual time units.) Also, P_S replaces P_i to help us remember that the probabilities are sequence probabilities. Our entropy relation (Eq. 26.3) in new clothing therefore becomes

$$H_{\Delta t} = \sum_{i=1}^{N_r} P_S \log(1/P_S) \tag{26.4}$$

in which N_r is the number of possible phase space routes.

Our example will have just two possible states, A and B. Figure 26.2 (mainly a reproduction of Fig. 6.7) shows that at "time 1" there are only two possible "routes," namely into bin A or into bin B. (And so entropy computations for time 1 are an exception to using sequence probabilities, because the only probabilities available are ordinary ones.) In Figure 26.2, the routes into bins A and B have ordinary probabilities of 0.60 and 0.40, respectively. For each route, Equation 26.4 says to compute $P\log(1/P)$, then add the two results. For bin A, $P\log_2(1/P) = 0.60\log_2(1/0.60) = 0.442$. For B, $P\log_2(1/P) = 0.40\log_2(1/0.40) = 0.529$. (For brevity, I'll drop the units of "bits" that actually go with each product.) According to Equation 26.4, entropy is the sum for all routes: $H_{\Delta t} = 0.442 + 0.529 = 0.971$.

Entropies for two or more successive time intervals are computed in the same way but with sequence probabilities. Each time gets its own entropy—its own sum for the several products of $P_S\log_2(1/P_S)$. Let's compute the entropy for "time 2" of Figure 26.2. To get to time 2, there are four possible routes and associated probabilities: AA ($P_S = 0.60 \times 0.70 = 0.420$), AB ($P_S = 0.60 \times 0.30 = 0.180$), BA ($P_S = 0.324$), and BB ($P_S = 0.076$). Using those sequence probabilities, we compute the quantity $P_S\log_2(1/P_S)$ for each route.

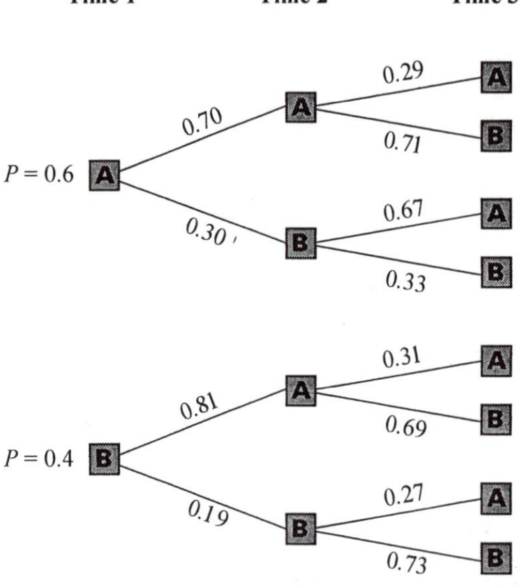

Figure 26.2 Branching pattern and hypothetical probabilities for two bins (A and B) over three time steps. Numbers along paths are conditional probabilities.

For instance, route AA has 0.420 $\log_2(1/0.420) = 0.526$. We then add up the four resulting numbers. That gives $H_{\Delta t} = 0.526+0.445+0.527+0.283 = 1.781$, the entropy for time 2.

Over three consecutive measurements with two possible states there are eight possible routes (AAA, AAB, etc.). For each of the eight sequence probabilities (0.122, 0.298, 0.121, etc. in the example of Fig. 26.2) we calculate $P_S \log_2(1/P_S)$, then add up the resulting eight values. That sum is the information entropy for time 3.

In practice, the computer estimates sequence probabilities from the basic data. We need the probability of every string that leads to each successive time. That is, any one sequence probability applies only to a unique route. For example, for time 2 we count the number of occasions the system followed each of the four possible routes. Then we add up the occurrences for all routes to get a grand total. Finally, we divide the number for each route by the grand total. That produces the sequence probabilities to use in computing the entropy for that time. Then we repeat that exercise for each subsequent time, in turn, to get its associated sequence probabilities.

The total number of required sequence probabilities can be very large. Therefore, we may well need a large dataset and much computer time. Efficient histogramming methods, such as a bin sort on a binary tree (Gershenfeld 1993), can reduce computer time.

Sequence probabilities in chaos analyses usually are based on lagged values of a single variable. That means pseudo phase space. As with attractor reconstruction, the two aspects that we can vary are the lag itself and the number of phase space dimensions.

- Secondly, K–S entropy represents a *rate*. The distinctive or indicative feature about a dynamical regime isn't entropy by itself but rather the entropy rate. A rate (actually an average rate) is any quantity divided by the time during which it applies. Hence, any computed entropy divided by the associated time gives an average entropy rate—an average entropy per unit time:

$$\begin{aligned} \textit{average entropy rate} &= H_{\Delta t}/time \\ &= \sum_{i=1}^{N_r} P_S \log(1/P_S)/(time). \end{aligned} \quad (26.5)$$

  Claude Shannon, in the classic treatise *The mathematical theory of communication* (Shannon & Weaver 1949: 55, theorem 5), seems to have been the first to articulate the concept of entropy rate.
- Thirdly, K–S entropy is a *limiting value*. For discrete systems or observations, two limits are involved, namely the entropy rate as time lengthens to infinity ($\lim_{t\to\infty}$) and bin size ε (width of classes, as in a histogram) shrinks to zero ($\lim_{\varepsilon\to 0}$).

  Based on the above three features, **K–S entropy** (H_{KS}) is the average entropy per unit time *at the limiting conditions of time going to infinity and of box size going to zero*. In symbols,

$$\begin{aligned} H_{KS} &= \lim_{\varepsilon\to 0}\lim_{t\to\infty} \text{ of entropy rate} \\ &= \lim_{\varepsilon\to 0}\lim_{t\to\infty} (H_{\Delta t}/time) \\ &= \lim_{\varepsilon\to 0}\lim_{t\to\infty} \left[\sum_{i=1}^{N_r} P_S \log(1/P_S)\right]/time. \end{aligned} \quad (26.6)$$

There are several ways to interpret K–S entropy. All of them stem from Table 26.1. For instance, K–S entropy measures the average amount of new information or knowledge gained per measurement or per unit time (Farmer 1982). In other words, it's the average temporal rate of production, creation or growth of information contained in a series of measurements (Fraser 1986, Atmanspacher & Scheingraber 1987, Gershenfeld 1988). Applied to predictions, K–S entropy is:

- the average amount of uncertainty in predicting the next n events (Young 1983)
- the average rate at which the accuracy of a prediction decays as prediction time increases, that is, the rate at which predictability will be lost (Farmer 1982)
- the average rate at which information about the state of the system is lost (Schuster 1988).

Estimating K–S entropy

Equation 26.6 tells us to do two general types of operations. It says to first arbitrarily choose a box size ε and, holding ε constant, see what happens to the entropy rate as time gets bigger and goes to infinity. That gives the so-called "inner limit" of Equation 26.6. The first steps or computations to make for that purpose are:

1. Choose a bin width ε.
2. Use the basic data to estimate sequence probabilities for all possible routes, for successive times.
3. For each time, compute entropy as

$$\sum_{i=1}^{N_r} P_S \log (1/P_S).$$

Possible routes that aren't represented all the way to the time in question have a probability of zero (based on the data on hand) and so can be neglected.

There are two ways to use those results to get the inner limit of Equation 26.6. One way is to divide each entropy by the associated time to get entropy rate, plot entropy rate versus time, and estimate the asymptotic entropy rate as time increases. That method takes many time events to reach the asymptotic limit, that is, to converge. In fact, for practical purposes it's computationally impossible in most cases because of the huge amount of data needed.

The other way needs fewer time steps to show the asymptotic limit (i.e. it converges faster). It uses an alternative relation that Shannon gave. That alternative expression (Shannon & Weaver 1949: 55, theorem 6) defines an entropy difference:

$$\textit{entropy difference} = t(H_t/t)-(t-1)(H_{t-1}/[t-1]) \tag{26.7}$$

where H_t is entropy as of time t, and $t-1$ refers to the observation just prior to time t. Within the first term on the right, namely $t(H_t/t)$, the two t's cancel out, leaving just H_t. In the same way, $t-1$ cancels out in the second product on the right, leaving just H_{t-1}. Equation 26.7 therefore reduces to simply

$$\textit{entropy difference} = H_t-H_{t-1}. \tag{26.8}$$

That result is nothing more than our friend, a first-differencing. It uses just the entropies themselves, not the entropy rates. At time 1, there isn't any entropy difference because there's no preceding entropy to subtract. At time 2, H_t-H_{t-1} is the entropy at time 2 minus the entropy at time 1; at time 3, H_t-H_{t-1} = entropy at time 3 minus that at time 2; and so on.

Shannon said that both the average entropy rate $H_{\Delta t}/t$ and the entropy difference H_t-H_{t-1}, when plotted against time, converge to the same asymptotic value of $H_{\Delta t}/t$. He also stated that H_t-H_{t-1} "is the better approximation." For our purposes, what's

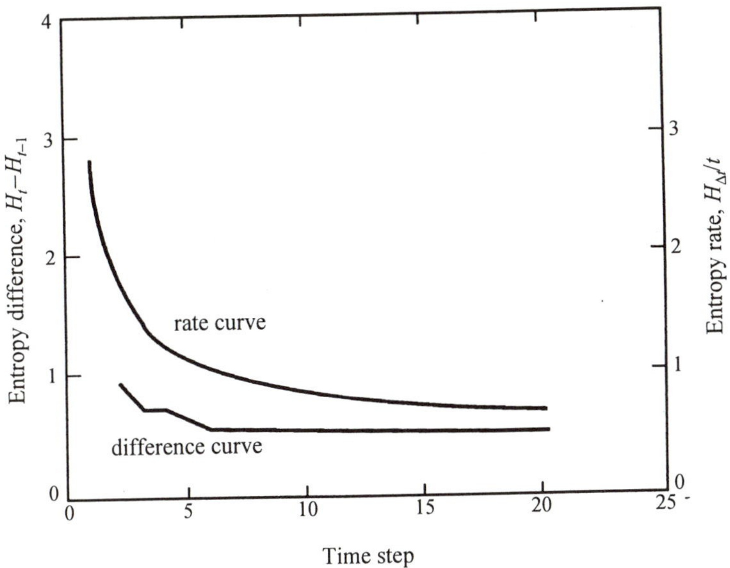

Figure 26.3 Decrease of entropy difference $H_t - H_{t-1}$ (the "difference curve") and entropy rate $H_{\Delta t}/t$ (the "rate curve") with time. Data are for logistic equation with k = 3.7 (hence chaotic regime), using 10 bins and 1 000 000 values.

important is that $H_t - H_{t-1}$ becomes asymptotic sooner (Fig. 26.3). It therefore requires fewer data, less computer power, and so on. So, in practice it's probably better to use Equation 26.8, plot $H_t - H_{t-1}$ versus time and take the asymptotic $H_t - H_{t-1}$ as the asymptotic $H_{\Delta t}/t$.

An asymptotic ($\lim_{t \to \infty}$) entropy rate applies only to the particular bin size chosen. The second major operation that Equation 26.6 requires is to repeat the entire procedure and get estimates of the inner limit for smaller and smaller bin sizes ($\lim_{\varepsilon \to 0}$). The formula for K–S entropy with this approach therefore is:

$$H_{KS} = \lim_{\varepsilon \to 0} \lim_{t \to \infty} (H_t - H_{t-1}). \tag{26.9}$$

In the next chapter I'll mention other ways to estimate K–S entropy.

Some features that influence computed values of entropy

As we've seen, computing entropy consists largely of just adding up a bunch of $P \log(1/P)$ values. However, anything that affects those P values can also affect the value of $P \log(1/P)$ and hence also the entropy. For instance, Figure 26.1 shows an example of how entropy can vary with the distribution of probabilities. Four other influential factors are the number (width) of bins, size of dataset, lag, and number of sequential events (or embedding dimension). Let's look at each of those:

- *Bin width (number of possible states)* Table 26.2 (cols 2 and 4) shows, for the case of equal probabilities, the different entropies that result from different partitionings or numbers of bins. Figure 26.4, based on the logistic equation with k = 3.7 and a lag of one, shows the different curves of $H_t - H_{t-1}$ versus time for different numbers of possible states (bin widths). Here there isn't a smooth progression toward an asymptotic limit as bin size decreases. Instead, for these particular data the relations for 2, 10, and 20 bins all seem to become asymptotic to roughly the same value of $H_t - H_{t-1}$ (about 0.47–0.48). (Extrapolating the values is somewhat uncertain.) However, the relation for 5 bins in Figure 26.4 becomes asymptotic to a noticeably lower value, namely about 0.43.

  Histogramming actually may not be the best procedure for estimating probabilities. Other methods (Ch. 6) are becoming popular.
- *Size of dataset* Sample size influences probabilities (and hence entropies) in that a short record (small dataset) probably won't adequately sample the system's long-term evolution. Probabilities (all types) for scanty datasets won't

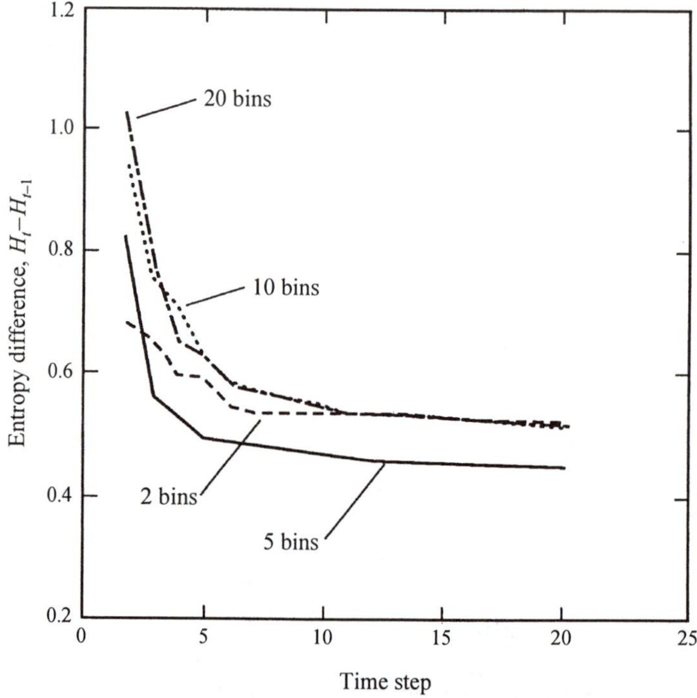

Figure 26.4 Change of entropy difference with time for different partitionings (numbers of bins). Data are for logistic equation with k = 3.7, using 1 000 000 values.

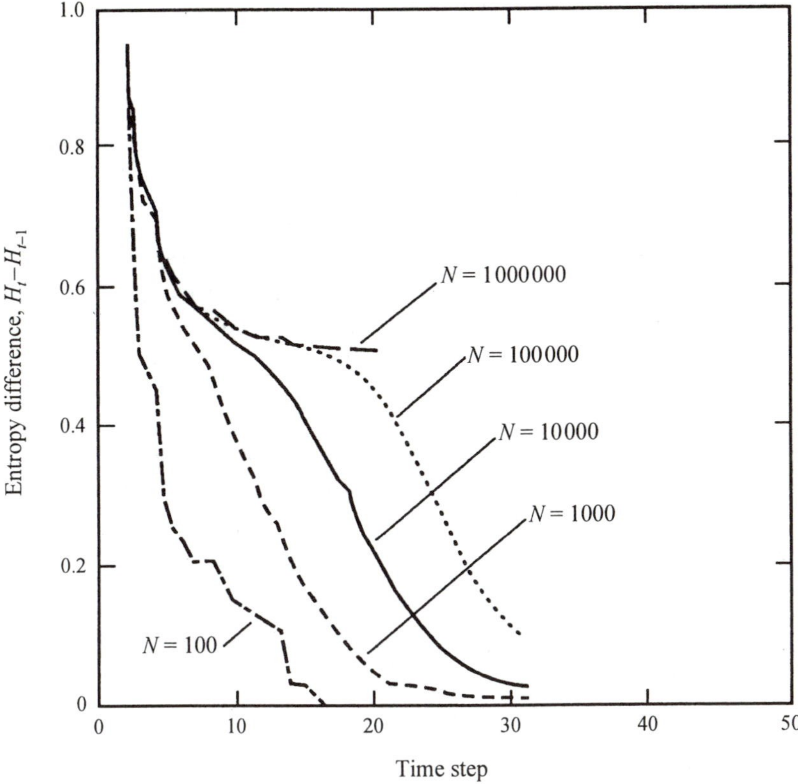

Figure 26.5 Effect of sample size on the relation between entropy difference $H_t - H_{t-1}$ and time. Data are for logistic equation with k = 3.7, using 10 bins.

be representative. They change as we collect more measurements. Some important bins may not get sampled at all.

Figure 26.5 shows some typical effects of sample size in the K–S entropy step of plotting $H_t - H_{t-1}$ against time. Data represent iterations of the logistic equation with k = 3.7, using ten bins and sample sizes ranging from 100 to 1 million (about the most my computer can handle). The relation for a dataset of 1 000 000 values seems to become asymptotic toward a constant, positive entropy difference of about 0.47–0.48 (as was also shown in Fig. 26.4). The relation for N = 100 000 seems to follow that same curve for about 15 time steps but then veers downward. Relations for datasets of 10 000 or less offer no hope (in this example) of indicating the asymptotic entropy difference. In fact, we'd have a final entropy difference of zero (rather than the correct value, possibly 0.47–0.48) for all cases except where we have at least a million observations. And that's for a relatively "clean" (noiseless) dataset.

- *Lag* Lag specifies the particular subset of data values used in estimating probabilities. Different subsets can lead to different sequence probabilities. For instance, using a lag of one, probabilities depend on the measured sequence x_t, x_{t+1}, x_{t+2}, and so on. Using a lag of two defines a different sequence of measurements, namely x_t, x_{t+2}, x_{t+4}, and so on. Sequence probabilities of getting the two different series of numbers very likely won't be the same, and therefore neither will the K–S entropy.
- *Embedding dimension (number of sequential measurements involved in the vector)* Sequence probabilities used in computing entropy (and hence the resulting value of K–S entropy) change as number of events (measuring occasions) increases. Most chaos calculations are in a context of pseudo (lagged) phase space, so they usually deal with lagged values of a single feature. Each additional lagged value attached to the vector is tantamount to another embedding dimension. That is, embedding dimension plays the same role as time or number of events. Each additional dimension or event brings an entire new set of phase space routes (Fig. 26.2) and sequence probabilities. Hence, K–S entropy can change with embedding dimension or time (Figs 26.3, 26.4, 26.5).

  For a given number of possible states, the number of routes increases geometrically with number of lag space dimensions, in a manner analogous to Equation 6.6. In the present context, Equation 6.6 says that the *number of routes* (here probabilities) = $states^{embedding\ dimension}$.

  That power-law relation plays a major role in the calculation of K–S entropy. Estimating the sequence probabilities for each of the various possible routes can require datasets numbering in the millions. Such large amounts of data are unrealistic for many of us. Consequently, researchers have pursued other ways to estimate K–S entropy. One proposal is to calculate and sum the positive Lyapunov exponents (based on the so-called Pesin theorem). Another is to do a mutual-information analysis, the subject of the next chapter. Both of those alternate approaches need further development, at least for applications to real-world data.

Identifying the dynamical regime

Let's now see how K–S entropy can distinguish between various types of data. We'll take the easy way and look only at noiseless data. In practice, of course, most data aren't noiseless. And therein lies the rub. One step in getting K–S entropy involves taking the limit as bin size goes to zero (Eq. 26.6). In practice, once bin size diminishes to the noise level (the error in the measurements), further decreases in bin size are meaningless. Our estimate of K–S entropy then becomes less accurate. The mutual-information approach discussed in the next chapter is an initial attempt to get around this problem.

The basic equation we'll use to estimate K–S entropy is Equation 26.9:

$$H_{KS} = \lim_{\varepsilon \to 0} \lim_{t \to \infty} (H_t - H_{t-1}).$$

H_t is entropy at time *t*, computed as

$$\sum_{i=1}^{N_r} P_S \log (1/P_S),$$

and H_{t-1} is the entropy at time $t-1$. Our recipe therefore is as follows:

1. Choose bin size ε.
2. For a given time step, estimate P_S for each route and compute H_{t-1} as the sum of $P_S \log(1/P_S)$ for all routes.
3. Repeat step 2 for the next time step and compute the entropy for that time, H_t.
4. Subtract to get the quantity $H_t - H_{t-1}$.
5. Repeat steps 2–4 for higher successive times.
6. See what happens to $H_t - H_{t-1}$ as time increases.
7. Repeat steps 1–6 for a smaller bin size, then a still smaller bin size, and so on, to see what happens as ε gets very small. (This step may not always be necessary.)

Fixed points

The logistic equation helps show the value of K–S entropy for various nonchaotic attractors. Let's start with ten compartments (step 1 from the above list). The logistic equation yields a fixed-point attractor when the control parameter k is less than 3. That is, for k<3 all iterations (once the transients die out) yield a constant fixed value. For instance, with k = 2.8 the fixed point is at $x^* = 0.643$ (Ch. 10). Excluding transients, the time series consists of a string of values of 0.643 (or of whatever value the fixed point has). Every route, in other words, consists of endless values of the same number. Sequence probabilities therefore are always 1.0. Log($1/P_S$) therefore is log(1), which is zero, so $P_S \log(1/P_S)$ is zero and entropies for all time steps are zero (steps 2–6). That takes care of the first limit in the K–S entropy equation.

Now for the second (final) limit, where bin size goes to zero (step 7). The single attractor value always falls in the same bin, for a given number of bins. The system is sure to go to that bin. All sequence probabilities therefore are always 1.0. In other words, using smaller and smaller bins has no effect. So, the K–S entropy—the limit as time goes to infinity and bin size goes to zero—is zero for a fixed-point attractor.

Period-two attractor

Now we'll increase k and enter the period-doubling zone. The simplest case is period two. As usual, we'll neglect transients. At k = 3.4, the system alternates between the two fixed values of $x^* = 0.452$ (bin boundaries 0.4 to 0.5, for a 10-bin partition) and $x^* = 0.842$ (bin boundaries 0.8 to 0.9). The first limit to get is the limit as time goes to infinity, for constant bin size. We'll use ten bins (step 1). The first calculations are for time 1. The two possible routes for time 1 are into bin 5 (x^* between 0.4 and 0.5) or into bin 9 (x^* within 0.8 to 0.9). At any arbitrary starting time, there's a 50 per cent chance (an ordinary probability P of 0.5) that the system is in either box. The quantity $P \log_2 (1/P)$ for each bin is $0.5 \log_2 (1/0.5)$ or 0.500. Summing the two values gives $H_{t-1} = 0.500+0.500 = 1.0$ (step 2).

Step 3 is to do the same computations over a duration of two time increments, thus getting H_t. Only two of our ten compartments (bins 5 and 9) are possible for our period-two attractor. Regardless of which bin the system goes to at time 1, it's sure (probability = 1) to go to the other at time 2. In other words, the only routes possible over two events are of the sort AB and BA. The sequence probability for each of those routes is 0.5 (for time 1) × 1 (for time 2), or 0.5. The two values of 0.5 for $P_S \log(1/P_S)$ sum to 1 (step 3). That is, entropy again is 1.0. Furthermore, all subsequent conditional probabilities are always 1.0 because, given one of the two attractor values, we know the value at the next time. Entropy therefore has a constant value of 1.0 for all times. Subtracting the entropy of any one time from that of the next time (step 4) gives an entropy difference of zero, at every time (steps 5 and 6). So, our first limit is zero.

The second limit (step 7) involves decreasing the bin size to very small. As with the fixed-point case, decreasing the bin size has no effect because the two attractor values always fall into just two known bins. K–S entropy therefore is zero for a period-two attractor.

Higher-periodicity attractors

Again let's start with bin sizes ε of 0.1 (step 1). A higher k of 3.52 produces a periodicity of four (with sequential x^* values of 0.879, 0.373, 0.823, and 0.512). That's a fixed succession. For instance, if an observation (say 0.373) is in bin 4 (x^* between 0.3 and 0.4), the system is certain to go next to bin 9 (x^* between 0.8 and 0.9), because the next attractor value in the sequence is 0.823. If in bin 6 (x^* between 0.5 and 0.6, such as for 0.512), the system is also certain to go next to bin 9, because the next attractor value in the sequence is 0.879. The probability for such certain paths is 1. If, however, we know only that an observation is in bin 9, we wouldn't know whether the actual value is 0.879 or 0.823. Hence, we have to assume a probability of 0.5 that the system goes next to bin 4 ($x^* = 0.373$) and also a probability of 0.5 that it goes instead to bin 6 ($x^* = 0.512$). The branching pattern going to time 2 therefore looks like that shown in Figure 26.6.

What are the sequence probabilities, entropies, and entropy differences for such data?

- At time 1, ordinary probabilities are 0.25, 0.25, and 0.50 that the system might be in bins 4, 6, or 9, respectively (Fig. 26.6). The three corresponding values of $P \log_2 (1/P)$ each compute to 0.500. Summing these gives an entropy of 1.500 (step 2).
- At time 2, there are four possible routes. If the system is in bins 4 (0.373) or 6 (0.512) at time 1, the routes to time 2 are certain (namely, to bin 9). Sequence probabilities for each of those two routes are $0.25 \times 1.0 = 0.25$. $P_S \log_2 (1/P_S)$ is 0.500 for each. The remaining possible bin at time 1 is bin 9 (possible values of 0.823 or 0.879). For each value, a route leads to time 2. The two possible routes from bin 9 each have a conditional probability of 0.5, so the associated values of P_S are each $0.50 \times 0.50 = 0.25$. $P_S \log_2 (1/P_S)$ for each of those two routes is 0.500. All four values of $P_S \log_2 (1/P_S)$ therefore are 0.500. The sum (the entropy) is 2.0 (step 3). Subtracting the entropy at time 1 (1.500) from that at time 2 (2.0) gives $2-1.5 = 0.5$ (step 4).

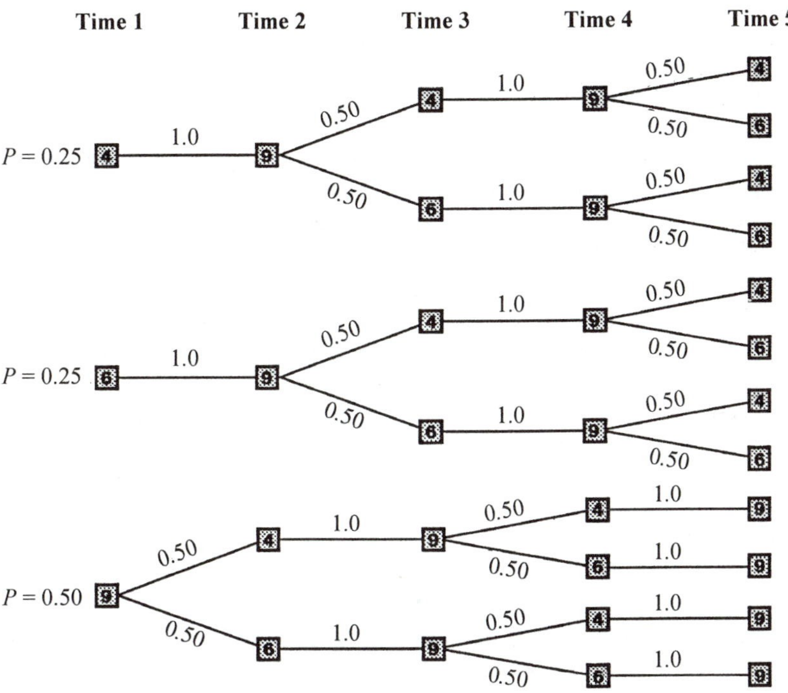

Figure 26.6 Branching patterns for five iterations of the logistic equation at period four (k = 3.52). Range of 0–1 was divided into 10 bins. Sequential attractor values are 0.373 (bin 4), 0.823 (bin 9), 0.512 (bin 6), and 0.879 (bin 9). Boxed numbers are bin numbers; numbers along individual routes are probabilities.

- Six possible routes lead to time 3 (Fig. 26.6). As always, the task is to determine sequence probabilities and associated values of $P_S \log_2 (1/P_S)$ for each route, then sum the results to get the entropy. Entropy here turns out to be 2.0 (step 3). Subtracting the preceding entropy (2) gives 2–2 = 0 (step 4). In fact, entropy for each succeeding time step also turns out to be 2.0 (step 5). Entropy differences therefore are zero at each time (step 6). Also, decreasing the bin size (step 7) again has no effect because of the fixed attractor values.

As with periodicities of two and four, K–S entropy also is zero for higher periodicities.

Summarizing, K–S entropy is zero for logistic-equation nonchaotic attractors. Nonchaotic systems don't evolve with time. The attractor consists either of one value (a point attractor), a group of values regularly repeated with time (periodic attractor or limit cycle), or a torus on which the frequencies are either periodic or quasiperiodic. Once we identify the attractor, we can reliably predict further results with virtually zero uncertainty. Entropy stays constant with time. The K–S entropy of zero means that we don't gain any new information over time.

Chaotic regime

We've already done some sample computations for the chaotic regime (Figs 26.3, 26.4, 26.5). A chaotic system continually evolves with time. Compared to any chosen time, the system at a later time is in an unpredictably different state. That unpredictable evolution provides a steady supply of new information. In other words, the K–S entropy for chaotic systems is some positive constant. For example, Figures 26.3, 26.4, and 26.5 show that the logistic equation's chaotic regime at k = 3.7 has a K–S entropy of about 0.47–0.48.

Even with unlimited amounts of noiseless data, extrapolating an asymptotic curve to estimate the limiting value is not easy. For instance, on Figures 26.3–5 it's no problem getting to within 0.1 of the approximate asymptotic value. However, we probably should question values in the literature that are reported to the nearest 0.01.

Random data

K–S entropy for random data can best be understood by looking at sample computations for the two limits in the K–S entropy equation (Eq. 26.9). (As usual, "random" here means that any determinism is essentially negligible. In addition, the treatment here assumes that the underlying probability distribution is uniform.) The first limit is time going to infinity. The second is bin size going to zero.

limit as time goes to infinity

Finding this first limit requires keeping the number of possible states constant. Let's use two states (step 1). For a uniform distribution, probabilities are equal of the system going to either state. At time 1, the ordinary probabilities are 0.5 for each state. $P \log_2 (1/P)$ for each state then is $0.5 \log_2 (1/0.5)$ or 0.500. Entropy H_{t-1} for time 1 is the sum of the two values of $P \log_2 (1/P)$, or 0.500+0.500, or 1.0 (step 2).

Time 2 and later times require sequence probabilities (P_S) in the entropy equation. There are four possible sequences or routes to time 2. Regardless of which of the two states the system went to at time 1, it has an equal chance of going to either state at time 2. In other words, each of the four routes is equally likely. Each sequence probability therefore is 0.50×0.50 or 0.25. $P_S \log_2 (1/P_S)$ for each of the four routes becomes $0.25 \log_2 (1/0.25)$ or 0.500. Adding the four values of 0.500, entropy $H_t = 2.0$ (step 3). Subtracting the entropy at time 1 (1.0) from that at time 2 (2.0) gives an entropy difference of 2.0–1.0 or 1 (step 4).

Step 5 is to move up one time step and make similar calculations. At time 3, there are eight possible routes, again all equally likely. Each sequence probability now is 0.50×0.50×0.50 = 0.125. $P_S \log_2(1/P_S)$ is $0.125 \log_2 (1/0.125)$ or 0.375. Adding up eight such values gives a new H_t of 3.0. Entropy difference now is entropy at time 3 minus entropy at time 2 = 3.0–2.0 = 1—exactly the same value as for the preceding time step.

Similar computations to include more successive observations yield the same result: at each time or iteration, $H_t - H_{t-1}$ is 1.0. Thus, for two states the limit of $H_t - H_{t-1}$ as time goes to infinity (step 6) is a constant, namely 1.0. The remaining job is to find out what happens to that constant as we increase the number of possible states (i.e. as bin size goes to zero) (step 7).

limit as bin size goes to zero

In the foregoing discussion, we used only two states and got $H_t - H_{t-1} = 1.0$ for any and all times. Now let's divide our interval of 0 to 1 into three possible states instead of two. For each time, computing the entropy and subtracting that of the preceding time again produces a constant value of $H_t - H_{t-1}$. However, the constant now is 1.585 instead of 1.0. Using four possible states produces a constant difference ($H_t - H_{t-1}$) of 2.0 for all times. In fact, each constant is the log (here base 2) of N_s, where N_s is the number of possible states. ($\text{Log}_2 2 = 1.0$; $\log_2 3 = 1.585$; $\log_2 4 = 2.0$; etc.) Hence, for uniformly distributed data, $H_t - H_{t-1} = \log N_s$.

Since $H_t - H_{t-1}$ = log of number of states, $H_t - H_{t-1}$ increases as long as the number of possible states increases, that is, as long as bin size ε decreases. The increase is at a steady, arithmetic rate as N_s increases geometrically. For example, as N_s progresses from 1 to 10 to 100, $H_t - H_{t-1}$ goes from 0 to 3.32 to 6.64, etc. Thus, $H_t - H_{t-1}$ doesn't become asymptotic to any constant. When the classes become infinitesimally narrow and approach zero in width, the number of possible states becomes infinitely large. $H_t - H_{t-1}$ then approaches infinity. Hence, for random data, K–S entropy is infinite.

Let's review briefly the relation between $H_t - H_{t-1}$ and time (number of iterations) for the various types of data. For pre-chaotic data, the relation is zero right from the start (time 1), so K–S entropy is zero. For chaotic data, the curve starts at a relatively high value of $H_t - H_{t-1}$ and decays by lesser and lesser amounts with time, becoming asymptotic toward some positive constant. Thus, K–S entropy is positive. For random and uniformly distributed data, the relation increases regularly (logarithmically) with number of states, so K–S entropy is infinite. In practice, noise, insufficient data, and computer capability complicate attempts to distinguish the three classes. On the other hand, K–S entropy for each category is different enough that it can sometimes be helpful.

Summary

"Entropy" has many meanings or interpretations. They include the proportion of energy available for doing work, disorder, probability of an event, uniformity of a distribution, uncertainty, randomness or unpredictability, freedom of choice, diversity, surprise, information, amount of information needed to specify the state of a system, and accuracy of data. Probabilities (relative frequencies) are the basic data used in calculating entropy or information; high probability implies a small amount of information, and vice versa. When all states of a system are equally probable, entropy increases as number of choices (phase space compartments) increases. When the states aren't all equally probable, entropy is

$$\sum_{i=1}^{N_s} P_i \log(1/P_i),$$

where P_i is the probability of the ith compartment. In that formula, entropy varies with number of compartments, distribution of probabilities among the compartments, and with value of control parameter.

Entropy also has various other general properties (see also Krippendorff 1986: 16ff.). For example, its value is always zero or positive. Entropy is zero (a minimum) when any bin has a probability of one (absolute certainty). It's a maximum for a uniform distribution (equally probable outcomes). Its value for that equally probable outcome case increases with number of possible states. Entropy doesn't care what the system or process is—it only evaluates a set of measured frequencies or probabilities. That is, it's a number that characterizes an ensemble of probabilities. That means, also, that it makes no *a priori* assumptions about the nature of the frequency distribution (that means it's nonparametric). An entropy is an average (a weighted sum). Finally, its generality has led to applications in many fields, including engineering, physics, chemistry, biology, sociology, Earth sciences, literature, art, journalism, television, and others.

Information entropy is always positive and finite. By itself, it doesn't identify chaos. To identify chaos, we need entropy rate (the average entropy per unit time) at the limiting conditions of time increasing toward infinity and of box size diminishing toward zero. That special case of entropy is called Kolmogorov–Sinai (or K–S) entropy (H_{KS}). It requires sequence probabilities. In symbols, K–S entropy

$$\begin{aligned} H_{KS} &= \lim_{\varepsilon\to 0}\lim_{t\to\infty} (H_{\Delta t}/time) \\ &= \lim_{\varepsilon\to 0}\lim_{t\to\infty} \left[\sum_{i=1}^{N_r} P_S \log(1/P_S) \right] /time \\ &= \lim_{\varepsilon\to 0}\lim_{t\to\infty} (H_t - H_{t-1}). \end{aligned}$$

K–S entropy quantifies how chaotic a dynamical system is: H_{KS} is zero for a deterministic system that isn't in chaos; it's a positive constant for a chaotic system; and it's infinite for a random process, at least for uniformly distributed data.

Chapter 27
Mutual information and redundancy

The idea of "information" (Ch. 6) is becoming very important in chaos theory. As long as there is some relation between two variables, one contains information about the other. Take, for example, the Christmas season. Sales of toys are highest by far at that time of year. Looking at a list of the volume of sales throughout the year, we can immediately recognize the Christmas season, even if we're not given any associated months of the sales. Measuring just one variable (either month or sales) gives *some* information about the other. (Low sales indicate that it's not Christmas time, and vice versa.) Similarly, the amount of water in a river (Ch. 9) typically rises to a peak in the springtime, then recedes to an annual low in late summer (at least where I live). A few sequential measurements of just one of the variables (either time of year or streamflow) give us some information about the other.

Mutual information, defined explicitly later in the chapter, appears in many fields. Like entropy, it's a commonly studied property of dynamical systems. Potential applications in chaos theory include estimating the optimum embedding dimension for attractor reconstruction (Fraser 1989a), estimating the accuracy of data (Fraser 1989a), indicating whether K–S entropy is zero or positive (and hence in revealing chaos) (Fraser 1989a, Paluŝ 1993), finding whether a time series is periodic, chaotic, or random, estimating the optimum lag to use in attractor reconstruction (Fraser 1989a, Pineda & Sommerer 1993), estimating how far into the future we can make reliable predictions, and testing data for nonlinearity (Paluŝ 1993). We'll discuss those research directions briefly, near the end of the chapter. However, most of the chapter explains the fundamentals of mutual information. The early part of the chapter establishes our foundation or tools. (We feel much more comfortable using a tool when we know how it's put together and what it's made from.) As with the entropy chapter, we'll rely very heavily on the probability and information material of Chapter 6.

Mutual information has several ingredients. Those ingredients are specialized but straightforward variants of entropy. The entropy or information discussed in the last chapter is sometimes called **self-entropy** or **self-information**, because it's the entropy of one system or variable. In other words, it's the information one system reveals about itself. This chapter extends our treatment of entropy from one system to two or more systems or dimensions. Arbitrarily, we'll couple them.

The "systems" can be anything—babies' heights and weights, dice, and so on. Also, coupling or combining systems doesn't imply anything about when we measured the data for each. We may have measured the systems at the same time (babies' heights and weights), at different times, or sequentially (a time series of one physical feature).

Joint and conditional entropies

Joint and conditional entropies, as defined in this chapter and as used in information theory, apply to *two* systems or dimensions; for more than two dimensions (discussed later in the chapter), terminology is still developing. For simplicity, we'll usually deal here with two systems. The first, system X, has discrete observations x_i; the second, system Y, has values y_j.

Chapter 6 discussed joint probability—the probability of seeing particular values of two or more variables in some sort of combination. In fact, we rolled a pair of dice (one white and one green, representing separate systems) and built a joint probability distribution. The joint probability distribution gave the probability of the joint occurrence of each possible combination of values of the two systems (Fig. 6.6). Since entropy is based solely on probability, the idea of joint probability easily applies to a so-called **joint entropy**—the average amount of information obtained (or uncertainty reduced) by individual measurements of two or more systems.

Having built a word concept of joint entropy, let's follow our standard procedure and express it in symbols. There are two cases. In the first, knowing value x from system X has no effect on the entropy of system Y. (In other words, individual observations x and y aren't mutually related.) In the second, knowing x affects our estimate of the entropy of system Y (and so x and y have some kind of association). For each case, the symbols (and hence the form of the definition) can be either those of probabilities or of entropies.

Mutually unrelated systems

Let's begin by talking about the first case, in which x and y aren't mutually related. We can express joint entropy in terms of probabilities or in terms of entropies. First we'll do the probability form, then the entropy form.

PROBABILITY FORM

In probability terms, the foundation for the symbol definition of joint entropy for two unassociated but coupled systems is Equation 6.17c. That is the general one-system equation derived in Chapter 6 for information (or entropy),

$$I_w = -\sum_{i=1}^{N_s} P_i \log P_i \,. \qquad \text{(6.17c, repeated)}$$

In that equation, I_w is information, N_s is the number of nonzero probabilities, and P_i is the ordinary probability of the *i*th bin. First we'll apply that general equation just to system X. Since it'll apply to just one system, we'll call it "self-entropy" instead of "information." We therefore replace symbol I_w with H_X (the self-entropy of system X) and the probability symbol P_i with $P(x_i)$ to indicate the probability of bin x_i. Equation 6.17c as rewritten for our system X then becomes

$$H_X = -\sum_{i=1}^{N_s} P(x_i) \log_2 P(x_i) \qquad (27.1)$$

in which the sum is over all bins that have at least one observation (and so have a nonzero probability).

Now for system Y. By analogy with Equation 27.1, we'll use $P(y_j)$ for the probability of bin y_j of system Y. The self-entropy H_Y for system Y then is

$$H_Y = -\sum_{j=1}^{N_s} P(y_j) \log_2 P(y_j). \qquad (27.2)$$

Now we want an equation for the joint entropy of the "combined" system of X and Y. Such an equation will have the same form as Equations 27.1 and 27.2. We only need to make two changes in Equations 27.1 and 27.2. First, the appropriate probabilities are no longer those of x_i or y_j alone. Instead, they are now joint probabilities—the chances associated with the joint occurrence of each possible combination of x and y. A statistician would say they are the joint probabilities that, for any given time or sequence, $x = x_i$ and $y = y_j$. The global symbol $P(x_i, y_j)$ represents the entire group of such joint probabilities. Thus, x_i and y_j are bin indices that identify any bin in the joint probability distribution.

The second change is that we now sum over all occupied bins in the joint probability distribution. Three of the more common ways of symbolizing that summation are

$$\sum_{i=1}^{N_s} \sum_{j=1}^{N_s}, \; \sum_i \sum_j, \text{ and } \sum_{ij}.$$

I'll use the first one.

With those two changes, we use the same format as Equations 27.1 and 27.2 to express a joint entropy in probability terms. The joint entropy of the two systems, $H_{X,Y}$, here is for mutually unrelated systems:

$$H_{X,Y} = -\sum_{i=1}^{N_s} \sum_{j=1}^{N_s} P(x_i, y_j) \log_2 P(x_i, y_j). \qquad (27.3)$$

Computing $H_{X,Y}$ in practice involves first preparing a joint-distribution table by empirically determining the joint probability of obtaining each possible combination of x and y, just as we did with the dice in Figure 6.6. We estimate that joint probability by looking at our basic data and counting the number of occurrences of each combination of classes of x and y, then dividing by the total number of all occurrences in the entire dataset. Next, Equation 27.3 says we need to calculate the quantity $P(x,y) \log_2 P(x,y)$ for every combination represented. Lastly, Equation 27.3 tells us to add up all those products.

The probabilities to use in computing joint entropy can be tricky. Although they are joint probabilities, they can be based on "simultaneous" associations or on sequential observations (Ch. 6). The most common situation in chaos theory is lag space. Observations x_i then become x_t, and observations y_j become x_{t+m}, where m is lag. In that case, joint probability $P(x_i,y_j)$ is the sequence variety of joint probability, $P(x_t,x_{t+m})$. That's the probability that one particular value of x follows another. It's comparable to successive-event routes AA, AB, BA, or BB, that we used as examples for two sequential time steps in Chapter 6. To compute the joint entropy, we have to estimate the sequence probability of each route or lag vector represented in the data. We do that by counting the number of occurrences of each particular route, then dividing each sum by the total number of occurrences of all observed two-event routes.

ENTROPY FORM

An alternate form shows that joint entropy for two mutually unrelated systems is the sum of their self-entropies. To see that, we'll just algebraically manipulate the probability form (Eq. 27.3) such that certain probability expressions within the restructured equation represent entropies.

Suppose you and I each flip a coin once, at the same time. There are four possible outcomes: HH, HT, TH, and TT, where the first letter (H for heads, T for tails) stands for your result and the second letter my result. With four possible outcomes, we know intuitively that, any time we flip our coins, each of the four possible outcomes has a 1 in 4 chance of happening. In other words, each outcome (HH, etc.) has a joint probability of 0.25. Let's express that in a more mathematical way. Assume that any flip has a probability of 0.5 of being heads. The joint probability that both of us get heads (HH) therefore is $0.5 \times 0.5 = 0.25$ (the same as that of the other three possible outcomes). In other words, the joint probability of any specified outcome, or $P(x_i,y_j)$, equals the ordinary probability of the one value $P(x_i)$ (here, that would be 0.5 that you get heads) times the ordinary probability of the other, $P(y_j)$ (here, 0.5 that I also get heads). Generalizing that multiplicative rule and writing it in symbol form:

$$P(x_i,y_j) = P(x_i)P(y_j). \tag{27.4}$$

Equation 27.4 lets us substitute the product $P(x_i)P(y_j)$ in place of $P(x_i,y_j)$ in Equation 27.3:

$$H_{X,Y} = -\sum_{i=1}^{N_s} \sum_{j=1}^{N_s} P(x_i)\ P(y_j) \log_2 [\ P(x_i)\ P(y_j)\ ]\,. \tag{27.5}$$

There's a log term on the far right. It has the form log(ab), in which a is $P(x_i)$ and b is $P(y_j)$. Since log(ab) = loga+logb (Appendix), we can alter that log term and rewrite Equation 27.5 as

$$H_{X,Y} = -\sum_{i=1}^{N_s} \sum_{j=1}^{N_s} P(x_i)\ P(y_j)\ [\log_2 P(x_i) + \log_2 P(y_j)\ ]\,. \tag{27.6}$$

The equation in this form says, among other things, to multiply each of the two log terms by $P(x_i)P(y_j)$. That's analogous to ab (c+d), which is abc+abd. Making that change in Equation 27.6:

$$H_{X,Y} = -\sum_{i=1}^{N_s} \sum_{j=1}^{N_s} [P(x_i)\ P(y_j) \log_2 P(x_i) + P(x_i)\ P(y_j) \log_2 P(y_j)]\,. \tag{27.7}$$

Everything to the right of the two summation signs is just the sum of two groups of products. Both summation signs apply to each of those two groups. We can show that by writing Equation 27.7 as follows:

$$\begin{aligned} {}_{X,Y} = &-\sum_{i=1}^{N_s} \sum_{j=1}^{N_s} P(x_i)\ P(y_j) \log_2 P(x_i) \\ &-\sum_{i=1}^{N_s} \sum_{j=1}^{N_s} P(x_i)\ P(y_j) \log_2 P(y_j)\,. \end{aligned} \tag{27.8}$$

The right-hand side of the equation now consists of two main parts, each involving

$$\sum_{i=1}^{N_s} \sum_{j=1}^{N_s}.$$

In that combined symbol,

$$\sum_{i=1}^{N_s}$$

applies to x values (as indicated by the subscript i), and

$$\sum_{j=1}^{N_s}$$

applies to y values (as indicated by the subscript j). Stating that same thing in symbols,

$$\sum_{i=1}^{N_s} \sum_{j=1}^{N_s} (x_i)(y_j) = \sum_{i=1}^{N_s} (x_i) \sum_{j=1}^{N_s} (y_j).$$

We therefore rewrite Equation 27.8 such that

$$\sum_{i=1}^{N_s}$$

more clearly applies to values involving x and

$$\sum_{j=1}^{N_s}$$

more clearly applies to values involving y:

$$H_{X,Y} = -\sum_{i=1}^{N_s} P(x_i) \log_2 P(x_i) \sum_{j=1}^{N_s} P(y_j) - \sum_{j=1}^{N_s} P(y_j) \log_2 P(y_j) \sum_{i=1}^{N_s} P(x_i). \quad (27.9)$$

The sum of the individual probabilities for any system is 1, so

$\sum_{i=1}^{N_s} P(x_i)$ is 1 and $\sum_{j=1}^{N_s} P(y_j)$ is 1.

Making those substitutions, Equation 27.9 reduces to

$$H_{X,Y} = -\sum_{i=1}^{N_s} P(x_i) \log_2 P(x_i) - \sum_{j=1}^{N_s} P(y_j) \log_2 P(y_j). \quad (27.10)$$

The first summed term is nothing more than the self-entropy H_X (Eq. 27.1). The second summed term is self-entropy H_Y (Eq. 27.2). Equation 27.10 therefore says simply that

$$H_{X,Y} = H_X + H_Y. \quad (27.11)$$

And that finishes our task. We've got those probabilities we started with expressed as entropies. Thus, the joint entropy of two mutually unaffiliated systems or of a pair of unrelated variables is just the sum of their self-entropies. In terms of uncertainty, the uncertainty of a joint event x,y is the uncertainty of event x plus the uncertainty of event y, as long as x and y have no mutual association.

Mutually related systems

Now for the second of our two main cases, namely the case where x and y have some kind of mutual relation. We can again express joint entropy in terms of either probabilities or entropies.

PROBABILITY FORM

Joint entropy as based on x,y pairs is a general concept that can apply to any two systems or variables, whether they're mutually affiliated or not. Equations 27.3 and 27.11 apply when they're totally disconnected. They might, on the other hand, have some kind of mutual association. For example, x might influence y. Even if the two processes aren't physically interdependent, the association means that measuring x gives some clue or information about the value of y. The probability associated with y_j therefore becomes a conditional probability—the probability of getting value y_j, given that value x_i has occurred (Ch. 6). In other words, when systems X and Y are mutually related we use a conditional probability for the second system (here Y), to reflect that affiliation.

Use of conditional probability means that $P(y_j)$ (the probability of getting the value y_j) for unrelated systems or variables (e.g. Eqs 27.2, 27.4) becomes $P(y_j|x_i)$ (the probability of getting y_j, given that x_i has occurred). So, Equation 27.4, namely $P(x_i,y_j) = P(x_i)P(y_j)$, changes to

$$P(x_i,y_j) = P(x_i)P(y_j|x_i). \tag{27.12}$$

As a result, the probability expression for joint entropy (Eq. 27.3) becomes

$$H_{X,Y} = -\sum_{i=1}^{N_s}\sum_{j=1}^{N_s} P(x_i, y_j|x_i) \log_2 P(x_i, y_j|x_i). \tag{27.13}$$

ENTROPY FORM

Deriving the entropy form, as before, is just a matter of straightforward algebra. Rewriting the joint entropy of Equation 27.6 so that it applies to the mutually affiliated case:

$$H_{X,Y} = -\sum_{i=1}^{N_s}\sum_{j=1}^{N_s} P(x_i)\ P(y_j|x_i)\ [\log_2 P(x_i) + \log_2 P(y_j|x_i)]. \tag{27.14}$$

The terms to the right of the summation signs in Equation 27.14 now consist of items having the form a (b+c). That's the same as ab+ac. Rewriting the equation that way while still applying the summations to everything:

$$H_{X,Y} = -\sum_{i=1}^{N_s}\sum_{j=1}^{N_s} P(x_i)\ P(y_j|x_i) \log_2 P(x_i)$$

$$-\sum_{i=1}^{N_s}\sum_{j=1}^{N_s} P(x_i)\ P(y_j|x_i)\log_2 P(y_j|x_i)\,. \tag{27.15}$$

The first pair of summation signs applies to the product of three items: $P(x_i)$ times $P(y_j|x_i)$ times $\log_2 P(x_i)$. Since the three items are multiplied, we can write them in any order. We'll write them as $P(x_i)\log_2 P(x_i)P(y_j|x_i)$. In that order, the first two probabilities involve x and the third involves y. With the product rearranged in that order, we now restate the entire first term so that each summation sign applies to its appropriate probabilities. The first summations to the right of the equals sign thereby become

$$-\sum_{i=1}^{N_s} P(x_i)\log_2 P(x_i)\sum_{j=1}^{N_s} P(y_j|x_i)\,.$$

Each of the two summations in that expression refers to a familiar quantity. For instance, the first summation

$$-\sum_{i=1}^{N_s} P(x_i)\log_2 P(x_i),$$

is nothing more than our friend H_X, the self-entropy of system X (Eq. 27.1). The expression therefore reduces to

$$H_X \times \sum_{j=1}^{N_s} P(y_j|x_i)\,.$$

The other summation in that expression is the sum of all the probabilities of y given x. Probabilities for any system sum to 1, so the latter sum is 1. The entire first term therefore is merely H_X times 1 or H_X. Inserting H_X in place of the first term in Equation 27.15 changes Equation 27.15 into

$$H_{X,Y} = H_X - \sum_{i=1}^{N_s}\sum_{j=1}^{N_s} P(x_i)\ P(y_j|x_i)\log_2 P(y_j|x_i)\,. \tag{27.16}$$

We can simplify that equation by rewriting the double summation. In keeping with their definitions, we'll associate the summation of i's, symbolized as

$$\sum_{i=}^{N_s},$$

with the x probabilities and the summation of j's

$$\sum_{j=1}^{N_s},$$

with the y probabilities. Rewriting the double summation that way gives

$$\sum_{i=1}^{N_s} P(x_i) \sum_{j=1}^{N_s} P(y_j|x_i) \log_2 P(y_j|x_i).$$

The first summation,

$$\sum_{i=1}^{N_s} P(x_i),$$

like all summations of probability distributions, is 1. The double summation therefore reduces to just

$$-\sum_{j=1}^{N_s} P(y_j|x_i) \log_2 P(y_j|x_i).$$

That term is a variation of the entropy of system Y as expressed in the last term of Equation 27.10. What's different about it now is that all y probabilities here are *conditional* probabilities, $P(y_j|x_i)$. For that reason, that revised entropy expression is a **conditional entropy**. A conditional entropy measures the average uncertainty in y from system Y, *given a measurement of* x from coupled system X. For two-dimensional lag space, conditional entropy measures the average uncertainty in x_{t+m}, given that we measured x_t at time t. (And again, the conditional probabilities in lag space are the probabilities that x_{t+m} will follow x_t.)

Let's use $H_{Y|X}$ as the symbol for the conditional entropy of Y,

$$-\sum_{j=1}^{N_s} P(y_j|x_i) \log_2 P(y_j|x_i).$$

Also, we'll retain $H_{X,Y}$ for the joint entropy of two systems (whether related or not). In terms of entropies, the joint entropy for two mutually associated systems (Eq. 27.16) then becomes

$$H_{X,Y} = H_X + H_{Y|X}. \tag{27.17}$$

The derivation is just as valid with the roles of the variables reversed. In that case, we symbolize the conditional entropy of X as $H_{X|Y}$. It represents the uncertainty in x as a result of knowing y. We then have

$$H_{Y,X} = H_Y + H_{X|Y}. \tag{27.18}$$

Equations 27.17 and 27.18 are both analogous to Equation 27.11 ($H_{X,Y} = H_X + H_Y$), the disassociated case. Equations 27.17 and 27.18 simply say that the joint entropy of two mutually related systems is the self-entropy of one system plus the conditional entropy of the other. In terms of uncertainty: the uncertainty of the joint event x,y is the sum of the uncertainty of x plus the uncertainty of y when we know x. Hence, for both the mutually unrelated and related cases joint entropy is the sum of two self-entropies or uncertainties—the uncertainty in x and the uncertainty in y.

Recapitulation

Let's summarize our two main accomplishments so far in this chapter.

1. We've developed four expressions for joint entropy. Two of them (a probability form and an entropy form) are for systems that have no mutual relation. The other two are counterpart forms for systems that are related. The probability form of joint entropy for two unassociated systems is Equation 27.3. The entropy form is Equation 27.11 (the sum of the two self-entropies). The probability form of joint entropy for two mutually associated systems is Equation 27.13. The entropy form is either of 27.17 or 27.18, that is, the self-entropy of one system plus the conditional entropy of the other.
2. We've stipulated what *conditional entropy* (e.g. $H_{Y|X}$) is. Our discussion actually revealed three alternative ways of writing or defining it:
   (a) In terms of probabilities, as the second of the two major terms in Equation 27.16:

$$H_{Y|X} = -\sum_{i=1}^{N_s} \sum_{j=1}^{N_s} P(x_i)\ P(y_j|x_i) \log_2 P(y_j|x_i). \tag{27.19}$$

   (b) In terms of probabilities, by replacing $P(x_i)P(y_j|x_i)$ in Equation 27.19 with its equal, $P(x_i,y_j)$, per Equation 27.12:

$$H_{Y|X} = -\sum_{i=1}^{N_s} \sum_{j=1}^{N_s} P(x_i,y_j) \log_2 P(y_j|x_i). \tag{27.20}$$

   (c) In terms of entropies, by rearranging Equation 27.17:

$$H_{Y|X} = H_{X,Y} - H_X. \tag{27.21}$$

This last equation says that the conditional entropy of system Y, given a measurement of system X, is the joint entropy of the two systems minus the self-entropy of system X. All three of the equations can also be written for $H_{X|Y}$, the conditional entropy of system X, given a measurement of y.

Mutual information

Two systems that have no mutual relation have no information about one another. Conversely, if they do have some connection, each contains information about the other. For instance, a measurement of x helps estimate y (or provides information about y, or reduces the uncertainty in y). Suppose now that we want to determine the amount (if any) by which a measurement of x reduces our uncertainty about a value of y. Two "uncertainties of y" contribute. First, y by itself has an uncertainty, as measured by the self-entropy H_Y. Secondly, there's an uncertainty of y given a measurement of x, as measured by the conditional entropy $H_{Y|X}$.

Conditional entropy $H_{Y|X}$ is a number that represents an amount of information about y. In particular, the basic uncertainty H_Y is lessened or partially relieved by an amount equal to $H_{Y|X}$. In symbols, that statement says that the overall decrease in uncertainty is $H_Y-H_{Y|X}$. The name for that difference or reduced uncertainty is *mutual information*, $I_{Y;X}$:

$$I_{Y;X} = H_Y - H_{Y|X}. \tag{27.22}$$

Using a semicolon in the subscript Y;X characterizes our symbol for *mutual* information. For *joint* probabilities or entropies, we used a comma, as in joint entropy $H_{X,Y}$. For *conditional* probabilities or entropies, we used a vertical slash, as in $H_{Y|X}$.

Alternate definitions

The technical literature and chaos theory also express mutual information in two other ways. As before, one is in terms of probabilities, the other in terms of entropies. This time we'll take the easy one (the entropy form) first.

ENTROPY FORM

The entropy form stems from Equation 27.22, $I_{Y;X} = H_Y - H_{Y|X}$. The last quantity on the right, conditional entropy $H_{Y|X}$, is defined by Equation 27.21 as $H_{X,Y}-H_X$. Substituting that difference in place of $H_{Y|X}$ in Equation 27.22:

$$\begin{aligned} I_{Y;X} &= H_Y - [H_{X,Y} - H_X] \\ &= H_Y + H_X - H_{X,Y}\,. \end{aligned} \tag{27.23}$$

In this form, mutual information is the sum of the two self-entropies minus the joint entropy.

We used the mutually associated case to derive Equation 27.23. However, Equation 27.23 also applies to the mutually unassociated case. For that case, mutual information turns out to be zero. To see that, we use the equation for the unassociated case, Equation 27.11 ($H_{X,Y} = H_X + H_Y$). Using that definition, we substitute H_X+H_Y in place of $H_{X,Y}$ in Equation 27.23. That gives $I_{Y;X} = H_Y + H_X -$

(H_X+H_Y), or $I_{Y;X}=0$. So, the mutual information of two independent systems is zero. In other words, one system tells us nothing about the other.

Incidentally, rearranging Equation 27.23 provides an alternate expression for joint entropy, $H_{X,Y}$:

$$H_{X,Y}=H_Y+H_X-I_{Y;X}. \tag{27.24}$$

Joint entropy for two systems or dimensions, whether mutually related or not, therefore is the sum of the two self-entropies minus the mutual information. (For two unrelated systems, mutual information $I_{Y;X}$ is zero. Eq. 27.24 then reduces to Eq. 27.11, $H_{X,Y}=H_Y+H_X$.)

PROBABILITY FORM

The probability form of mutual information relieves us of having to compute H_Y, H_X, and $H_{X,Y}$ (or $H_{Y|X}$ per Eq. 27.22) individually. It's a more economical form, making mutual information easier and faster to calculate. The procedure in deriving it is easy. We'll do it with Equation 27.23, which is $I_{Y;X}=H_Y+H_X-H_{X,Y}$. First, we rewrite H_Y and H_X in terms of probabilities. We do so in such a way that, insofar as possible, those probabilities consist of the same kinds of probability-terms as $H_{X,Y}$. Then we simplify the resulting expression of $H_Y+H_X-H_{X,Y}$ (Eq. 27.23) by factoring out some common terms. (And so it's our trusty three-step procedure for developing symbol definitions. We've got the concept that mutual information is the sum of the two self-entropies minus the joint entropy. Next we express that in probability symbols. Finally, we simplify.)

Let's begin with H_Y (Eq. 27.2):

$$H_Y = -\sum_{j=1}^{N_s} P(y_j)\log_2 P(y_j). \tag{27.2, repeated}$$

The crux of that equation is $P(y_j)$. Equation 6.8 defines $P(y_j)$ as

$$\sum_{j=1}^{N_s} P(x_i,y_j).$$

Substituting that into the *first* $P(y_j)$ term in Equation 27.2:

$$H_Y = -\sum_{i=1}^{N_s}\sum_{j=1}^{N_s} P(x_i,y_j)\log_2 P(y_j). \tag{27.25}$$

That gives an expression for H_Y that's longer and more cumbersome than Equation 27.2, but there's a reason for such madness. The reason is that H_Y now begins with

$$-\sum_{i=1}^{N_s}\sum_{j=1}^{N_s} P(x_i,y_j),$$

just like $H_{X,Y}$ (Eq. 27.3) does.

Next we do the same for H_X. That is, we take Equation 6.7, namely

$$P(x_i) = \sum_{j=1}^{N_s} P(x_i, y_j)$$

and substitute the right-hand side for the *first* $P(x_i)$ term in Equation 27.1:

$$_{X} = -\sum_{i=1}^{N_s} P(x_i) \log_2 P(x_i) \qquad \text{(27.1, repeated)}$$

$$= -\sum_{i=1}^{N_s} \sum_{j=1}^{N_s} P(x_i, y_j) \log_2 P(x_i) \quad . \qquad (27.26)$$

Rewriting H_Y and H_X in that fashion puts all three of H_Y, H_X, and $H_{X,Y}$ into the common form of

$$-\sum_{i=1}^{N_s} \sum_{j=1}^{N_s} P(x_i, y_j) \log_2 P(\text{A}),$$

where A represents certain bins and varies for each of the three expressions (Eqs 27.25, 27.26, and 27.3, respectively). Now we substitute Equations 27.25 (H_Y), 27.26 (H_X), and 27.3 ($H_{X,Y}$) into Equation 27.23:

$$I_{Y;X} = H_Y + H_X - H_{X,Y} \qquad \text{(27.23, repeated)}$$

$$= \text{Equation 27.25} + \text{Equation 27.26} - \text{Equation 27.3}$$

$$= -\sum_{i=1}^{N_s} \sum_{j=1}^{N_s} P(x_i, y_j) \log_2 P(y_j)$$

$$+ \left[-\sum_{i=1}^{N_s} \sum_{j=1}^{N_s} P(x_i, y_j) \log_2 P(x_i) \right]$$

$$- \left[-\sum_{i=1}^{N_s} \sum_{j=1}^{N_s} P(x_i, y_j) \log_2 P(x_i, y_j) \right].$$

That completes step two—expressing our concept in terms of symbols. The final step is to simplify. In the lengthy expression just created, the double summation

$$\sum_{i=1}^{N_s} \sum_{j=1}^{N_s}$$

and the multiplication by $P(x_i, y_j)$ are operations that apply to all three entropies. Hence, we can factor out those common ingredients and place them in front. It's like writing –ab –ac +ad as a (–b –c +d). That shortens our equation to

$${}_{\mathrm{Y;X}} = \sum_{i=1}^{N_s} \sum_{j=1}^{N_s} P(x_i,y_j)\,[-\log_2 P(x_i) - \log_2 P(y_j) + \log_2 P(x_i,y_j)\,]$$

in which, for convenience, I have switched the order of the first two log terms. (And so we've done a lot of simplifying or condensing already!)

We can do more. One of the rules of logarithms is that –log a = log(1/a). Applying that rule to the first two terms within the brackets,

$${}_{\mathrm{Y;X}} = \sum_{i=1}^{N_s} \sum_{j=1}^{N_s} P(x_i,y_j)\left[\log_2\frac{1}{P(x_i)} + \log_2\frac{1}{P(y_j)} + \log_2 P(x_i,y_j)\right].$$

We can now finish the entire escapade by using another log rule to combine the log terms within the brackets. That rule of logarithms is that log a+log b+log c = log(abc). The brackets include the summation of three such log terms. Multiplying those three log terms gives $\log_2[P(x_i,y_j)]/[P(x_i)P(y_j)]$. Uniting the log terms in that fashion brings the entire expression for mutual information to

$$I_{\mathrm{Y;X}} = \sum_{i=1}^{N_s} \sum_{j=1}^{N_s} P(x_i,y_j)\log_2\frac{P(x_i,y_j)}{P(x_i)\;P(y_j)}\,. \qquad (27.27)$$

That step completes our derivation. It puts mutual information into a condensed symbol form, using probabilities. The final result (Eq. 27.27) doesn't look like $H_{\mathrm{Y}}+H_{\mathrm{X}}-H_{\mathrm{X,Y}}$ (Eq. 27.23), but it is. (That's what we started out with, several paragraphs ago.)

The log term in Equation 27.27, namely $\log_2[P(x_i,y_j)/P(x_i)P(y_j)]$, involves a ratio. The i and j subscripts tell us to consider each bin individually, in turn, in the joint probability distribution. The numerator of the ratio is the joint probability of a given bin, $P(x_i,y_j)$. The denominator is the product of the two marginal probabilities. That is, it's the product of $P(x_i)$ (the marginal probability of state x_i) times $P(y_j)$ (the marginal probability of state y_j). Also, Equation 27.27 says to multiply the log of that ratio by $P(x_i,y_j)$. We therefore end up using each joint probability $P(x_i,y_j)$ twice in the overall calculations pertaining to each bin—first as the numerator in the ratio and then again as the multiplier. Finally, once we have the product of $P(x_i,y_j)$ times $\log_2[P(x_i,y_j)/P(x_i)P(y_j)]$ for every observed bin in the joint probability distribution, Equation 27.27 says to add up all those products.

Mutual information is a global measure in that it uses probabilities that we have measured over the entire attractor. (We have at least measured those probabilities over our entire dataset, and we hope and assume those data represent the entire attractor.)

In lag space, variable x_i becomes x_t and y_j becomes x_{t+m}. The joint probabilities $P(x_i,y_j)$ of Equation 27.27 then become sequence probabilities, $P(x_t,x_{t+m})$. That's the estimated probability that value x_{t+m} follows value x_t in the time series. Equation 27.27 requires an estimate of that probability for every observed phase space route. Also, $I_{Y;X}$ as computed with Equation 27.27 has the by-now-familiar basic "$P \log P$" form. As such, it's a weighted average value—a weighted average mutual information. (Abarbanel et al. 1993 and some other authors in fact call it "average mutual information." For brevity, however, I'll just call it mutual information.)

Figure 19.9 dealt with estimating optimum lag in attractor reconstruction by the method of the first minimum in mutual information. Either Equation 27.27 or 27.23 is the equation to use for that purpose.

Interpretations

Figure 27.1 helps clarify the general concepts of self-entropy, conditional entropy, joint entropy, and mutual information for two systems. Circles in the figure represent self-entropies H_X and H_Y. The first case (part (a) of the figure) is that in which x and y aren't related. The joint entropy is the sum of areas H_X and H_Y (per Eq. 27.11). Neither of those entropies contains information about the other. Mutual information therefore is zero (the smallest possible value).

As an aside, let's take a short paragraph to see how a mutual information of zero can occur in terms of probabilities. Equation 27.27 says that zero mutual information can only come about when $\log_2[P(x_i,y_j)/P(x_i)P(y_j)] = 0$ for every bin in the joint probability distribution. That log term is 0 for any one bin when the numerator of the log term equals the denominator, that is, when $P(x_i,y_j) = P(x_i)P(y_j)$ for that bin. The reason is that when the two are equal, their ratio is 1, and the log of 1 is 0. In other words, if the joint probability $P(x_i,y_j)$ equals the product of the marginal probabilities $P(x_i)P(y_j)$ for every bin, the two systems are statistically independent. Mutual information then is zero.

Now suppose that the two systems or variables have some sort of loose association. That means that measuring one variable gives some information about (perhaps even a rough approximation of) the other. Figure 27.1b shows the same two self-entropies as Figure 27.1a. The amount of overlap of the two self-entropies indicates the degree of affiliation between systems. That overlap zone represents the mutual information, $I_{Y;X}$. The finite value of mutual information (or existence of an overlap zone) means that measuring just one variable (i.e. without measuring the other) reduces our uncertainty in the other. For instance, whereas our uncertainty in y was a maximum (H_Y) in the unrelated case (Fig. 27.1a), it decreases to the conditional entropy $H_{Y|X}$ (the uncertainty in y after knowing x) for the mutually related case. The magnitude of that lesser uncertainty ($H_{Y|X}$) is the area of the hachured overlap zone in Figure 27.1b. The amount of reduction in uncertainty is the size of the overlap zone. In symbols, that's $H_Y - H_{Y|X}$, and it's the mutual information (Eq. 27.22). The same ideas also apply to x.

Joint entropy in Figure 27.1b is the area within the outer perimeter of the combined systems. Any of Equations 27.17, 27.18, or 27.24 gives its magnitude.

The largest possible value of mutual information (not shown diagrammatically

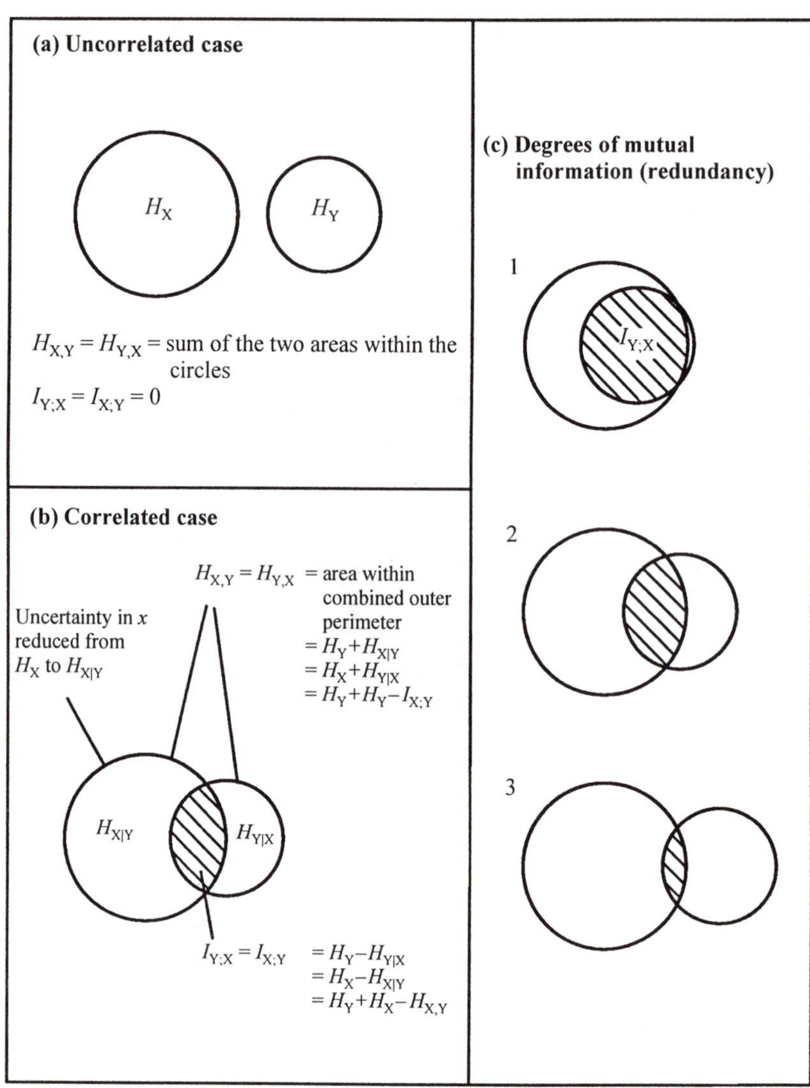

Figure 27.1 Sketches showing relations between self-entropy, joint entropy, conditional entropy, and mutual information (redundancy), for two systems, X and Y. (a) Uncorrelated case. (b) Correlated case (in which knowing a value y reduces the uncertainty in x, and vice versa). Modified and expanded from Cover & Thomas (1991: fig. 2.2). (c) Differing amounts of mutual information or redundancy. Part 1 has a high value of mutual information (high correlation between x and y). Part 3 has a low value of mutual information (weak correlation between x and y).

in Fig. 27.1) occurs when x completely determines y and vice versa. For that case, as a few trials on a joint distribution grid show, two results are certain. First, the sums of the marginal probabilities of the two systems are the same, so the self-entropies of the two systems are the same (Eqs 27.1 or 27.2). Secondly, the sum of the joint probabilities

$$\sum_{i=1}^{N_s} \sum_{j=1}^{N_s} P(x_i y_j),$$

equals the sum of the marginal probabilities for either system, for example

$$\sum_{i=1}^{N_s} P(x_i).$$

Hence, the joint entropy per Equation 27.3 equals the self-entropy of either system. For instance, $H_{X,Y} = H_X$. When that occurs, we can substitute H_X in place of $H_{X,Y}$ in the general relation $I_{Y;X} = H_Y + H_X - H_{X,Y}$ (Eq. 27.23). That leaves just $I_{X;Y} = H_Y$. (Also, $H_{X,Y} = H_Y$, so $I_{Y;X} = H_X$.) So when one system uniquely determines the other, mutual information is just the self-entropy of one system.

Overall, then, mutual information ranges from a minimum of zero (no relation between the two systems) to a maximum of the self-entropy of either system (which occurs at the limiting condition of one system uniquely determining the other).

With the above background, let's look at three ways to interpret mutual information:

- One interpretation is in terms of uncertainty. Mutual information is the amount by which a measurement of one variable reduces the uncertainty in another (Fig. 27.1b). In other words, it's the reduction in the uncertainty of one random variable as a result of knowing the other (Cover & Thomas 1991: ch. 2).
- A second interpretation is in terms of information. Three ways to restate the general idea are as follows. Mutual information is:
  - A quantitative measure (in bits) of the average amount of common information between two variables. The overlap zones in Figure 27.1c, discussed below, demonstrate this concept. This interpretation shows that we can write our symbol for mutual information either as $I_{Y;X}$ or $I_{X;Y}$, as in the legends of Figures 27.1a,b.
  - A measure of the average amount of information contained in one system, process, or variable about another. That is, it's the average amount of information x gives about y. If mutual information is relatively minor, x doesn't contain much information about y. In fact, a mutual information of zero means we can't find out a thing about y by measuring x. In other words, x and y are independent (Fig. 27.1a). Conversely, where mutual information is large, x contains much information about y.
  - The information contained in a process or variable (e.g. H_Y) minus the

information contained in that variable when we've measured another variable (e.g. $H_{Y|X}$) (per Eq. 27.22).

- From a third point of view, mutual information measures the degree of association between two processes or random variables. For instance, a high value for mutual information (large reduction in uncertainty) implies that x and y have a close mutual affiliation (Fig. 27.1c, part 1). Conversely, a low value for mutual information indicates that x and y have only a weak affiliation (Fig. 27.1c, part 3). In fact, a value of zero for mutual information means that x and y are independent of one another (Fig. 27.1a). (The notion of the degree of association between two variables again shows why we can symbolize mutual information alternatively as $I_{Y;X}$ or $I_{X;Y}$.)

All of the above interpretations can be applied to lag space. In the context of lag space, mutual information is:

- The amount (in bits) by which a measurement of x_t reduces the uncertainty of x_{t+m}.
- The amount of information that one measurement gives about a later measurement of the same variable (Fraser & Swinney 1986). Here Equation 27.27 gives the average mutual information between observations at time t and those at time $t+m$. In other words, it's the average amount of information about x_{t+m} that we get by making observation x_t.
- A measure of the degree to which knowing x at time t specifies x at time $t+m$. In other words, it measures predictability—how well we can predict x at time $t+m$, given that we measured x at time t.

Redundancy

Within chaos theory, **redundancy** (explained below) is nothing more than the extension of mutual information to three or more dimensions. Even then, some chaologists don't use the term "redundancy" that way and prefer, instead, to use "mutual information" for any number of dimensions. As usual, terminology is still developing. In any case, we'll develop our tools in the rest of this chapter.

First, here's how the idea of "redundancy" applies. Suppose x and y have some kind of direct (straight-line) relation and we know the equation of the line. Measuring one variable then tells us the other exactly. In that case, measuring the second variable as well as the first is a waste of time. It merely duplicates information we already got by measuring the first variable. Measuring the second variable in that case is unnecessary, superfluous, or *redundant*. Thus, when measurements of one system provide complete information about the other (i.e. reduce the uncertainty in the other to zero), then mutual information is a maximum. At the same time, all measurements of the second system are wholly redundant. That is, both mutual information and *redundancy* are a maximum.

Now let's go to the other extreme, namely where there's no relation at all between the two variables. Measuring *x* for unrelated systems (Fig. 27.1a) provides absolutely no information at all about *y*. Mutual information then is zero, and there's nothing redundant (zero redundancy) in measuring the second variable as well as the first. So, as before, we can think of mutual information and redundancy as the same thing.

For intermediate conditions, *x* provides some but not all information about *y* (and vice versa). Mutual information then is at some intermediate value, and measurements of the second variable might be described as partly redundant (Fig. 27.1b).

To recapitulate, what we've done in the last three paragraphs is lay a general foundation between mutual information and the idea of redundancy. Two systems that have a perfect one-to-one relationship have maximum mutual information. All measurements of the second system in that case are completely redundant (have maximum redundancy). For partially affiliated systems, mutual information is less. Measurements of the second system then are partly redundant. Finally, with unrelated systems, mutual information is zero. Measurements of the second system therefore aren't the least bit redundant (redundancy also is zero).

Figure 27.1c summarizes those concepts graphically. In part 1 of the figure, two systems or variables are closely related. Measuring one variable enables us to estimate the other quite well, without even having to measure that other variable. Measuring the second variable in addition to measuring the first is largely redundant. The relatively large area taken up by the mutual information on the sketch (i.e. the large value of mutual information) reflects that high degree of redundancy. In part 2 of the figure, the interrelation between systems is less. Measurements of the second variable then are still redundant to some extent, but not as much as in part 1. The lesser area represented by the mutual information indicates that lesser redundancy. Finally, in part 3, measuring one variable gives only a vague idea of the other. In that case, there's very little redundancy associated with measuring the second variable in addition to the first. Mutual information also is small.

The point here is that, for the two-system case, mutual information measures (and equals) redundancy. Any value for mutual information also is a quantitative measure of redundancy. Redundancy thereby joins company with entropy, information, and mutual information as a numerical measure of an ensemble of probabilities.

Derivations of redundancy

ORIGINAL DEFINITIONS

Original definitions of redundancy came from information theory, long before chaos came along. Those definitions aren't exactly the same as the one used in chaos theory. In other words, the idea of redundancy as mutual information isn't the only concept of "redundancy" in the literature. However, most schools of thought, within both information theory (e.g. Shannon & Weaver 1949: 56) and chaos theory (e.g. Gershenfeld & Weigend 1993: 48), define redundancy in ways that are closely related to the "chaotic" ones. For background, let's take a quick look at how information theory usually defines redundancy. There are two ways. Each can be derived from the other.

The first definition is:

redundancy = the difference between some reference entropy and an actual or observed entropy. (27.28)

Depending on the circumstances or problem, we can choose different definitions for "actual entropy" and "reference entropy." For instance, for one system or variable, redundancy might be the difference between the maximum possible entropy (the reference entropy) and the actual or self-entropy, say H_X. Maximum possible entropy occurs when the probability of every bin is the same; in that case it equals $\log N_s$, where N_s is the number of possible states or bins. Redundancy by the above definition then would be the difference between the reference entropy (here $\log N_s$) and self-entropy (here H_X), or $(\log N_s)-H_X$.

The second definition is a standardized version of the first definition. The standardization is done by dividing the basic difference by the reference entropy. That puts all redundancies within the range of 0 to 1. In the example just used, standardized redundancy becomes definition one divided by the reference entropy, or $(\log N_s-H_X)/\log N_s$. That ratio, according to the rules of basic algebra, is $(\log N_s/\log N_s)-(H_X/\log N_s)$, or $1-(H_X/\log N_s)$. Thus, the second definition of redundancy as used in information theory is:

redundancy = 1–(*actual entropy/reference entropy*).

The ratio of the actual entropy to the reference entropy is a type of **relative entropy** (the ratio of one entropy to another).

CHAOS THEORY DEFINITION IN TERMS OF ENTROPIES

Now we advance to chaos theory definitions of redundancy. In that context, as mentioned above, "mutual information" and "redundancy" are essentially the same thing. However, following Fraser (1986), I'll use "mutual information" only for *two* (and only two) variables, systems, or dimensions. "Redundancy," on the other hand, will refer to *three or more* variables, systems, or dimensions. That is, redundancy will be the generalization of mutual information from two dimensions to three or more dimensions.

Further, we'll usually apply "redundancy" to lag space. Let's now generalize our equation for mutual information (Eq. 27.23) so that it expresses redundancy (that is, it applies to three or more dimensions) and applies to lagged values of one feature (x). In this section we'll do it in terms of entropies, in the next in terms of probabilities.

Mutual information is the starting point. By definition, it involves two dimensions. It's just the sum of the two self-entropies minus the joint entropy (Eq. 27.23):

$$I_{X;Y} = H_X + H_Y - H_{X,Y}. \quad \text{(27.23, repeated)}$$

For lag space, system X becomes X_t and system Y becomes X_{t+m}. To avoid the inconvenience of having a "sub-subscript," we'll write $H(X_t)$ for H_{X_t} (the self-entropy of system X_t) and make similar changes for the other symbols. With those modifications, Equation 27.23 as applied to lag space becomes

$$I(X_t;X_{t+m}) = H(X_t)+H(X_{t+m})-H(X_t,X_{t+m}).$$

In getting our new definition (redundancy), we follow our standard three-step procedure for defining a new quantity. The first step is to conceptualize. Here the concept is that redundancy (the multidimensional version of mutual information) is the sum of three or more self-entropies minus the joint entropy.

Step two is to express that idea in symbols. The modified version of Equation 27.23 that we just wrote is for just two dimensions (X_t, X_{t+m}). Now we want to generalize that relation to D dimensions, where D is three or more. Embedding dimension D represents D axes in lag space or D lagged measurements of x. The last member of the group of entropies then corresponds to $H(X_{t+(D-1)m})$, as explained in the vector chapter. For instance, the symbol for mutual information, $I(X_t;X_{t+m})$, (spoken as "I of X sub t and X sub t plus m") when generalized to redundancy R in that way, becomes $R(X_t,X_{t+m}, \ldots ,X_{t+(D-1)m})$. Similarly, the entire symbol definition for redundancy, based on our modified version of Equation 27.23, becomes:

$$R(X_t,X_{t+m}, \ldots ,X_{t+(D-1)m}) = H(X_t)+H(X_{t+m})+\ldots+H(X_{t+(D-1)m}) \\ -H(X_t,X_{t+m}, \ldots , X_{t+(D-1)m}).$$

So, the sum of the self-entropies is symbolized by $H(X_t)+H(X_{t+m})+\ldots+ H(X_{t+(D-1)m})$. Similarly, the joint entropy that we're subtracting from that sum is $H(X_t,X_{t+m}, \ldots ,X_{t+(D-1)m})$.

Finally, step three of our standard procedure is to simplify and condense. We can do that by letting

$$\sum_{i=1}^{N} H(x_i)$$

represent the sum of the self-entropies. That is,

$$\sum_{i=1}^{N} H(x_i) = H(X_t)+H(X_{t+m})+\ldots+H(X_{t+(D-1)m}).$$

In that global summation symbol, N is the total number of self-entropies, $H(X_i)$ is the group of self-entropies, and $i = t, t+m, \ldots, t+(D-1)m$. That step condenses our symbol definition for redundancy in terms of entropies to:

$$R(X_t, X_{t+m}, \ldots, X_{t+(D-1)m}) = \left[\sum_{i=1}^{N} H(x_i)\right] - H(X_t, X_{t+m}, \ldots, X_{t+(D-1)m}). \quad (27.29)$$

CHAOS THEORY DEFINITION IN TERMS OF PROBABILITIES

Another common chaos theory definition of redundancy is in terms of probabilities. Here we'll again follow our three-step procedure but only as a first stage, namely to get a definition applicable to two systems or variables. Then we'll generalize it to lag space and the multidimensional case.

For step one (the concept), we invoke the above idea that redundancy is a reference entropy minus an actual entropy (Eq. 27.28). Step two is to express that in probability symbols. For that purpose, we use a reference entropy based on marginal probabilities (Ch. 6):

$$-\sum_{i=1}^{N_s}\sum_{j=1}^{N_s} P(x_i, y_j) \log_2 [P(x_i)P(y_j)].$$

The actual or observed entropy is the joint entropy,

$$-\sum_{i=1}^{N_s}\sum_{j=1}^{N_s} P(x_i, y_j) \log_2 P(x_i, y_j). \qquad \text{(Eq. 27.3, repeated)}$$

Redundancy R then is

R = *reference entropy* minus *actual entropy*

$$= -\sum_{i=1}^{N_s}\sum_{j=1}^{N_s} P(x_i, y_j) \log_2 [P(x_i)P(y_j)] - \left[-\sum_{i=1}^{N_s}\sum_{j=1}^{N_s} P(x_i, y_j) \log_2 P(x_i, y_j)\right].$$

The third and final step is to simplify. The actual entropy (the expression on the far right) in that equation is a negative quantity that's subtracted from something. That makes it positive, so we rewrite the preceding equation by putting the actual entropy first:

$$R = \sum_{i=1}^{N_s}\sum_{j=1}^{N_s} P(x_i, y_j) \log_2 P(x_i, y_j) - \sum_{i=1}^{N_s}\sum_{j=1}^{N_s} P(x_i, y_j) \log_2 [P(x_i)P(y_j)].$$

Common features of the two entropies on the right-hand side of the equation are

$$\sum_{i=1}^{N_s}\sum_{j=1}^{N_s} P(x_i, y_j).$$

In other words, the expression is comparable to ab–ac. That's a(b–c). Thus:

$$R = \sum_{i=1}^{N_s} \sum_{j=1}^{N_s} P(x_i,y_j) \{ \log_2 P(x_i,y_j) - \log_2 [P(x_i)P(y_j)] \}.$$

Since $\log a - \log b = \log(a/b)$,

$$R = \sum_{i=1}^{N_s} \sum_{j=1}^{N_s} P(x_i, y_j) \log_2 \frac{P(x_i, y_j)}{P(x_i) P(y_j)}. \qquad \text{(27.27, repeated)}$$

That expression for "redundancy" is a bit of a misnomer because it applies only to two systems, as indicated by the x's and y's. In fact, I've labeled it "Equation 27.27" because it's the same as Equation 27.27 for mutual information. Incidentally, the log term is another relative entropy, in this case the ratio of the entropy of the joint probability distribution to that of the "product" probability distribution. The products are those of the marginal probabilities (cf. Cover & Thomas 1991: 18, eq. 2.28).

Let's now generalize the simplified equation we just got to *lag space* and three or more dimensions. Lag space means we're using time-sequential values of one variable. That in turn means we need sequential probabilities. For lag space, we reinterpret the symbols in the usual way: system X becomes X_t, system Y becomes X_{t+m}, and the final system in our group is $X_{t+(D-1)m}$; for actual observations, x_i becomes x_t and y_j becomes x_{t+m}; we indicate data from additional "systems" or "variables" by additional lags, up to $x_{t+(D-1)m}$; and our former summation over all possible values of i and j translates to summing over all represented phase space routes, N_r. Incorporating those symbols lets us generalize Equation 27.27 to D dimensions as follows:

$$R(X_t, X_{t+m}, \ldots, X_{t+(D-1)m}) \qquad (27.30)$$

$$\sum_{i=1}^{N_r} P(x_t, x_{t+m}, \ldots, x_{t+(D-1)m}) \log_2 \frac{P(x_t, x_{t+m}, \ldots, x_{t+(D-1)m})}{P(x_t)\ P(x_{t+m}) \ldots P(x_{t+(D-1)m})}.$$

That equation is our goal—redundancy for lag space, in terms of probabilities.

The joint probability $P(x_t, x_{t+m}, \ldots, x_{t+(D-1)m})$ that occurs twice in Equation 27.30 is deceptive. It's a global symbol that can encompass a *very large* number of probabilities. Each variable or dimension (x_t, x_{t+m}, etc.) represents an additional "event" or measurement in the time sequence. Number of events in any specified route or sequence greatly increases the number of sequence probabilities we need to estimate, according to Equation 6.6: $N_r = N_s^n$, where N_r is number of phase space routes, N_s is number of possible states, and n is number of events. We need a sequence probability for every route, or at least for those routes that occur at least once. Therefore, redundancy for more than two or three events or embedding dimensions usually requires estimating many thousands of joint (sequence) probabilities. That, in turn, means the dataset (time series) must be of colossal length.

Incremental redundancy and its interpretations

The definitions given above form the basis for the interpretations of redundancy that I'll mention here. Those definitions generally stem from or follow Fraser's (1989a) ideas. However, some concepts and definitions aren't yet standardized because the field is relatively young. For example, Gershenfeld & Weigend (1993: 48) define and interpret redundancy differently from Fraser.

Redundancy as defined in Equation 27.30 quantifies the average amount of common information among several systems or variables (Paluŝ 1993). In the context of Figure 27.1, for instance, we'd be looking at the overlap zone of not just two spheres but three or more spheres. If the variables are lagged measurements of one time series (e.g. x_t, x_{t+1}, x_{t+2}), redundancy is the number of bits that are redundant in a vector (Fraser & Swinney 1986), that is, redundant in the vector's components as a group.

Chaos theory sometimes focuses on the amount of change in the redundancy of a vector when we increase the vector's dimension by one (e.g. from two dimensions to three), at a constant lag. That quantity is just the difference between the two successive redundancies. That's the redundancy at the higher dimension minus that at the lower. (And so it's really just another version of first-differencing, discussed in earlier chapters.) Some authors symbolize it as $R_{D+1}-R_D$, others as R_D-R_{D-1}. Fraser (1989a) coined the term **marginal redundancy** for each such difference. I'm going to call it **incremental redundancy**. (Like "fractal dimension," terminology for mutual information and redundancy can be inconsistent from one author to another.)

One way to interpret incremental redundancy when dealing with lag space is in terms of information. Incremental redundancy quantifies the average amount of information that several successive measurements of x give about the next x (Paluŝ 1993: 390). It's the average amount of information that n sequential observations contain about observation $n+1$. (In these interpretations, each measurement gets a phase space dimension, so there are D dimensions providing information about dimension or observation number $D+1$.)

Another lag space view of incremental redundancy is in terms of prediction. Here incremental redundancy is a quantitative measure of the average number of bits that several sequential measurements of x can predict about the next x. That is, it estimates the number of bits that the earlier components of a vector can forecast about the last component. Hence, it reflects predictability—how well the last member of a measurement sequence can be predicted from its predecessors. The minimum value of incremental redundancy (zero) means that the data are useless in regard to predictions; we might just as well throw a dart at the wall or guess any number within the range of possible values. In contrast, a high value means close association and reliable predictions.

Typical values

Because redundancy depends strictly on probabilities, anything that affects those probabilities affects the value of redundancy. Five such influential factors are the number of possible states (bin size), number of sequential events (embedding dimension), lag, distribution of probabilities, and size of dataset. The last two—distribution of probabilities and size of dataset—are fixed once we've measured the time series. Hence, in analyzing a given time series for redundancy, we can regulate only the first three—number of possible states (the original partitioning), embedding dimension, and lag. Varying each of those three factors yields a different set of probabilities and hence a new value for redundancy. Therefore, each value of redundancy that we calculate pertains only to a particular combination of bin size, lag, and embedding dimension.

Suppose we choose a particular bin width or partitioning. Then redundancy varies only with lag and embedding dimension. The usual procedure is to choose a lag and compute redundancy for each of several successive embedding dimensions. In keeping with the philosophy of mutual information and redundancy, we need at least two measurements (two embedding dimensions) to compute a meaningful redundancy. (Redundancy for an embedding dimension of one is zero.) The first meaningful embedding dimension to use is two. At lag one, that means x_t and x_{t+1}. So, we'd first go methodically through the entire dataset to estimate the probabilities. Then we'd compute redundancy (actually mutual information, since $D = 2$). After that, we compute redundancy at that same lag (one) for three embedding dimensions (x_t, x_{t+1} and x_{t+2}). Then for lag one at four embedding dimensions, five embedding dimensions, and so on. Data limitations may prevent computations of redundancy for more than about five or ten embedding dimensions, but it's worth doing as many as possible.

Once we've computed redundancy for successive embedding dimensions at lag one, the next step is to go to the next lag (lag two) and again compute redundancy for various successive embedding dimensions. Then repeat for lags three, four, and so on. Typically, we advance lag by increments of one until lag reaches about 10 or 15, then by increments of 5 thereafter. Again, data limitations may prevent calculations at lags of more than 30 or 40.

To show what some typical values look like, Table 27.1 lists redundancies and incremental redundancies based on the x values of 9000 iterations of the Hénon equations (Eq. 13.11). Parameter values used in the equations were a = 1.4 and b = 0.3. Those values produce a chaotic attractor. The table includes computations for four embedding dimensions at lags 1–15, then continues by lags of 5 through lag 50. At a given lag, incremental redundancy is the amount by which the redundancy changes with each successively higher embedding dimension. That is, it's the redundancy at a particular embedding dimension minus that at the preceding embedding dimension. For instance, at lag one, incremental redundancy for an embedding dimension of 4 is the redundancy at $D = 4$ (having a value of 21.04, listed in col. 5 of Table 27.1) minus the redundancy at $D = 3$ (value 14.31, in col.

Table 27.1 Redundancies and incremental redundancies computed from 9000 iterations of the Hénon equations (lagged values of x only; y values not involved); redundancy calculations provided by Andrew Fraser.

1	2	3	4	5	6	7	8
	Redundancies				Incremental redundancies		
	Embedding dimension				Embedding dimension		
Lag	1	2	3	4	2	3	4
1	0.00	6.43	14.31	21.04	6.43	7.88	6.73
2	0.00	5.76	12.27	16.89	5.76	6.51	4.62
3	0.00	4.80	9.97	12.14	4.80	5.17	2.17
4	0.00	4.17	7.39	8.26	4.17	3.22	0.87
5	0.00	3.03	5.35	5.42	3.03	2.32	0.07
6	0.00	2.38	3.76	3.60	2.38	1.38	–0.16
7	0.00	1.83	2.61	2.32	1.83	0.78	–0.29
8	0.00	1.43	1.93	1.67	1.43	0.50	–0.26
9	0.00	1.03	1.24	1.01	1.03	0.21	–0.23
10	0.00	0.65	0.72	0.62	0.65	0.07	–0.10
11	0.00	0.52	0.59	0.65	0.52	0.07	0.06
12	0.00	0.44	0.47	0.44	0.44	0.03	–0.03
13	0.00	0.28	0.31	0.40	0.28	0.03	0.09
14	0.00	0.16	0.25	0.37	0.16	0.09	0.12
15	0.00	0.09	0.11	0.16	0.09	0.02	0.05
20	0.00	0.00	0.04	0.02	0.00	0.04	–0.02
25	0.00	0.00	0.02	0.07	0.00	0.02	0.05
30	0.00	0.01	0.00	0.06	0.01	–0.01	0.06
35	0.00	0.01	0.03	0.09	0.01	0.02	0.06
40	0.00	0.00	0.00	0.06	0.00	0.00	0.06
45	0.00	0.00	0.03	0.11	0.00	0.03	0.08
50	0.00	0.00	0.01	0.03	0.00	0.01	0.02

4). That difference, $R_{D+1}-R_D$, is 6.73 (col. 8). The calculations are very crude; for instance, negative incremental redundancies, as columns 7 and 8 show for some greater lags, aren't possible theoretically.

Applications

Chaologists have suggested several applications of redundancy and incremental redundancy. Fraser's (1989a) work is the basis for the following idealized descriptions. The potential applications we'll discuss are:

- estimating optimal embedding dimension
- assessing the accuracy of the basic data
- estimating Kolmogorov–Sinai entropy
- distinguishing between periodic, chaotic and random data
- estimating optimal lag
- assessing predictability.

Each application carries certain qualifications that we needn't go into here. Also, all applications are exploratory and need further development to be easily and reliably usable (cf. Grassberger et al. 1991, Martinerie et al. 1992, Pineda & Sommerer 1993).

EMBEDDING DIMENSION

Two successive measurements of *random* data have no mutual relation. One value carries no information about the next. Their mutual-information values or redundancies are zero. The difference between two such redundancies (i.e. the incremental redundancy) therefore also is zero. This holds true regardless of embedding dimension (D) or lag. So, for any and all embedding dimensions with random data, any measurement of x is useless for predicting the latest x within a series (the latest component of a vector). (I'll sometimes speak of incremental redundancy as the difference between two successive redundancies, with symbol ΔR. At other times, I'll speak of it as the average number of bits that several sequential measurements of x can predict about the next x, that is, about the last component of a series of measurements.)

Data (chaotic or not) based on an underlying rule, on the other hand, have some association (apparent or not) between successive measurements of x. Such measurements therefore carry some mutual information or redundancy about each other. For these kinds of data, redundancy for any series of measurements (any embedding dimension) increases as we add another measurement (another embedding dimension), for a fixed lag. For example, three successive values (x_t, x_{t+m}, x_{t+2m}) are more redundant than two, four are more redundant than three, and so on. In other words, computed values of redundancy R increase with increase in embedding dimension, for such data (at fixed lag). Incremental redundancy in those cases is positive (>0), at least in theory.

One or two measurements of x (i.e. low embedding dimensions) may not give us a very good prediction about the last component of our string. In other words, incremental redundancies at small embedding dimensions tend to be small, for a fixed lag. Predictability of the last member of a series of measurements (the incremental redundancy) increases rather rapidly at first, as we incorporate more measurements or components into the vector, that is, as we increase the embedding dimension (Fig. 27.2). Further increases in the number of components bring a lesser and lesser gain in predictability of that last member. Eventually the incremental redundancy becomes approximately constant. A constant incremental redundancy means that further increases in embedding dimension don't improve the ability of the sequence of measurements to predict the last member of the string. Hence, there isn't any advantage to using an even larger embedding dimension. (The overall pattern is similar to that between the correlation dimension and embedding dimension, as we saw in Fig. 24.6.)

The best embedding dimension to use for attractor reconstruction with this approach is the one at which incremental redundancy no longer increases significantly, at that lag and measuring accuracy. One way to estimate that embedding

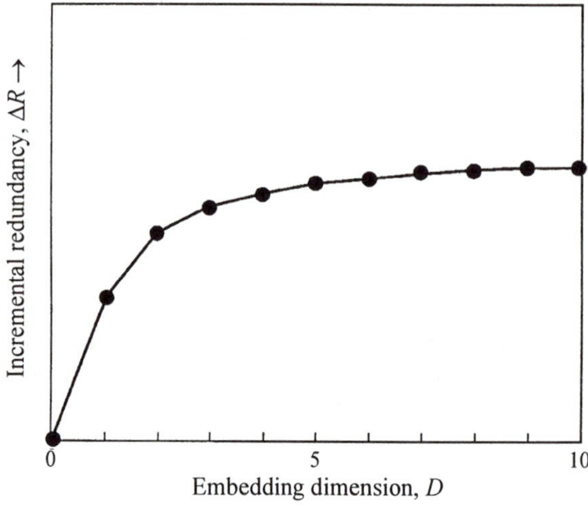

Figure 27.2 Hypothetical change of incremental redundancy with embedding dimension, at a fixed lag.

dimension is from a graph of incremental redundancy as a function of embedding dimension, with lag as a third variable. (A simplified version of such a graph is Fig. 27.2, where we analyzed only one lag.)

A second type of graph for estimating an optimum embedding dimension is a plot of incremental redundancy as a function of lag, with embedding dimension as a third variable. Figure 27.3 shows hypothetical examples. Figure 27.4 (from Paluŝ 1993) shows a similar plot for real data—laboratory measurements of fluctuations in a far-infrared laser. On such graphs, relations for successively higher embedding dimensions become closer and closer to one another. When they become very close to one another, incremental redundancy at a given lag stays about the same (nearly constant) for still larger embedding dimensions (Fig. 27.2). In other words, there's little advantage in going to a larger embedding dimension. The optimum embedding dimension as estimated from graphs such as Figures 27.3 and 27.4 therefore is the smallest one for which the plotted relations get relatively close to one another. That's a subjective decision on our part.

ACCURACY

Figures 27.3 and 27.4 show that the relations of incremental redundancy versus lag not only become closer as embedding dimension increases; they also become straighter. In fact, at small lags the lines converge toward some limiting perfectly straight "asymptotic accumulation line" as D goes to infinity. That asymptotic accumulation line, shown as a dashed line on Figure 27.3, is important.

Straight lines have the form $y = \mathrm{c} + \mathrm{b}x$, where y is the quantity plotted on the ordinate, x is on the abscissa, c is the intercept, and b is the slope. For our case, y is incremental redundancy ΔR and x is delay or lag, m. The straight line's equation therefore is $\Delta R = \mathrm{c} + \mathrm{b}m$. Ordinarily, we would fit a straight line by some rigorous analysis of the data to find the parameters b and c. Our data unfortunately are noisy, less plentiful than we might prefer, and probably won't let us examine conditions

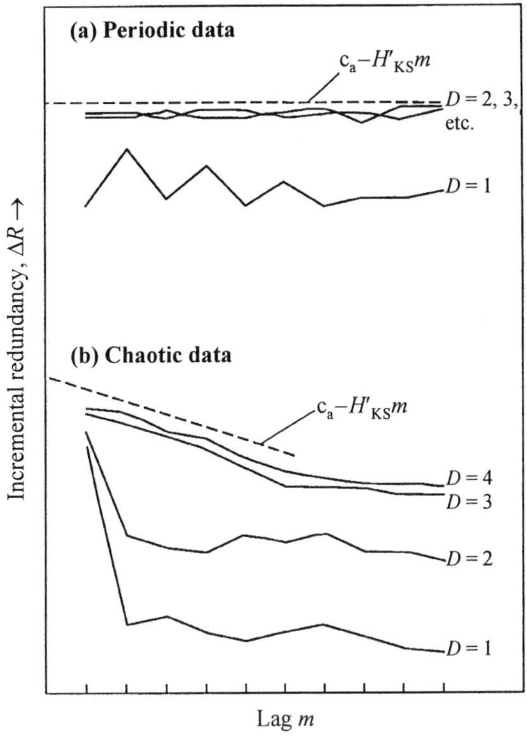

Figure 27.3 Idealized sketch of incremental redundancy as a function of lag.

of D nearing infinity. Therefore, we can't compute b and c directly. As a result, we have to fit the straight asymptotic accumulation line by eye. To do that, we first compute the relation for ΔR as a function of lag for larger and larger embedding dimensions D. Then, when those relations get very close to one another on the graph, we make a best guess as to where their limit at very large D (the straight asymptotic accumulation line) will plot, and we draw it in by eye. (For chaotic data,

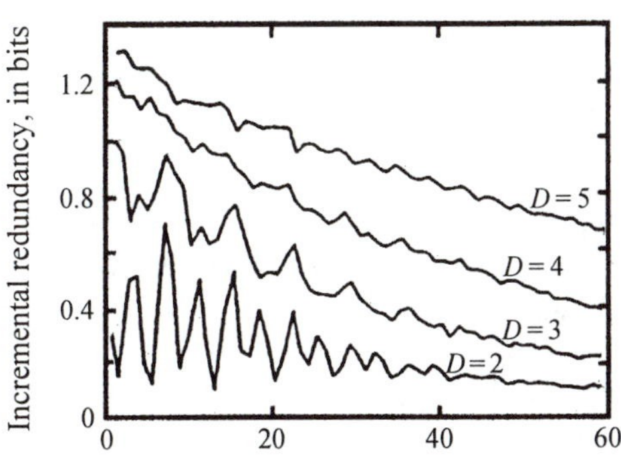

Figure 27.4 Incremental redundancy as a function of lag for measurements of fluctuations in far-infrared lasers (adapted from Paluŝ 1993, with permission from Addison–Wesley–Longman Publishing Co.).

as in Fig. 27.3b, those straight sections only last over small lags. The asymptotic accumulation lines for chaotic data therefore apply only to those small lags.) Finally, we measure the values of slope b and intercept c directly from the graph. For instance, to get the intercept we extrapolate our straight line back to where it intersects the vertical axis ($m = 0$) and read the value of ΔR.

Let's look further at the intercept, c. In our present context, it's the value of incremental redundancy at a lag of zero, at an infinitely large embedding dimension. In other words, it reflects the number of bits of information that several measurements of x predict about the last of a group of such measurements, at that lag. Naturally, we wouldn't be interested in a lag of zero, but we might be interested in a lag very close to zero. Suppose we make several measurements very close together in time and average them. Then the only difference between that average and the next measurement (also taken almost immediately) will be attributable to noise. That noise represents the *accuracy* of that latest measurement. So, the incremental redundancy at a lag of zero (or very nearly zero), that is, the value of the intercept c, in a sense reflects the accuracy of our last measurement. (This concept of accuracy works in practice only for data that aren't very accurate.)

To help us keep that potential role of the intercept in mind, I'll relabel it c_a, in which subscript "a" stands for accuracy. The equation for the asymptotic accumulation line then becomes $\Delta R = c_a + bm$.

KOLMOGOROV–SINAI (K–S) ENTROPY

The standard definition of K–S entropy, written in terms of sequence probabilities (Eq. 26.6), doesn't work well with noisy data. Fraser (1989a) attempted to avoid that defect by developing an alternate definition based on incremental redundancies. The plot of incremental redundancy versus lag (e.g. Fig. 27.3) exemplifies his method. We'll label K–S entropy estimated in this way as H'_{KS}.

An esoteric mathematical theory of dynamical systems (not discussed here), when applied to a plot of incremental redundancies at successive lags, leads to defining the slope of the asymptotic accumulation line on the plot as the *negative* of K–S entropy, or $-H'_{KS}$. In other words, that quantity replaces slope b in the latest version of our straight-line equation. Thus, the equation of the asymptotic accumulation line (Fig. 27.3) is:

$$\Delta R = c_a - H'_{KS} m. \tag{27.31}$$

Estimating K–S entropy with this approach is just a matter of measuring the slope of the straight asymptotic accumulation line and then reversing the sign. Slope as always is an ordinate distance divided by an abscissa distance. Here the ordinate distance Δy is $\Delta R_{m+1} - \Delta R_m$, and the abscissa distance Δx is $(m+1)-m$, where ΔR_{m+1} and ΔR_m are incremental redundancies at lags $m+1$ and m, respectively. Again, all of this discussion of the asymptotic accumulation line applies only in the limit where embedding dimension D approaches infinity. The symbols $D \to \infty$ refer to those conditions. Hence, the slope of the straight line is:

$$\text{slope} = \lim_{D\to\infty} (\Delta y/\Delta x)$$

or

$$\lim_{D\to\infty} \left[\frac{(\Delta R_{m+1} - \Delta R_m)}{(m+1) - m} \right].$$

Our estimate of K–S entropy (H'_{KS}) is the negative of that slope. So

$$H'_{KS} = \lim_{D\to\infty} -\frac{\Delta y}{\Delta x}$$

$$= \lim_{D\to\infty} -\left[\frac{\Delta R_{m+1} - \Delta R_m}{(m+1) - m} \right]. \quad (27.32)$$

This method at present isn't particularly precise. However, it can be useful for identifying zero versus nonzero K–S entropy, as discussed in the next two paragraphs.

NATURE OF TIME SERIES

Although K–S entropy as estimated from a plot of incremental redundancy versus lag isn't accurate, certain tendencies at high embedding dimensions on that plot can characterize or help distinguish between periodic, chaotic, and random data. Periodic data have a K–S entropy of zero, as calculated with Equation 27.32. When H'_{KS} is zero, Equation 27.31 ($\Delta R = c_a - H'_{KS}m$) reduces to just $\Delta R = c_a$ at all lags. In other words, the relation between incremental redundancy and lag (Fig. 27.3a) is a horizontal straight line at a positive and constant ordinate value of c_a. Usually, the overall trend with lag is only generally horizontal and can include periodic spikes (not shown here). An asymptotic accumulation line for such data also is roughly horizontal.

For chaotic data, K–S entropy H'_{KS} is positive. Equation 27.31 then is $\Delta R = c_a - (+H'_{KS})m = c_a - H'_{KS}m$. That equation says that, on the plot of ΔR versus lag, we get a straight line sloping downward (Fig. 27.3b).

Finally, as discussed earlier, incremental redundancies for random data are virtually zero, regardless of lag.

LAG

I mentioned above and in Chapter 19 that mutual information (Eqs 27.23 or 27.27) might help to indicate an optimum lag. (Mutual information by our definition means two dimensions, such as x_t and x_{t+m}.) That approach involved plotting the relation for mutual information versus lag, at a chosen embedding dimension, and taking the "best" lag to be the one corresponding to the first minimum in the plotted relation (e.g. Fig. 19.9).

Redundancy, including incremental redundancy, also can possibly help us estimate an optimum lag, for a given dataset and embedding dimension, D. With this approach, the optimum lag is the one for which the string of D measurements provides the maximum amount of useful information. "Useful information" here is a special expression involving redundancy but with an assumed small-scale noise factor removed. The method is as follows. First, some definitions (Fraser 1989a):

- the quantity $H'_{KS}m$ = the information lost to small scales in the time between any two successive measurements
- the quantity $(D-1)H'_{KS}m$ = the information lost to small scales over the entire string
- the quantity Dc_a-R = the total amount of information known over the entire string.

Using those estimates, an expression for the maximum amount of useful information is just a matter of algebraic manipulation, as follows. We define "useful information" as the total amount known minus that lost to small scales, where both amounts are taken over the string of D values. That useful information is definition 3 above minus definition 2, or (Dc_a-R) minus $(D-1)\, H'_{KS}m$. It's a function of lag m. We'll symbolize that function as $f_1(m)$. Stating that function as equal to the useful information (i.e. neglecting any proportionality constants):

$$f_1(m) = Dc_a-R-(D-1)H'_{KS}m.$$

Moving Dc_a to the left-hand side:

$$f_1(m)-Dc_a = -R-(D-1)H'_{KS}m.$$

Dividing everything by $D-1$:

$$\frac{f_1(m) - Dc_a}{D-1} = \frac{-R}{D-1} - H'_{KS}m\,.$$

Now we'll call the left-hand side $f_2(m)$. It's our new version of "useful information." The optimum value of lag is the lag that, for the chosen embedding dimension, gives the largest value for $f_2(m)$, that is, the largest value for the quantity $[-R/(D-1)]-H'_{KS}m$. For any $D\geq 2$, the value of m that maximizes $f_1(m)$ also maximizes $f_2(m)$, since D and c_a are fixed. That's true even if we add a constant to $f_2(m)$. For reasons that will soon become clear, we'll therefore add the constant c_a to $f_2(m)$, thereby obtaining what we'll call $f_3(m)$. So

$$\begin{aligned} f_3m &\equiv c_a+f_2(m) \\ &= c_a + \left[\frac{-R}{D-1} - H'_{KS}m\right] \\ &= c_a - H'_{KS}m - \frac{R}{D-1}\,. \end{aligned} \tag{27.33}$$

Equation 27.33 is our final expression for "useful information." We take the optimum lag to be the one for which Equation 27.33 yields the largest value.

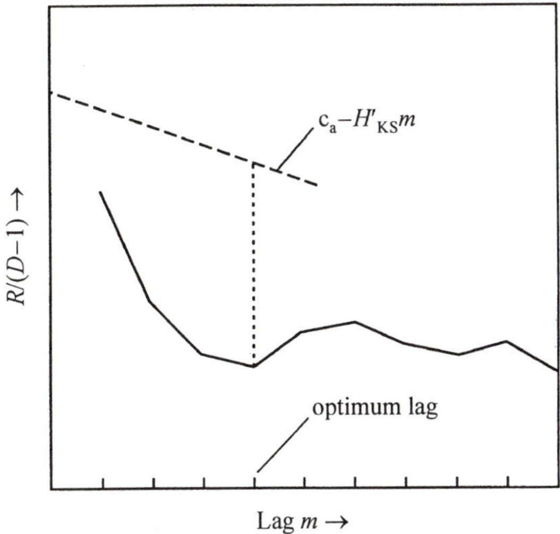

Figure 27.5 Sketch showing method of estimating optimum lag, for one embedding dimension.

Writing "useful information" in the form of Equation 27.33 puts it in familiar quantities or relations. There are actually two such relations. Both are functions of lag m, as the left-hand side of Equation 27.33 indicates. One relation consists of lag as a function of the first two terms, $c_a - H'_{KS}m$. That's the straight asymptotic accumulation line (incremental redundancy at the limit of infinitely large D, per Eq. 27.31). The other relation is the remaining term, $R/(D-1)$, as a function of lag. As we'll see in a moment, $R/(D-1)$ is the average redundancy. Equation 27.33 says to subtract that average redundancy from the value of the accumulation line, at a given lag, in order to get a value for useful information.

Figure 27.5 is a hypothetical plot of $R/(D-1)$ against lag, for a given embedding dimension (the one we have chosen by whatever means). The same graph also includes the asymptotic accumulation line (incremental redundancy ΔR as D approaches infinity). Why can we plot that relation for ΔR on a graph on which the ordinate scale is $R/(D-1)$? Because in the limit of D approaching infinity the two are equal. That is, in the limit of D approaching infinity, $\Delta R = R/(D-1)$. Table 27.2 shows why that's true.

The first column in the table is embedding dimension D. The second column lists a series of incremental redundancies that asymptotically approach a limit (here 5.0) with increase in embedding dimension (in accordance with Fig. 27.2). Column 3 lists the associated total redundancies R. Finally, the last column is $R/(D-1)$.

Incremental redundancies, as in Table 27.2, are like the vertical distances from one floor to the next in a building. Summing those separate distances (those incremental redundancies) to any floor gives the elevation or total height of that floor (analogous to redundancy) above the street. In an alternate way, we'd get the distance from any one floor to the next (incremental redundancy) by subtracting their redundancies, that is, by subtracting the elevation of the upper floor from the elevation of the one below it.

Table 27.2 Hypothetical redundancies and associated values.

D	ΔR	R	$R/(D-1)$
1	1.0	–	–
2	3.0	1.0	1.0
3	4.0	4.0	2.0
4	4.5	8.0	2.67
5	4.75	12.5	3.125
6	4.875	17.25	3.45
7	4.9375	22.125	3.6875
8	4.96875	27.0625	3.866
9	4.984375	32.03125	4.004
10	–	37.015625	4.113

Besides the asymptotic increase of incremental redundancies, there are two other important features in the table. One is that the values of $R/(D-1)$ also approach a limit of 5.0 (although much more slowly than the incremental redundancies do). So, in the limit of D approaching infinity, both ΔR and $R/(D-1)$ will have the same value (here 5.0). In other words, in the limit of D approaching infinity, $\Delta R = R/(D-1)$. That's why we can plot the asymptotic accumulation line on the graph of $R/(D-1)$ versus m for the special condition of D nearing infinity. The second feature is that, to get an average value for any R, there are $D-1$ values that contribute. (There isn't any redundancy for a dimension of 1, that is, when we have only one measurement by itself.) So, the last column, $R/(D-1)$, lists average total redundancies.

As mentioned, the best lag is taken to be the one for which Equation 27.33 yields the largest value. That's the lag at which we get the largest amount of useful information, according to Equation 27.33. That value is biggest when $R/(D-1)$, subtracted from $c_a - H'_{KS}m$, gives the largest number. Graphically, that's the lag for which we have the greatest vertical distance between the asymptotic accumulation line (or $c_a - H'_{KS}m$) and $R/(D-1)$ (the other plotted relation on Fig. 27.5).

PREDICTABILITY

Incremental redundancies quantify how well several sequential measurements of x predict the next x, at that lag and embedding dimension. For instance, say incremental redundancy at the optimum embedding dimension stays roughly constant with lag (as happens with deterministic, nonchaotic data, as in Fig. 27.3a). That means predictability about the next x remains constant at some positive value, regardless of the number of measurements included in the vector. In contrast, a downward-sloping relation (such as for chaos, as in Fig. 27.3b) means that our ability to predict the next x from D measurements decreases with lag.

A typical goal in regard to predictability is to find the smallest embedding dimension that provides optimum predictive power, for the given accuracy of the measurements and lag. That dimension is indicated by the proximity of the relations for successive embedding dimensions on the graph of incremental redundancy versus lag (e.g. Fig. 27.3), as mentioned earlier. When those relations get close to the asymptotic accumulation line, predictive ability has nearly reached the noise scale. Further increases in embedding dimension then are fruitless; no more predictive power can be gained.

The same types of graphs also indicate approximate limits of predictability in terms of lag. If the general trend of the ΔR-versus-m relation is roughly horizontal (e.g. Fig. 27.3a), then lag doesn't affect predictive power. If, instead, incremental redundancy decreases with lag (Fig. 27.3b at small lags) at a large embedding dimension, then we're losing predictive power as we increase lag. That decay of ability to predict typifies chaos, although it's not unique to chaotic systems (Ellner 1991). Predictability at that resolution becomes essentially zero when the relation falls to the vicinity of the abscissa (an incremental redundancy near zero).

In addition to the possible uses of mutual information and redundancy discussed above, other applications are being looked at as well. For instance, Paluŝ (1993) proposes a way to use redundancies to test for nonlinearity in a time series.

Summary

Mutual information extends the idea of entropy to two systems. Entropy thereby becomes a joint entropy—the average amount of uncertainty reduced (or information acquired) by measurements of two or more systems. Joint entropy for two mutually unrelated systems is the sum of their self-entropies. Joint entropy for two mutually related systems, in contrast, is the self-entropy of one system plus the conditional entropy of the other. Mutual information is the reduction in the uncertainty of a value of one system as a result of knowing a value of the other system. In terms of entropies, it's the sum of the two self-entropies minus the joint entropy. That's often expressed in one condensed equation, in terms of probabilities. Other interpretations of mutual information are that it's a measure of:

- the average amount of information contained in one system about another
- the degree of association between two processes or variables
- the amount of information that one measurement gives about a later measurement of the same variable.

Redundancy as used in chaos theory extends the idea of entropy to three or more systems. It's the mutual information of three or more systems or dimensions. As such, it's still the sum of the self-entropies minus the joint entropy. As with mutual information, it's often defined by a condensed equation in terms of probabilities. It quantifies the average amount of information common to several systems or variables. Incremental redundancy (sometimes called marginal redundancy) is the amount of increase in redundancy as a vector's embedding dimension is increased from D to $D+1$. Incremental redundancy also is the average amount of information that several successive measurements of x give about the next x. Alternatively, it's a quantitative measure of the average number of bits that several sequential measurements of x can predict about the next x.

Refinements or improvements are needed in regard to practical applications of mutual information, redundancy, and incremental redundancy. Even so, those quantities seem to have some potential in estimating optimum embedding dimension, data accuracy, K–S entropy, nature of a time series, appropriate lag, predictability, and the presence of nonlinearity.

Epilogue

Phase space portraits, correlation dimensions, Lyapunov exponents, K–S entropy, and other measures, are beautiful and potentially useful tools. However, they still need some development to be trustworthy in practice. For example, Caputo & Atten (1987: 1311) comment that there's no reliable way to estimate Lyapunov exponents. Eubank & Farmer (1990: 171) say that "algorithms for computing dimension, Lyapunov exponents, and metric entropy are notoriously unreliable. They produce a number. But . . . it is very difficult to know a priori how much to trust the number . . . Many incorrect conclusions have been reached by naive application of these methods." Paul Rapp, lecturing at a 1992 US national conference on chaos, put it as follows: "These methods are not robust against misapplication. They fail in a particularly pernicious way. Rather than simply failing to produce a result, they can produce [plausible but totally] spurious results. Even when applied rigorously, care must be exercised when interpreting results."

The main reasons for those notorious practical difficulties are:

- Small and/or unrepresentative datasets.
- *Noise* The typical indicator of chaos was developed through numerical experiments with virtually unlimited amounts of noiseless data. In other words, the methods work best (and indeed depend) on clean, accurate data with a large number (say, thousands or millions) of observations. We can generate such voluminous and virtually noise-free datasets in computer experiments and sometimes in the laboratory. However, they're mighty scarce in the real world.
- *Number of important variables* The methods were developed with (and therefore work best on) low-dimensional systems. Identifying chaos in high-dimensional systems is much more difficult than in low-dimensional systems (Glass & Mackey 1988: 42).
- *Questionable fulfillment of basic assumptions* (e.g. deterministic systems; transients no longer relevant) Such assumptions may not hold for a real-world system (Glass & Kaplan 1993).

For all of these reasons, the very important practical aspect of identifying chaos in real data is still in relative infancy.

Because of the problems just mentioned and others, no single criterion by itself usually is enough or reliable for determining chaos (Wolf et al. 1985, Glass & Mackey 1988, Prichard & Price 1992, Provenzale et al. 1992). Instead, try several different tests or approaches. Most of these involve identifying and characterizing the chaotic attractor. Another advantage to doing several tests is that you learn more about the system, even if it turns out not to be chaotic or if you can't determine whether it's chaotic.

On another note, one important message I hope you've absorbed from this book is not to be intimidated by the mysterious, impressive-sounding names given to analytical tools in science and mathematics. Analyses or concepts with imposing names can be quite simple and straightforward. *D*-dimensional phase space, invariant probability distribution of attractor, autocorrelation function, K–S entropy, Lyapunov exponents, correlation dimension, mutual information, redundancy, and the many other associated ideas really aren't all that difficult once we pick them apart and see what they consist of.

We've come a long way, you and I. Take a second and think back on what you knew about chaos when you started this book . . . but only a second. Now that you're up to this rung of the ladder, it's on to the next. It's the climb that's fun. Hope to see you along the way.

Appendix

Selected laws of powers, roots and logarithms

$0^c = 0$ (c being positive)

$0^{-c} = \infty$ (c being negative), since $0^{-c} = 1/0^c = 1/0 = \infty$

$1^c = 1$ (regardless of whether c is positive or negative)

$x^0 = 1$ (if x doesn't equal 0)

$x^1 = x$

$x^{-c} = 1/x^c = (1/x)^c$

$x^{-1} = 1/x$

$(xy)^c = x^c y^c$

$x^c(x^d) = x^{(c+d)}$

$x^c/x^d = x^{(c-d)}$

$(x/y)^c = x^c/y^c$

$(x^c)^d = x^{cd}$

$-(\log x) = \log(x^{-1}) = \log(1/x)$

$\log x = -[\log(1/x)]$

$\log(x^c) = c\log x$

$\log(xy) = \log x + \log y$

$y = cx^d$ transformed into logs: $\log y = \log c + d(\log x)$

$\log_a x = \log_b x / \log_b a$

Hence, if a = 2 and b = e: $\log_2 x = \log_e x / \log_e 2 = \log_e x / 0.69315 = 1.443 \log_e x$.

Glossary

accuracy (a) A numerical *measure* of how close an approximation is to the truth; (b) correctness, in the sense of lack of bias.

adaptive histogramming Any histogramming procedure in which bin width is allowed to vary (usually according to local densities in the data).

adaptive kernel density estimator A *probability*-estimation technique consisting of the *kernel density estimator* but with the variation that bin width changes according to local density of data points.

affine Produced by a *transformation* in which the *variables* or *axes* haven't all been scaled by the same factor.

algorithm A recipe, plan, sequence of steps, *set* of instructions, list of rules, or set of mathematical equations used to solve a problem, usually with the aid of a computer. Some of the many possible forms of an algorithm include a loosely phrased group of text statements, a diagram called a flowchart, or a computer program.

almost-periodic orbit An *orbit* that comes closer and closer to repeating itself with time.

amplitude (a) The extreme range of a fluctuating quantity, such as a pendulum, tidal *cycle*, etc.; (b) the maximum difference between the value of a *periodic* oscillation and its *mean*. In "next-amplitude" plots, an amplitude is simply a *variable*'s local maximum or peak value of each oscillation, within a sequence of oscillations.

aperiodic (a) Not repeating with time, that is, lacking *periodicity* or *quasiperiodicity* (tantamount to having an infinite *period*); (b) occurring as a *transient* or pulse only once over infinite time.

arithmetic mean Same as *mean* (the sum of all data values divided by the number of such values).

attractor The *phase space* point or *set* of points representing the various possible *steady-state* conditions of a *system*; in other words, an equilibrium *state* or group of states to which a *dynamical system converges*.

attractor reconstruction See *reconstruction of attractor*.

autocorrelation Correlation of a *variable* at one time with itself at another time. Also called *serial correlation*.

autocorrelation coefficient A dimensionless numerical indicator (calculated as *autocovariance* divided by *variance*) of the extent to which the measurements of a *time series* are mutually related.

autocorrelation function The *spectrum* or entire series of *autocorrelation coefficient*s for a *time series*.

autocorrelation time The time required for the *autocorrelation function* to drop to

1/e (= 1/2.718 = 0.37).

autocovariance A *measure* of the degree to which the observations of a *time series* are related to themselves, numerically calculated as:

$$\frac{1}{N}\sum_{t=1}^{N-m}(x_t-\bar{x})\,(x_{t+m}-\bar{x})\,.$$

autonomous (a) Self-governing, propagating from within; (b) independent of time.

axis (a) One of a group of mutually perpendicular reference lines passing through the *origin* of a graph; (b) a line about which a body or group of bodies can rotate.

axis vector *Principal axis.*

band-pass filter A *filter* that passes only those frequencies that fall within a desired range or band.

basic wave *Fundamental wave.*

basin of attraction The group of all possible *phase space* points that can evolve onto a given *attractor.*

basis A *set* of nonparallel and linearly independent *vectors.*

bifurcation (a) In general, a branching into parts or into connected segments (usually from one segment to two); (b) any abrupt change in the qualitative form of an *attractor* or in a *system*'s *steady-state* behavior, as one or more *parameters* are changed.

bifurcation-rate scaling law An equation that estimates the critical *parameter*-value at which a particular *bifurcation* occurs, within a series of bifurcations in *period-doubling.*

bifurcation diagram A graph showing all possible solutions (excluding *transients*) to an equation that relates a *variable* to a *control parameter.* Such a graph usually is drawn specifically to show *bifurcations* (*period-doublings*) of the variable.

bifurcation point A critical *parameter* value at which two or more branches of *system*-behavior emerge.

binary Having two equally likely, *mutually exclusive* possibilities.

binary digit Either of the two *digits*, conventionally 0 and 1, used in a *binary system* of reading numbers or of naming numbers; as such, it's the smallest amount of *information* that can be stored in a computer.

binary system A *system* that is set up or operates on *binary* principles.

bit (a) Contraction of "*binary digit*;" (b) either of the digits 0 or 1; (c) the unit of *information* when logs are taken to the base 2, that is, a *measure* of *information* equal to the decision content of one of two *mutually exclusive* and equally likely values or states (sometimes also called a shannon).

box-counting dimension *Similarity dimension* as obtained from a grid of boxes.

box dimension Same as *box-counting dimension.*

broadband spectrum A *frequency-domain* (*Fourier-analysis*) plot revealing no outstanding *periodicity.*

butterfly effect *Sensitive dependence on initial conditions.* The "butterfly" name stems from the theoretical possibility that a very slight change in the *state* of a *system* (such as a butterfly flapping its wings) might create slightly different "initial" conditions and thereby influence the long-term resulting pattern (the weather), even in some other part of the world.

byte (a) The amount of memory space needed to store one character on a computer (normally eight *bits*); (b) a string of *binary* elements operated on or treated as a unit.

Cantor set A *fractal* obtained by dividing a line into equal subparts, deleting one or more of those parts, and repeating that process indefinitely.

capacity A *dimension* D_c defined only at the limit where the scaling object approaches length zero; written in equation form as:

$$D_c = \lim_{\varepsilon \to 0} (\log N / \log [1/\varepsilon]).$$

cellular automaton A mathematical construction consisting of a *system* of entities, called cells, whose temporal evolution is governed by a *set* of rules and whose behavior over time becomes, or at least may appear, highly complex.

chain rule A rule that tells how to differentiate a *function* of a function, that is, how to find the *derivative* of a composite function. The name stems from the fact that the various functions fit together like a chain. For example, if y is a function of t, and t is a function of x, then the *chain rule* says that $(dy/dx) = (dy/dt)(dt/dx)$.

chaologist A person who studies *chaos*.

chaology The study of *chaos*.

chaos (a) Sustained and *random*-like long-term evolution that satisfies certain special mathematical criteria and that happens in *deterministic*, *nonlinear*, *dynamical systems*; (b) largely unpredictable long-term evolution occurring in a deterministic, nonlinear dynamical *system* because of *sensitivity to initial conditions*.

chaos theory The principles and operations underlying *chaos*.

chaotic attractor An *attractor* that shows *sensitivity to initial conditions* (*exponential divergence* of neighboring trajectories) and that, therefore, occurs only in the chaotic domain.

characteristic exponent *Lyapunov exponent.*

circle map Any of a class of *nonlinear difference equations* that *map* a point on a circle to another point on the circle (Middleton 1995).

classical decomposition The breaking down of a *time series* into its component constituents of *trend*, *seasonality*, *cyclicity*, and *noise*.

coarse graining Categorizing or *partitioning*, such as when reducing a *set* of observations of a *continuous* variable to a stream of *discrete* symbols.

coefficient (a) A constant number or symbol prefixed as a multiplier to a *variable* or unknown quantity (and hence a *constant* as opposed to a variable); (b) a dimensionless, numerical *measure* of a *set* of data, for example, an *autocorrelation coefficient*.

complex number Any number of the form a+jb, where a and b are real numbers and j is an *imaginary number* (the square root of −1).

complexity A type of *dynamical* behavior in which many independent agents continually interact in novel ways, spontaneously organizing and reorganizing themselves into larger and more complicated patterns over time.

conditional entropy The uncertainty in a measurement of *variable* y from *system* Y, given a value of variable x from coupled system X.

conditional probability The likelihood that a particular *state* or event will happen, given that one or more other specified states or events have taken place previously.

conservative Not losing energy (keeping instead a constant area or volume in *phase space*), with time.

constant A quantity that doesn't vary under specified conditions.

continuous Defined at all values of the given (independent) *variable*.

continuous random variable A *random variable* that can take on any value over a prescribed *continuous* range.

control parameter A controllable *constant* or quantity, different values of which can produce different cases or outcomes of an experiment or equation.

converge To approach a finite limiting value.

coordinate (noun) (a) One of a *set* of numbers that locate a point in space; (b) the *axis* of a graph, as in "coordinate axes" or "coordinate *system*."

coordinate vector A *vector* whose starting point is at the *origin* (0,0) of the graph. The word "coordinate" for such vectors is often dropped for convenience.

correlation dimension An exponent in a *power law* (i.e. slope of straight line) on a logarithmic plot of *correlation sum* (as the *dependent variable*) versus the radius ε of an encompassing circle, sphere, or *hypersphere*.

correlation exponent *Correlation dimension.*

correlation integral *Correlation sum.*

correlation sum A normalized total number of pairs of points within a circle or sphere of radius ε, obtained by dividing that total sum by the total number of points on the *attractor*.

correlogram A plot of *autocorrelation coefficient* (on the ordinate) versus *lag*, on arithmetic scales.

crisis An abrupt, discontinuous change in a *chaotic attractor* as a *parameter* is varied, characterized by either destruction of the chaotic attractor or its expansion to a much larger *interval* of x.

cubic polynomial A *polynomial* in which the highest *power* to which any *variable* is raised is 3.

cubic spline A smoothly varying, cubic-*polynomial* curve fitted between two data points, often used to interpolate "data" values between the two measured points.

cycle A series of events or observations that occur in a fixed sequence and return to the original *state* and which then repeat themselves in a regular pattern.

cyclicity Same as *periodicity*, that is, the tendency for any pattern to repeat itself over fixed *intervals* of time.

data-adaptive Said of methods of analysis in which the operations are applied to classes or subgroups that are defined differently from one to another, according to specified peculiarities of the data (e.g. number of observations in a class).

decomposition (a) The numerical expression of a quantity in terms of its simpler components; (b) see *classical decomposition*.

degrees of freedom The number of independent quantities (*variables* or *parameters*) or pieces of *information* that must be specified to define the *state* of a *system* completely. The field of statistics also has several other less common meanings.

delay-coordinate method Same as *time-delay method.*

delay method Same as *time-delay method.*

delay time (a) A delay, in time units, usually accompanied by an *integer parameter* to account for the units; (b) in some authors' usage, same as *lag* time.

delay vector Same as *lag vector*.

density Same as *probability density function*.

density estimation The estimation of probabilities.

density function Same as *probability density function*.

dependent variable The output *variable* of a *function*, as determined by the values of the *independent variables*.

derivative (a) Generally, a value that derives from, comes from, or is traceable to a particular source, such as a point on a curve; (b) in mathematics, the slope (rate of change) of the line that is tangent to a point on a curve.

deterministic (a) Completely and exactly specified (at least to within measuring *accuracy*) by one or more mathematical equations and a given initial condition; (b) said of a *system* whose past and future are completely determined by its present *state* and any new input.

deterministic fractal A *fractal* that looks exactly the same at all levels of detail.

detrend To transform trended data in such a way that they have an approximately constant *mean* and *variance* with time, so that the *trend* is removed.

diffeomorphic Smooth and *invertible*.

diffeomorphism A differentiable *mapping* that has a differentiable inverse.

difference equation A recurrence equation based on changes that occur at *discrete* times and solved by *iteration*. An example is the *logistic equation*. In mathematical jargon, a difference equation is an equation in which a difference operator is applied to a *dependent variable* one or more times. (A difference operator is a twofold mathematical operation in which (1) a small increment is added to the *independent variable*, and then (2) the original value of the independent variable is subtracted.)

difference plot A graph on which the *coordinates* of each plotted point are successive differences. Each successive difference is the difference between an observation x_t and some later observation (on the abscissa) and the difference between x_{t+m} and a similarly lagged observation (on the ordinate).

differencing The filtering of a *time series* by subtracting each value, in turn, from a subsequent (lagged) value. See also *first-differencing*, *second-differencing*, and *difference plot*.

differential equation An equation expressing a relationship between a *function* and one or more of its *derivatives* and that therefore is based on changes that occur continuously.

digit A code character, such as a number from 0 to 9 or a letter of the alphabet.

digital In numerical form, as in (a) calculation by numerical methods or by *discrete* units or (b) representation of data by numerical digits or discrete units.

digitize (a) To approximate by discrete samplings; (b) to put data into digital (numerical) notation (as for use in a digital computer).

dimension Generally, a magnitude measured in a particular direction, as on the axes of a graph. In *chaos theory*, "dimension" is used in any of several variations of that general meaning, such as (a) each *axis* of a *set* of mutually perpendicular axes in Euclidean space; (b) the number of *coordinates* needed to locate a point in space; (c) the maximum number of *variables* of a *system*; (d) any of various quantitative, topological measures of an object's *complexity*, and other variations.

discrete Defined or occurring only at specified values.

discrete Fourier transform (DFT) A mathematical operation that transforms a series of *discrete*, equally spaced observations measured over a finite range of time into a discrete, *frequency-domain spectrum*. Also called a finite *Fourier transform*.

discrete random variable A *random variable* that only takes on certain specified outcomes or values, with no possible outcomes or values in between.

discretize To extract equally spaced (in time) *discrete* values from a continuous *time series*.

dissipative system A *system* that loses energy with time. Evidence of energy loss with time includes an irreversible evolution toward an asymptotic or limiting condition over time and a decrease of *phase space* area or volume, with time.

distance formula The simple equation, based on the *Pythagorean theorem*, for finding the straight-line distance c between two points in N_d-dimensional space:

$$c = ([x_2-x_1]^2+[y_2-y_1]^2+[z_2-z_1]^2+\ldots+[w_2-w_1]^2)^{0.5}$$

where w is the N_dth *variable* of the group of N_d variables.

distribution function A mathematical operation that gives the proportion of members in a sample or *population* having values less than or equal to any given value. See also *frequency distribution* and *probability distribution.*

dot product A *scalar* quantity for two *vectors* that have a common *origin*, obtained either by (a) multiplying their *x coordinates* (x_1x_2), their *y* coordinates (y_1y_2), etc. for any other *dimensions* involved, and then summing those coordinate products ($x_1x_2+y_1y_2$+etc.), or (b) multiplying the vectors' lengths by the cosine of their included angle.

dynamical Changing with time.

dynamical system (a) Anything that moves or that evolves in time; (b) any process or *model* in which each successive *state* is a *function* of the preceding state.

dynamical-systems theory (a) The study of phenomena that vary with time; (b) a language that describes the behavior of moving or evolving systems, especially as affected by external *control parameters*.

dynamics (a) That branch of physics (mechanics) that deals with forces and their relation to the motion and sometimes the equilibrium of bodies; (b) the pattern of change or growth of an object or phenomenon.

embedding The preparation of a *pseudo phase space* graph to reconstruct a *system*'s dynamics (*attractor*), using successively lagged values of a single *variable*.

embedding dimension The total number of separate *time series* (consisting of the original series and subgroups obtained by lagging that series) used in a *pseudo phase space* plot or in a more rigorous mathematical analysis.

emergent Said of phenomena or *systems* in which new, increasingly complex levels of order appear over time.

entropy (a) A *measure* of unavailable energy (thermodynamics), degree of disorder or disorganization, *probability* (in inverse proportion), uncertainty, randomness, variety of choice, surprise, or *information*; (b) a quantity,

$$\sum_{i=1}^{N} P_i \log (1/P_i)$$

computed for a *discrete random variable* whose ith outcome has *probability* P.

equation of motion An equation in which time is the *independent variable*.

equilibrium point Same as *fixed point*.

ergodic (a) The property whereby statistical measures of an ensemble don't change with time and, in addition, all statistics are *invariant* from one *time series* to another within the ensemble; (b) said of a *system* for which spatial or ensemble averages are equal to time averages (meaning that time averages are independent of starting time and that most points visit every region of *phase space* with about equal *probability*); (c) said of a *trajectory* if it comes back arbitrarily close to itself after some time; (d) the property whereby averages computed from a data sample *converge* over time to ensemble averages (i.e. statistics of all initial states ultimately lead to the same *set* of statistics). Eubank & Farmer (1990) mention additional usages of the word and *state* that "there is no universally accepted definition of the word 'ergodic'."

ergodic hypothesis (a) The *dynamical* theory that says that, in the limit as the number of observations goes to infinity, a time average equals a space average (i.e. the theory says that the *point* that represents the *state* of the *system* spends, in each *phase space* compartment, an amount of time proportional to the volume of that compartment); (b) the *dynamical* theory that says that, for a system in statistical equilibrium, all accessible states are equally probable, so that the system passes rapidly through all of them.

ergodic theory (a) The mathematical study of the long-term average behavior of *dynamical systems*; (b) the study of *measure-preserving transformations*; (c) *ergodic hypothesis*.

error The difference between a quantity and an estimate of the quantity.

Euclidean dimension The common or standard notion of "*dimension*" whereby a point has dimension zero, a line dimension one, a surface dimension two, and a volume dimension three.

exponential (a) In a general sense, relating to powers (exponents); (b) referring to a rate of change that's proportional to a *constant* raised to a *power*, where the power includes the *independent variable* (see *exponential divergence* and *exponential equation*); (c) referring to a specific mathematical series known as an "exponential series."

exponential divergence Temporal separation of two adjacent trajectories according to an exponential law, that is, by a straight-line relation between the log of separation distance (as the *dependent variable*) and time.

exponential equation (exponential function) An equation relating a *dependent variable* to some *constant* raised to a *power*, where the power includes the *independent* (given) *variable*. Examples are $y = \mathrm{c}^x$ and $y = \mathrm{ac}^{\mathrm{b}x}$, where x and y are variables and a, b, and c are constants. An exponential equation plots as a straight line on semilog paper, with the log *scale* being used for y (or y/a) and the arithmetic scale for x.

false nearest neighbor A point in lagged space that is close to another point only because the *embedding dimension* is too low.

fast Fourier transform (FFT) Any member of a family of computer *algorithms* for calculating the various real and imaginary parts of the *discrete Fourier transform* (DFT) efficiently and quickly in about $N\log_2 N$ operations.

feedback That part of the output that returns to serve as input again, in a temporal process.

Feigenbaum constant Same as *Feigenbaum number*.

Feigenbaum number A universal *constant* (4.6692 . . .), discovered in the mid-1970s by Mitchell Feigenbaum, that represents the rate (in the limit where number of periods, n, becomes infinite) at which new periods appear during the *period-doubling* route to *chaos*. The Feigenbaum number (also known as the *Feigenbaum constant*) is defined as:

$$\lim_{n\to\infty} (k_n - k_{n-1}) / (k_{n+1} - k_n) = 4.6692\ldots$$

fidelity A *measure* of how closely linked one measurement is to its predecessor.

fiducial Referring to something used as a standard of reference for measurement or calculation. Examples: "fiducial point," *fiducial trajectory*.

fiducial trajectory A *trajectory* used as a reference trajectory from which to compute orbital gaps and the *Lyapunov exponent*.

filter (linear filter) Any mathematical technique or operator that systematically changes an input series, such as a *time series* or *power spectrum*, into a new (output) series that has certain desired qualities. Some purposes of *filters* are to (a) eliminate *periodicity* or *trend*; (b) reduce or eliminate *noise*; (c) suppress high or low frequencies; and (d) remove *autocorrelation*.

final state *Point attractor*.

first derivative The initial *derivative* of a *function.*

first-difference plot A graph on which the *coordinates* of each plotted point are first-differenced data, in the form of the difference between x_{t+m} and x_t on the abscissa and the difference between x_{t+2m} and x_{t+m} on the ordinate, in which m is *lag*.

first-difference transformation A *transformation* performed by *first-differencing* the data.

first-differencing Calculation of the difference between a measured value (x_t) and a lagged measurement (x_{t+m}), for all values in a *time series*. This is the most common form of *differencing.*

first harmonic A wave that has the same *frequency* as the *fundamental wave*, in *Fourier analysis*.

first principal axis The *principal axis* that is stretched the most (or reduced the least) in the *phase space* evolution of an arbitrarily designated element.

first-return map A (pseudo) *phase space model* of *Poincaré-section* data, giving the value of some *variable* as related to its preceding value at that section.

fixed point (a) In *discrete* processes, a single *phase space* point that is its own *iterate* (a point for which $x_{t+1} = x_t$); (b) in *continuous* processes, a *constant*, time-independent solution to a *differential equation*. Also called *equilibrium point* or *steady state*. See also *stable fixed point* and *unstable fixed point.*

fixed-point attractor *Point attractor.*

flip bifurcation *Period-doubling.*

flow (a) A *set* of *differential equations*; (b) a *phase space trajectory* or bundle of trajectories obtained by solving a set of differential equations.

folding A topologist's explanation of how a particular range of output values results from (a) two different ranges of input values during *iteration*; or (b) continued *phase space* evolution of a chaotic *trajectory* as it reaches the limiting value of the *variable*(s) and then rebounds or deflects back onto the *attractor.*

forced oscillator A device to which extra energy is periodically added, by some external means.

fork-width scaling law An equation that estimates the width between the two parts of any particular *bifurcation* (fork), within a series of bifurcations in *period-doubling.*

Fourier analysis A mathematical technique for uniquely describing a *time series* in terms of the *frequency-domain* characteristics of its *periodic* constituents.

Fourier coefficient (a) A numerical value reflecting the strength of a particular constituent wave relative to that of other constituent waves; (b) one of the *coefficient*s needed to express a *function* formally in terms of its *Fourier series.*

Fourier cosine series An equation for getting wave height for a particular *harmonic* by adding the cosines of angles associated with constituent *wavelengths*, in *Fourier analysis.*

Fourier integral A mathematical expression that extends the *Fourier series* to the more general situation of an infinitely long *period* by decomposing a continuous *time series* into *sinusoidal* constituents at all *frequencies* and merging the *variances* into a *continuous* distribution of variances.

Fourier series An equation that describes a *periodic time series* in terms of the cosines and/or sines of constituent *harmonics* and their associated *coefficients.*

Fourier sine series An equation for getting wave height for a particular *harmonic* by adding the sines of angles associated with constituent *wavelengths*, in *Fourier analysis.*

Fourier transform A mathematical *frequency-domain* characterization of a *time series*, consisting of constituent *amplitudes* and *phases* at each *frequency.*

fractal (a) A pattern that repeats the same design and detail or definition over a broad range of *scale*; (b) a *set* of points whose *dimension* is not a whole number (Lorenz 1993).

fractal dimension A generic term for any *dimension* (e.g. *similarity dimension, capacity, Hausdorff–Besicovich dimension, correlation dimension*, etc.) that can take on a non-*integer* value.

fractional dimension *Fractal dimension.*

frequency (a) In physics, the number of repeating *wavelengths, periods*, or *cycles* in a unit of time; (b) in statistics, the number of observations or individuals in a class; (c) also in statistics, a *probability* or proportion (*relative frequency*).

frequency analysis Same as *Fourier analysis.*

frequency distribution A list of class *intervals* or values and their associated number of observations (*frequencies*). The frequencies or number of observations in each class are often normalized to range from 0 to 1 and are often assumed equal to probabilities.

frequency domain The representation of *time series* data in terms of their *frequencies* (or of some other wave characteristic) and respective *variances.*

frequency locking *Frequency* adjustment by an oscillator in response to some *periodic* stimulus.

frequency spectrum A plot of the distribution of calculated *powers* as a function of *frequency*, as obtained in a *Fourier analysis.* Also known as *power*-density *spectrum, power spectrum*, or spectral density.

function (a) An output *variable* or *dependent variable* whose value is uniquely determined by one or more input (*independent*) *variables*; (b) an equation or relation between two groups A and B such that at least one member of group A is matched with one member of group B (a "single-valued" *function*) or is matched with two or more members of group B (a "multivalued" function). A single-valued function, for instance, is $y = 3x$; for any value of x, there is only one value of y, and vice versa. A multivalued function is $y = x^2$; for $y =$ 4, x can be +2 or –2, that is, one value of y is matched with two values of x. Many authors use "function" to mean single-valued function.

fundamental frequency The *frequency* of the selected *basic wave* in a *Fourier analysis.*

fundamental wave A wave (usually the longest available, or the length of the *time series*) chosen as a standard or reference wave in *Fourier analysis.*

fundamental wavelength The *wavelength* of the selected *basic wave* in a *Fourier analysis.*

geometric progression A sequence of terms whose successive members differ by a constant ratio or multiplier. (Also known as a *geometric series.*)

geometric series A series of numbers in which the ratio of each member to its predecessor is the same throughout the sequence. Also known as a *geometric progression.* Example: 1, 2, 4, 8, 16, 32, etc., in which the ratio of any number to its predecessor is 2.

golden mean The unique value obtained by sectioning (a) a straight line such that the ratio of the shorter segment to the longer segment equals the ratio of the longer segment to the total length, or (b) the sides of a rectangle such that the ratio of their difference to the smaller equals the ratio of the shorter to the longer.

Gram–Schmidt orthogonalization Same as *orthogonalization.*

Hamiltonian system A *system* with no friction.

harmonic (a) As an adjective, expressible in terms of sine and cosine *functions*; (b) as a noun, any component of a *periodic* quantity having a *frequency* that's an *integer* multiple of a given *fundamental frequency*. In a *Fourier analysis*, for example, the *first harmonic* is the *fundamental wave*; the *second harmonic* is the constituent wave having a frequency twice that of the fundamental wave; the third harmonic is the constituent wave that has a *frequency* three times that of the fundamental wave; etc.

harmonic analysis The *frequency-domain* description of a *time series*, especially of a *periodic function*, by summing sine and cosine functions (its constituent harmonics).

harmonic number The number corresponding to a particular *harmonic*. For example, the harmonic number of the *second harmonic* is two, that of the third harmonic is three, etc.

Hausdorff dimension That critical *dimension* D_H at which the computed value of the *measure* $M\varepsilon$ changes abruptly from zero to infinity (or from infinity to zero), using the relation $M\varepsilon = N\varepsilon^{D_H}\varepsilon^{D-D_h}$.

Hausdorff–Besicovich dimension Same as *Hausdorff dimension.*

Heaviside function A simple mathematical *function* or number that is zero if a specified expression is negative (less than zero) and 1 if it is positive (zero or greater).

Hénon map A *phase space* plot of iterates of the equations $x_{t+1} = y_t + 1 - ax_t^2$ and $y_{t+1} = bx_t$ where a and b are *constants*.

high-pass filter A *filter* that lets high frequencies pass and blocks low frequencies.

histogram A bar diagram showing the *frequency distribution* of a *variable*.

homoclinic orbit An *orbit* that is asymptotic to a *fixed point* or *periodic* orbit and that emanates from the same point or orbit (Lorenz 1993).

homoclinic point The *fixed point* from which a *homoclinic orbit* emanates and which it subsequently approaches (Lorenz 1993).

Hopf bifurcation An abrupt increase (by one) in the number of fundamental frequencies of a *system*, caused by the increase of a *control parameter* past a critical value. Commonly applied in the quasiperiodic route to *chaos*. Named for the work done on the subject by mathematician Eberhard Hopf in 1942.

hypercube (a) An imaginary *phase space* zone or subspace of four or more dimensions and characterizable by a length; (b) the multidimensional analog of a cube.

hyperspace Space of more than three dimensions.

hypersphere (a) An imaginary *phase space* zone or subspace of four or more dimensions and characterizable by a length (e.g. radius); (b) the multidimensional analog of a sphere.

identity line (or identity map) A 45° straight line on an arithmetic-*scale* two-*coordinate* graph, representing the relation $y = x$.

imaginary number Any number consisting of a real number times the square root of minus one.

incremental redundancy The amount of change in the *redundancy* of a *vector* when we increase its *dimension* by one, at a constant *lag*.

independent variable (a) An input number to a *function*; (b) a *variable* unaffected by the value taken on by other variables; (c) a variable that an experimenter deliberately manipulates, to find its effect on some other quantity (the latter being the dependent variable).

information A numerical *measure* of (a) knowledge or content of any statement, (b) how much is learned when the contents of a message are revealed or (c) the uncertainty in the outcome of an experiment to be done.

information dimension The slope of a straight line on a semilog plot of information, here defined as I_ε (*dependent variable*, on arithmetic *scale*) versus $1/\varepsilon$ (log scale), where ε is characteristic size of measuring device and

$$I_\varepsilon = \sum_{i=1}^{N_s} P_i \log_2 (1/P_i).$$

information entropy A *measure* devised by Shannon (Shannon & Weaver 1949; called by him simply "entropy") for the amount of *information* in a message, and identical in equation form to *thermodynamic entropy*.

information theory The formal, standard treatment of *information* as a mathematically definable and measurable quantity.

initial conditions Values of variables at the beginning of any specified time *period*.

inner product Same as *dot product*.

integer A whole number with no decimal or fractional part.

integral (a) A sum obtained by adding a group of separate elements; (b) the result of mathematically integrating a *function* or an equation.

intermittency A complex *steady-state* behavior (often a route to *chaos*) consisting of orderly *periodic* motion (regular oscillations, with no *period-doubling*) interrupted by occasional bursts of *chaos* or *noise* at uneven *intervals*.

interpolation The process of estimating one or more intermediate values between two known values.

interval (a) The length of time between successive events; (b) a *set* of real numbers that fall between two end-member real numbers.

invariant (a) Independent of the particular *coordinates*, that is, unaffected by a change in coordinates or by a particular *transformation*, such as a change from an original *phase space* to a time-delay reconstruction; (b) remaining forever in (never escaping from) a particular region of *phase space*; (c) unchanged by the *system*'s *dynamics* over time.

invariant manifold A collection of *phase space* trajectories, none of which ever leave the collection.

invariant measure (a) A measurable property that's unchanged by transformations; (b) a measurable feature that doesn't change with time, that is, doesn't change under the action of the *dynamics*; (c) a *probability-distribution function* describing the long-time likelihood of finding a *system* in a particular zone of *phase space*.

invariant probability distribution The *frequency distribution* that is approached as time goes to infinity. Also known as (an *attractor*'s) *probability distribution*, *probability density*, *probability density distribution*, *invariant measure*, and other combinations of these terms.

invertible (a) Having an inverse; (b) having a unique successor or predecessor, or in other words, capable of being solved uniquely either forwards or backwards in time. Example: capable of indicating either x_t or x_{t+m}, given the other.

isotropic Independent of direction.

iterate (a) As a verb, to repeat an operation over and over, often with the aim of coming closer and closer to a desired result; (b) as a noun, a value calculated by any mathematical process of successive approximation.

iteration (a) Any process of successive approximation; (b) repeated application of a mathematical procedure, with the outcome of one solution fed back in as input for the next; (c) each successive step of an iterative process.

joint entropy The average amount of *information* obtained (or average amount of uncertainty reduced) by individual measurements of two or more systems.

joint probability The *probability* that two specified events (usually independent events) will happen together. "Together" doesn't necessarily mean "simultaneously."

joint probability distribution A list, table, or *function* for two or more *random variables*, giving the *probability* of the joint occurrence of each possible combination of values of those variables.

K–S entropy *Kolmogorov–Sinai entropy.*

kernel density estimator A *probability*-estimation technique in which a fixed, local *probability distribution* (kernel) is applied to the neighborhood centered on each datum point.

Koch snowflake A geometric pattern formed from a straight line by applying a particular, constant, repeated generating procedure. The most common generating procedure is to replace the middle third of the line with an equilateral triangle (also known as "*von Koch snowflake*").

Kolmogorov–Sinai (K–S) entropy Entropy (based on sequence probabilities) per unit time in the limits where time goes to infinity and bin size goes to zero. Also known as source entropy, entropy of the source, *measure-theoretic* entropy, *metric-invariant* entropy, and *metric entropy.*

lag The basic time *interval* or amount of offset between any two values being compared, within a *time series.*

lag-m autocorrelation coefficient *Autocorrelation coefficient.*

lag-m serial correlation coefficient *Autocorrelation coefficient.*

lag space A special but very common type of *pseudo phase space* in which the axes or dimensions represent successive values of the same feature (x) separated by a constant time *interval.*

lag vector A plotted point as defined by successively lagged values of some *variable* in reconstructed (pseudo) *phase space.*

lagged phase space See *lag space.*

limit cycle A self-sustaining phase space loop that represents *periodic* motion. Hence, it's a *periodic attractor.*

line spectrum *Periodogram.*

linear (a) Pertaining to lines, usually straight ones; (b) having no *variable* raised to a *power* other than one (see *linear equation*).

linear equation An equation having a straight line for its graph, that is, an equation in which the *variables* are raised to the first *power* only (also called a *polynomial* equation of the first degree). Examples of linear equations are $\mathrm{a}x+\mathrm{b}=0$, $y=\mathrm{c}+\mathrm{b}x$, and $\mathrm{a}x+\mathrm{b}y+\mathrm{c}z=0$, where x, y, and z are variables and a, b, and c are parameters.

linear filter See *filter.*

linear function A mathematical relationship in which the *variables* appear only in the first degree, multiplied by *constants*, and combined only by addition and subtraction.

linear interpolation *Interpolation* based on straight-line relations between observations.

linear system A *system* in which all observations of a given *variable* plot as a straight line (arithmetic scales) against observations of a second (or lagged) variable. In *lag space*, for example, all observations then are said to be "linearly related" to observations at a later time.

local Lyapunov exponent The *exponential* rate of *trajectory* convergence or *divergence* in a local region of an *attractor.*

logarithm An exponent that is the *power* to which a base number (usually 10, 2, or e) is raised to give another number.

logarithmic equation An equation of the form $\mathrm{a}^y=\mathrm{c}x^{\mathrm{d}}$, in which a, c, and d are *constants.*

Such an equation plots as a straight line on semilog paper (with x on the log *scale* and y on the arithmetic scale).

logistic (a) In mathematics and statistics, referring to a special type of so-called growth curve (an expression that specifies how the size of a *population* varies with time) (however, that curve isn't the "logistic equation" of *chaos theory*; see *logistic equation*); (b) in a general sense, skilled in computation; (c) in military usage: referring to the provision of personnel and material.

logistic equation (a) Historically and generally, the relation $x_{t+1} = \mathrm{k}/(1+\mathrm{e}^{\mathrm{a}+\mathrm{b}x})$ where the *constant* b is less than zero; (b) in *chaos theory*, the relation $x_{t+1} = \mathrm{k}x_t(1-x_t)$. Both equations are iterative types that are popular in biology as models for *population* growth.

low-pass filter A *filter* that lets low frequencies pass and blocks high frequencies.

Lyapunov characteristic exponent *Lyapunov exponent.*

Lyapunov characteristic number *Lyapunov number.*

Lyapunov exponent The average of many local *exponential* rates of convergence or divergence of adjacent trajectories, expressed in *logarithms* and measured over the entire *attractor*. As such, it reflects the average rate of expansion or contraction of neighboring trajectories with time.

Lyapunov number The number whose *logarithm*, to a given base, equals the *Lyapunov exponent*.

manifold (a) In general, an object consisting of many diverse elements; (b) in mathematics, a class with subclasses (e.g. a plane is a two-dimensional manifold of points because it is the class of all its points); (c) any smooth geometric object (point, curve, surface, volume, or multidimensional counterpart); (d) in the nonchaotic domain, an *attractor*; (e) the basic space of a *dynamical system*.

map (a) A *function*, mathematical *model*, or rule specifying how a *dynamical system* evolves; (b) a synonym for correspondence, function, or *transformation*; (c) the mathematical process of taking one point to another. In the latter sense, a map tells how x will go, by a *discrete* step, to a new value of x. More than one *variable* can be involved. Common forms of a map are an iterative equation and a graph; in graphical form, a map shows a historical sequence of values.

mapping (a) A series of *iterations* of a *map*; (b) a *dynamical system* whose *variables* are defined only at *discrete* times; (c) a *function*, correspondence, or *transformation*.

marginal distribution The complete distribution (summing to 1.0) of probabilities for any chosen *variable* within a joint distribution.

marginal probability The *probability* of getting a particular value of one *variable* within a joint distribution, regardless of the value of the other variable.

marginal probability distribution Same as *marginal distribution*.

marginal redundancy Same as *incremental redundancy*.

marginally stable Tending to keep a *perturbation* at about its original magnitude, over time.

mathematical fractal An object having the property that any suitably magnified part looks exactly like the whole.

mean Arithmetic average (sum of values divided by the number of values).

measure (noun) (a) The size or quantity of something expressed in terms of a standard unit; (b) a *scalar* associated with a *vector* and indicating its magnitude and sense but not its orientation; (c) the *probability* of finding a value of a *variable* within a particular domain.

measure-preserving transformation A one-to-one *transformation* made such that the measure of every *set* and of its inverse image agree.

measure-theoretic entropy Same as *Kolmogorov–Sinai (K–S) entropy*.

metric (noun) (a) In general, a standard of measurement (e.g. "there is no metric for joy"); (b) in mathematics, a way of specifying values of a *variable* or positions of a point (e.g. "a Euclidean metric"); (c) also in mathematics, a differential expression of distance in a generalized *vector* space; (d) (adjective) referring to measurement or to the meter (100 centimeters).

metric entropy Same as *Kolmogorov–Sinai (K–S) entropy*.

metrical Quantitative.

microstate A *phase space* compartment, possible outcome, or solution.

model A simplified representation of a real phenomenon, in other words, a stripped-down or uncomplicated description or version of a real-world process. Models can be classified into physical (*scale*), mathematical, analog, and conceptual models.

monotonic Always increasing or always decreasing, that is, pertaining to a *continuous* line along which the slope keeps the same sign at all points.

moving average A transformed *time series* in which each value in the original time series is replaced by the *arithmetic mean* of that value and a specified, constant number of adjacent values that come before and after it.

multifractal Having different fractal scalings (*dimension* values) at different times or places, on an *attractor*.

multitaper spectral analysis A type of *Fourier analysis* that optimally combines *information* from *orthogonal* tapered estimates to minimize leakage and bias in the spectral estimate at each *frequency*.

mutual information (a) The amount by which a measurement of one *variable* reduces the uncertainty in another; (b) the quantity of *information* one *system* or variable contains about another; (c) a *measure* of the degree to which two processes or *random variables* are mutually related; (d) the amount of surprise or predictability associated with a new measurement.

mutually exclusive A statistical term meaning that only one of various possible outcomes can occur at a time.

nat The unit of measurement for the separation (or convergence) rate of two neighboring trajectories when logs are taken to base 2.

natural fractal An object having the property that any suitably magnified part looks approximately like the whole, the differences being minor, negligible and ascribable to chance.

natural measure A *measure* or observation that is a convergent time-average value.

natural probability measure *Invariant measure*.

nearest neighbor A *pseudo phase space* datum point that plots close to another point, for a particular *embedding dimension*.

next-amplitude map A one-dimensional *first-return map* that uses only the high values of successive oscillations (the successive maxima) of the *time series*.

next-amplitude plot *Next-amplitude map*.

noise (a) Any unwanted disturbance superimposed on useful data and tending to obscure their *information* content; (b) unexplainable variability or fluctuation in a quantity with time; (c) in a general sense, anything that impedes communication. Such disturbance or variability may be *random* fluctuations, reading *errors*, analytical errors, sampling errors, and other factors.

nonautonomous Time-dependent.

noninvertible Not capable of being uniquely solved backwards in time, for example, not able to indicate x_t from a given value of x_{t+m}.

nonlinear Not having a straight-line relationship, that is, referring to a response that isn't directly (or inversely) proportional to a given *variable*.

nonlinear dynamics The study of motion that doesn't follow a straight-line relation, that is, the study of nonlinear movement or evolution. As such, nonlinear dynamics is a broad field that includes *chaos theory* and many mathematical tools used in analyzing complex temporal phenomena.

nonlinear equation An equation in *variables* x and y which cannot be put in the form $y = c+bx$, where b and c are *coefficients*.

nonlinear system A *system* in which the observations of a given *variable* do not plot as a straight line (on arithmetic scales) against observations of a second (or lagged) variable.

nonmonotonic Pertaining to a *continuous* line along which the slope changes sign.

nonparametric Not involving assumptions about specific values of parameters or about the form of a distribution. ("Parameter" in this case usually refers to the statistical definition, namely numerical characteristics of a *population*.)

nonperiodic Same as *aperiodic*.

nonstationary Said of a *time series* for which (a) a *moving average* isn't approximately *constant* with time or (b) the *mean* and *variance* aren't approximately constant with time.

non-uniformity factor (NUF) The *standard deviation* of the local rates of convergence or divergence of neighboring trajectories, sampled over the entire *attractor*.

norm The magnitude of a *vector*.

normal distribution A special type of symmetrical (bell-shaped) and *continuous frequency distribution*, the graph or curve of which is given by a particular general equation (not reproduced here).

normalization The process of adjusting or converting one or more values to some standard *scale*. The standard scale for a group of values usually is from 0 to 1. The conversion then consists of dividing each value of the original dataset by some maximum reference quantity, such as the greatest value in the dataset or a theoretical maximum value. A *vector* is normalized by dividing it by its magnitude, yielding a so-called *unit vector*. *Probability distributions* are normalized by changing the *variable* so that the distribution has a *mean* of 0 and a *variance* of 1.

null hypothesis A hypothesis that supposes no significant difference between a statistic for one group and the same statistic for another.

observable (noun) A physical quantity that can be measured.

one-dimensional map An equation (in the form of either a written statement or a graphical plot) that gives the value of a *variable* as a *function* of its value at one or more earlier times. The typical expression is x_{t+1} as some function of x_t.

orbit The path through space taken by a moving body or point. Examples: (a) a *trajectory* that completes a circuit (as, "the Moon orbits the Earth"); (b) a trajectory or chronological sequence of states as represented in *phase space*.

ordinary probability A nonstandard term used in this book to mean the likelihood of getting a specified *state* at a particular time.

origin A reference point in ordinary space or *phase space*. Most often, it's the point at which all *variables* have a value of zero.

orthogonal (a) Perpendicular or normal (having to do with right angles); (b) unrelated or independent; (c) said of elements having the property that the product of any pair of them is zero.

orthogonalization A procedure for realigning two or more nonorthogonal vectors into a *set* of an equal number of mutually *orthogonal* vectors, all of which have the same *origin.*

orthonormal Said of axes or vectors that are mutually perpendicular and of normalized (unit) length.

orthonormalization The process of reducing mutually *orthogonal* vectors to unit length.

parabola The curve or path of a moving point that remains equidistant from a *fixed point* (the focus) and from a fixed straight line (the directrix).

parallelogram law A graphical method for adding two vectors, whereby their starting points are placed together to form two sides of a parallelogram, the two opposite sides then are drawn in, and the sum or *resultant* is given by the diagonal drawn from the two starting points to the opposite corner.

parameter (a) In physics, a controllable quantity kept *constant* in an experiment as a *measure* of some influential or driving environmental condition; (b) in mathematics, an arbitrary constant in a mathematical expression and that can be changed to provide different cases of the phenomenon represented; (c) in mathematics, a special *variable* in terms of which two or more other variables can be written; (d) in statistics, a numerical characteristic of a *population* (such as, for example, the *arithmetic mean*).

parametric Involving parameters.

parametric equations In math, a *set* of equations that express quantities in terms of the same set of *independent variables* (called parameters).

partition (a) As a noun, the collection of possible outcomes of an experiment; (b) as a noun, a compartment (bin, cell, etc.) or group of compartments within *phase space*; (c) as a verb, to subdivide a phase space or the range of values of a *variable* into a *set* of *discrete*, connected subintervals.

partitioning The process of dividing a dataset into classes or groups such that any one observation is a member of only one class.

percentage-of-moving-average method Same as *ratio-to-moving-average method.*

period The amount of time needed for a *system* to return to its original *state*, that is, the time required for a regularly repeating phenomenon to repeat itself once.

period-doubling A process whereby increases in a *control parameter* in certain iterated equations produce trajectories made up of a successively doubled number of *attractor* values (e.g. 2, 4, 8, 16, etc. *attractor* values), eventually leading to *chaos*.

periodic Regularly repeating. The repetition can be exact (in a pure mathematical sense) or approximate (as with virtually all measured data).

periodic attractor An *attractor* consisting of a finite *set* of points that form a closed loop or *cycle* in *phase space*.

periodic toroidal attractor A *torus* on which the composite *trajectory's* motion repeats itself regularly, i.e., on an exact or *integer* basis.

periodic points Points (values) that are members of a *cycle* (and hence that recur periodically).

periodicity (a) The number of measured observations or *iterations* to a *cycle* or *period*; (b) the quality of recurring at a definite *interval* (in other words, repetition of a given pattern over fixed intervals).

periodogram A graph showing the relative strengths (usually variances) of constituent waves at their *discrete* frequencies. Discrete *wave periods* can be used in place of frequencies and are probably the origin of the name.

perturbation (a) A difference between two neighboring observations, at a given time; (b) an intentional displacement of an observation, at any given time; (c) a deliberate change (usually slight) in one or more parameters of an equation.

perturbation vector *Principal axis.*

phase (a) The stage that a *dynamical system* is in at any particular time; (b) the fraction of a *cycle* through which a wave has passed at any instant.

phase angle The starting angle or reference point within a wave *cycle*, from which a process begins.

phase diagram Same as *phase portrait.*

phase locking The beating in harmony or resonating together of many individual oscillators.

phase portrait A *phase space* plot of a *system*'s various possible conditions, as shown by one or more trajectories.

phase space An abstract mathematical space in which *coordinates* represent the *variables* needed to specify the *phase* (or *state*) of a *dynamical system* at any time.

phase space portrait Same as *phase portrait.*

phase space reconstruction See *reconstruction of phase space.*

pitchfork bifurcation The spawning of two *equilibrium points* from one *equilibrium point*, as in *period-doubling.*

Poincaré map Same as *return map.*

Poincaré return map Same as *return map.*

Poincaré section A slice or cross section in *phase space*, cutting through an *attractor* transverse to the *flow*. The section shows dots that represent the *trajectory*'s intersections with the plane or slice.

point A *state* of a *system* (values of its *variables* at a given time).

point attractor A single *fixed point* in *phase space*, representing the final or equilibrium *state* of a *system* that comes to rest with the passage of time.

polynomial An algebraic expression involving two or more summed terms, each term consisting of a constant multiplier and one or more *variables* raised to nonnegative *integer* powers. (In some mathematics texts, a polynomial can have one term.)

population Any group of items that is well defined (has a common characteristic).

power (a) The number of times, as indicated by an exponent, in which a given number (say, x) is used when being repeatedly multiplied by itself (e.g. $x^3 = x \cdot x \cdot x$, and we say x is raised to the third power); (b) a term used in *Fourier analysis* as a synonym for *variance.*

power equation An equation of the form $y = ax^c$, where x and y are *variables* and a and c are *constants*. A power equation plots as a straight line on log paper. See also *power law.*

power law Any equation in which the *dependent variable* varies in proportion to a specified *power* of the *independent variable*. Synonymous with *power equation.*

power spectrum The ensemble of variances (powers) calculated in a *Fourier analysis* and plotted against their respective wave characteristic (*frequency*, *wavelength*, *period*, or *harmonic number*).

precision (a) In general, the *accuracy* with which a calculation is made; (b) reproducibility, as with repeated samplings.

principal axis Any of a body's two or three axes of structural symmetry that were mutually perpendicular before deformation.

probability (a) Limiting *relative frequency*; (b) the ratio of the number of times an event occurs to the number of trials; (c) the encoding of all that we know about the likelihood that a particular event will happen, with the encoding expressed as a number between zero (no chance that the event will take place) and 1 (certainty that the event will happen); (d) the formal study of chance occurrences.

probability density Same as *probability density function.*

probability density curve Same as *probability density function.*

probability density distribution Same as *probability density function.*

probability density function The limiting *relative frequency density* as sample size becomes infinitely large and bin width goes to zero, for a *continuous random variable.*

probability distribution A list of the possible outcomes of an experiment and their associated probabilities, that is, a table or *function* that assigns to each possible value of a *random variable* the *probability* of that value's occurrence. Also known (possibly depending on whether the *variable* is discrete or *continuous*) as *distribution function,* statistical distribution, *probability density function, density function,* or *frequency* function.

projection The image of a geometric object or *vector* superimposed on some other vector. As such, the projection is a new vector and is called the projection of the first vector onto the second.

pseudo phase space An imaginary graphical space in which the first *coordinate* represents a physical feature and the other coordinates represent lagged values of that feature.

Pythagorean theorem The theorem that, in a right triangle, the square of the length of the hypotenuse equals the sum of the squared lengths of the other two sides.

quadratic Of the second degree, that is, involving one or more unknowns (*variables*) that are squared but none that are raised to a higher *power.*

quadratic map An iterated equation (and sometimes also the *phase space* plot of those iterates) in which x_{t+1} is given as some *function* of x_t^2.

quasiperiodic toroidal attractor A *torus* on which the composite *trajectory*'s motion almost but not exactly repeats itself regularly.

quasiperiodicity (a) A *dynamical* state characterized by the superposition of two or more *periodic* motions (characteristic frequencies); (b) a route to *chaos* caused by the simultaneous occurrence of two periodicities whose frequencies are out of *phase* (not commensurate) with one another.

random (a) Based strictly on a chance mechanism (the luck of the draw), with negligible *deterministic* effects (the definition used in this book); (b) disorganized or haphazard; (c) providing every member an equal chance of selection; (d) unlikely repeatability of any observation; (e) unpredictability of individual events to within any reasonable degree of certainty; (f) difficult to compute.

random process A process based on *random* selection, as in a process in which new observations depend strictly on chance, or in which every possible observation has an equal chance of selection.

random variable (a) A *variable* whose value cannot be foretold, except statistically; (b) a variable that takes on different values at different random events (experiments or trials).

ratio-to-moving-average method A *classical decomposition* technique for isolating the value of [*seasonality* times *random* deviations] by dividing the value of the *variable* by the *moving-average* value, at each sequential time.

reconstruction dimension The *embedding dimension* in which an *attractor* is *reconstructed.*

reconstruction of attractor The graphical or analytical recreation of the *topology* or essence of a multidimensional *attractor* by analyzing lagged values of one selected *variable*.

reconstruction of phase space A pseudo (lagged) *phase space* plot in two or three dimensions, made with the hope of seeing an *attractor* (if there is one). Also known as *phase space reconstruction, state space reconstruction, phase portrait reconstruction, trajectory* reconstruction, and similar expressions.

recursive Referring to any process or *function* in which each new value is generated from the preceding value. (Hence, iterative.) Examples: (a) the *logistic equation* of *chaos theory*; (b) forecasting techniques in which each successive forecast incorporates the preceding forecast.

reductionism The notion that the world is an assemblage of parts.

redundancy The multidimensional generalization of *mutual information.*

regression A mathematical *model* stating how a *dependent variable* varies with one or more *independent variables*.

relative entropy A ratio of two entropies (typically an actual *entropy* to a reference entropy), reflecting the magnitude of one relative to the other.

relative frequency The ratio of number of occurrences of an outcome to total number of possible occurrences.

relative frequency density The ratio of *relative frequency* to class width.

relative frequency distribution A list of relative frequencies, for a particular process or *system*.

renormalization A mathematical scaling technique consisting of rescaling a physical *variable* and transforming a *control parameter* such that the properties of an equation at one *scale* can be related to those at another scale, and the properties at the limiting scale (infinity) can be determined.

reorthonormalization A procedure for again realigning all vectors to be mutually perpendicular and then making each *vector* of unit length.

residual The difference between an observed value and the value predicted by a *model* fitted to the general dataset.

resolution (a) The act of breaking something up into its constituent elements; (b) the act of determining or rendering distinguishable the individual parts of something; (c) the size of the biggest element in a *partitioning*, such as the length of a box or radius of a sphere.

resultant A *vector* that is the sum of two or more other vectors.

return map A rule, *function*, or *model* that tells where the *trajectory* will next cross a particular one-dimensional *Poincaré section*, given the location of a previous crossing at that section. "Locations" along the line are described in terms of distance (L) along the line from some arbitrary *origin* on the line. Commonly, a *first-return* (or first-order-return) *map* is a model of L_{n+1} as a function of L_n for the Poincaré section; a *second-return* (or *second-order-return*) *map* is a model of L_{n+2} versus L_n; and so on.

Richardson plot A logarithmic graph of a coastline's measured length (on the ordinate) versus the associated ruler length.

robust Resistant to, or steady under, *perturbation.*

saddle (saddle point) An *unstable steady-state phase space* condition (*fixed point, limit cycle*, etc.), the *flow* near which is *stable* in some directions (*converges* toward the fixed point) and unstable in others (moves away from the fixed point).

scalar A number representing a magnitude only, as might be indicated on a simple *scale*.

scalar product Same as *dot product*.

scalar time series An ordinary *time series*, in which each successive measurement is recorded along with its time or order of measurement (and, hence, synonymous with the general meaning of *time series*).

scale (a) A sequence of collinear marks, usually representing equal steps or increases, used as a reference in measuring; (b) the proportion that a *model* bears to whatever it represents (e.g. a scale of one centimeter to a kilometer); (c) to reduce or enlarge according to a fixed ratio or proportion; (d) to determine a quantity at some order of magnitude by using data or relationships known to be valid at other orders of magnitude.

scale invariance The property whereby an object's size cannot be determined or estimated without additional *information* because the item looks like any of a family of objects of different size but similar appearance.

scale length Length of tool or device used (usually by successive increments) to estimate an object's size.

scaling ratio A proportion (and hence a value less than 1.0), used in scaling, and equal to that fraction of the original object which the new (smaller) object represents as measured in the direction of any *topological dimension*.

scaling region A straight-line relation (the slope of which is the *correlation dimension*) on a plot of log of *correlation sum* versus log of measuring radius.

Schmidt orthogonalization Same as *orthogonalization*.

seasonality Same as *periodicity* (the tendency for any pattern to repeat itself over fixed *intervals* of time). In this book, the fixed intervals of time for seasonality are arbitrarily taken as one year or less, whereas periodicity is used as a generic or general term and can have any fixed interval.

second derivative The *derivative* of the *first derivative* of a *function*.

second-differencing The computation of differences of differenced data.

second harmonic A wave that has twice the *frequency* as the *fundamental wave*, in *Fourier analysis*.

second-order return map *Second-return map*.

second-return map A *pseudo phase space* plot of *Poincaré-section* data on which each observation at the section is plotted against its second return to that section.

self-affine That property whereby an object is reproduced by a *transformation* involving different scaling factors for different dimensions or *variables*.

self-entropy (a) The *entropy* for one *system* or *variable*; (b) the average amount of *information* gained by making a measurement of (or provided by) a *random process* about itself.

self-information Same as *self-entropy*.

self-organization The spontaneous, self-generated occurrence of some kind of pattern or structure from an (usually) orderless *dynamical system*.

self-organized criticality The property whereby a *dynamical system* naturally evolves toward a critical *state* (a condition where the *system* undergoes a sudden change). Various *perturbations* of the system at that critical state provoke different responses that follow a *power law*.

self-similarity The property whereby any part of an object, suitably magnified, looks mostly like the whole.

self-similarity dimension *Similarity dimension*.

sensitive dependence on initial conditions (a) The quality whereby two slightly different values of an input *variable* evolve to two vastly different trajectories; (b) the quality whereby two initially nearby trajectories diverge exponentially with time.

sensitivity to initial conditions *Sensitive dependence on initial conditions*.

separatrix A boundary between regions of *phase space*, such as a boundary between two basins of attraction in a *dissipative system*.

sequence probability A nonstandard term used in this book to mean the *joint probability* of certain successive events in time and representing the *probability* that a given sequence of events will take place.

serial correlation Same as *autocorrelation*.

set A group of points or values that have a common characteristic or rule of formation.

similarity dimension A *dimension* D defined as $D = \log N/\log(1/r)$, used in transforming figures into similar figures of different sizes.

sine wave A wave corresponding to the equation $y = \sin x$, or more generally, $y = A \sin (x+\phi)$, where A is *wave amplitude* and ϕ is *phase angle*.

singular spectral analysis An alternative to *Fourier analysis* that uses empirical *basis functions* formed by a principal component analysis of the embedded *phase space* data matrix instead of the sine/cosine *basis* functions.

singular system analysis A *phase space reconstruction* technique in which *orthonormal* reconstruction axes near each *point* x_t are a maximal set of linearly independent vectors that are derived from the local distribution of points near x_t by *singular value decomposition*.

singular value decomposition *Singular system analysis*.

sink A *phase space* point toward which all nearby points *flow*.

sinusoid (a) The sine curve ($y = \sin x$); (b) any curve derivable from the sine curve by multiplying by a *constant* or by adding a constant; (c) any curve shaped like a sine curve but with different *amplitude*, *period*, or intercepts with the *axis*.

sinusoidal (a) Relating to, and shaped like, a sine curve; (b) describable by a sine or cosine equation.

smooth Said of a *frequency distribution* that is *continuous* or, in mathematical terms, is differentiable at every point.

smoothing (a) In general, reduction or removal of discontinuities or fluctuations in data; (b) more specifically, application of a *low-pass filter*, that is, any mathematical technique that combines two or more observations to reduce variability and to highlight general trends.

smoothing parameter Bin width, usually as required in methods for estimating *probability* densities.

soliton Solitary wave.

source A *phase space* point away from which all nearby points *flow*.

space-filling curve A curve passing through every point in a space of two or more dimensions.

spectral analysis The calculation and interpretation of a *spectrum*, especially in terms of the *frequency-domain* characteristics of a *time series*. Generally used synonymously with *Fourier analysis*.

spectrum A collection of all the frequencies, *wavelengths*, or similar components of a complex process, showing the distribution of those components according to their magnitudes.

spline (a) A flexible strip used in drafting to form—and to help draw—a curve between two *fixed points*; (b) a smoothly varying curve between two data points, often fitted by a *cubic polynomial* and used to interpolate "data" values between the two measured points.

stable Tending to dampen *perturbations* or initial differences, over time.

stable fixed point A type of *attractor* in the form of a *phase space* point that attracts all neighboring points, that is, a point to which all iterates beginning from some other point *converge*.

stable manifold The set of points that asymptotically approach a given point, over time.

standard deviation A descriptive statistic that reflects the spread of a group of values about their *mean*, usually computed as:

$$\left[\frac{1}{N}\sum_{i=1}^{N}(x_i-\bar{x})^2\right]^{0.5}.$$

As such, it's the square root of the average squared deviation, or the square root of the *variance*.

standard phase space A term used in this book for the usual *phase space*, the *coordinates* for which represent the measured and different physical features that are to be plotted on the associated graph or *phase portrait.*

standardization A *transformation* done by subtracting the *mean* from each observation of a dataset and then dividing each such difference by the *standard deviation*. Such a transformation, for example, removes *seasonality*. It converts data of different units to a common or standard *scale*, namely a scale for which the mean is zero and the *standard deviation* is one. The new scale measures the transformed *variable* in terms of the deviation from its mean, in units of standard deviation, and is dimensionless.

state (a) The condition of a *system* or values of its *variables*, at any one time; (b) an arbitrarily defined subrange (usually with specified numerical boundaries) that one or more variables of a system can be in at one particular time.

state space Same as *phase space*.

state space reconstruction See *reconstruction of phase space*.

state space vector Same as *lag vector*.

stationary Time-*invariant*, that is, (a) lacking a *trend*; or (b) (more rigidly), keeping a constant *mean* and *variance* with time; or (c) (more formally still), said of a randomlike process whose statistical properties are independent of time. ("Independence of time" is often a matter of the particular timescale being used.)

statistical self-similarity The property whereby any suitably magnified part of an object looks approximately like the whole, the differences being minor, negligible and ascribable to chance.

steady state (a) A condition that doesn't change with time; (b) the *state* toward which the *system*'s behavior becomes asymptotic as time goes to infinity. The associated equation gives a constant solution.

stochastic (a) Having no regularities in behavior; (b) characterizable only by statistical properties, thus involving randomness and *probability* as opposed to being mainly *deterministic*; (c) *random*; (d) developed according to a probabilistic *model*.

strange attractor (a) Same as *chaotic attractor*; (b) an *attractor* having such geometrical features as *fractal dimension*, *Cantor-set* structure, and so on.

stretching (a) A topologist's interpretation of either (1) the amplification of a certain range of input values to a larger range of output values during *iteration* or (2) the *phase space exponential* divergence of two nearby chaotic trajectories; (b) a *transformation* of the form $x' = ax$ and $y' = ay$, where $a>1$.

stroboscopic map A lag-time *phase space model* in which data are taken at equal time *intervals*.

subharmonic bifurcation *Period-doubling*.

successive-maxima plot Same as *next-amplitude map*.

surface of section *Poincaré section*.

surrogate data Artificially generated data that mimic certain selected features of an observed *time series* but that are otherwise *stochastic*.

system (a) An arrangement of interacting parts, units, components, *variables*, etc. into a whole; (b) a group, series, or sequence of elements, often in the form of a chronological dataset.

tangent bifurcation That special situation on a *one-dimensional map* whereby the *function* is tangent to the *identity line*.

theoretic (a) Restricted to theory; (b) lacking verification.

thermodynamic entropy A *measure* conceived by Clausius for describing the unavailable energy of a closed system such as a heat-producing engine, and subsequently modified for application to other thermodynamic systems.

time-delay method A lag-time analytical technique that uses data of just one measured physical feature (x) (regardless of any other features that may have been measured), whereby x_t is compared to one or more lagged subseries, often with the aim of reconstructing an *attractor*.

time domain The representation of *time series* data in their raw or unaltered form.

time series (a) A chronological list or plot of the values of any *variable* or variables and their time or order of measurement; (b) in a narrow and more rare sense, a *set* of values that vary randomly with time.

topological dimension An *integer* that reflects the *complexity* of a geometric continuum, equal to 1+the *Euclidean dimension* of the simplest geometric object that can subdivide that continuum. The topological dimension usually has the same value as the Euclidean dimension.

topology (a) That branch of geometry that deals with the properties of figures that are unaltered by imposed deformations or cumulative transformations; (b) the branch of mathematics that studies the qualitative properties of spaces, as opposed to the more delicate and refined geometric or analytic properties.

toroidal attractor *Torus*.

torus (a) A three-dimensional ring- or doughnut-shaped surface or solid generated by rotating a circle about any *axis* that doesn't intersect that circle; (b) an *attractor* that represents two or more *limit cycles*. Also called a toroid.

trajectory (a) A path taken by a moving body or point (and hence an *orbit*); (b) a sequence of measured values or list of successive states of a *dynamical system*; (c) a solution to a *differential equation*; (d) graphically, a line on a *phase space* plot, connecting points in chronological order.

transformation (a) A change in the numerical description or *scale* of measurement of a *variable* (in other words, a *filter*); (b) a change in position or direction of the axes of a coordinate *system*, without altering their relative angles; (c) a *mapping* between two spaces.

transient An early, atypical observation or temporary behavior in a *system* when first activated that dies out with time.

translation A shifting (*transformation*) of the axes of a coordinate *system* while keeping the same orientation of those axes. For example, say a is the amount that the x *axis* is shifted and b is the amount that the y axis is shifted; then the *origin* of new axis x' is at $x+a$ and the origin of new axis y' is at $y+b$; any point plotted on the new *coordinates* is located at $x'=x-a$ and $y'=y-b$.

trend (a) A systematic change, prevailing tendency, or general drift (in *chaos*, usually a prevailing direction of plotted points over some *period* of time); (b) a nonconstant *mean*.

triangle law A graphical method of adding two vectors whereby the starting point of the second *vector* is placed at the terminal point of the first vector, the sum or *resultant* being given by the new vector drawn from the starting point of the first vector to the terminal point of the second.

truncate (a) To shorten a number by keeping only the first few (significant) *digits* and discarding all others; (b) to approximate an infinite series by a finite number of terms; (c) to exclude sample values that are greater (or less) than a specified constant value.

two-dimensional map A pair of equations, in each of which x_t and y_t together are used to yield x_{t+m} (with the first equation) and y_{t+m} (with the second equation).

unit vector A *vector* having a magnitude of 1. It's usually obtained by dividing a vector by its length (magnitude).

universal Typical of entire classes of systems.

universality The property whereby many seemingly unrelated systems or equations behave alike in a particular respect, so that they can be grouped into a class by their generic behavior.

unstable Tending to amplify *perturbations* or initial differences, over time.

unstable fixed point A *fixed point* from which successive iterates move farther and farther away.

unstable manifold The *set* of points that exponentially diverge from a given point, over time.

unstable orbit (unstable trajectory) A *trajectory* for which, arbitrarily close to any input value, there's another possible input which gives rise to a vastly different trajectory.

variable A characteristic or property that can have different numerical values.

variable-partition histogramming Same as *adaptive histogramming*.

variance A *measure* of the spread or dispersion of a group of values about their *mean*, specifically the average squared deviation from the mean, or

$$\frac{1}{N}\sum_{i=1}^{N}(x_i-\bar{x})^2.$$

vector A directed straight line, that is, a straight line representing a quantity that has both magnitude and direction, drawn from its starting point (point of *origin*) to its terminal point.

vector array The list of paired values that make up a *vector time series*.

vector time series A listing or series of measurements of a *variable* in which each value is associated not with its time or order of measurement explicitly but rather with its value at some constant *lag* time later.

von Koch snowflake *Koch snowflake*.

wave amplitude A *measure* of the maximum height of a wave, taken either as the vertical height from trough to crest or as half the vertical height from trough to crest.

wave frequency See *frequency*.

wavelength The distance from any point on a wave to the equivalent point on the next wave.

wave period The time needed for one full *wavelength* or *cycle*.

weight A factor by which some quantity is multiplied and that reflects that quantity's relative importance.

weighted sum A sum obtained by adding weighted quantities.

weighting The multiplication of each item of data by some number that reflects the item's relative importance.

white noise Data that are mutually unrelated. For instance, an observation made at one time has no relation to observations made at earlier times. Such *noise* might, for example, be generated from independent, identically distributed observations of a *variable*. The label "white" comes from the analogy with white light and means that all possible *periodic* oscillations or frequencies are present with equal *power* or *variance*.

window A selected subrange or *interval* of values.

References

Abarbanel, H. D. I. 1992. Local and global Lyapunov exponents on a strange attractor. In *Nonlinear modeling and forecasting*, M. Casdagli & S. Eubank (eds), 229–47. Redwood City, California: Addison-Wesley.

Abarbanel, H. D. I., R. Brown, J. B. Kladtke 1990. Prediction in chaotic nonlinear systems: methods for time series with broadband Fourier spectra. *Physical Review A* **41**(4), 1782–807.

Abarbanel, H. D. I., R. Brown, M. B. Kennel 1991a. Variation of Lyapunov exponents on a strange attractor. *Journal of Nonlinear Science* **1**, 175–99.

— 1991b. Lyapunov exponents in chaotic systems: their importance and their evaluation using observed data. *International Journal of Modern Physics B* **5**(9), 1347–75.

— 1992. Local Lyapunov exponents computed from observed data. *Journal of Nonlinear Science* **2**, 343–65.

Abarbanel, H. D. I., R. Brown, J. J. Sidorowich, L. S. Tsimring 1993. The analysis of observed chaotic data in physical systems. *Reviews of Modern Physics* **65**(4), 1331–92.

Abarbanel, H. D. I & U. Lall 1996. Nonlinear dynamics of the Great Salt Lake: system identification and prediction. *Climate Dynamics* **12**, 287–97.

Abraham, R. H. & C. D. Shaw 1982. *Dynamics—the geometry of behavior*, part 1: *Periodic behavior*. Santa Cruz, California: Aerial Press.

— 1983. *Dynamics—the geometry of behavior*, part 2: *Chaotic behavior*. Santa Cruz, California: Aerial Press.

Abraham, N. B., A. M. Albano, N. B. Tufillaro 1989. Complexity and chaos. In *Measures of complexity and chaos*, N. B. Abraham, A. B. Albano, A. Passamante, P. E. Rapp (eds), 1–27. New York: Plenum.

Albano, A. M., A. I. Mees, G. C. de Guzman, P. E. Rapp 1987. Data requirements for reliable estimation of correlation dimensions. In *Chaos in biological systems*, H. Degn, A. V. Holden, L. F. Isen (eds), 207–220. New York: Plenum.

Aleksic, Z. 1991. Estimating the embedding dimension. *Physica D* **52**, 362–8.

Amato, I. 1992. Chaos breaks out at NIH, but order may come of it. *Science* **256**(26 June), 1763–4.

Anonymous 1994. Balancing broomsticks. *The Economist* **331**(7869), 85–6.

Atmanspacher, H. & H. Scheingraber 1987. A fundamental link between system theory and statistical mechanics. *Foundations of Physics* **17**(9), 939–63.

Atmanspacher, H., H. Scheingraber, W. Voges 1988. Global scaling properties of a chaotic attractor reconstructed from experimental data. *Physical Review A* **37**(4), 1314–22.

Barcellos, A. 1984. The fractal geometry of Mandelbrot. *College Mathematics Journal* **15**, 98–114.

Bergé, P., Y. Pomeau, C. Vidal 1984. *Order within chaos*. New York: John Wiley.

Berryman, A. A. 1991. Chaos in ecology and resource management: what causes it and how to avoid it. In *Chaos and insect ecology*, J. A. Logan & F. P. Hain (eds), 23–38. Information Series 91–3, Virginia Agricultural Experiment Station and Virginia Polytechnic Institute and State University.

Berryman, A. A. & J. A. Millstein 1989. Are ecological systems chaotic—and if not, why not. *Trends in Ecology and Evolution* **4**(1), 26–8.

Bloomfield, P. 1976. *Fourier analysis of time series: an introduction*. New York: John Wiley.

Borowitz, S. & L. A. Bornstein 1968. *A contemporary view of elementary physics*. New York: McGraw-Hill.

Bracewell, R. N. 1978. *The Fourier transform and its applications*, 2nd edn. New York: McGraw-Hill.

Briggs, J. & F. D. Peat 1989. *The turbulent mirror*. New York: Harper & Row.

Briggs, K. 1990. An improved method for estimating Lyapunov exponents of chaotic time series. *Physics Letters A* **151**(1–2), 27–32.

Brigham, E. O. 1988. *The fast Fourier transform and its applications*. Englewood Cliffs, New Jersey: Prentice-Hall.

Broomhead, D. S. & G. P. King 1986. Extracting qualitative dynamics from experimental data. *Physica D* **20**(2/3), 217–36.

Broomhead, D. S., R. Jones, G. P. King 1987. Topological dimension and local coordinates from time series data. *Journal of Physics A* **20**, L563–L569.

Bryant, P. 1992. Computation of Lyapunov exponents from experimental data. In *Proceedings of the first experimental chaos conference*, S. Vohra, M. Spano, M. Shlesinger, L. Pecora, W. Ditto (eds), 11–23. Singapore: World Scientific.

Bryant, P., R. Brown, H. D. I. Abarbanel 1990. Lyapunov exponents from observed time series. *Physical Review Letters* **65**(13), 1523–6.

Burrough, P. A. 1984. The application of fractal ideas to geophysical phenomena. *The Institute of Mathematics and its Applications* **20**(March/April), 36–42.

Çambel, A. B. 1993. *Applied chaos theory: a paradigm for complexity*. Boston: Academic Press.

Campbell, D. 1989. Introduction to nonlinear phenomena. In *Lectures in the sciences of complexity*, D. L. Stein (ed.), 3–105. Redwood City, California: Addison-Wesley.

Campbell, D., J. Crutchfield, D. Farmer, E. Jen 1985. Experimental mathematics: the role of computation in nonlinear science. *Communications of the Association for Computing Machinery* **28**(4), 374–84.

Caputo, J. G. & P. Atten 1987. Metric entropy: an experimental means for characterizing and quantifying chaos. *Physical Review A* **35**(3), 1311–16.

Casdagli, M. 1989. Nonlinear prediction of chaotic time series. *Physica D* **35**, 335–56.

Caswell, W. E. & J. A. Yorke 1986. Invisible errors in dimension calculations: geometric and systematic effects. See Mayer-Kress (1986: 123–36).

Cipra, B. 1993. If you can't see it, don't believe it. *Science* **259**(January 1), 26–7.

Collet, P. & J-P. Eckmann 1980. *Iterated maps on the interval as dynamical systems*. Basel: Birkhäuser.

Cover, T. M. & J. A. Thomas 1991. *Elements of information theory*. New York: John Wiley.

Cremers, J. & A. Hübler 1987. Construction of differential equations from experimental data.

Zeitschrift für Naturforschung, Series A **42**(8), 797–802.

Crutchfield, J. P., J. D. Farmer, N. H. Packard, R. S. Shaw 1986. Chaos. *Scientific American* **255**(6), 46–57.

Crutchfield, J. P. & B. S. McNamara 1987. Equations of motion from a data series. *Complex Systems* **1**, 417–52.

Crutchfield, J. P. & K. Young 1989. Inferring statistical complexity. *Physical Review Letters* **63**(2), 105–108.

Cvitanovic, P. (ed.) 1984. *Universality in chaos*. Bristol: Adam Hilger.

Davis, J. C. 1986. *Statistics and data analysis in geology*, 2nd edn. New York: John Wiley.

Denbigh, K. G. & J. S. Denbigh 1985. *Entropy in relation to incomplete knowledge*. Cambridge: Cambridge University Press.

Devaney, R. L. 1988. Fractal patterns arising in chaotic dynamical systems. See Peitgen & Saupe (1988: 137–67).

Ditto, W. L. & L. M. Pecora 1993. Mastering chaos. *Scientific American* **269**(August), 78–84.

Dvorák, I. & J. Klaschka 1990. Modification of the Grassberger–Procaccia algorithm for estimating the correlation exponent of chaotic systems with high embedding dimension. *Physics Letters A* **145**(5), 225–31.

Eckmann, J-P. & D. Ruelle 1985. Ergodic theory of chaos and strange attractors. *Reviews of Modern Physics* **57**(3) (part 1), 617–56.

— 1992. Fundamental limitations for estimating dimensions and Lyapunov exponents in dynamical systems. *Physica D* **56**, 185–7.

Eckmann, J-P., S. Oliffson-Kamphorst, D. Ruelle, S. Ciliberto 1986. Liapunov exponents from time series. *Physical Review A* **34**(6), 4971–9.

Elliott, D. F. & K. R. Rao 1982. *Fast transforms: algorithms, analyses, applications*. Orlando, Florida: Academic Press.

Ellner, S. 1991. Detecting low-dimensional chaos in population dynamics data: a critical review. In *Chaos and insect ecology*, J. A. Logan & F. P. Hain (eds), 63–90. Information Series 91–3, Virginia Agricultural Experiment Station and Virginia Polytechnic Institute and State University.

Essex, C. & M. A. H. Nerenberg 1990. Fractal dimension: limit capacity or Hausdorff dimension? *American Journal of Physics* **58**(10), 986–8.

— 1991. Comments on "deterministic chaos: the science and the fiction" by D. Ruelle. *Proceedings of the Royal Society of London A* **435**, 287–92.

Eubank, S. & J. D. Farmer 1990. An introduction to chaos and randomness. In *1989 lectures in complex systems*, E. Jen (ed.), 75–190. Redwood City, California: Addison-Wesley.

Farmer, J. D. 1982. Information dimension and the probabilistic structure of chaos. *Zeitschrift für Naturforschung*, Series A **37**, 1304–325.

Farmer, J. D., E. Ott, J. A. Yorke 1983. The dimension of chaotic attractors. *Physica D* **7**, 153–80.

Farmer, J. D. & J. J. Sidorowich 1987. Predicting chaotic time series. *Physical Review Letters* **59**(8), 845–8.

— 1988a. Exploiting chaos to predict the future and reduce noise. In *Evolution, learning, and cognition*, Y. C. Lee (ed.), 277–330. Singapore: World Scientific.

— 1988b. Predicting chaotic dynamics. In *Dynamic patterns in complex systems*, J. A. S. Kelso, A. J. Mandell, M. F. Shlesinger (eds), 265–92. Singapore: World Scientific.

Feder, J. 1988. *Fractals*. New York: Plenum.

Feigenbaum, M. J. 1978. Quantitative universality for a class of nonlinear transformations. *Journal of Statistical Physics* **19**(1), 25–52.

— 1979. The universal metric properties of nonlinear transformations. *Journal of Statistical Physics* **21**(6), 669–706.

— 1980. Universal behavior in nonlinear systems. *Los Alamos Science* **1**, 4–27.

— 1993. The transition to chaos. In *Chaos: the new science*, J. Holte (ed.), 45–53. Lanham, Maryland: University Press of America.

Firth, M. 1977. *Forecasting methods in business and management*. London: Edward Arnold.

Fisher, A. 1985. Chaos: the ultimate asymmetry. *Mosaic* **16**(1), 24–33.

Fokas, A. S. 1991. Some remarks on integrability. *Nonlinear Science Today* **1**(3), 6–11.

Ford, J. 1983. How random is a coin toss? *Physics Today* **36**(4), 40–47.

— 1989. What is chaos, that we should be mindful of it? In *The new physics*, P. Davies (ed.), 348–72. Cambridge: Cambridge University Press.

Fowler, T. & D. Roach 1991. Dimensionality analysis of objects and series data. In *Nonlinear dynamics, chaos and fractals*, G. V. Middleton (ed.), 59–81. Short Course Notes 9, Geological Association of Canada, Toronto.

Fraedrich, K. 1986. Estimating the dimensions of weather and climate attractors. *Journal of the Atmospheric Sciences* **43**(5), 419–32.

Fraser, A. M. 1986. Using mutual information to estimate metric entropy. See Mayer-Kress (1986: 82–91).

— 1989a. Information and entropy in strange attractors. *Institute of Electrical and Electronics Engineers (IEEE) Transactions on Information Theory* **35**(2), 245–62.

— 1989b. Reconstructing attractors from scalar time series: a comparison of singular system and redundancy criteria. *Physica D* **34**, 391–404.

Fraser, A. M. & H. L. Swinney 1986. Independent coordinates for strange attractors from mutual information. *Physical Review A* **33**(2), 1134–40.

Froehling, H., J. P. Crutchfield, D. Farmer, N. H. Packard, R. Shaw 1981. On determining the dimension of chaotic flows. *Physica D* **3**, 605–617.

Fujisaka, H. 1983. Statistical dynamics generated by fluctuations of local Lyapunov exponents. *Progress of Theoretical Physics* **70**(5), 1264–75.

Gatlin, L. L. 1972. *Information theory and the living system*. New York: Columbia University Press.

Gershenfeld, N. 1988. An experimentalist's introduction to the observation of dynamical systems. In *Directions in chaos* (vol. 2), B-L. Hao (ed.), 310–84. Singapore: World Scientific.

— 1993. Information in dynamics. In *Workshop on physics and computation, PhysComp '92*, D. Matzke (ed.), 276–80. Los Alamitos, California: Institute of Electrical and Electronics Engineers (IEEE) Computer Society Press.

Gershenfeld, N. A. & A. S. Weigend 1993. The future of time series: learning and understanding. See Weigend & Gershenfeld (1993: 1–70).

Glass, L. & M. C. Mackey 1988. *From clocks to chaos: the rhythms of life*. Princeton, New Jersey: Princeton University Press.

Glass, L. & D. T. Kaplan 1993. Complex dynamics in physiology and medicine. See Weigend & Gershenfeld (1993: 513–28).

Gleick, J. 1987. *Chaos: making a new science*. New York: Penguin.

Godfray, H. C. J. & S. P. Blythe 1990. Complex dynamics in multispecies communities. *Philosophical Transactions of the Royal Society of London A* **330**, 221–33.

Goldman, S. 1953. *Information theory*. New York: Prentice-Hall.

Goodchild, M. F. & D. M. Mark 1987. The fractal nature of geographic phenomena. *Annals of the Association of American Geographers* 77(2), 265–78.

Grassberger, P. & I. Procaccia 1983a. Characterization of strange attractors. *Physical Review Letters* **50**(5), 346–9.

— 1983b. Measuring the strangeness of strange attractors. *Physica D* **9**, 189–208.

— 1984. Dimensions and entropies of strange attractors from a fluctuating dynamics approach. *Physica D* **13**, 34–54.

Grassberger, P., T. Schreiber, C. Schaffrath 1991. Nonlinear time sequence analysis. *International Journal of Bifurcation and Chaos* **1**(3), 521–47.

Grebogi, C., E. Ott, J. A. Yorke 1982. Chaotic attractors in crisis. *Physical Review Letters* **48**(22), 1507–510.

— 1983. Crises, sudden changes in chaotic attractors, and transient chaos. *Physica D* **7**, 181–200.

— 1987. Chaos, strange attractors, and fractal basin boundaries in nonlinear dynamics. *Science* **238**(30 October), 632–8.

Grebogi, C., E. Ott, S. Pelikan, J. A. Yorke 1984. Strange attractors that are not chaotic. *Physica D* **13**, 261–8.

Greenside, H. S., A. Wolf, J. Swift, T. Pignataro 1982. Impracticality of a box-counting algorithm for calculating the dimensionality of strange attractors. *Physical Review A* **25**(6), 3453–6.

Grossman, S. & S. Thomae 1977. Invariant distributions and stationary correlation functions of one-dimensional discrete processes. *Zeitschrift für Naturforschung* **32a**(12), 1353–63.

Härdle, W. 1991. *Smoothing techniques (with implementation in S)*. Berlin: Springer.

Havstad, J. W. & C. L. Ehlers 1989. Attractor dimension of nonstationary dynamical systems from small datasets. *Physical Review A* **39**(2), 845–53.

Hayles, N. K. 1990. Self-reflexive metaphors in Maxwell's demon and Shannon's choice. In *Literature and science: theory and practice*, S. Peterfreund (ed.), 209–237. Boston: Northeastern University Press.

Hénon, M. 1976. A two-dimensional mapping with a strange attractor. *Communications in Mathematical Physics* **50**, 69–77.

Herivel, J. 1975. *Joseph Fourier: the man and the physicist*. Oxford: Oxford University Press.

Hoffman, B. 1975. *About vectors*. New York: Dover.

Hofstadter, D. R. 1981. Metamagical themas. *Scientific American* **245**(5), 22–43.

— 1985. *Metamagical themas: questing for the essence of mind and pattern*. New York: Basic Books.

Holden, A. V. (ed.) 1986. *Chaos*. Princeton: Princeton University Press.

Holden, A. V. & M. A. Muhammad 1986. A graphical zoo of strange and peculiar attractors. See Holden (1986: 15–35).

Holzfuss, J. & G. Mayer-Kress 1986. An approach to error-estimation in the application of dimension algorithms. See Mayer-Kress (1986: 114–22).

Horgan, J. 1992. Claude E. Shannon. *Institute of Electrical and Electronics Engineers (IEEE) Spectrum* **29**(4), 72–5.

Hunt, B. R. & J. A. Yorke 1993. Maxwell on chaos. *Nonlinear Science Today* **3**(1), 1–4.

Izenman, A. J. 1991. Recent developments in nonparametric density estimation. *Journal of the American Statistical Association* **86**(413), 205–224.

Jackson, E. A. 1991. *Perspectives of nonlinear dynamics*, vol. 1. Cambridge: Cambridge University Press.

Jenkins, G. M. & D. G. Watts 1968. *Spectral analysis and its applications*. San Francisco: Holden-Day.

Jensen, R. V. 1987. Classical chaos. *American Scientist* **75**(March–April), 168–81.

Johnson, B. R. & S. K. Scott 1990. Period doubling and chaos during the oscillatory ignition of $CO+O_2$ reaction. *Journal of the Chemical Society, Faraday Transactions* **86**, 3701–705.

Kac, M. 1983. What is random? *American Scientist* **71**(4), 405–406.

Kachigan, S. K. 1986. *Statistical analysis*. New York: Radius.

Kadanoff, L. P. 1983. Roads to chaos. *Physics Today* **36**(12), 46–53.

Kanasewich, E. R. 1981. *Time sequence analysis in geophysics*. Edmonton: University of Alberta Press.

Kaplan, D. T. 1994. Exceptional events as evidence for determinism. *Physica D* **73**, 38–48.

Kaplan, D. T. & L. Glass 1992. Direct test for determinism in a time series. *Physical Review Letters* **68**(4), 427–30.

Kennel, M. B., R. Brown, H. D. I. Abarbanel 1992. Determining embedding dimension for phase space reconstruction using a geometrical construction. *Physical Review A* **45**(6), 3403–411.

Kennel, M. B. & S. Isabelle 1992. Method to distinguish possible chaos from colored noise and to determine embedding parameters. *Physical Review A* **46**(6), 3111–18.

Kolmogorov, A. N. 1959, On entropy per unit time as a metric invariant in automorphisms. *Proceedings of the Academy of Sciences of the USSR* **124**(4), 754–5 (in Russian).

Kostelich, E. J. & J. A. Yorke 1990. Noise reduction: finding the simplest dynamical system consistent with the data. *Physica D* **41**, 183–96.

Kovalevskaya, S. V. 1978. *A Russian childhood* [translated and edited by Beatrice Stillman, with an analysis of Kovalevskaya's mathematics by P. Y. Kochina]. Berlin: Springer.

Kreyszig, E. 1962. *Advanced engineering mathematics*. New York: John Wiley.

Krippendorff, K. 1986. *Information theory: structural models for qualitative data*. Los Angeles: Sage.

Lall, U. & M. Mann 1995. The Great Salt Lake: a barometer of low-frequency climatic variability. *Water Resources Research* **31**(10), 2503–515.

Laskar, J. 1989. A numerical experiment on the chaotic behaviour of the solar system. *Nature* **338**(6212), 237–8.

Lauwerier, H. A. 1986a. One-dimensional iterative maps. See Holden (1986: 39–57).

— 1986b. Two-dimensional iterative maps. See Holden (1986: 58–95).

Lewin, R. 1992. *Complexity—life at the edge of chaos*. New York: Macmillan.

Liapounoff, M. A. 1949. *Problème général de la stabilité du mouvement*. Princeton, New Jersey: Princeton University Press.

Lichtenberg, A. J. & M. A. Lieberman 1992. *Regular and chaotic dynamics*, 3rd edn. Berlin: Springer.

Liebert, W. & H. G. Schuster 1989. Proper choice of the time delay for the analysis of chaotic time series. *Physics Letters A* **142**(2–3), 107–111.

Liebert, W., K. Pawelzik, H. G. Schuster 1991. Optimal embeddings of chaotic attractors from topological considerations. *Europhysics Letters* **14**(6), 521–6.

Liebovitch, L. S. & T. Toth 1989. A fast algorithm to determine fractal dimensions by box counting. *Physics Letters A* **141**(8–9), 386–90.

Longley, P. A. & M. Batty 1989. Fractal measurement and line generalization. *Computers and Geosciences* **15**(2), 167–83.

Lorenz, E. N. 1963. Deterministic nonperiodic flow. *Journal of the Atmospheric Sciences* **20**(2), 130–41.

— 1976. Nondeterministic theories of climatic change. *Quaternary Research* **6**, 495–506.

— 1987. Deterministic and stochastic aspects of atmospheric dynamics. In *Irreversible phenomena and dynamical systems analysis in geosciences*, C. Nicolis & G. Nicolis (eds), 159–79. Norwell, Massachusetts: Reidel.

— 1991. Dimension of weather and climate attractors. *Nature* **253**(19 September), 241–3.

— 1993. *The essence of chaos*. Seattle: University of Washington Press.

Machlup, F. 1983. Semantic quirks in studies of information. In *The study of information: interdisciplinary messages*, F. Machlup & U. Mansfield (eds), 641–71. New York: John Wiley.

Makridakis, S., S. C. Wheelwright, V. E. McGee 1983. *Forecasting: methods and applications*, 2nd edn. New York: John Wiley.

Mandelbrot, B. B. 1967. How long is the coast of Britain: statistical self-similarity and fractional dimension. *Science* **156**(5 May), 636–8.

— 1983. *The fractal geometry of nature*. New York: W. H. Freeman.

Mañé, R. 1981. On the dimension of the compact invariant sets of certain nonlinear maps. In *Dynamical systems and turbulence*, D. A. Rand & L-S. Young (eds), 230–42 [Lecture Notes in Mathematics 898]. Berlin: Springer.

Martinerie, J. M., A. M. Albano, A. I. Mees, P. E. Rapp 1992. Mutual information, strange attractors, and the optimal estimation of dimension. *Physical Review A* **45**(10), 7058–7064.

May, R. M. 1974. Biological populations with nonoverlapping generations: stable points, stable cycles, and chaos. *Science* **186**(15 November), 645–7.

— 1976. Simple mathematical models with very complicated dynamics. *Nature* **261**(June 10), 459–67.

— 1987. Chaos and the dynamics of biological populations. *Proceedings of the Royal Society of London A* **413**, 27–44.

May, R. M. & G. F. Oster 1976. Bifurcations and dynamic complexity in simple ecological models. *The American Naturalist* **110**(974), 573–99.

Mayer-Kress, G. (ed.) 1986. *Dimensions and entropies in chaotic systems*. Berlin: Springer.

— 1987. Application of dimension algorithms to experimental chaos. In *Directions in chaos* (vol. 1), B-L. Hao (ed.), 122–47. Singapore: World Scientific.

McAuliffe, K. 1990. Get smart: controlling chaos. *OMNI* **12**(5), 43–8, 86–92.

Mende, W., H. Herzel, K. Wermke 1990. Bifurcations and chaos in newborn infant cries. *Physics Letters A* **145**(8–9), 418–24.

Middleton, G. V. 1995. Appendix I: chaossary—a short glossary of chaos. In *Nonlinear dynamics and fractals—new numerical techniques for sedimentary data*, G. V. Middleton, R. E. Plotnick, D. M. Rubin (eds), 133–58. Short Course 36, Society for Sedimentary Geology (SEPM), Tulsa, Oklahoma.

Moon, F. C. 1992. *Chaotic and fractal dynamics*. New York: John Wiley.

Morrison, F. 1988. On chaos. *EOS, Transactions of the American Geophysical Union* **69**(25), 668–9.

Mullin, T. 1993. A dynamical systems approach to time-series analysis. In *The nature of chaos*, T. Mullin (ed.), 23–50. New York: Oxford University Press.

Mundt, M. D., W. B. Maguire II, R. R. P. Chase 1991. Chaos in the sunspot cycle: analysis and prediction. *Journal of Geophysical Research* **96**(A2), 1705–716.

Murray, J. D. 1991. Mathematics—biology—nonlinearity. *Nonlinear Science Today* **1**(3), 1–5.

Nerenberg, M. A. H. & C. Essex 1990. Correlation dimension and systematic geometric effects. *Physical Review A* **42**(12), 7065–7074.

Nese, J. M. 1989. Quantifying local predictability in phase space. *Physica D* **35**, 237–50.

Nicolis, J. S., G. Mayer-Kress, G. Haubs 1983. Non-uniform chaotic dynamics with implications to information processing. *Zeitschrift für Naturforschung* **38a**, 1157–69.

Nychka, D., S. Ellner, A. R. Gallant, D. McCaffrey 1992. Finding chaos in noisy systems. *Journal of the Royal Statistical Society*, Series B **54**(2), 399–426.

Olsen, L. F. & H. Degn 1985. Chaos in biological systems. *Quarterly Review of Biophysics* **18**(2), 165–225.

Orford, J. D. & W. B. Whalley 1987. The quantitative description of highly irregular sedimentary particles: the use of the fractal dimension. In *Clastic particles—scanning electron microscopy and shape analysis of sedimentary and volcanic clasts*, J. R. Marshall (ed.), 267–80. New York: Van Nostrand Reinhold.

Osborne, A. R. & A. Provenzale 1989. Finite correlation dimension for stochastic systems with power-law spectra. *Physica D* **35**, 357–81.

Oseledec, V. I. 1968. A multiplicative ergodic theorem: Ljapunov characteristic numbers for dynamical systems. *Transactions of the Moscow Mathematical Society* **19**, 197–231 (in English).

Ottino, J. M., F. J. Muzzio, M. Tjahjadi, J. G. Franjione, S. C. Jana, H. A. Kusch 1992. Chaos, symmetry, and self-similarity: exploiting order and disorder in mixing processes. *Science* **257**(7 August), 754–60.

Packard, N. H., J. P. Crutchfield, J. D. Farmer, R. S. Shaw 1980. Geometry from a time series. *Physical Review Letters* **45**(9), 712–16.

Paluŝ, M. 1993. Identifying and quantifying chaos by using information-theoretic functionals. See Weigend & Gershenfeld (1993: 387–413).

Parker, T. S. & L. O. Chua 1989. *Practical numerical algorithms for chaotic systems*. Berlin: Springer.

Paulos, J. A. 1990. *Innumeracy—mathematical illiteracy and its consequences*. New York: Vintage.

Peitgen, H-O. & D. Saupe (eds) 1988. *The science of fractal images*. Berlin: Springer.

Peitgen, H-O., H. Jürgens, D. Saupe 1992. *Chaos and fractals: new frontiers of science*. Berlin: Springer.

Percival, I. 1989. Chaos: a science for the real world. *New Scientist* **124**(1687), 42–7.

Peters, E. E. 1991. *Chaos and order in the capital markets*. New York: John Wiley.

Peterson, I. 1988. In the shadows of chaos. *Science News* **134**(3 December), 360–61.

Pielou, E. C. 1969. *An introduction to mathematical ecology*. New York: John Wiley.

Pineda, F. J. & J. C. Sommerer 1993. Estimating generalized dimensions and choosing time delays: a fast algorithm. See Weigend & Gershenfeld (1993: 367–85).

Pippard, B. 1982. Instability and chaos: physical models of everyday life. *Interdisciplinary Science Reviews* **7**(2), 92–101.

Plotnick, R. E. & K. Prestegaard 1993. Fractal analysis of geologic time series. In *Fractals in geography*, N. S. Lam & L. de Cola (eds), 193–210. Englewood Cliffs, New Jersey:

Prentice-Hall.
Pomeau, Y. & P. Manneville 1980. Intermittent transition to turbulence in dissipative dynamical systems. *Communications in Mathematical Physics* **74**, 189–97.
Pool, R. 1989a. Is it chaos, or is it just noise? *Science* **243**(6 January), 25–8.
— 1989b. Ecologists flirt with chaos. *Science* **243**(20 January), 310–13.
— 1989c. Is it healthy to be chaotic? *Science* **243**(3 February), 604–607.
— 1989d. Quantum chaos: enigma wrapped in a mystery. *Science* **243**(17 February), 893–5.
— 1989e. Is something strange about the weather? *Science* **243**(10 March), 1290–93.
— 1989f. Chaos theory: how big an advance? *Science* **243**(7 July), 26–8.
— 1990a. Fractal fracas. *Science* **249**(27 July), 363–4.
— 1990b. Putting chaos to work. *Science* **250**(2 November), 626–7.
Prichard, D. 1994. The correlation dimension of differenced data. *Physics Letters A* **191**(3–4), 245–50.
Prichard, D. & C. P. Price 1992. Spurious dimension estimates from time series of geomagnetic indices. *Geophysical Research Letters* **19**(15), 1623–6.
— 1993. Is the AE index the result of nonlinear dynamics? *Geophysical Research Letters* **20**(24), 2817–20.
Priestley, M. B. 1981. *Spectral analysis and time series* [2 volumes]. London: Academic Press.
Provenzale, A., L. A. Smith, R. Vio, G. Murante 1992. Distinguishing between low-dimensional dynamics and randomness in measured time series. *Physica D* **58**, 31–49.

Ramirez, R. W. 1985. *The FFT: fundamentals and concepts*. Englewood Cliffs, New Jersey: Prentice-Hall.
Ramsey, J. B. & H-J. Yuan 1989. Bias and error bars in dimension calculations and their evaluation in some simple models. *Physics Letters A* **134**(5), 287–97.
Rapp, P. E. 1992. Dynamical analysis of biological systems: a case for guarded optimism. In National Institutes of Health, workshop on "The head and heart of chaos: nonlinear dynamics in biological systems," Executive Report (15–16 June), 43–6.
Rapp, P. E., T. R. Bashore, J. M. Martinerie, A. M. Albano, I. D. Zimmerman, A. I. Mees 1989. Dynamics of brain electrical activity. *Brain Topography* **2**(1–2), 99–118.
Rapp, P. E., A. M. Albano, T. I. Schmah, L. A. Farwell 1993. Filtered noise can mimic low-dimensional chaotic attractors. *Physical Review E* **47**(4), 2289–97.
Rasband, N. S. 1990. *Chaotic dynamics of nonlinear systems*. New York: John Wiley.
Rayner, J. N. 1971. *An introduction to spectral analysis*. London: Pion.
Read, P. L., M. J. Bell, D. W. Johnson, R. M. Small 1992. Quasi-periodic and chaotic flow regimes in a thermally driven, rotating fluid annulus. *Journal of Fluid Mechanics* **238**, 599–632.
Richardson, L. F. 1961. The problem of contiguity: an appendix to *Statistics of deadly quarrels*. *General Systems Yearbook* **6**(1), 139–87.
Robinson, A. L. 1982. Physicists try to find order in chaos. *Science* **218**(5 November), 554–6.
Rössler, O. E. 1976. An equation for continuous chaos. *Physics Letters* **57A**(5), 397–8.
— 1979. An equation for hyperchaos. *Physics Letters* **71A**, 155–7.
Rothman, T. 1991. *A physicist on Madison avenue*. Princeton, New Jersey: Princeton University Press.
Rowlands, G. & J. C. Sprott 1992. Extraction of dynamical equations from chaotic data. *Physica D* **58**, 251–9.
Ruelle, D. 1990. Deterministic chaos: the science and the fiction. *Proceedings of the Royal*

Society of London A **427**, 241–8.
— 1991. *Chance and chaos*. Princeton, New Jersey: Princeton University Press.
— 1994. Where can one hope to profitably apply the ideas of chaos? *Physics Today* **47**(7), 24–30.

Salas, J. D., J. W. Delleur, V. Yevjevich, W. L. Lane 1980. *Applied modeling of hydrologic time series*. Littleton, Colorado: Water Resources Publications.
Sangoyomi, T. B., U. Lall, H. D. I. Abarbanel 1996. Nonlinear dynamics of the Great Salt Lake: dimension estimation. *Water Resources Research* **32**(1), 149–59.
Sano, M. & Y. Sawada 1985. Measurement of the Lyapunov spectrum from a chaotic time series. *Physical Review Letters* **55**(10), 1082–1085.
Sauer, T., J. A. Yorke, M. Casdagli 1991. Embedology. *Journal of Statistical Physics* **65**(3/4), 579–616.
Saupe, D. 1988. Algorithms for random fractals. See Peitgen & Saupe (1988: 71–136).
Schaffer, W. M. & M. Kot 1985. Do strange attractors govern ecological systems? *BioScience* **35**(6), 342–50.
— 1986. Chaos in ecological systems: the coals that Newcastle forgot. *Trends in Ecology and Evolution* **1**(3), 58–63.
Schaffer, W. M., S. Ellner, M. Kot 1986. Effects of noise on some dynamical models in ecology. *Journal of Mathematical Biology* **24**, 479–523.
Schaffer, W. M., L. F. Olsen, G. L. Truty, S. L. Fulmer 1990. The case for chaos in childhood epidemics. In *The ubiquity of chaos*, S. Krasner (ed.), 138–66. Washington DC: American Association for the Advancement of Science.
Schaffer, W. M., G. L. Truty, S. L. Fulmer 1992. *Dynamical software professional–user's manual and introduction to chaotic systems*, vol. 2 (version II.2). Tucson: Dynamical Systems [distributed by Academic Computing Specialists, Salt Lake City, Utah].
Schneider, E. D. 1988. Thermodynamics, ecological succession, and natural selection: a common thread. In *Entropy, information, and evolution: new perspectives on physical and biological evolution*, B. H. Weber, D. J. Depew, J. D. Smith (eds), 107–138. Cambridge, Massachusetts: Massachusetts Institute of Technology Press.
Schuster, H. G. 1988. *Deterministic chaos: an introduction*, 2nd edn. Weinheim, Germany: VCH.
Scott, D. W. 1992. *Multivariate density estimation*. New York: John Wiley.
Seydel, R. 1988. *From equilibrium to chaos*. Amsterdam: Elsevier.
Shannon, C. E. & W. Weaver 1949. *The mathematical theory of communication*. Urbana–Champaign: University of Illinois Press.
Shaw, R. 1984. *The dripping faucet as a model chaotic system*. Santa Cruz, California: Aerial Press.
Shinbrot, T. 1993. Chaos: unpredictable yet controllable? *Nonlinear Science Today* **3**(2), 1–8.
Shinbrot, T., C. Grebogi, E. Ott, J. A. Yorke 1993. Using small perturbations to control chaos. *Nature* **363**(3 June), 411–17.
Shumway, R. H. 1988. *Applied statistical time series analysis*. Englewood Cliffs, New Jersey: Prentice-Hall.
Silverman, B. W. 1986. *Density estimation for statistics and data analysis*. London: Chapman & Hall.
Sinai, Ya. 1959. On the concept of entropy in a dynamical system. *Proceedings of the Academy of Sciences of the USSR* **124**(4), 768–71 (in Russian).

Smith, L. A. 1988. Intrinsic limits of dimension calculations. *Physics Letters A* **133**(6), 283–8.

Stewart, H. B. 1984. *The geometry of chaos*. Brookhaven Lecture Series 209, Report BNL-51892, Brookhaven National Laboratory.

Stewart, I. 1989a. *Does God play dice?: the mathematics of chaos*. Cambridge, Massachusetts: Basil Blackwell.

— 1989b. Portraits of chaos. *New Scientist* **124**(1689), 42–7.

Stoop, R. & J. Parisi 1991. Calculation of Lyapunov exponents avoiding spurious elements. *Physica D* **50**, 89–94.

Sussman, G. J. & J. Wisdom 1992. Chaotic evolution of the solar system. *Science* **257**(3 July), 56–62.

Swinney, H. L. 1986. Experimental observations of order and chaos. In *Nonlinear dynamics and chaos: geometrical methods for engineers and scientists*, J. M. T. Thompson & H. B. Stewart (eds), 332–49. New York: John Wiley.

Takens, F. 1981. Detecting strange attractors in turbulence. In *Dynamical systems and turbulence*, D. A. Rand & L-S. Young (eds), 366–81 [Lecture Notes in Mathematics 898]. Berlin: Springer.

— 1985. On the numerical determination of the dimension of an attractor. In *Dynamical systems and bifurcations*, B. L. J. Braaksma, H. W. Broer, F. Takens (eds), 99–106. Berlin: Springer.

Tashman, L. J. & K. R. Lamborn 1979. *The ways and means of statistics*. New York: Harcourt Brace Jovanovich.

Theiler, J. 1986. Spurious dimension from correlation algorithms applied to limited time-series data. *Physical Review A* **34**(3), 2427–32.

— 1987. Efficient algorithm for estimating the correlation dimension from a set of discrete points. *Physical Review A* **36**(9), 4456–62.

— 1990a. Estimating fractal dimension. *Journal of the Optical Society of America A* **7**(6), 1055–1073.

— 1990b. Statistical precision of dimension estimators. *Physical Review A* **41**(6), 3038–3051.

Theiler, J., B. Galdrikian, A. Longtin, S. Eubank, J. D. Farmer 1992. Using surrogate data to detect nonlinearity in time series. In *Nonlinear modeling and forecasting*, M. Casdagli & S. Eubank (eds), 163–88. Redwood City, California: Addison-Wesley.

Theiler, J. & S. Eubank 1993. Don't bleach chaotic data. *Chaos* **3**(4), 771–82.

Theiler, J., P. S. Linsay, D. M. Rubin 1993. Detecting nonlinearity in data with long coherence times. See Weigend & Gershenfeld (1993: 429–55).

Thompson, J. M. T. & H. B. Stewart 1986. *Nonlinear dynamics and chaos: geometrical methods for engineers and scientists*. New York: John Wiley.

Tredicce, J. R. & N. B. Abraham 1990. Experimental measurements to identify and/or characterize chaotic signals. In *Lasers and quantum optics*, L. M. Narducci, E. J. Quel, J. R. Tredicce (eds), 148–98. Singapore: World Scientific.

Tribus, M. & E. C. McIrvine 1971. Energy and information. *Scientific American* **224**(3), 179–88.

Troutman, B. M. & G. P. Williams 1987. Fitting straight lines in the Earth sciences. In *Use and abuse of statistical methods in the earth sciences*, W. B. Size (ed.), 107–128. New York: Oxford.

Tsonis, A. A. 1989. Chaos and unpredictability of weather. *Weather* **44**, 258–63.

Tsonis, A. A., G. N. Triantafyllou, J. B. Elsner, J. J. Holdzkom II, A. D. Kirwan Jr 1994. An investigation of the ability of nonlinear methods to infer dynamics from observables. *Bulletin of the American Meteorological Society* **75**(9), 1623–633.

Tufillaro, N. B., T. Abbott, J. Reilly 1992. *An experimental approach to nonlinear dynamics and chaos*. Redwood City, California: Addison-Wesley.

Vassilicos, J. C., A. Demos, F. Tata 1993. No evidence of chaos but some evidence of multifractals in the foreign exchange and the stock markets. In *Applications of fractals and chaos*, A. J. Crilly, R. A. Earnshaw, H. Jones (eds), 249–65. Berlin: Springer.

Vautard, R. & M. Ghil 1989. Singular spectrum analysis in nonlinear dynamics, with applications to paleoclimatic time series. *Physica D* **35**, 395–424.

Voss, R. F. 1988. Fractals in nature: from characterization to simulation. See Peitgen & Saupe (1988: 21–70).

Waldrop, M. M. 1992. *Complexity: the emerging science at the edge of order and chaos*. New York: Simon & Schuster.

Wastler, T. A. 1963. *Application of spectral analysis to stream and estuary field surveys, 1. individual power spectra*. Public Health Service Publication 999-WP-7, US Department of Health, Education, and Welfare, Cincinnati.

Wayland, R., D. Bromley, D. Pickett, A. Passamante 1993. Recognizing determinism in a time series. *Physical Review Letters* **70**(5), 580–82.

Wegman, E. J. 1988. On randomness, determinism and computability. *Journal of Statistical Planning and Inference* **20**, 279–94.

Weigend, A. S. 1994. Paradigm change in prediction. *Philosophical Transactions of the Royal Society of London A* **348**(1688), 405–420.

Weigend, A. S. & N. A. Gershenfeld (eds) 1993. *Time-series prediction: forecasting the future and understanding the past*. Reading, Massachusetts: Addison-Wesley.

West, B. J. 1990. *Fractal physiology and chaos in medicine*. Singapore: World Scientific.

Whitney, H. 1936. Differentiable manifolds. *Annals of Mathematics* **37**(3), 645–80.

Wilkes, M. V. 1966. *A short introduction to numerical analysis*. Cambridge: Cambridge University Press.

Wolf, A. 1983. Simplicity and universality in the transition to chaos. *Nature* **305**(15 September), 182–3.

— 1986. Quantifying chaos with Lyapunov exponents. In *Chaos*, A. V. Holden (ed.), 273–90. Princeton, New Jersey: Princeton University Press.

Wolf, A., J. B. Swift, H. L. Swinney, J. A. Vastano 1985. Determining Lyapunov exponents from a time series. *Physica D* **16**, 285–317.

Wonnacott, T. H. & R. J. Wonnacott 1984. *Introductory statistics for business and economics*, 3rd edn. New York: John Wiley.

Wright, J. 1984. Method for calculating a Lyapunov exponent. *Physical Review A* **29**(5), 2924–7.

Yorke, J. A., C. Grebogi, E. Ott, L. Tedeschini-Lalli 1985. Scaling behavior of windows in dissipative dynamical systems. *Physical Review Letters* **54**(11), 1095–1098.

Young, L-S. 1983. Entropy, Lyapunov exponents, and Hausdorff dimension in differentiable dynamical systems. *Institute of Electrical and Electronics Engineers (IEEE) Transactions on Circuits and Systems* **CAS–30**(8), 599–607.

Yule, G. U. 1927. On a method of investigating periodicities in disturbed series, with special reference to Wolfe's sunspot numbers. *Philosophical Transactions of the Royal Society of London A* **226**, 267–98.

Zeng, X., R. Eykholt, R. A. Pielke 1991. Estimating the Lyapunov-exponent spectrum from short time series of low precision. *Physical Review Letters* **66**(25), 3229–32.

Zeng, X., R. A. Pielke, R. Eykholt 1992. Extracting Lyapunov exponents from short time series of low precision. *Modern Physics Letters B* **6**(2), 55–75.

— 1993. Chaos theory and its applications to the atmosphere. *Bulletin of the American Meteorological Society* **74**(4), 631–44.

Selected further reading

Baker, G. L. & J. P. Gollub 1990. *Chaotic dynamics: an introduction.* Cambridge: Cambridge University Press.

Devaney, R. L. 1990. *Chaos, fractals, and dynamics: computer experiments in mathematics.* Menlo Park, California: Addison-Wesley.

— 1992. *A first course in chaotic dynamical systems: theory and experiment.* Reading, Massachusetts: Addison-Wesley.

Hall, N. (ed.) 1991. *Exploring chaos—a guide to the new science of disorder.* New York: Norton.

Hao, B-L. (ed.) 1984. *Chaos.* Singapore: World Scientific.

Hayles, N. K. (ed.) 1991. *Chaos and order—complex dynamics in literature and science.* Chicago: University of Chicago Press.

Hilborn, R. C. 1994. *Chaos and nonlinear dynamics.* New York: Oxford University Press.

Marek, M. & I. Schreiber 1991. *Chaotic behavior of deterministic dissipative systems.* Cambridge: Cambridge University Press.

Ott, E. 1993. *Chaos in dynamical systems.* Cambridge: Cambridge University Press.

Peterson, I. 1988. *The mathematical tourist—snapshots of modern mathematics.* New York: Freeman.

Strogatz, S. H. 1994. *Nonlinear dynamics and chaos.* Reading, Massachusetts: Addison-Wesley.

Tsonis, A. A. 1992. *Chaos—from theory to applications.* New York: Plenum.

Index